Fueling Development
ENERGY TECHNOLOGIES FOR DEVELOPING COUNTRIES

Cover photo credit: James P. Blair, © National Geographic Society

CONGRESS OF THE UNITED STATES
OFFICE OF TECHNOLOGY ASSESSMENT

333,79
F953

Recommended Citation:

U.S. Congress, Office of Technology Assessment, *Fueling Development: Energy Technologies for Developing Countries*, OTA-E-516 (Washington, DC: U.S. Government Printing Office, April 1992).

For sale by the U.S. Government Printing Office
Superintendent of Documents, Mail Stop: SSOP, Washington, DC 20402-9328
ISBN 0-16-036185-0

Foreword

This report examines the delivery of energy services in developing countries and how the United States can help to improve these energy services while minimizing environmental impacts. OTA examines the technologies and policies that will enable more efficient use of energy and the most promising new sources of energy supply. This assessment was requested by the Senate Committee on Governmental Affairs; the House Committee on Energy and Commerce and its Subcommittee on Energy and Power; the Subcommittees on Human Rights and International Organizations and on Africa of the House Committee on Foreign Affairs; the Subcommittee on International Development, Finance, Trade, and Monetary Policy of the House Banking Committee; and individual members of the Senate Committee on Environment and Public Works, the House Select Committee on Hunger, and the Congressional Competitiveness Caucus.

Such extensive congressional interest is certainly warranted. American economic, political, and environmental self-interest lies in developing healthy relationships with these present and potential trading partners and allies. Furthermore, the developing world will require our close attention for decades to come. Based on present trends, 90 percent of the expected world population growth over the next 30 years will occur in the countries of Africa, Asia, and Latin America, bringing their population to almost 7 billion. Energy will play an indispensable role in raising the economic status of these people.

This report builds on the analysis presented in the interim report of this assessment, *Energy in Developing Countries*, which OTA published in 1991. The assessment, as a whole, takes a broad look at energy options and opportunities in the many countries of the developing world. It finds that there are many valuable lessons that developing countries can learn from the prior experiences of the United States and other industrial countries.

In the course of this assessment, OTA drew on the experience of many organizations and individuals. In particular, we appreciate the generous assistance of our distinguished advisory panel and workshop participants, as well as the efforts of the project's contractors. We would also like to acknowledge the help of the many reviewers who gave their time to ensure the accuracy and comprehensiveness of this report. To all of the above goes the gratitude of OTA, and the personal thanks of the staff.

JOHN H. GIBBONS
Director

Advisory Panel—Energy, Technology, and the Environment in Developing Countries

Harry G. Barnes, Jr., *Chairman*
Critical Languages and Area Studies Consortium

Irma Adelman
Department of Economics
University of California, Berkeley

Jeffrey Anderson
Institute of International Finance

Edward S. Ayensu
Pan-African Union for Science and Technology

Gerald Barnes
General Motors

Mohamed T. El-Ashry
World Bank

Eric Helland-Hansen
United Nations Development Programme

Carl N. Hodges
Environmental Research Laboratory
University of Arizona

Donald Jost
Sun Refining and Marketing Co.

Amory Lovins
Rocky Mountain Institute

Mohan Munasinghe
World Bank

Henry Norman
Volunteers in Technical Assistance

Waafas Ofosu-Amaah
WorldWIDE Network

R.K. Pachauri
Tata Energy Research Institute

D. Louis Peoples
Hagler-Bailly, Inc.

Gabriel Sanchez-Sierra
Latin-American Energy Organization (OLADE)

Kirk R. Smith
Environment and Policy Institute
East-West Center

Irving Snyder
Dow Chemical Co.

Thomas B. Stoel, Jr.
Private Consultant
Washington, DC

Jack W. Wilkinson
Sun Refining and Marketing Co.

Robert H. Williams
Center for Energy and Environmental Studies
Princeton University

Lu Yingzhong
Professional Analysis Inc. (PAI)

Montague Yudelman
World Wildlife Fund and The Conservation Foundation

Eugene W. Zeltmann
General Electric

Observer

David Jhirad
Office of Energy, United States Agency for International Development

NOTE: OTA appreciates and is grateful for the valuable assistance and thoughtful critiques provided by the advisory panel members. The panel does not, however, necessarily approve, disapprove, or endorse this report. OTA assumes full responsibility for the report and the accuracy of its contents.

OTA Project Staff for Fueling Development

Lionel S. Johns, *Assistant Director,*
OTA Energy, Materials, and International Security Division

Peter D. Blair, *Energy and Materials Program Manager*

Project Staff

Joy Dunkerley, *Project Director*

Samuel F. Baldwin, *Assistant Project Director*

Paul S. Komor, *Analyst*

Sharon Burke, *Research Analyst*   Daniel Kumi, *Summer Intern*

Administrative Staff

Lillian Chapman, *Office Administrator*

Linda Long, *Administrative Secretary*

Phyllis Aikens, *Secretary*

OTA Contributors

Rosina Bierbaum   Roger Chin   Nina Goldman
Gretchen Kolsrud   Karen Larsen   Howard Levenson
Edward MacDonald   Robin Roy

Contractors

Jack Casazza and Herb Limmer
Casazza, Schultz and Associates, Inc.

Russell deLucia
deLucia and Associates, Inc.

Ashok Desai
National Council of Applied Economic Research

K.G. Duleep and Sergio Ostria
Energy and Environmental Analysis, Inc.

Ahmad Faruqui and Greg Wikler
Barakat, Howard & Chamberlin

Howard Geller
American Council for an Energy Efficient Economy

Theodore J. Gorton
Petroleum Consultant

Donald Hertzmark
Independent Consultant

H. Mike Jones
Independent Consultant

Virendra Kothari
Energy and Environmental Analysis, Inc.

Eric Larson
Princeton University

Karin Lissakers
Columbia University

Arjun Makhijani
Institute for Energy and Environmental Research

Peter Meier, John Lee and Mujahid Iqbal
International Development and Energy Associates

Henry Peskin
Edgevale Associates, Inc.

Judith M. Siegel
Judith M. Siegel and Associates

Vaclav Smil
University of Manitoba

Workshop Participants

Technology Transfer to Developing Countries
September 13, 1989

Robert H. Annan
Committee on Renewable Energy Commerce
    and Trade
U.S. Department of Energy

Mark Bernstein
Center for Energy and the Environment
University of Pennsylvania

James H. Caldwell
Independent Consultant
Washington, DC

David Curry
U.S. Department of the Treasury

Ghazi Darkazali
Spire Corp.

Robert Ford
U.S. Department of State

Les Garden
U.S. Department of Commerce

Michael Greene
Board on Science and Technology in Development
National Academy of Sciences

Erik Helland-Hansen
United Nations Development Programme

David Jhirad
Office of Energy
U.S. Agency for International Development

Linda Ladis
U.S. Export Council for Renewable Energy

Mark Levine
Lawrence Berkeley Laboratory

Frances Li
U.S. Department of State

Janice Mazer
U.S. Department of Commerce

Mohan Munasinghe
The World Bank

D. Lou Peoples
Hagler, Bailly, Inc.
    (formerly Bechtel Power Corp.)

Paul Schwengels
U.S. Environmental Protection Agency

Joseph Sedlak
Volunteers in Technical Assistance

James Sullivan
U.S. Agency for International Development

Franklin Tugwell
Environmental Enterprises Assistance Fund
(formerly Winrock International Institute
    for Agricultural Development)

Jack Vanderryn
U.S. Agency for International Development

Frank Vukmanic
U.S. Department of Treasury

Timothy Weiskel
Rockefeller Foundation

Gerald T. West
Overseas Private Investment Corp.

Thomas Wilbanks
Oak Ridge National Laboratory

John W. Wisniewski
Export-Import Bank

Environmental Problems and Priorities in Developing Countries
April 19, 1990

Michael Adler
U.S. Environmental Protection Agency

Lutz Baehr
Center for Science and Technology in Development
United Nations

Leonard Berry
Florida Atlantic University

Al Binger
Conservation Foundation/Biomass Users
    Network

Janet Welsh Brown
World Resources Institute

Lalanath de Silva
Environmental Foundation, Ltd.
Sri Lanka

Clarence Dias
International Center for Law in Development

Paul Dulin
Associates in Rural Development

(continued on next page)

John J. Gaudet
Africa Bureau
U.S. Agency for International Development

Robert Goodland
Environment Division
Latin American Division
World Bank

Lupe Guinand
BIOMA
Venezuela

Robert Ichord
Energy & Natural Resources Division
Asia Near East Bureau
U.S. Agency for International Development

Kari Keipi
Environmental Protection Division
Inter-American Development Bank

Ananda Krishnan
Center for Science and Technology in Development
United Nations

Russell Mittermeier
Conservation International

Hind Sadek
WorldWide

Paul Schwengels
U.S. Environmental Protection Agency

Thomas Stoel, Jr.
Natural Resources Defense Council

Energy for Transportation in the Developing World
September 25, 1990

David Greene
Oak Ridge National Laboratory

Martin Bernard
Argonne National Laboratory

Mia Birk
International Institute for Energy Conservation (IIEC)

K.G. Duleep
Energy and Environmental Analysis, Inc.

Asif Faiz
The World Bank

Fred Moavenzadeh
Center for Construction Research and Education
Massachusetts Institute of Technology

Sergio Ostria
Energy and Environmental Analysis, Inc.

Setty Pendakur
School of Community and Regional Planning
University of British Columbia

Gabriel Roth
Private Consultant
Washington, DC

Kumares C. Sinha
Department of Civil Engineering
Purdue University

NOTE: OTA appreciates and is grateful for the valuable assistance and thoughtful critiques provided by the workshop participants. The workshop participants do not, however, necessarily approve, disapprove, or endorse this report. OTA assumes full responsibility for the report and the accuracy of its contents.

Outside Reviewers

Michael L.S. Bergey
Bergey Windpower Co.

Martin Bernard
Argonne National Laboratory

Mia Birk
International Institute for Energy Conservation

Chaim Braun
United Engineers and Constructors

Bob Chronowski
Alternative Energy Development, Inc.

Kurt Conger
American Public Power Association

W.D. Craig
Bechtel Power Corp.

K.G. Duleep
Energy and Environmental Analysis, Inc.

Gautam Dutt
Independent Consultant

Elaine Evans
National Hydropower Association

Asif Faiz
The World Bank

Willem Floor
The World Bank

Gerald Foley
Panos Institute (London)

Howard Geller
American Council for an Energy Efficient Economy

David Greene
Oak Ridge National Laboratory

Toni Harrington
Honda North America, Inc.

Allen R. Inversin
National Rural Electric Cooperative Association

H. Michael Jones
Independent Consultant
Washington, DC

John Kadyszewski
Winrock International Institute for Agricultural Development

Lars Kristofferson
Stockholm Environment Center

Eric Larson
Princeton University

Barry McNutt
U.S. Department of Energy

Denise Mauzerall
Harvard University

Steven Meyers
Lawrence Berekeley Laboratory

Fred Moavenzadeh
Center for Construction Research and Education
Massachusetts Institute of Technology

Steve Nadel
American Council for an Energy Efficient Economy

William Obenchain
U.S. Department of Energy

Jim Ohi
Solar Energy Research Institute

Tim Olson
California Energy Commission

Sergio Ostria
Energy and Environmental Analysis, Inc.

Ralph Overend
Solar Energy Research Institute

S. Padmanabhan
U.S. Agency for International Development

Setty Pendakur
School of Community and Regional Planning
Univeristy of British Columbia

David Pimentel
Cornell University

Steve Plotkin
Energy and Materials Program
Office of Technology Assessment

Glenn Prickett
Natural Resources Defense Council

Marc Ross
University of Michigan

Gabriel Roth
Independent Consultant
Washington, DC

Chris Rovero
Oak Ridge Associated Universities

Gabriel Sanchez Sierra
Latin-American Energy Organization (OLADE)

Judith M. Siegel
Judith M. Siegel and Associates

D.P. Sengupta
Department of Electrical Engineering
Indian Institute of Science

Kumares C. Sinha
Department of Civil Engineering
Purdue University

Scott Sklar
U.S. Export Council for Renewable Energy

Stratos Tavoulareas
Office of Energy
United States Agency for International Development

Michael Totten
International Institute for Energy Conservation

Sergio Trindade
SET International, Ltd.
(formerly United Nations Center for Science and Technology for Development)

Xin Hau Wang
United Nations
Centre for Science and Technology for Development

NOTE: OTA appreciates and is grateful for the valuable assistance and thoughtful critiques provided by the reviewers. The reviewers do not, however, necessarily approve, disapprove, or endorse this report. OTA assumes full responsibility for the report and the accuracy of its contents.

Contents

	Page
Chapter 1: Overview	3
Chapter 2: Energy and the Developing Countries	17
Chapter 3: Energy Services: Residential and Commercial	47
Chapter 4: Energy Services: Industry and Agriculture	91
Chapter 5: Energy Services: Transport	145
Chapter 6: Energy Conversion Technologies	179
Chapter 7: Energy Resources and Supplies	231
Chapter 8: Issues and Options	261
Appendix A: Capital and Life Cycle Costs for Electricity Services	289
Appendix B: Electricity Supply Technologies	304
Appendix C: Conversion Factors, Abbreviations, and Glossary	312
Index	315

Chapter 1
Overview

Photo credit: U.S. Agency for International Development

CONTENTS

	Page
SUMMARY	3
ENERGY ISSUES IN DEVELOPING COUNTRIES	3
IMPROVING EFFICIENCIES IN RESIDENTIAL, COMMERCIAL, INDUSTRIAL, AND TRANSPORTATION ENERGY USE	5
The Potential	5
The Example of Electricity	5
Impediments to the Adoption of Energy Efficient Technology	8
Policy Responses	9
IMPROVING ENERGY SUPPLY EFFICIENCY	9
Conversion Technologies	9
Primary Energy Supplies	10
The Benefits	11
Impediments to Adoption	11
ISSUES AND OPTIONS FOR THE UNITED STATES	12

Chapter 1
Overview

SUMMARY

Developing countries need energy to raise productivity and improve the living standards of their populations. Traditionally, developing countries have addressed their energy needs by expanding the supply base with little attention to the efficiency of energy use. This approach is now, however, raising serious financial, institutional, and environmental problems. The magnitude of these problems underlines the need for improving the efficiency with which energy is currently used and produced in developing countries.

OTA finds that there are major opportunities—drawing on currently available or near-commercial technologies—for improving efficiencies throughout the energy systems of developing countries. These technologies promise to save energy, diminish adverse environmental impacts, reduce life cycle costs to consumers, and lower systemwide capital costs. Despite these advantages, efficient technologies may not be rapidly adopted unless technology transfer is improved and policies and procedures in donor agencies and the developing countries themselves are designed to remove impediments to their adoption. There are already promising signs of greater attention to removal of such barriers, but much remains to be done.

The way in which developing countries meet their energy needs directly relates to a number of U.S. policy concerns. International political stability depends on steady broad-based economic growth in the developing countries, which in turn requires economic and reliable energy services. The developing countries are of growing importance in global energy markets and global environmental issues: these countries are projected to account for over one-half of the increase in global energy consumption over the next three decades with corresponding increases in their emissions of carbon dioxide, a major greenhouse gas. Sharply rising demand for oil from the developing countries contributes to upward pressure on international oil prices. Developing country debt, often energy related, affects the stability of U.S. and international banking systems. At the same time, developing countries offer the United States important trade opportunities in their large and expanding markets for energy technologies.

The United States has a number of programs influencing the diffusion of energy efficient and renewable energy technologies in developing countries. The United States could increase its influence by providing greater leadership in technology transfer, including research, development and demonstration, information dissemination, and training. The United States could also promote the adoption of more energy efficient technologies by supporting policy changes in both lending agencies and developing countries. Finally, the United States could set a good example of energy efficient behavior at home.

ENERGY ISSUES IN DEVELOPING COUNTRIES

Commercial energy consumption in developing countries is projected to triple over the next 30 years, driven by rapid population growth and economic development. Even assuming continued declines in fertility rates, the population of the developing world will increase by nearly 3 billion—to almost 7 billion—over the next three decades, stimulating a sharp increase in demand for energy services. Securing higher living standards for this growing population requires rapid economic growth, further increasing the demand for energy services. This demand is augmented by structural changes inherent in the development process, especially urbanization; the building of the commercial, industrial and transportation infrastructure; and the substitution of commercial for traditional fuels. Demand is further augmented by the rapid rise in demand for consumer goods—lights, refrigerators, TVs—stimulated by their lowered real cost, improved availability, and frequently subsidized energy prices. Even though per capita commercial energy consumption will remain well below the levels of the industrial countries, the more rapid population and economic growth in developing countries means that their share of global commercial energy consumption would rise—from 23 percent today to a projected 40 percent in 2020.

Efforts to supply energy on this scale face serious financial, operational, and environmental constraints. Capital intensive electricity generating stations and petroleum refineries already account for a large part of all public investment budgets in developing countries, with electric utilities taking the lion's share. Yet, annual electricity sector investments would have to double to provide supplies at projected growth rates. A large part of the investment in energy facilities and in the fuel to operate them must be paid for in foreign exchange, already under pressure in many countries. In addition, there is often a shortage of local currency to pay for energy development due to inadequate revenues from existing operations. The high cost of developing national energy infrastructures and of importing energy to support growing energy demands could, in some cases, slow overall economic growth.

The energy supply sector in many developing countries also experiences a wide range of operational problems. These problems raise questions about the ability of energy supplies to expand rapidly even if financial resources were available. Finally, the environmental impacts of rapid supply expansion could be substantial. The production and use of both commercial and traditional fuels contribute to the accelerating rates of environmental degradation now occurring in many developing countries. Energy trends in developing countries are also of global environmental concern. These countries are already important contributors to greenhouse gas emissions from fossil fuel use, accounting for one-quarter of annual global energy sector carbon dioxide emissions. Deforestation and the emission of other greenhouse gases, such as methane and NO_x, further raise the share of developing countries in total global greenhouse gas emissions. Although per capita levels of greenhouse emissions from energy use are much lower than in the industrial countries, the developing countries' rapid population and economic growth will increase their share of total emissions in the future.

The magnitude of these problems underlines the need for improving the efficiency of energy use and production in developing countries. Improved efficiencies can moderate the expansion of energy systems while still providing the energy services needed for development. Energy efficiencies vary in the developing world but, on average, appear to be much lower than in the industrial countries.

While the wide differences in technical efficiencies in reasonably standardized operations, such as cooking, steel making, and electricity generation, suggest that dramatic improvements in efficiencies are possible, factors other than technology also play an important role in improving efficiencies. Thus, an energy system may seem technically "inefficient" when, in many cases, users and producers are acting rationally given the framework of resources, incentives, and disincentives within which they make their decisions. One of the reasons that poor households use inefficient traditional fuels is that they lack the financial means to buy more efficient technologies and fuels. Many industrial processes are inefficient due to antiquated machinery and erratic fuel supplies of uncertain quality. The policy environment that determines this pattern of incentives and disincentives is crucial to the adoption of new technologies.

The way in which developing countries provide their energy services is important to the United States for a number of reasons:

- **International Political Stability.** Steady broad-based economic growth in the developing countries is a prerequisite for long-term international political stability. The provision of economic and reliable energy services plays a key role in securing such economic growth.
- **Humanitarian Concerns.** Humanitarian and equity concerns have long been a core element of U.S. foreign relations with developing countries. Helping developing countries to meet their energy needs can play an important role in assisting low income groups.
- **Trade and Competitiveness.** With the large trade deficits of recent years and the growing internationalization of the economy, the United States has little choice but to pay close attention to export markets. Many of these will be in the developing countries. The electricity sector of developing countries alone is projected by the World Bank to need a capital investment of nearly $750 billion during the 1990s. Similarly, there will be large markets in consumer products such as automobiles, refrigerators, and air conditioners. The United States faces intense competition in the increasingly important markets for energy efficient manufacturing processes and consumer products.
- **Global Environmental Issues.** Regional and global environmental issues such as acid rain,

ozone depletion, and global warming are strongly related to energy production and use. These issues are becoming of increasing concern to developing and industrial countries.

- **Global Oil Markets.** The World Energy Conference projects that developing countries will account for about 90 percent of the increase in world oil consumption between 1985 and 2020. This will put significant upward pressure on oil markets and could lead to both higher prices and greater volatility, with corresponding impacts on U.S. inflation, balance of trade, and overall economic performance.
- **Global Financial Markets.** High levels of developing country indebtedness (a significant portion of which was incurred in building the energy sector) affect global capital markets and the global banking system. This contributes to the instability of the U.S. and international money and banking systems.

IMPROVING EFFICIENCIES IN RESIDENTIAL, COMMERCIAL, INDUSTRIAL, AND TRANSPORTATION ENERGY USE

The Potential

This OTA study shows that it is possible to improve the low technical efficiencies with which energy is produced, converted, and used in developing countries through the adoption of proven cost effective technologies. On the demand side, these include efficient lights, refrigerators, cars and trucks, industrial boilers, electric motors, and a variety of new manufacturing processes for energy intensive industries such as steel and cement. Moreover, numerous technologies at various stages of RD&D and commercialization can further increase the efficiency of delivering these energy services. Widespread adoption of these technologies could achieve substantial energy savings, while still providing the energy services needed for development. Capturing these energy savings would help environmental quality, and ease the burden of high import bills for the many developing countries that import most of their energy supplies.

Improved technologies are not limited to the modern urban sectors of developing countries. Technology can also provide energy services for the vast majority of the population of the developing world that lives outside or on the margin of the modern economy. In rural and poor urban households, more efficient biomass stoves could both reduce fuel use and cut back the hazardous smoke emissions that are a potentially significant contributor to ill-health among women and young children. Simple motor driven systems for pumping water or grinding grain can dramatically reduce the burden on women who now spend several hours per day performing these physically demanding tasks. Energy efficient pumps, fertilizers, and mechanical traction (e.g., rototillers and small tractors) can improve agricultural productivity. Technology also can boost the efficiency, quality, and productivity of traditional small scale industry, which accounts for one-half to three-quarters of manufacturing employment in many developing countries and is an important source of income for the rural and urban poor.

The Example of Electricity

The large potential for energy efficient technologies can be illustrated in the electricity sector. Electricity demand is rising rapidly, by 10 percent or more annually in many developing countries. Matching this increase in demand with corresponding increases in electricity generation may not be feasible for many developing countries due to their high indebtedness and already strained budgets. Even if it were possible, such rates of increase would imply a substantial increase in electricity's share of development budgets, to the detriment of other important development expenditures.

This OTA study, based on conservative estimates,[1] shows that for a wide range of electricity using services—cooking, water heating, lighting, refrigeration, air conditioning, electronic information services, and industrial motor drive—overall electricity savings of nearly 50 percent are possible with currently available[2] energy efficient technologies (see figure 1-1). Further, these technologies would provide financial savings to individual con-

---

[1] See app. A for details of assumptions and calculations.

[2] The one technology included in the scenario that is not currently generally available is the high efficiency refrigerator. Refrigerators with efficiencies as high or higher have been demonstrated and are for sale in small lots, however.

Figure 1-1—Electricity, Life Cycle Cost, and Capital Cost Savings of High Efficiency End Use Equipment in the Electric Sector

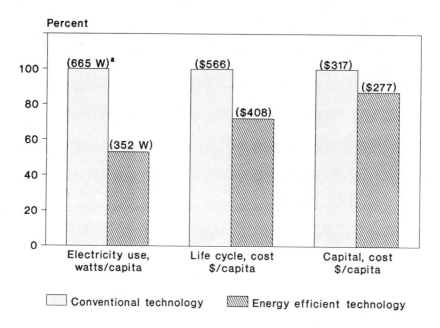

This figure shows that society-wide (including both utility and end user investments) electricity savings of about 47 percent, life cycle cost savings of 28 percent, and capital cost savings to society of about 13 percent are possible by investing in energy efficient equipment rather than the standard equipment most commonly purchased today. The assumed mix of energy services and activity levels corresponds to a Western European or American standard of living. Primary energy savings are about 2 percent less than suggested here because of the substitution of gas cooking for part of the electric cooking in the high efficiency case.

The estimates provided here are probably substantial underestimates of the capital and life cycle cost advantage of energy efficient equipment. The costs of energy supply were substantially underestimated—system performance, for example, was assumed to be higher than that achieved in virtually any full service developing country power system. On the end use side, costs were generally overestimated and savings underestimated. For example, many lower cost energy efficient alternatives—such as insulation for buildings to reduce heat gain and consequent air conditioner loads—were not included; synergisms between energy efficient equipment were not considered (e.g., high efficiency equipment gives off less heat, reducing the load on air conditioners); and the total cost of these improvements was allocated to efficiency alone whereas many of the benefits—and often the basis for making the investment—are unrelated to energy use (e.g., users may invest in improved motor drives to better control manufacturing processes, and at the same time realize substantial energy savings). Finally, various external costs such as that of the pollution associated with energy supply equipment were not included.

Details of the calculation and a sensitivity analysis are provided in appendix A at the back of this report.

[a] W = watt

SOURCE: U.S. Congress, Office of Technology Assessment, 1992.

sumers over the lifetime of the equipment of more than 25 percent.

It is commonly believed, however, that widespread adoption of energy efficient technologies will not occur, in large part because of their perceived high capital costs. This is an important consideration for poor, heavily indebted countries. The argument that the capital costs of energy efficient equipment are too high depends critically, however, on the frame of reference for considering capital costs. This OTA study shows that when all the systemwide financial costs are accounted for, energy efficient equipment usually can provide the same energy services to the Nation at a **lower** installed capital cost than less efficient equipment (see figure 1-1). The estimates presented here suggest that over 10 percent of the initial capital investment in the

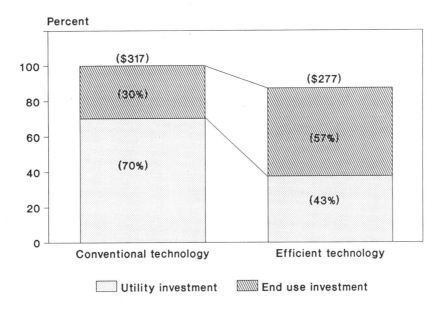

Figure 1-2—Allocation of Capital Investment Between End User and Utility

This figure shows that the capital costs of high efficiency end use equipment fall largely on the end user (but the end user still saves money over the equipment's lifetime) while the required capital investment by the utility drops substantially (and the total capital investment to society is reduced—figure 1-1). This is the basis for the belief that end use equipment has a higher initial capital cost—it does cost more for the end user. Redistributing these capital costs so as to reduce the burden felt by the end user (who is typically extremely sensitive to initial capital costs) could thus allow greater penetration of these highly advantageous technologies.
SOURCE: U.S. Congress, Office of Technology Assessment, 1992.

electric sector could be avoided by investing in currently available high-efficiency equipment. With projected "business as usual" investments in the electric utility supply sector in developing countries expected to total $750 billion during the 1990s, energy efficiency could save substantial capital.

Capital could be saved because the higher initial cost of efficient end use equipment is usually outweighed by the savings realized from building fewer power plants. Taking advantage of opportunities to install energy efficient equipment is particularly important in developing countries because of the rapid growth in stocks of energy using equipment and the high share of total investment budgets devoted to increasing energy supplies.

The perception that energy efficient equipment has a high capital cost results from the way institutions account for capital costs of energy supply and end use equipment. Consumers who purchase end use equipment see the increase in capital cost of more efficient designs, but not the decrease in capital costs as fewer power plants are built in order to provide a given level of energy service.

Though energy efficiency could result in substantial capital savings, the allocation of investment between energy supply and end use would need to change dramatically. Capital investment in electricity supply (borne by the utility) would be cut nearly in half while that in end use equipment (currently borne by the end user) would increase by about two-thirds (see figure 1-2). Such a reallocation of expenditure patterns is likely to impede adoption of efficient end use equipment. To overcome these barriers, institutional changes are needed to enable decisionmaking on a systemwide basis, and to initiate innovative mechanisms that focus the financial resources of utilities on the adoption of efficient end use equipment. A powerful tool for achieving such a systems approach is Integrated Resource Planning (IRP—see ch.3), in which energy efficiency investments are explicitly included as an alternative to capacity expansion. This methodology is supported by many utilities and public utility commissions in the United States.

Impediments to the Adoption of Energy Efficient Technology

Despite the multiple benefits of these energy efficient technologies, many have not yet been widely used because of pricing policies, a variety of market failures, institutional impediments, and technical barriers.

Pricing Policies

Government pricing policies frequently discourage the adoption of energy efficient equipment. Energy pricing policies have several objectives: the efficient allocation of resources, social objectives, reasonable returns to producers, substitution between fuels for national security or environmental reasons, and industrial competitiveness. Social objectives, including the desire to ensure that household fuels are price-stable and affordable by the poor, play a major role in pricing policies in developing countries. Economic objectives, notably the desire to encourage key strategic development sectors, are reflected in policies that promote rural electrification or that keep diesel prices low. As a result, energy prices in developing countries, particularly for electricity and—in many oil-producing countries—for petroleum-based fuels, are often set by the government at well below market value. For example, the average cost of electricity to consumers in developing countries is just 60 percent of the cost of producing it.

Policies that keep key energy prices low, however, can bias the choice of technology away from energy efficiency. Low prices also may result in revenues that cannot cover the costs of supplying the energy, leading to a decline in quality and availability of energy supplies. Though designed to help the poorer part of the community, energy subsidies may in fact largely benefit the more affluent classes who are the heaviest users of commercial fuels.

Market Failures

The diffusion of technology within the developing countries is also impeded by a variety of market failures. In the residential/commercial sector, for example, consumers are extremely sensitive to the first costs of equipment, and in many cases are not even the purchasers of the equipment they use. The contractor who builds the office or the landlord who rents the housing often purchase the appliances used in the building and base their choice of appliance on lowest first cost rather than life cycle operating costs. The tenants, not the owners, must pay the cost of operating this inefficient equipment.

In the industrial sector, plant managers' efforts to improve energy efficiency can be impeded by frequent brownouts or blackouts; lack of foreign exchange to purchase critical components not available locally; and lack of skilled engineers and managers. Each of these factors can be particularly disruptive to the utilization of energy efficient technologies.

Moreover, energy efficiency is often of secondary interest to industrial firms. Energy is just one, and often a minor, component of overall corporate strategy to improve profitability and competitiveness. Energy must compete with other factors—the financial return, the quality and quantity of product produced, the timeliness and reliability of the production equipment, and the flexibility of the equipment—when investment choices are made and scarce time of skilled manpower allocated. In many countries, local industries are protected from competition and therefore have less incentive to lower costs through the introduction of energy efficient equipment.

Further, because of poor transport infrastructure, scarcity of capital, and limited ability to bear risk, manufacturing plants are often much smaller in scale than their industrial country counterparts. Small scale manufacturing plants, however, are typically less energy efficient and require greater capital investment per unit output to realize the energy savings that can be achieved in large plants.

In the transport sector, higher first costs of energy efficient autos, trucks, and motorcycles often deter consumers from buying them, despite being cost effective on a life cycle basis. The available infrastructure may strongly influence the choice of technology as well as the efficiency of its use. Poor land use planning or ineffective controls, for example, may result in urban sprawl, which in turn leads to reliance on personal rather than mass transport. A high degree of reliance on personal transport can then cause congestion, lower average speeds, and reduce efficiencies, subsequently increasing demands for capital-intensive highways.

Supply Biases

In the public sector, the multilateral development banks and bilateral donor agencies influence the amount and type of technology transferred through their lending and technical assistance activities. The bulk of technology transfer activities in these agencies has been directed to large scale supply oriented projects—notably major electricity generating facilities including hydroelectricity. The activities and approach of these agencies do not appear to be well adapted to the special needs of transferring conservation and small renewables technologies. The private sector, too, is accustomed to large scale conventional supply projects.

Technical Barriers

The energy efficient equipment developed in the industrial countries may not be suited to conditions in developing countries. Considerable adaptation may be needed, but manufacturers may not have adequate funds available for the necessary RD&D—especially when supplying companies are small and developing country markets are limited and expensive to access. There are also technology gaps. Technologies important to the rural populace, the majority of the population in developing countries, get relatively little attention.

Policy Responses

A number of policies might be adopted throughout the technology transfer process to help overcome these barriers. These include: increased attention to technology adaptation; increased training in energy efficient end use or improved supply technologies; reforming energy pricing policies to reflect the full costs to the Nation of supplying energy; taxation of inefficient equipment; improved information to consumers; financial incentives (e.g., tax relief or low-cost loans) to encourage production or purchase of energy efficient equipment; and efficiency standards. Reforms in energy pricing policies can encourage the purchase of energy efficient equipment, but prices alone are usually insufficient to overcome the strong nonprice and institutional barriers to improved efficiency. Energy efficiency standards, innovative financial mechanisms, or other policies might be used to reinforce price signals.

Changes in regulation of the electricity supply industry to create closer links between suppliers and users could overcome a major institutional barrier to the adoption of efficient appliances. Decisionmaking in the key electricity sector is strongly biased in favor of supply expansion and typically does not take a "systems" view (see ch.3) of electricity service. One way to encourage this to happen is through the use of Integrated Resource Planning (IRP—see ch.3) in utility decisionmaking. IRP includes both energy efficiency and energy supply options in decisions about how to provide energy services. This contrasts sharply with current procedures that focus on supply options only. Though straightforward in concept, IRP can, however, be difficult to implement. Supporting changes in utility regulation would be required, as well as substantial complementary education, training, demonstration, and monitoring of results.

IMPROVING ENERGY SUPPLY EFFICIENCY

Even with substantial end use efficiency gains, growing populations and increased economic activity in developing countries will require expansion of energy supplies. Here, too, available technologies can improve the efficiency of supplying and converting energy into useful forms and of developing domestic resources. Improved technologies also can moderate the environmental effects of energy production.

Conversion Technologies

A number of technologies have the potential to improve the efficiency and performance of the energy conversion sector. Given low operational efficiencies in the electricity sector, technologies for plant rehabilitation, life extension, system interconnection, and improvements in transmission and distribution systems often offer higher returns to capital investment than do new plants. Improvement of existing systems would have the added advantage of putting into place a more efficient framework for future capacity expansion. By the same token, failure to improve the existing system will be detrimental to efficient supply expansion in the future, regardless of technology.

For capacity expansion, fluidized-bed combustion (FBC) has greater tolerance for the low-quality coal often used in developing countries and can reduce SO_2 and NO_x emissions. Advanced gas turbines, particularly when used in a combined-cycle or steam-injected mode—promise to be one of the most attractive new technologies for electricity generation in the developing world. Operating efficiencies can be high, and because they are small and modular with short construction lead times, gas turbines are also suited to private power producers with limited capital.

The relatively small, modular, and safer nuclear power technologies under development may extend the market for nuclear power in developing countries. Their cost and performance characteristics, however, are uncertain. The technical skill requirements, relatively high cost, and ongoing concerns about waste disposal may also continue to limit the use of nuclear power in developing countries. Another important concern is the potential for nuclear weapons proliferation, which has been greatly heightened by the recent disclosure of the relatively advanced status of Iraq's nuclear weapons program.

Development of the extensive hydroelectric resources in many countries has already provided large amounts of power. Large scale hydroelectric development remains problematic, however, due to the economic and social impacts of large scale projects, and, in many cases (as in Africa), initial lack of markets. Smaller scale hydro could overcome some of these problems.

Stand alone renewables, such as wind turbines and photovoltaics, are now cost competitive with diesel generators and grid extension in many situations. Although the role of renewables remains very limited today, it could expand substantially in the future as their costs continue to decline. These technologies are of particular importance in providing high-quality energy sources to rural areas. Geothermal energy is also available for a select few countries. A number of technologies offer promise for the conversion of biomass into improved liquid and gaseous fuels, if cost effective and sustainable biomass supplies can be assured (see below). Increased conversion of biomass is particularly promising where there are captive residues, such as in agroindustry or the forest products industry. Much more efficient use of the biomass now consumed is possible as well, through modern bioenergy conversion and utilization technologies.

Cost effective retrofits could improve refinery efficiencies, and new technologies could permit small scale and relatively low cost conversion of methane to methanol.

Primary Energy Supplies

China, India, and several other countries have large coal resources. China and India, the two biggest users, plan to expand their coal use considerably. Coal as presently produced and used in these countries has severe environmental impacts such as emissions of acid rain precursors and greenhouse gases, and mining damage to the land and water. Large scale use also poses major logistical problems when coal fields are distant from major users. Substantial improvements in coal mining in India and China could be achieved through increased use of high-performance mining and beneficiation technologies currently used in other countries.

The developing world possesses only limited crude oil reserves, with a reserves/production ratio of 26 years compared with a worldwide ratio of 43 years. These reserves are concentrated in a few countries, including Mexico and Venezuela. One-half of the developing countries have no discovered recoverable reserves. The industry consensus is that the oil reserves likely to be proved in developing countries will be relatively small. The development of such fields, while traditionally unattractive to the major oil companies, is important for the developing countries themselves—especially the poorer ones. A number of recent technical developments in oil exploration and development may reduce risks and costs, thus making small field development more attractive than before, and facilitating the entry of smaller oil companies in developing countries.

These technologies also could aid exploration for natural gas, a fairly versatile fuel and one that is environmentally less damaging than coal or oil—both locally and globally. The reserves/production ratio for natural gas in the developing world is about 88 years—much more favorable than crude oil. Natural gas resources are more widely dispersed among the developing countries than oil. In several countries, large amounts of developed natural gas are currently flared because of the lack of infrastructure and organized markets.

Biomass is an important energy resource that could be better used in most countries. In the technical arena, several advances have been made to improve plant productivity in recent years. Fast growing species, and intercropping and multiple species opportunities have been identified, physiological knowledge of plant growth processes has been improved, and there have been breakthroughs in manipulation of plants through biotechnology. Crop residue densification increases their energy content and in some cases can thus reduce transport and handling costs.

The Benefits

These new energy supply technologies offer numerous potential advantages:

- **Modular, Small Scale, and Short Lead Times.** Several of these new supply technologies, notably advanced gas turbines and many types of renewable energy technologies, are small and modular, and can therefore better match demand growth both in size and construction time. Conventional coal-fired power plants or petroleum refineries generally are constructed in very large units with long lead times—as much as 10 years or more. The large incremental size of energy supply systems makes it difficult to match supply to demand. The long lead times result in large quantities of capital being tied up in a project for years before the project becomes productive. Both of these factors raise costs and increase risks.
- **Plant Reliability and Performance.** Developing countries experience frequent service curtailments, including blackouts, brownouts, and sharp power surges. Unreliable service means that industries and offices are unable to operate, production is lowered, and raw materials are wasted. Losses can be considerable. Lost industrial output caused by shortages of electricity in India and Pakistan is estimated to have reduced Gross Domestic Product (GDP) by about 1.5 to 2 percent. Many consumers—residential, commercial, and industrial—are obliged to invest in a variety of equipment—voltage boosters, standby generators, storage batteries, kerosene lamps—in order to minimize the impact of disrupted supplies. Improved operating and maintenance procedures as well as a variety of new technologies could improve plant reliability and performance.
- **Rural Energy.** Most of the population of developing countries live in rural areas, the great majority in poverty and without access to the services that could increase their productivity and improve their standard of living. Several of the smaller scale technologies (modern biomass energy and decentralized renewables) can bring high quality fuels to rural areas and thus promote rural development and employment.
- **Environmental Benefits.** Different types of energy supply may have different environmental impacts over various time periods. For example, a coal-fired power plant generates air emissions today with both near-term acid rain and potentially longer term global warming impacts, while a nuclear plant has the potential (though the probability is low) for catastrophic release of radiation at any time. This, plus the general lack of knowledge of the environmental impacts, makes the comparison of different types of energy supply difficult.

  With these reservations, this analysis suggests that among conventional systems, natural gas generally has the least adverse environmental impacts. Increased emphasis on natural gas could reduce land disturbances, air and water pollution, and occupational health hazards compared with coal, and could avoid some of the problems of large hydro and nuclear. Modern biomass systems (properly handled) could also reduce environmental impacts compared to coal or other conventional fuels. Decentralized renewables generate less air pollution, and, because of their small scale incremental nature, may avoid some of the environmental problems of large scale energy projects. There is, however, little experience with large scale use of decentralized technologies to serve as a basis for firm decisionmaking.
- **Foreign Exchange Saving.** New technologies that develop local energy resources will reduce energy imports—which currently account for over 50 percent of export earnings in several of the poorest countries.
- **Employment.** Decentralized renewables could stimulate employment. Even though developing countries would, at least initially, import much of the technology, installation and servicing would create jobs. Production of biomass could create rural employment.

Impediments

In practice, a variety of factors impede the use of these energy supply technologies, including problems in the energy sector itself, such as a lack of adequate technology.

institutional structure and procedures. The operating efficiency of electricity generating equipment and oil refineries in many developing countries is often substantially below that of industrialized countries with similar technology. A number of factors lie behind this poor performance: official interference in day-to-day management of the utilities; overstaffing but shortages of trained manpower; lack of standardization of equipment; limited system integration and planning; distorted pricing structures; obligations to provide parts of the population with electricity at less than cost; shortages of foreign exchange to buy spare parts; and regulatory frameworks that discourage competition. The effective deployment of technology will therefore depend on addressing these related financial, policy and institutional issues.

Institutional issues are also important for accelerated oil and gas development. Despite promising geological prospects, hydrocarbon exploration activity (density of wells drilled) in the developing countries is much lower than the world average, and is concentrated in countries where resources have already been developed. In most developing countries (with the exception of a few large-population countries, such as Mexico and Venezuela), investment in petroleum exploration and development is carried out almost exclusively by international oil companies. Thus, the fiscal and contractual arrangements between country and company need to include appropriate incentives. These incentives have traditionally been biased in favor of large, low cost fields and may need to be changed to encourage the development of small higher cost fields. Gas development faces additional obstacles. Unlike oil markets, gas markets must be developed concurrently with the resource, adding to the start-up costs and complexity of gas projects. In the past, this has been a formidable obstacle, but new technologies, such as small high-performance gas turbines, could greatly help market development. Even so, gas sold in local markets does not directly generate the foreign exchange needed to repatriate profits to foreign investors.

The introduction of modern biomass energy industries encounters different problems. These include inadequate research, development, and demonstration; direct or indirect subsidies to other fuels that may discourage investment; high infrastructure (notably roads) costs; and credit, which may not be as readily available for modern biomass supply systems as for other more conventional supplies such as coal- or oil-fired electricity.

The development of large scale, cost effective, sustainable biomass feedstocks also faces uncertainties. Data on the extent of forest area and the annual increment of forest growth are sparse and unreliable, and little is known about the impacts of intensive biomass development on soils and other environmental assets. Improvements in forest management are notoriously difficult to achieve, and the introduction of high-yield energy crops together with the necessary improvements in agricultural practice will require long-term sustained efforts. Finally, policies to promote bioenergy could create competition for land between energy crops for the rich or food crops for the poor.

ISSUES AND OPTIONS FOR THE UNITED STATES

Developing countries have been slow to adopt improved energy end use and supply technologies, despite the potential advantages, due to the technical, institutional, and economic barriers described above. Developing countries are, however, demonstrating interest in seeking alternative ways of meeting the demand for energy services, despite the difficulties of changing entrenched systems. Increased attention is being given, for example, to politically sensitive questions such as energy price reform, improved management, and operations efficiency in state-owned energy supply industries. Several developing countries are taking steps to encourage private investment in electricity and in oil and gas. Many countries have developed capable energy resource and policy institutions. Progress is also being made on the environmental front. Although much of the current environmental focus of developing countries is on local rather than global conditions, these countries also participate in international resource and environmental protection policy discussion and treaties.

There is also evidence of change in donor institutions. The bilateral donor agencies and the multilateral development banks (of which the most influential is the World Bank) are beginning—often under pressure from Congress and nongovernmental organizations—to incorporate environmental planning into their projects, to develop energy conservation projects, and to encourage a larger role for the private sector. This momentum for change offers a timely opportunity for U.S. initiatives.

A substantial number of U.S. agencies already has programs that influence the diffusion of improved energy technologies in developing countries. These include the Agency for International Development (AID), the Trade and Development Program, the Departments of Energy, Commerce, and the Treasury, the Overseas Private Investment Corporation, the Export-Import Bank, the Small Business Administration, and the U.S. Trade Representative. The United States exercises additional influence through membership in international organizations, notably the World Bank and the regional development banks and the United Nations programs. A number of U.S. based industry groups and nongovernmental organizations is also active in this field. These agencies and organizations cover a wide range of technology transfer and diffusion activities—research, development, and demonstration; project loans and grants; education, training, and technical assistance; information services; policy advice; support for exports and private investment; and others.

In one sense, the United States already has a considerable policy infrastructure in place for promoting energy technologies in developing countries. This policy infrastructure has, however, focused on supply development, including renewables in recent years, and is only now beginning to accept efficiency as an important theme. Further, given the large number of programs and activities, the question arises of their cohesion and cooperation to ensure maximum effectiveness. Efforts have been made to coordinate at least some aspects of the work of the different agencies and organizations through both formal and informal channels. For example, in the Renewable Energy Industry Development Act of 1983, Congress initiated a multiagency committee called the Committee on Renewable Energy Commerce and Trade (CORECT) to promote trade in U.S. renewable energy technologies. Now could be a good time to examine the extent to which this model of coordination could be applied to other relevant areas, such as energy efficiency and the environment.

Despite the large number of programs and wide range of activities, the current level of U.S. bilateral aid for energy is small. USAID grants and other assistance in the energy sector total about $200 million per year compared with multilateral development bank (MDBs) annual energy loans of $5 billion. The small scale of U.S. bilateral assistance for energy suggests that the sums available are currently and will continue to be used to greatest effect by:

- using limited U.S. bilateral grant monies to promote technical assistance and institution building;
- including technology transfer in broader bilateral policy discussions, such as debt negotiations;
- influencing the activities of the multilateral development banks;
- cooperating with other bilateral donors, lending agencies, and private voluntary organizations (PVOs); and
- encouraging a wider role for the private sectors of both the industrial and developing countries.

In this context, the analysis presented below leads to a series of broad policy options for Congress to consider. In particular, the following policy areas merit priority consideration:

- devoting additional attention to energy efficiency, the environmental impacts of energy developments, and the energy needs of the rural and urban poor in current bilateral and multilateral lending programs;
- encouraging energy price reform in developing countries to stimulate the adoption of energy efficient equipment and to help finance needed supply expansion;
- providing technical assistance to support Integrated Resource Planning (IRP) and associated regulatory reform to guide investments in energy supply and end use projects, particularly in the electricity sector—this could help developing countries secure the savings potential illustrated in figure 1-1;
- providing technical assistance, information, and training in other potentially significant areas, such as environmental protection, appliance and industrial equipment efficiency, utility management, and transportation planning;
- assisting in institution building, especially for technology research, development, adaptation, testing, and demonstration;
- encouraging private sector (from both the United States and developing countries) participation in energy development;

- expanding U.S. trade and investment programs—U.S. energy related exports and investment have traditionally been an important channel for energy technology transfer to developing countries;
- making sure that the United States sets a good example for the rest of the world in energy efficiency and environmental protection.

Congress has already taken action in several of these areas—support for IRP, efficient energy pricing, and consideration of environmental impacts of projects. In these cases, the primary question may be the effectiveness of existing interventions rather than the need to take additional action.

Efficient energy technologies often reduce systemwide capital investment as well as life cycle operating costs. Under these conditions, redirecting capital funds from supply expansion to energy efficiency projects in the MDBs and other financing institutions would free resources for additional investment in energy services or other pressing development needs. Even if all these savings were reinvested in the energy sector, however, the rapid rise in demand for energy services will require substantially more investment than that projected to be available. The MDBs and other bilateral and multilateral financial institutions will need, therefore, to continue providing high levels of support, while at the same time supporting actions (e.g., debt negotiations, macroeconomic reform, and privatization) to encourage increased private sector participation.

Several of the options for accelerating the adoption of energy efficient technology imply an increase in bilateral assistance. While increases in bilateral aid run counter to efforts to control budget expenditures, the share of bilateral aid (particularly in AID) devoted to energy is low in relation to: total bilateral expenditures; the share of energy in the aid efforts of other donors; and the potential importance of developing countries as markets for U.S. exports of improved energy technologies. Some redistribution of expenditures could be considered. The geographical distribution of existing AID energy expenditures, concentrated in the Near East, may not adequately reflect the totality of U.S. policy interests.

The diffusion of improved energy technologies in developing countries and the ways in which the United States can accelerate this diffusion is a complicated process involving a number of agencies and institutions. The principle behind these actions is, however, simple: energy is not used for its own sake, but for the services it makes possible. If energy policy concentrates on the best way to provide these services, rather than automatically encouraging increased energy supplies, opportunities expand dramatically. Though it is not easy for institutions designed for supply expansion to change to a service orientation, their interest is growing and the stakes are high. This analysis provides strong evidence that such a change could release billions of dollars that could provide energy services to those who would otherwise be without, or to finance the many other economic and social goals of developing countries. The United States can make an important contribution to this change in policy through its bilateral programs, its influence on MDB programs, and through cooperation with other donors. In so doing, the United States could develop closer partnerships with the developing countries themselves, who are increasingly looking for new, more efficient ways to meet the rapidly growing demand for energy services.

Chapter 2
Energy and the Developing Countries

Photo credit: U.S. Agency for International Development

Contents

	Page
INTRODUCTION	17
ANALYTICAL FRAMEWORK	19
DEVELOPING AND INDUSTRIALIZED COUNTRIES	21
PATTERNS OF ENERGY USE AND SUPPLY IN DEVELOPING COUNTRIES	22
Energy Use	22
Energy Supply	25
Developing Countries in World Energy	26
TRENDS IN ENERGY DEMAND IN DEVELOPING COUNTRIES	27
Factors Increasing Energy Demand	27
Difficulties in Meeting Energy Demands	29
ENERGY, ECONOMIC GROWTH, AND ENVIRONMENTAL QUALITY	31
Energy and Economic Development	31
Energy and the Traditional Economy	33
Energy and the Environment	37
Greenhouse Gases and Global Climate Change	39
CONCLUSION	43

Boxes

Box	Page
2-A. The OTA Study in Context	18
2-B. Highlights of the Intergovernmental Panel on Climate Change 1990 Scientific Assessment	41

Chapter 2
Energy and the Developing Countries

INTRODUCTION[1]

Energy use in the developing countries has risen more than fourfold over the past three decades and is expected to continue increasing rapidly in the future. The increase in the **services** that energy provides is necessary and desirable, since energy services are essential for economic growth, improved living standards, and to provide for rising populations. But the traditional way of meeting these energy service needs—primarily by increasing **energy supplies** with little attention to the efficiency with which energy is used—could cause serious economic, environmental, and social problems. For many of the developing countries, much of the additional energy needed will be supplied by imported oil, thus further burdening those countries already saddled with high oil import bills. Similarly, building dams or powerplants to meet higher demands for electricity could push these nations even deeper into debt. Energy production and use in developing countries contributes to local and regional environmental damage, and, on the global scale, accounts for a substantial and rising share of greenhouse gas emissions.

The way in which developing countries provide their energy services is important to the United States for a number of reasons:

- **International Political Stability.** Steady broad-based economic growth in the developing countries is a prerequisite for long-term international political stability. The provision of economic and reliable energy services plays a key role in securing such economic growth.
- **Humanitarian Concerns.** Humanitarian and equity concerns have long been a core element of U.S. foreign relations with developing countries. Assisting developing countries to meet their energy needs can play an important role in helping low income groups.
- **Trade and Competitiveness.** With the large trade deficits of recent years and the growing internationalization of the economy, the United States has little choice but to pay close attention to export markets. Many of these will be in the developing countries. The developing countries are potential markets for U.S. manufacturers through direct sales, joint ventures, and other arrangements. The electric power sector of developing countries alone is projected by the World Bank to need a capital investment of nearly $750 billion during the 1990s. There will similarly be large markets in consumer products such as automobiles, refrigerators, air conditioners, and many other goods. The United States faces intense competition in the increasingly important markets for energy efficient manufacturing processes and consumer products.
- **Global Environmental Issues.** Regional and global environmental issues such as acid rain, ozone depletion, and global warming are strongly related to energy production and use. These issues are becoming of increasing concern in developing and industrial countries.
- **Global Oil Markets.** The World Energy Conference projects that developing countries will account for about 90 percent of the increase in world oil consumption between 1985 and 2020. This will put significant upward pressure on oil markets and could lead to both higher prices and greater volatility, with corresponding impacts on U.S. inflation, balance of trade, and overall economic performance. With business as usual, U.S. demand for imported oil is also likely to increase dramatically over the same time period.
- **Global Financial Markets.** High levels of developing country indebtedness (a significant portion of which was incurred in building the energy sector) affect global capital markets and the global banking system. This contributes to the instability of the U.S. and international money and banking systems.

This report (see box 2-A) evaluates ways of better providing energy services for development. The analysis examines technologies (hardware, but also the knowledge, skills, spare parts, and other infrastructure) that permit energy to be used and supplied

---

[1] This chapter draws heavily on the interim report of this project, U.S. Congress, Office of Technology Assessment, *Energy in Developing Countries*, OTA-E-486 (Washington, DC: U.S. Government Printing Office), January 1991.

> ### Box 2-A—The OTA Study In Context
>
> The eight Congressional committees and subcommittees that requested or endorsed this study asked OTA to examine: the potential of new energy technologies to contribute to economic growth while protecting the environment in the developing countries, and the role of U.S. policy in encouraging the rapid adoption of these technologies.
>
> OTA responded to this request in two documents, an interim report, *Energy in Developing Countries* released in January 1991, and the present final report *Fueling Development: Energy Technologies for Developing Countries*. The interim report examines how energy is currently supplied and used in the developing countries and how energy is linked with economic and social development and the quality of the environment. A major finding was that despite the low level of energy consumption in developing countries, energy was often produced, converted, and used inefficiently.
>
> The present report, *Fueling Development*, builds on the findings of the interim report by examining the role of technology in improving the efficiency with which energy is supplied and used. This final report incorporates part of the interim report, but readers with more specific interest in patterns of energy demand and supply and its relation to economic growth and environmental quality are referred to the interim report.
>
> The emphasis in the current report is on improving energy efficiencies in using and supplying energy. The emphasis on efficiency should not be taken to minimize the importance of economic structure on energy consumption. Variations in economy wide energy/GDP ratios between countries, and differing experiences over time, suggest that differences in economic structure do have an important role to play in explaining differences in energy consumption. These differences are more often related to economic and social policies, and country specific conditions and endowments, however, than to technology—the focus of this study.
>
> We deliberately take a conservative approach to technology, examining only those that are at present commercially available or are expected to become commercial soon. This is not to discount the remarkable improvements in efficiency now on the drawing board, or the new energy supplies whose commercial development would have revolutionary consequences for all countries of the world. Instead, it recognizes that there is much that can be done here and now in the developing countries using established technologies, and that poor countries cannot be expected to take technological risks, especially when they are not necessary.
>
> Given the immediate pressing needs of the citizens of the developing world for energy services, and the vast potential for improving energy efficiencies with known technologies, it seemed more appropriate to concentrate on relatively near-term responses. Many of the options—improving efficiencies and moving to cleaner energy and renewables—will nevertheless contribute to the solution of longer-term issues such as global warming. The basis for our economic calculations of the cost effectiveness of energy efficient technologies is also conservative, designed to under- rather than over-estimate the cost advantages of energy efficient equipment.

more efficiently, and the institutional and policy mechanisms that determine their rate of adoption. Based on this analysis, the role of the U.S. Government in promoting the transfer and accelerated adoption of improved energy technologies to developing countries is examined.

Developing countries vary widely in their socio-economic development; their patterns of energy use and supply; and the linkages between their energy use, economic development, and environmental quality. Awareness of these factors is an important first step in developing policies that can effectively respond to the wide range of conditions found in developing countries. For example, the problems—energy or otherwise—faced by a relatively rich and developed country such as Brazil are different from those faced by a poor country like Ethiopia, as are the resources available for their solution. An appreciation of these differences is necessary for the realistic assessment of energy technologies.

As background to the subsequent analysis, this chapter therefore briefly introduces five sets of issues. First, it describes OTA's analytic approach. Second, it examines who the developing countries are and how they differ from the industrialized countries and from each other. Third, it reviews how energy is used and supplied in these countries. Fourth, it examines the trends in energy use that are leading to the developing countries' growing share in world energy consumption. Finally, it explores how energy is linked with economic growth and with environmental quality. Subsequent chapters examine the extent to which more efficient and cost effective energy use and supply systems can provide

> Despite our focussed approach, we were obliged, due to the breadth of the subject and congressional interest in our considering the whole group of vastly disparate developing countries, to be selective in our coverage. As our interim report pointed out, most of the energy used in developing countries today is in the residential sector, for cooking and lighting, and the industrial sector for process heat and electric drive. We therefore emphasize these sectors and services in our energy service analysis. The third major sector, transport, is quantitatively less important than the others, but merits attention because of its reliance on oil, and its contribution to urban air pollution. Perhaps more than the other sectors, however, improvements in the overall efficiency of the transport sector in developing countries depend primarily on transport and urban planning—rather outside the scope of this report—than pure technology, though here again there is major scope for improvement. The services provided by electricity warrant particular attention for a number of reasons; the major share of the electricity sector in development budgets; the current financial and operational problems faced by the sector; the large bilateral and multilateral aid component in electricity development that provides the opportunity for leveraging sectoral change; and the large and costly increases in generation capacity that could be avoided through end-use efficiency improvements.
>
> Oil and gas exploration also are examined in some detail because of the heavy burden imposed by oil imports for many countries, and the major environmental advantages of gas as a fuel. Because most of the population of the developing world live in rural areas, we examine the potential for providing the energy needed for rural and agricultural development—decentralized forms of electricity generation based on locally available renewable resources, and superior forms of fuel based on biomass.
>
> Many of these issues are common to all developing countries. Certain groups of countries, however, command particular attention. India and China receive particular emphasis because of the weight of their total energy consumption, their intensive coal use, and the high energy intensities of their economies. The countries of Africa, though their commercial energy consumption is very small in global and even in developing country terms, need special attention because of their acute poverty and often declining living standards. Latin America receives special mention as well, due to an especially heavy debt burden and strong traditional economic ties to the United States.
>
> Our treatment of policy options is conditioned by three factors. First, it was not possible to cover in consistent detail the content of the many U.S. and Multilateral Development Bank activities affecting the transfer of improved energy technology to developing countries. Second, many of these programs have changed since this project was started, and are continuing to change. Third, Congress has already addressed or is addressing many of the energy issues in connection with its interest in environmental quality. Thus the options identify policy issues rather than providing solutions. Each of the areas identified deserves detailed study before specific options could be justified.
>
> SOURCE: Office of Technology Assessment, 1992.

the energy services needed for development, thus moderating increases in energy consumption.

ANALYTICAL FRAMEWORK

The analysis presented in this OTA study has three important features. First, the analysis focuses on the services energy provides, rather than on simply increasing energy supplies. The reason for this approach is simple. Energy is not used for its own sake, but rather for the services it makes possible—cooking, water heating, cooling a house, heating an industrial boiler, transporting freight and people. Further, there may be many different means of providing a desired service, each with its own costs and benefits. For example, transport can be provided in a number of ways—by bicycle, motorcycle, car, bus, light rail, aircraft. The consumer chooses among these according to such criteria as cost, comfort, convenience, speed, and even aesthetics. Within these consumer constraints, a more efficient car may be preferable to increasing refinery capacity in order to reduce capital and/or operating costs, or because of environmental benefits. More than just engineering and economics must be considered, including social, cultural, and institutional factors. Such factors are more readily included in a services framework than in a conventional energy supply analysis. A flow diagram comparing the analytical frameworks for energy services versus energy supplies is shown in figure 2-1.

Second, within this services framework the changes in how energy is used are traced from traditional rural areas to their modern urban counterparts. This progression from the traditional rural to the modern

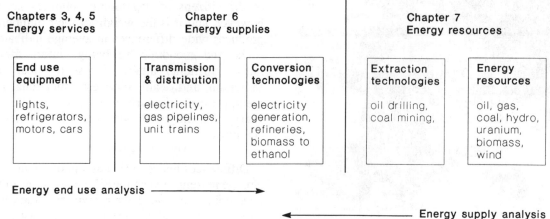

Figure 2-1—The OTA Analytical Framework

SOURCE: Office of Technology Assessment, 1992.

urban captures well the dynamics of energy use and how it can be expected to change in the future.

Third, the entire system needed to provide energy services—from the energy resource to the final energy service, including production, conversion, and use—is examined as a whole. This is done in order to show the total costs and consequences to society, as well as to the individual, of providing particular services and how they might be provided more effectively in terms of financial, environmental, and other costs. For example, increased lighting services might be met by increasing the amount of electricity generated, by increasing the use of more efficient light bulbs and reflectors, or by a combination of the two, perhaps in conjunction with daylighting techniques. A systems approach permits the comparison of efficiency and supply options in achieving the desired end.

The energy services analysis explicitly identifies potential institutional problems by highlighting the gap between what is and what could be. The analysis thereby recognizes that technology adoption and use is embedded in an institutional framework that provides both incentives and disincentives to users, and largely determines which and how technologies will be used.

This approach has a number of implications both for the way technology is used now and for the adoption of new technologies in the future. Thus, the energy sector in many developing countries is frequently characterized as "inefficient" in the sense that more energy is used to provide a given service or output than is usual in industrial countries. In a wider context, however, taking into account the many other relevant factors (financial, infrastructural, managerial, and institutional) the technology may well be used to the best of human ability and often with considerable ingenuity and resourcefulness. In many cases, although energy appears from the outside to be used inefficiently, energy users may be acting logically given the framework of incentives and disincentives within which they make their decisions. It follows, therefore, that the adoption of a new technology will depend not only on the intrinsic superiority of the technology itself but also on whether institutional factors favor its adoption. The policy environment is of crucial importance to the adoption of new technologies.

---

[2]This report largely follows the definition of "developing" countries—low-and middle-income countries—used by the World Bank, including all of the countries of Africa, Latin America, and Asia, excluding Japan. Saudi Arabia, Kuwait, and the United Arab Emirates are not included by virtue of their high per capita income. The World Bank does, however, include as developing countries some East and West European countries, such as Poland, Hungary, Yugoslavia, Greece, and Turkey, that qualify as developing countries by virtue of their income levels, but, due to their integration with industrial economies of East and West Europe, do not share other characteristics of underdevelopment, and are therefore not included in this OTA report. Where group averages of general economic and social indicators are reported directly from the *World Development Report 1989*, these countries are included in the total. In more detailed analysis, they are excluded. While every effort is made to adhere to these definitions, it is not always possible, especially when other sources of data with slightly different definitions are used. See, for example, World Bank, *World Development Report 1989* (New York, NY: Oxford University Press, 1989), Table 1, World Development Indicators, pp. 164-165.

Photo credit: Appropriate Technologies, International.

DEVELOPING AND INDUSTRIALIZED COUNTRIES

It is difficult to clearly define the group of "developing nations."[2] If all the countries in the world are ranked by any widely used development index (e.g., GDP per capita), there is no distinct gap in the series that would suggest that those above were "rich" countries, and those below "poor." Further, as is readily observable, some citizens of developing countries have standards of living as high as, or higher than, many citizens of industrial countries. Indeed, the differences between the rich and poor in virtually any developing country are usually more dramatic than the average differences between developing and industrial countries. Most of the citizens of developing countries are poor, however, and it is the weight of their numbers that produces wide differences in average indicators of social and economic wellbeing—infant mortality rates; nutritional intake; access to clean water, sanitation, and health services; educational attainment; per capita income; life expectancy; per capita energy consumption[3]—between the rich and poor countries.

Differences between rich and poor countries also emerge in trends over time. The economies of many developing countries have grown more rapidly than those of industrial countries over the past century, but much of this increase has been absorbed by population growth—leaving per capita Gross Domestic Product (GDP) well below the levels of the industrial countries.[4]

The developing countries have, nonetheless, made rapid progress in improving the quality of life for their citizens, lagging the industrial countries by no more than a generation or two in reducing infant mortality rates (see figure 2-2A) or increasing average life expectancy at birth. Today, only a few countries have lower life expectancies at birth than did the United States in 1900.[5] Substantial gains have also been achieved in providing access to education and in improving other aspects of the quality of life. These improvements have been realized by developing countries at a much earlier point in economic development (as measured by per-capita GDP) than was achieved by today's industrial countries as they developed economically (see figure 2-2B). For example, China and Korea achieved infant mortality rates less than 50 per 1,000 live births at GDPs of about $1,500 per capita (corrected for $1990 Purchasing Power Parity (PPP)). The United States did not achieve these rates of

---

[3] United Nations Development Program, *Human Development Report 1991* (New York, NY: Oxford University Press, 1991).

[4] For example, a sample of six large Latin American and nine Asian countries found overall annual GDP growth rates between 1900 and 1987 of 3.8 percent in Latin America and 3.2 percent in Asia, compared to 2.9 percent for the Organization for Economic Cooperation and Development (OECD) countries. (The Latin American and Asian countries, however, were starting from per capita incomes of between one-fourth and one-half that of the OECD countries.) Population growth rates for the Latin American, Asian, and OECD countries however, were 2.2 percent, 1.8 percent, and 0.9 percent, respectively, leaving per capita GDP growth rates of 1.7 percent, 1.3 percent, and 2.1 percent. The annual difference of 0.4 to 0.8 percent in per capita GDP results in an income differential of 40 to 200 percent over the 87 year time period. See Angus Maddison, *The World Economy in the 20th Century* (Paris: Organization for Economic Cooperation and Development, 1987).

[5] The U.S. life expectancy at birth in 1900 was 47.3 years. Today, countries with similar life expectancies include: Mozambique (48), Ethiopia (47), Chad (46), Malawi (47), Somalia (47), Bhutan (48), Burkina Faso (47), Mali (47), Guinea (43), Mauritania (46), Sierra Leone (42), Yemen Arab Republic (47), Senegal (48). The weighted average life expectancy at birth for the low-and middle-income developing countries is 63 years. See World Bank, *World Development Report 1991* (New York, NY: Oxford University Press, 1991).

Figure 2-2A—Infant Mortality Rate Versus Calendar Year for the United States, Brazil, China, India, Korea, and Nigeria

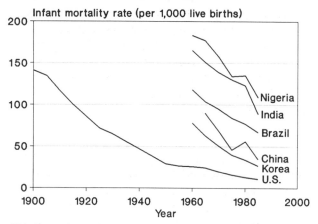

This figure shows that these developing countries lag the United States in reducing infant mortality rates by just 1 to 2 generations or less.

SOURCES: U.S. Department of Commerce, Bureau of the Census, "Historical Statistics of the United States: Colonial Times to 1970" (Washington, DC: U.S. Government Printing Office); Bureau of the Census, "Statistical Abstract of the United States, 1990," (Washington, DC: U.S. Government Printing Office); World Bank, "World Development Report" (New York, NY: Oxford University Press), various years.

infant mortality until its per capita GDP was over $7,000.

Pioneering advances in science and technology, agriculture, medicine, sanitation, and other fields have been important contributors to these improvements in the quality of life. Wide differences in these quality of life indicators between countries with similar per capita GDPs, however, also emphasize the important role of social policies and other factors in realizing such gains.[6]

Wide differences in economic and social development also exist both within and among developing countries. A generation of exceptionally fast economic growth in the Newly Industrializing Countries (NICs), principally in Asia, combined with the slow growth or economic stagnation in some other countries, principally in Africa, has widened the gap among developing countries (see figure 2-3). Per-capita incomes in the upper middle-income developing countries (e.g., Brazil, Argentina, Algeria, Venezuela, and Korea) are almost 7 times higher than in the low-income countries.

The income differential reflects important differences in economic structure. In the upper middle-income countries, industry has a much larger share in total output and agriculture a much lower share. India and China are exceptions, with atypically large shares of industry for their relatively low incomes. The share of the total population living in urban areas is much lower in the low-income countries. In several African countries only about 10 percent of the total population are urban dwellers, whereas in countries like Brazil, Argentina, and Venezuela, levels of urbanization (about 80 percent of the population live in towns) are similar to those in the industrial countries.

PATTERNS OF ENERGY USE AND SUPPLY IN DEVELOPING COUNTRIES

Energy Use

The wide variations in social and economic conditions between developing and industrialized countries are also reflected in their energy use. At the level of final consumption,[7] consumers in developing countries used about 45 exajoules (43 quads) of commercial energy and 16 EJ (15 quads) of traditional fuels, or 6l EJ (58 quads) total in 1985 (see table 2-1). In comparison, consumers in the United States used about 52 EJ (49 quads) of energy (again, not including conversion losses). On a per capita basis, people in the United States used an average of 215 Gigajoules (GJ, or 204 million Btu) each in 1985, compared to 15, 20, and 35 GJ per capita (14, 19 and 33 million Btu) in Asia, Africa, and Latin America, respectively (see table 2-2).

These wide variations in energy use are also seen among developing countries. In the upper middle-income developing countries, per capita annual **commercial** energy consumption is 12 times higher

---

[6]This theme is developed further in a recent publication by the United Nations Development Programme, *Human Development Report 1990* (New York, NY: Oxford University Press, 1990).

[7]Note that this does not include conversion losses. Conversion losses are the energy consumed in converting raw energy (such as coal and crude oil) to forms (electricity, gasoline) that can be used by consumers. The largest conversion loss is in the electricity sector. Total conversion losses in the United States account for about 30 percent of total primary energy consumption. Electricity losses account for about 80 percent of that total or one-quarter of total energy consumption.

Figure 2-2B—Infant Mortality Rates Versus Per Capita GDP Corrected for Purchasing Power Parity (PPP) for the United States, Brazil, China, India, Korea, and Nigeria

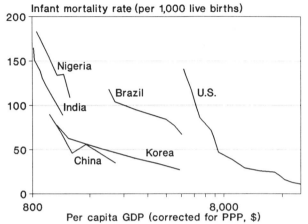

This figure shows that these developing countries have been able to reduce their infant mortality rates at a much earlier point in their economic development as measured by per capita income than did the United States during its economic development. This was made possible by using the scientific, medical, agricultural, public health, and other advances pioneered by the industrial countries. Note that the per capita GDPs have been smoothed—otherwise the curves shown would reverse themselves during serious recessions—so that the curves are monotonic along the GDP axis.

SOURCE: Adapted from figure 2-2A. GDP data corrected by Purchasing Power Parity (PPP) are from: Angus Maddison, "The World Economy in the 20th Century" (Paris, France: Organization for Economic Cooperation and Development, 1989); and from Robert Summers and Alan Heston, "A New Set of International Comparison of Real Product and Price Levels Estimates for 130 Countries, 1950 to 1985," *The Review of Income and Wealth*, vol. 34, No. 1, March 1988, pp. 1-25.

than in the low-income countries.[8] China and India differ from the other low-income countries, with per capita consumption of commercial energy more than 3 times higher. Per capita consumption of **traditional** biomass fuels, on the other hand, is generally higher in the poorest countries, depending on the biomass resources available.[9]

Energy use is typically divided according to end use sector: residential, commercial, industrial, agricultural, and transport. Tables 2-1 and 2-2 list four of these. For commercial fuels only, industry is the largest single end use sector in the developing countries and accounts for about half of the total. In comparison, industry accounts for about one-third of total energy use in the United States. If traditional fuels are included, the residential/commercial sector in developing countries often uses as much or more energy than the industrial sector. Transport accounts for about 10 percent of all energy use in Asia, but for 20 to 30 percent in both Africa and Latin America.

The principal energy services demanded in developing countries today are cooking and industrial process heat (see table 2-3—note that unlike tables 2-1 and 2-2, these values include conversion losses). These services account for 60 percent of total energy use in Brazil, India, and China. Services provided by electricity, such as lighting, appliances, and motor drive, account for about 12 percent of the total, but their share is rising rapidly. The rapid increase in demand for electricity and its high cost explain the emphasis on the electricity sector in this report.

The traditional rural economy relies heavily on biomass—wood, crop residues, animal dung—for cooking or heating; and on human and animal muscle power to grind grain, haul water, plant and harvest crops, transport goods, run cottage industries, and meet other needs for motive power. These forms of energy are extremely limited in output and efficiency. Hauling water from the village well, for example, can take 30 minutes to 3 hours per household each day. The same amount of water could instead be pumped by a motor and piped to the home at a direct cost in electricity of typically less than a penny per day.

As incomes grow and access to improved fuels become more reliable, people are widely observed to shift from traditional biomass fuels to cleaner, more convenient purchased fuels such as charcoal, kerosene, LPG or natural gas, and electricity to meet their energy needs. Modern mechanical drive (electric motors, diesel engines, etc.), in particular, substitutes for human and animal muscle power throughout the economy and allows dramatic increases in the output and the productivity of labor.

A correspondingly wide range of technologies are currently used to provide energy services in developing countries. For example, cooking technologies include stoves using fuelwood, charcoal, kerosene, liquid petroleum gas, natural gas, and electricity, among others, all with different cost and performance characteristics. These technologies vary widely

---

[8] World Bank, op. cit., footnote 2.

[9] Brazil, despite its relatively high income, uses substantial quantities of biomass fuels in modern applications, such as charcoal for steelmaking and ethanol for cars. This contrasts with the use of biomass in the poorer countries, as a cooking fuel using traditional technologies.

Figure 2-3—Per Capita GDP Corrected for Purchasing Power Parity (PPP) Versus Calendar Year for Eight Industrial and Developing Countries.

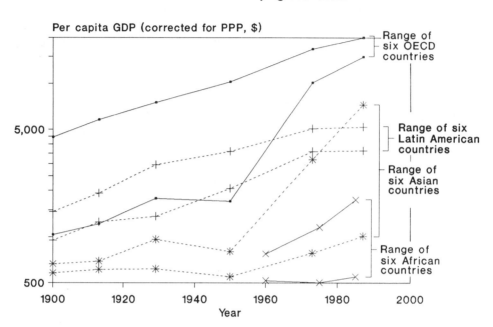

This figure shows the change over time in per capita income for the low and high income countries within each of four sets of six countries for Africa, Asia, Latin America, and OECD Groups, or 24 countries overall. Note the large overlap in per capita incomes during the early part of this century between some of today's industrial countries and some developing countries. Over time, today's industrial countries made institutional reforms, invested heavily in human and physical capital, and made other changes, resulting in their incomes converging at a high level. The relative success of Latin American, Asian, and African countires in making similarly successful institutional reforms, human and capital investments, controlling population growth, and making other changes is indicated by their per capita GDPs over time.

SOURCE: Adapted from Angus Maddison, "The World Economy in the 20th Century," (Paris, France: Organization for Economic Cooperation and Development, 1989); and Robert Summers and Alan Heston, "A New Set of International Comparisons of Real Product and Price Levels Estimates for 130 Countries, 1950 to 1985," *The Review of Income and Wealth*, vol. 34, No. 1, pp. 1-25, 1988.

in their energy efficiency. In an open fire, for example, only about 17 percent of the energy contained in fuelwood goes into cooking. In contrast, in the "modern" gas stove, about 60 percent of the energy contained in the gas is used in cooking. The preponderance of traditional stoves and fuels in many developing countries suggests opportunities for increasing efficiencies and therefore providing more cooking services from the same amount of energy.

Differences in efficiencies in providing energy services are also observed in the industrial sector—industrial process heat, electric or mechanical drive, and other processes. For example, the two largest developing country energy consumers, India and China, currently use roughly twice as much energy to produce a tonne of steel in their integrated iron and steel plants as is used in integrated plants in the United States and Japan.[10]

Despite these differences in aggregate indicators, there are strong similarities in energy use among developing countries within their rural and urban sectors. Energy use in traditional villages throughout the developing world is fairly similar in terms of quantity used, source (biomass, muscle power), and services provided (cooking, subsistence agriculture). At the other end of the scale, energy use by the economically well off is also reasonably similar between developing countries. Energy use by this group in the developing countries is also similar to industrial country energy use in terms of quantity used (to within a factor of 2 or 3), source (oil, gas, coal, electricity), and services provided (electric lighting and appliances, industrial goods, private automobiles, etc.). The large differences between

---

[10]Note that there are wide variations between individual plants, with key plants in India and China performing much better than these averages suggest. There are also differences in product mix which modify these estimates. See ch. 4 for details.

Table 2-1—Total Delivered Energy by Sector, in Selected Regions of the World, 1985 (exajoules)[a]

Region	Residential/commercial		Industry		Transport		Total		Total energy
	Commercial fuels	Traditional fuels[b]	Commercial fuels	Traditional fuels[b]	Commercial fuels	Traditional fuels[b]	Commercial fuels	Traditional fuels[b]	
Africa	1.0	4.0	2.0	0.2	1.5	NA	4.4	4.1	8.5
Latin America	2.3	2.6	4.1	0.8	3.8	NA	10.1	3.4	13.5
India and China	7.3	4.7	13.0	0.2	2.0	NA	22.2	4.8	27.1
Other Asia	1.9	3.2	4.0	0.4	1.9	NA	7.8	3.6	11.3
Developing countries	12.5	14.5	23.1	1.6	9.2	NA	44.5	15.9	60.4
United States	16.8	NA	16.4	NA	18.6	NA	51.8	NA	51.8

NA = Not available or not applicable.
NOTES: This is delivered energy and does not include conversion losses from fuel to electricity, in refineries, etc. The residential and commercial sector also includes others (e.g., public services, etc.) that do not fit in industry or transport. Traditional fuels such as wood are included under commercial fuels for the United States.
[a] Exajoule (10^{18} Joules) equals 0.9478 Quads. To convert to Quads, multiply the above values by 0.9478.
[b] These estimates of traditional fuels are lower than those generally observed in field studies.
SOURCE: U.S. Congress, Office of Technology Assessment, *Energy in Developing Countries*, OTA-E-486 (Washington, DC: U.S. Government Printing Office, January 1991) p. 49.

Table 2-2—Delivered Energy Per Capita by Sector in Selected Regions, 1985
(Gigajoules[a]—includes traditional fuels)

Region	Residential/commercial	Industry	Transport	Total
Africa	11.8	5.2	3.5	20.5
Latin America	12.7	12.5	9.7	34.9
India and China	6.7	7.3	1.1	15.1
Other Asia	7.2	6.2	2.7	16.1
United States	69.8	68.5	77.5	215.8

NOTE: These estimates do not include conversion losses in the energy sector, and underestimate the quantity of traditional fuels used compared to that observed in field studies.
[a] Gigajoule (GJ) is 10^9 Joules and equals 0.9478 million Btu.
SOURCE: U.S. Congress, Office of Technology Assessment, *Energy In Developing Countries*, OTA-E-486 (Washington, DC: Government Printing Office, January 1991) p. 49.

developing countries are then in large part due to the relative share of traditional villagers and the economically well off in the population, and in the forms and quantities of energy used by those who are making the transition between these two extremes. These broad similarities within specific population sectors imply that it is possible to make generalizations about technology that are applicable to a wide range of otherwise disparate countries.

Energy Supply

Biomass fuels are probably[11] the largest single source of energy in developing countries (table 2-4), providing one-third of the total. Coal and oil are the next largest, providing 26 and 28 percent. Primary electricity (mainly hydro) and natural gas account for 7 percent each. Compared with the industrial world, the share of oil and gas is much smaller and the share of biomass fuels larger.

The relative shares of these energy sources in the overall energy supply mix vary significantly across different regions and countries, due in part to unequal endowments of energy resources. Coal supplies about half of the energy requirements for developing countries in Asia, due largely to high levels of coal consumption in China and India. Oil is the major source of commercial primary energy for most countries of the developing world, India and China being the notable exceptions. Natural gas supplies a relatively small fraction of energy in the developing world, despite a more abundant resource base compared with oil.

Overall, the developing world produces more energy than it consumes. There are, however, large disparities between countries. While many countries have some energy resources, three-quarters of the developing countries depend on imports for part or all of their commercial energy supplies (table 2-5).

---

[11] The term ''probably'' is used because data on biomass fuels are unreliable but tend on the whole to underestimate the amounts used.

Table 2-3—Per Capita Energy Use by Service in Selected Countries (Gigajoules[a])

	Brazil	China	India	Kenya	Taiwan	U.S.A.
Residential	6.2	11.7	5.5	16.9	8.9	64.9
cooking	5.3	8.5	5.0	16.4	4.7	3.5
lighting	0.3	0.4	0.5	0.5	0.7	NA
appliances	0.6	NA	0.05	NA	3.1	13.0[b]
Commercial	1.5	0.7	0.26	0.4	4.2	45.2
cooking	0.4	NA	0.13	0.24	1.9	NA
lighting	0.5	NA	0.05	0.16	0.8	7.2
appliances	0.6	NA	0.07	NA	1.5	NA
Industrial	19.4	13.8	4.1	4.8	39.2	94.1
process heat	17.5	10.2	2.7	NA	NA	55.8
motor drive	1.6	3.6	1.3	NA	NA	20.4
lighting	0.1	NA	0.05	NA	NA	NA
Transport	13.3	1.2	1.3	2.7	11.5	80.8
road	12.0	0.2	0.8	1.8	10.1	66.7
rail	0.2	0.7	0.4	0.2	0.1	2.0
air	0.7	NA	0.1	0.7	0.7	11.3
Agriculture	2.1	1.8	0.6	0.5	2.6	2.5
Total	43.4	27.0	11.7	25.6	67.7	288.0

NA = Not available or not applicable.
[a]Gigajoule (GJ) is 10^9 Joules and equals 0.9478 million Btu.
[b]This is the combined total for appliances and lighting.
SOURCE: U.S. Congress, Office of Technology Assessment, *Energy In Developing Countries*, OTA-E-486 (Washington, DC: U.S. Government Printing Office, January 1991) p. 49.

Table 2-4—1987 Primary Energy Supplies (exajoules[a])

	Coal	Oil	Gas	Primary electricity	Total commercial	Biomass	Total energy
World	88.7	104.6	58.2	33.0	284.5	36.9	321.3
Industrial	63.5	77.0	51.7	26.6	218.7	5.5	224.2
Developing	25.2	27.7	6.5	6.4	65.7	31.3	97.1
Share of Industrial countries	72%	74%	89%	81%	77%	15%	70%
Share of developing countries	28%	26%	11%	19%	23%	85%	30%

NOTE: As in table 2-1, the values reported for developing country biomass are too low. Field surveys indicate that biomass accounts for roughly one-third of the energy used in developing countries.
[a]Exajoule (10^{18} Joules) equals 0.9478 Quads. To convert to Quads, multiply the above values by 0.9478.
SOURCE: Adapted from World Energy Conference, *Global Energy Perspectives 2000-2020*, 14th Congress, Montreal 1989 (Paris: 1989).

Oil imports can be a considerable strain on already tight foreign exchange budgets. In several countries, particularly in Africa and Central America, oil imports represent over 30 percent of foreign exchange earnings from exports.

As noted above, a well established energy transition takes place as development proceeds. Biomass is the primary energy supply for traditional villages and is normally used in its raw form with virtually no processing. When rural populations migrate to urban areas to look for seasonal or full time employment, they continue to use traditional biomass fuels. As incomes increase, however, people are gradually able to purchase processed fuels (when available) that are more convenient, efficient, and cleaner. This shift from traditional biomass fuels to purchased fuels changes the structure of the energy supply industry. As development takes place, an increasing amount of processing takes place, notably in the share of fossil fuels converted into electricity. This means that the conversion sector—electric utilities, refineries, etc.—becomes more important as development proceeds.

Developing Countries in World Energy

The developing countries play an important role in world energy consumption, accounting for about 30 percent of global energy use, including both commercial and traditional energy (table 2-4). Several developing countries—China, India, Mexico,

Table 2-5—Energy Import Dependence in Developing Countries

Country income group	Number of countries in group[a]	Number of energy exporters	Number of energy importers	High importers (70-100%)	Medium importers (30-70%)	Low importers (0-30%)
Low-income	38	4	34	29	3	2
China and India	2	1	1	0	0	1
Lower middle-income	30	10	20	15	3	2
Upper middle-income	10	6	4	2	1	1
Total	80	21	59	46	7	6

[a] Includes all countries for which import dependence data are available.
SOURCE: Adapted from World Bank, *World Development Report 1989* (New York, NY: Oxford University Press, 1989).

Brazil, and South Africa—are among the world's top 20 commercial energy consumers. China alone accounts for almost 10 percent of the world's total commercial energy use.

Three countries—China, India, and Brazil—together account for about 45 percent of total developing country consumption of both commercial and biomass fuels. And these countries plus four more—Indonesia, Mexico, Korea, and Venezuela—account for 57 percent of the total. At the other end of the scale are a large number of small countries that, combined, account for only a small part of global consumption. Concerns about global energy use and its implications focus attention on the large consumers, but the energy needs of the small developing nations, though of lesser importance to global totals, are critical to their development prospects.

The developing countries are becoming increasingly important actors in global energy. Their share of global commercial energy consumption has risen sharply in recent years (figure 2-4), from 17 percent of global commercial energy in 1973 to over 23 percent now. Despite their much lower levels of per capita commercial energy consumption (figure 2-5), rapid population and economic growth has meant that developing countries accounted for one-half of the total *increase* in global commercial energy consumption since 1973.

The rising share of the developing countries in global commercial energy consumption is widely predicted to continue. The World Energy Conference projects an increase in their share to 40 percent by 2020, and similar results are also found in a large number of other studies.[12] Again, due to rapid population and economic growth, the developing countries are projected to account for almost 60 percent of the global *increase* in commercial energy consumption by 2020. China alone accounts for over one-third of this increase. These rising shares are sufficiently large to have a major impact on world energy markets. Despite the more rapid rate of growth in energy consumption in developing countries, however, per capita consumption of commercial energy will continue to be far below the levels in industrial countries.

TRENDS IN ENERGY DEMAND IN DEVELOPING COUNTRIES

Factors Increasing Energy Demand

Factors contributing to the rapidly rising energy consumption in developing countries include population growth, economic growth and structural change, and declining real costs of consumer goods.

Population Growth

Over the next three decades the population of the developing world is projected to increase by nearly 3 billion—to almost 7 billion total—while that of

---

[12] An analysis of projections of global commercial energy consumption over the next 20 years can be found in Allan S. Manne and Leo Schrattenholzer, *International Energy Workshop: Overview of Poll Responses* (Stanford University International Energy Project, July 1989). This analysis reports the results and assumptions of over 100 projections of global energy consumption and production, and provides the means of the different studies. Not all studies report results for all regions. The coverage is nonetheless a comprehensive indicator of how energy forecasters view the future. They suggest that the developing countries' share could rise to over one-third by 2010. Longer term projections in general arrive at similar conclusions. For example, the Environmental Protection Agency's *Emissions Scenarios* document, prepared by the Response Strategies Working Group of the Intergovernmental Panel on Climate Change (IPCC), Appendix Report of the Expert Group on Emissions Scenarios (RSWG Steering Committee, Task A), April 1990, concludes that, over a wide range of scenarios, the share of developing countries (Centrally Planned Asia, Africa, Middle East, and South and East Asia) will increase from a 1985 reference level of 23 to between 40 and 60 percent of global energy in 2100, and that this group of developing countries would account for between 60 and 80 percent of the total increase in energy consumption over this period. Further, developments in the Third World define much of the difference between the low and high growth scenarios.

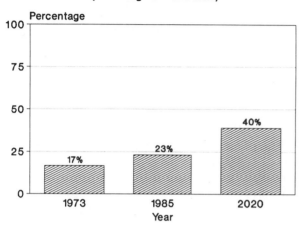

Figure 2-4—Commercial Energy Consumption, 1973, 1985, and 2020 (developing nation energy consumption as a percentage of world total)

SOURCE: World Energy Conference, Global Energy Perspective 2000-2020, 14th Congress, Montreal 1989 (Paris, 1989).

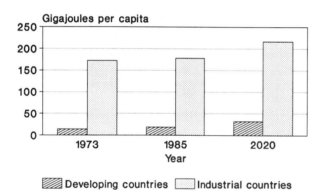

Figure 2-5—Per Capita Commercial Energy Consumption, 1973, 1985, and 2020

SOURCE: World Energy Conference, Global Energy Perspective 2000-2020, 14th Cong., Montreal 1989 (Paris, 1989).

Organization for Economic Cooperation and Development (OECD) countries[13] will increase by only 100 million—to 850 million total. Even assuming continued decreases in fertility rates (the number of children expected to be born to a woman during the course of her life), the population of these countries[14] could reach 10 billion or more in 2100 (figure 2-6). Developing countries would then account for 88 percent of the global population. The increase in population alone in developing countries would account for a 75 percent increase in their commercial energy consumption by 2025 even if per capita consumption remained at current levels.[15]

Economic Growth

Securing higher living standards for this rising population requires rapid economic growth, further increasing the demand for energy services. If energy consumption were to increase in proportion to economic growth (ignoring the enormous potential for improvements in the efficiency of both supplying and using energy), then an average annual gross rate of economic expansion of 4.4 percent (including both economic and population growth) in the developing countries—as projected by the World Energy Conference—would represent more than a fourfold increase in economic activity and commercial energy consumption between now and 2020. The demand for energy services could be increased even more by structural changes inherent in the development process, including:

- **Urbanization.** Urban populations in developing countries are projected to continue rising rapidly, by more than 100 million additional people annually during the 1990s. This rapid growth in urban populations results in rising transportation energy needs as food and raw materials, and finished products are hauled longer distances and as personal transport needs grow.
- **Substitution of commercial for traditional fuels.** Traditional biomass fuels such as wood, crop residues, and animal dung remain today the primary source of energy for more than 2 billion people, but there is a strong preference for commercial fuels as soon as they become available and affordable;
- **Increased use of energy intensive materials.** Developing countries have a large demand for energy intensive material such as steel and cement needed to build commercial, industrial, and transportation infrastructures (see figure 2-7).

---

[13]The OECD countries are Australia, Austria, Belgium, Canada, Denmark, Finland, France, Germany, Greece, Iceland, Ireland, Italy, Japan, Luxembourg, the Netherlands, New Zealand, Norway, Portugal, Spain, Sweden, Switzerland, Turkey, the United Kingdom, and the United States.

[14]Rudolfo A. Bulatao, Eduard Bos, Patience W. Stephens, and My T. Vu, *Europe, Middle East, and Africa (EMN) Region Population Projections, 1989-90 Edition* (Washington, DC: World Bank, 1990), table 9.

[15]A more detailed analysis of the factors driving population growth is given in, U.S. Congress, Office of Technology Assessment, *Energy in Developing Countries*, OTA-E-486 (Washington, DC: U.S. Government Printing Office, January 1991).

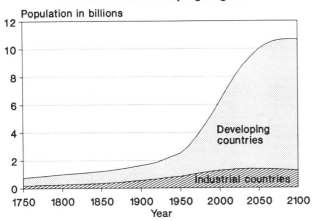

Figure 2-6—World Population Growth 1750-2100 in Industrial and Developing Regions

SOURCE: Thomas Merrick, Population Reference Bureau, "World Population in Transition" *Population Bulletin*, vol. 41, No. 2, April 1986, update based on United Nations 1989 projections.

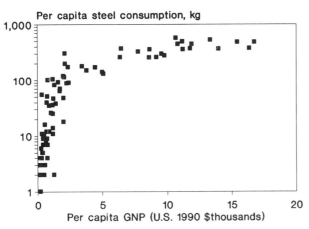

Figure 2-7—Per Capita Steel Consumption Versus GNP for Various Countries

The saturation of the steel market at higher income levels as national infrastructures are developed is readily seen in linear or logarithmic plots. It is shown here in a semi-log plot so as to better display both low-end and high-end data. Each data point represents a country.

SOURCE: U.S. Congress, Office of Technology Assessment, *Energy in Developing Countries*, OTA-E-486 (Washington, DC: U.S. Government Printing Office, January 1991).

Accelerated Consumer Demand

Demand for energy services is further augmented by rapidly rising demand for a wide range of energy-using appliances. Modern manufacturing techniques and improved materials have sharply lowered the real cost of consumer goods—radios, refrigerators, television—in recent years. For example, the real cost of refrigerators in the United States has decreased by a factor of 5 since 1950 (see figure 2-8A). Global distribution systems have also increased the accessibility of these appliances. People in developing countries can thus purchase these goods at a far earlier point in the development cycle (as measured by per capita GDP) than did people in today's industrial countries (see figure 2-8B). Further, as women in developing countries increasingly enter the formal workforce, the demand for (and the means to purchase) labor- and time-saving household appliances such as refrigerators (to store perishable foods and thus reduce the frequency of grocery shopping) can be expected to grow dramatically. The increase in demand for appliances is further stimulated by frequently subsidized electricity prices.

The increase in consumer appliances is already creating an explosive demand for energy both directly to power these goods and indirectly to manufacture and distribute them. A recent review of 21 of the largest developing countries in Asia, Latin America, and Africa found electricity use to be growing faster in the residential than in other sectors in all but four. Annual growth rates in residential electricity use averaged about 12 percent in Asian countries examined, 10 percent in African countries, and 5 percent in Latin American countries.[16] The rapidly increasing use of these appliances has a strong impact on the electric power infrastructure due to the additional demand placed on systems that are typically already short of capacity. Further, much of the residential demand comes at peak times—the most expensive power to generate.

Difficulties in Meeting Energy Demands

Meeting these rising demands for energy services through the traditional strategy of expanding supplies from large scale conventional energy systems faces large problems—financial, institutional, and environmental.

Financial Constraints

Capital intensive electricity generating stations and petroleum refineries already account for a large part of all public investment budgets in developing countries (see table 2-6). Yet according to the U.S.

---

[16]Stephen Meyers et al., "Energy Efficiency and Household Electric Appliances in Developing and Newly Industrialized Countries," Draft Report No. LBL-29678, Lawrence Berkeley Laboratory, October 1990.

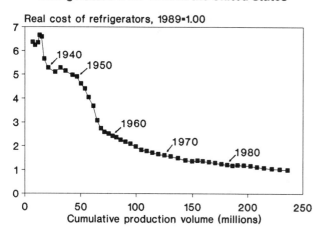

Figure 2-8A—Reduction in the Real Cost of Refrigerators Over Time in the United States

Over the past 40 years, the real price of refrigerators has dropped by almost a factor of 5. For developing countries, such price reductions would allow households to invest in refrigerators at a much earlier point in time than was the case for the United States and other industrialized countries at a similar level of development.

SOURCE: U.S. Congress, Office of Technology Assessment, *Energy in Developing Countries*, OTA-E-486 (Washington, DC: U.S. Government Printing Office, January 1991).

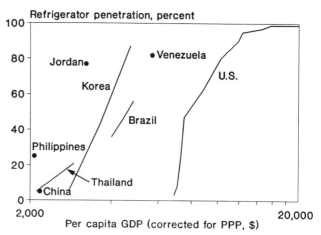

Figure 2-8B—Nationwide Refrigerator Penetration Versus Per Capita GDP Corrected for PPP

This figure shows for selected developing countries that a much higher percentage of households own refrigerators than did households in the United States at a similar level of economic development, as measured by purchasing power parity (PPP) corrected per capita incomes in 1990$. Thus, countries like Jordan, Korea, and Venezuela have refrigerator penetration rates of around 80 percent with per capita incomes of $3,200 to $5,700, respectively. In comparison, refrigerator penetration rates of 80 percent did not occur in the United States until per capita incomes were $10,000.

SOURCE: GDP data is from: Angus Maddison, "The World Economy in the 20th Century" (Paris, France: Organization for Economic Cooperation and Development, 1989); and Robert Summers and Alan Heston, "A New Set of International Comparisons of Real Product and Price Levels Estimates for 130 Countries, 1950-1985," *The Review of Income and Wealth*, vol. 34, pp. 1-25, 1988. Refrigerator penetration data for the United States is from: Donald W. Jones, "Energy Use and Fuel Substitution in Economic Development: What Happened in Developed Countries and What Might be Expected in Developing Countries?" Oak Ridge National Laboratory, ORNL-6433, August 1988; Refrigerator penetration data for the various developing countries shown is from: S. Meyers, et al., "Energy Efficiency and Household Electric Appliances in Developing and Newly Industrialized Countries," Lawrence Berkeley Laboratory, LBL-29678, December 1990. Note that the per capita GDPs have been smoothed—otherwise the curves shown would reverse themselves during serious recessions—so that the curves are monotonic along the GDP axis.

Agency for International Development (AID), annual power sector investments would have to double—to $125 billion annually[17]—to provide adequate supplies. This would take up a large share of the entire projected annual increase in the combined gross national product (GNP) of the developing countries, leaving little for other pressing development needs.

In the past, about one-half of all investments in energy supply have been in the form of foreign exchange. High levels of debt now make it difficult for many developing countries to increase their borrowing from abroad, and capital is likely to remain tight.[18] A continuing source of difficulty is capital outflow or "Capital Flight" from many developing countries, particularly the heavily indebted middle income countries.[19] Similarly, there is often a shortage of local currency to pay for energy development due to inadequate revenues from existing operations. Out of concern for the rural and urban poor and to aid development of key sectors such as agriculture, energy prices—including that of kerosene, diesel, and electricity—are often kept too low to finance the expansion of new facilities.

Institutional Constraints

The power sectors in developing countries frequently experience a wide range of institutional problems, including excessive staffing, inadequate

---

[17]This can be compared to the World Bank estimate of $75 billion annually that is being planned for in developing countries.

[18]Marcus W. Brauchli, "Capital-Poor Regions Will Face World of Tighter Credit in the '90s," *Wall Street Journal*, June 5, 1991, p. A8.

[19]Rigmar Osterkamp, "Is There a Transfer of Resources From Developing to Industrial Countries," *InterEconomics*, Sept./Oct. 1990, p.242-247; Glennon J. Harrison, "Capital Flight and Highly Indebted Countries: An Analytical Survey of the Literature," Congressional Research Service, Library of Congress, March 21, 1991; Glennon J. Harrison, "Capital Flight: Problems Associated with Definitions and Estimates," Congressional Research Service, Library of Congress, March 21, 1991.

Table 2-6—Estimated Annual Energy Investment as a Percentage of Annual Total Public Investment During the Early 1980s

Over 40%	40-30%	30-20%	20-10%	10-0%
Argentina	Ecuador	Botswana	Benin	Ethiopia
Brazil	India	China	Egypt	
Colombia	Pakistan	Costa Rica	Ghana	
Korea	Philippines	Liberia	Jamaica	
Mexico	Turkey	Nepal	Morocco	
			Nigeria	
			Sudan	

SOURCE: Mohan Munasinghe, *Electric Power Economics* (London: Butterworths, 1990), p. 5.

management, weak planning, poor maintenance, deficient financial monitoring, and few incentives to improve efficiency of operation. This raises questions about the ability of this key sector to continue expanding rapidly even if financial resources were available.

Environmental Constraints

On the one hand, modern energy technologies can substitute cleaner modern fuels for smoky traditional biomass fuels; can improve the productivity of traditional agriculture and thus slow the expansion of cultivated lands into tropical forests or other fragile lands; or can power environmental control systems such as sewage treatment. In these cases, modern energy technologies help improve environmental quality. On the other hand, fossil fuel combustion in modern industry, transport, and electricity generation causes air pollution—already often higher in developing country cities than in most industrial country cities—as well as contaminate water supplies and land. Energy production, such as hydro-electric or coal mine development, can cause the loss of agricultural land and displace local populations.

Energy use in developing countries is also of global environmental concern (see below for further discussion). For example, their share of world emissions of carbon dioxide—the most significant greenhouse gas—from the burning of fossil fuel is projected to rise from 25 percent to 44 percent in 2025.[20] If the carbon dioxide impacts of tropical deforestation are included, the developing countries' share of world carbon dioxide emissions rises significantly.

ENERGY, ECONOMIC GROWTH, AND ENVIRONMENTAL QUALITY

Energy and Economic Development

A two-way linkage exists between energy and economic development. The process of economic development strongly influences the amount and type of energy needed. At the same time, developments in the energy sector affect economic growth.

This linkage raises a potential dilemma. On the one hand, the rapid rates of economic growth necessary to provide rising standards of living for growing populations requires sharp increases in energy services. The high cost of providing these services through the conventional route of supply expansion could, however, divert an excessive amount of available investment funds to energy, to the extent of limiting economic growth itself. This possibility is highly undesirable given current low and, in some cases, declining living standards in developing countries.

On the other hand, the inability to supply needed energy services can frustrate economic and social development. In many countries, unreliability and poor quality of energy supplies lead to large costs to the economy through wasted materials, slowdown or stoppage of operations, and investment in standby equipment.

Improving energy efficiencies could help solve this dilemma. The analysis presented here suggests that, on a systems basis, energy efficient equipment can usually provide energy services at lower initial capital costs. Efficiency improvements could therefore increase the capital resources available for

---

[20]Intergovernmental Panel on Climate Change (IPCC), "Emissions Scenarios," Report of the Expert Group on Emissions Scenarios, April 1990.

investment in social and economic development. The adoption of energy efficient technology could also save foreign exchange, a major constraint in developing countries. On the supply side, improved operating procedures and new technologies may well improve the reliability of energy supplies, and thus reduce the heavy economic losses caused by blackouts and brownouts.

While the major focus of this report is on technologies to improve energy efficiencies in both energy use and supply, the broader issues of the energy implications of different development strategies also need to be considered. A detailed analysis of these issues is beyond the scope of this report, but some general considerations may serve to indicate the importance of the energy sector in broader development goals and strategies.

The traditional route to economic development has been through rapid industrialization, motivated either by "import substitution" or "export promotion" considerations. In both cases, the rising share of industry in total output and the associated sharp increase in urban populations has led to a rapid rise in commercial energy consumption.[21] Neither strategy appears to be inherently more energy intensive than the other. Import substitution strategies tend to begin with low energy-intensity assembly type operations, but as they expand to include domestic manufacture of previously imported components—including in some cases metal fabrication—average energy intensities rise.

Conversely, energy intensities associated with export promotion, especially exports based on mining and mineral fuel exploitation, start high, but fall as countries integrate forward to capture the higher value added. The export promotion strategy, however, may have indirect benefits on energy efficiency as the need to compete in foreign markets usually results in the more efficient allocation of resources throughout the economy. Furthermore, export promotion provides the foreign exchange earnings needed to pay for imports of energy efficient technologies. To the extent that export promotion strategies lead to more rapid rates of economic growth, greater opportunities to introduce new, more efficient energy technology are provided.

Changes in the types of products produced within the industrial sector, however, can have a sizable impact on the energy intensity of a country. While the energy content of most industrial products is quite small, a few products—chemicals, plastics, steel, paper, and cement—are conspicuously energy intensive, and a change in the distribution of industrial output between these two categories can impact an economy's overall energy intensity. One-third of the 40 percent decline in the U.S. energy/GDP ratio between 1972 and 1985, for example, is attributed to structural change in the industrial sector, notably the relative decline of steel making; the remainder is due to efficiency improvements. There are similar examples in the developing countries. About half of the post-1979 decline in energy intensity in China (which fell by 30 percent between the late 1970s and the late 1980s) can be ascribed to limits on the expansion of heavy industries and the promotion of light manufactures (e.g., textiles, consumer electronics, processed foodstuffs, and plastics).[22] In the era of cheap oil supplies, countries without domestic energy resources were able to develop energy intensive industries. In recent years, however, countries have been generally reluctant to develop heavy industry based on imported energy.

In adopting industrialization strategies, the developing countries are broadly following the path of the older industrial countries. This raises the possibility of "leapfrogging"—taking a more efficient route to economic and social improvement than that followed by the industrial countries, whose progress was marked by trial and error and constrained by the need to develop technologies where none existed before. To some extent leapfrogging is already taking place. For example, much steel industry development in the industrial countries took place before the invention of current energy efficient techniques. As demand for steel has been slow, many of these countries have had difficulty modern-

---

[21]Joy Dunkerley et al., *Energy Strategies for Developing Nations* (Washington, DC: Johns Hopkins University Press, 1981).

[22]See U.S. Congress, Office of Technology Assessment, *Energy Use and the U.S. Economy* (Washington, DC: U.S. Government Printing Office, June 1990); U.S. Department of Energy, Energy Information Administration, *Monthly Energy Review*, DOE/EIA-0035(91/05 (Washington, DC: May 1991); Ministry of Energy, People's Republic of China, "Energy in China" 1990. The energy intensity went from 13.36 tonnes coal equivalent per 10,000 yuan (tce/10^4 Y) in 1980 to 9.48 (tce/10^4 Y) in 1989 (constant yuan). Total (primary) energy consumption still increased, however, from 603 million tonnes coal equivalent (Mtce) in 1980 to 969 Mtce in 1989, but the economy grew at twice that rate. See also Vaclav Smil, "China's Energy: A Case Study," contractor report prepared for the Office of Technology Assessment, April 1990.

izing their industries. In contrast, the more recently developed steel industries in South Korea and Taiwan are based on near state-of-the-art technology and are more energy efficient.

Although rapid industrialization contributed to unprecedented rises in standards of living in the 1950s, 1960s, and 1970s, and for some developing countries even in the 1980s, there is dissatisfaction in many countries with some of the side effects—high unemployment as job creation fails to keep pace with rising populations, unmanageable urbanization, foreign debt, growing dependence on food imports, persistent poverty, environmental degradation. Further, industrialization based on export promotion is threatened by rising protectionism. Together, these factors are leading to consideration of alternative development paths, giving greater emphasis to agriculture[23] or to light, rather than heavy, manufacturing as sources of economic growth.

In such a strategy, it is argued that agricultural development would generate mutually reinforcing linkages with the industrial sector. Thus, increases in farm output would require inputs from the industrial sector in the form of fertilizers, pumps, tractors, and other equipment. The growth of agricultural processing industries would add to these demands from the industrial sector, and also create jobs in rural areas. Higher rural incomes would, in turn, generate additional demand for the products of the industrial sector. This strategy, relying more on raising rural incomes than past approaches, could have favorable impacts on income distribution.

Such a strategy could have important implications for both overall energy intensity and the forms of energy needed. Economies in which agriculture accounts for a large share of total economic activity, tend to have a lower energy intensity.[24] Reliable supplies of efficient, modern fuels for agricultural processing, operating pump sets, farm transport, and operating domestic and workshop appliances would be needed in rural areas. The importance of rural electrification to rural development has long been recognized, and major investments have been made

Photo credit: U.S. Department of Agriculture

Many developing-country farmers rely on muscle power. Modernization of agricultural techniques usually requires improved energy supplies.

in rural electrification projects, mainly grid extension. Cost and technical improvements in a wide range of small scale, decentralized technologies based on renewable forms of energy (see ch. 6) now offer, in many situations, a more cost effective and sustainable approach to rural electrification. The rising demand for liquid fuels and gases stemming from accelerated rural development could potentially be met through the development of a modern biomass fuels industry (see ch. 6) which simultaneously could increase farm and rural industry employment and income.

The possibilities of alternative strategies for economic and social development, and their energy implications, underline the need to include energy considerations in development planning. Whatever development path is chosen, the energy supply sector is critical for economic development.

Energy and the Traditional Economy

Two-thirds of the developing world's population—some 2.5 billion people—live in rural areas[25] with low standards of living based largely on low-resource farming. This type of farming is characterized by high labor requirements, low productivity

---

[23]Irma Adelman "Beyond Export-Led Growth," *World Development*, vol. 12, No. 9 pp. 937-949, 1984; John P. Lewis and Valeriana Kallab (eds.), Overseas Development Council, "Development Strategies Reconsidered" (New Brunswick, NJ: Transaction Books, 1986).

[24]To cite an admittedly extreme example: the energy intensity of Denmark is one-third that of Luxembourg—despite their very similar levels of income and social development—because Denmark has a large (though highly energy intensive) agricultural sector, while Luxembourg's small manufacturing sector is dominated by the steel industry.

[25]World Bank, *World Development Report 1989*, op. cit., footnote 2.

per hectare and, because of the marginal subsistence, strong risk aversion. Rural populations have little access to commercial fuels and technologies and only limited connection with the modern economy. Biomass fuels satisfy the heating and cooking needs of these populations, and muscle power largely provides for their agricultural, industrial, and transportation energy needs. Although these energy sources provide crucial energy services at little or no direct financial cost, they generally have low efficiencies and limited output and productivity levels.

In many areas, biomass supplies are diminishing due to a host of factors, including population growth and the expansion of agricultural lands, commercial logging, and fuelwood use. The poorest rural peoples often have limited access to even these resources and, therefore, must spend longer periods of time foraging for fuel sources—exacerbating their already difficult economic position.

Traditional villages are complex, highly interconnected systems that are carefully tuned to their environment and the harsh realities of surviving on meager resources.[26] Villages are largely closed systems. The biomass that is used for fuel is part of a system that provides food for humans, fodder for animals, construction materials, fiber for ropes, and even traditional medicines. Similarly, the bullock that pulls a plow also provides milk, meat, leather, and dung for fertilizer or fuel. Changes in any one part thus affect other elements of village life. Changes in agricultural practices, for example, change the amount and type of energy supplies available. In turn, energy sector developments, such as rural electrification, can have major impacts on agricultural practice and income distribution. Making changes in rural systems frequently proves difficult due to the large risks that changes can pose to populations living on the margin of subsistence.

Several factors affect the linkages between energy and the economic and social development of rural economies:

Seasonality

The seasons affect every aspect of rural life: the availability of food, fuel, and employment; the incidence of disease; and even the rates of fertility and mortality.[27] Labor requirements for planting are seasonally peaked to take advantage of limited rainfall and other favorable growing conditions.[28] Labor requirements to harvest crops are similarly peaked. Thus, while there may be a large labor surplus during most of the year, labor shortages occur during the critical planting and harvesting seasons. Studies of African agriculture indicate that labor is "the major scarce resource in food production."[29] Modern equipment could reduce the high labor demands during planting and harvesting.[30]

Although agriculture demands very high levels of labor during the peak seasons, during the remainder of the year rural areas experience serious underemployment. In turn, this seasonal unemployment in rural areas propels a large amount of both seasonal and permanent migration to urban areas.[31] In Africa and Asia, where the migrants are mostly men,[32] more of the burden for subsistence crop production is shifted to the women who stay behind. Migration to cities increases pressure on forests, because urban dwellers generally *purchase* their wood supplies, which are likely to be derived from cutting whole trees, rather than the gathering of twigs and branches more typical of rural foragers.

---

[26] See M.B. Coughenour et al, "Energy Extraction and Use in a Nomadic Pastoral Ecosystem," *Science*, vol. 230, No. 4726, Nov. 8, 1985, pp. 619-625; J.S. Singh, Uma Pandey, and A.K. Tiwari, "Man and Forests: A Central Himalayan Case Study," *AMBIO*, vol. 13, No. 2, 1984, pp. 80-87; Amulya Kumar N. Reddy, "An Indian Village Agricultural Ecosystem—Case Study of Ungra Village. Part II. Discussion," *Biomass*, vol. 1, 1981, pp. 77-88.

[27] Robert Chambers, Richard Longhurst, and Arnold Pacey (eds.), *Seasonal Dimensions to Rural Poverty* (London: Frances Pinter Publishers, Ltd., and Totowa, New Jersey: Allanheld, Osmun and Co., 1981); Robert Chambers, "Rural Poverty Unperceived: Problems and Remedies," *World Development*, vol. 9, 1981, pp. 1-19.

[28] Robert Chambers, Richard Longhurst, and Arnold Pacey (eds.), *Seasonal Dimensions to Rural Poverty*, op. cit., footnote 27, pp. 10-11.

[29] Jeanne Koopman Henn, "Feeding the Cities and Feeding the Peasants: What Role for Africa's Women Farmers?" *World Development*, vol. 11, No. 12, 1983, pp. 1043-1055.

[30] Prabhu Pingali, Yves Bigot, and Hans P. Binswanger, *Agricultural Mechanization and the Evolution of Farming Systems in Sub-Saharan Africa* (Baltimore, MD: Johns Hopkins University Press for the World Bank, 1987).

[31] Michael P. Todaro, *Economic Development in the Third World* (New York, NY: Longman, Inc., 1977); Gerald M. Meier, *Leading Issues in Economic Development*, 4th ed. (New York, NY: Oxford University Press, 1984); Scott M. Swinton, *Peasant Farming Practices and Off-Farm Employment in Puebla, Mexico* (Ithaca, NY: Cornell University, 1983).

[32] Michael P. Todaro, *Economic Development in the Third World*, op. cit., footnote 31, pp. 192-193. Note that in Latin America more women than men now migrate.

The seasons also affect the availability and usability of renewable energy resources. During the rainy season, wood is less easily obtained and more difficult to burn than during the drier months. In areas heavily dependent on crop residues for fuel, shortages at the end of the dry season can force the use of noxious weeds as substitutes, particularly by the very poor.[33] Correspondingly, in mountainous areas or elsewhere with large seasonal temperature variations, fuel demands can increase significantly during the winter.[34]

Inequities in Resource Distribution and Access

In regions where biomass fuel supplies are limited—particularly those with dry climates and/or high population densities—rural people may travel long distances to collect fuel for domestic use, as much as 20 miles round trip in some areas under special conditions. More generally, when wood is scarce they rely on crop wastes, animal dung, or other materials as substitutes. Estimates of time spent foraging range as high as 200 to 300 persondays per year per household in Nepal.[35] Foraging is also heavy work. In Burkina Faso, typical headloads weigh 27 kg (60 pounds).[36] In many regions, women and children do most of the fuel collection.

Despite these heavy burdens, villagers often prefer to invest their capital and labor in technologies for income-producing activities rather than in fuel-conserving stoves or tree-growing efforts.[37] Reasons for this investment preference include lack of cash income, the ability to minimize wood use or to switch to alternative fuels when wood becomes scarce,[38] conflicts over ownership of land or trees, and easy access to common lands. In addition, villagers often carry out fuelwood collection in conjunction with other tasks, such as walking to and from the fields or herding animals. In this case, collecting biomass resources may prove less burdensome than it appears.[39]

To the village user, the immediate value of these fuels outweighs their potential long-term environmental costs.[40] In India, for example, a ton of cow dung applied to the fields produces an estimated increase in grain production worth $8, but if the dung is burned, it eliminates the need for firewood worth $27 in the market.[41] The diversion of crop residues previously used as soil enhancers to fuel use, however, can over a long period of time lead to a loss in soil fertility. Local fuel shortages often have their most serious impacts on the most vulnerable groups. Rural landless and/or marginal farmers may have little access to fuel supplies, especially when the market value of biomass rises.[42]

The Role of Women and Children

Women are particularly affected by biomass fuel availability as they shoulder the burden of most domestic tasks, including foraging for fuelwood and cooking. In many areas they also perform much of

---

[33] Varun Vidyarthi, "Energy and the Poor in an Indian Village," *World Development*, vol. 12, No. 8, 1984, pp. 821-836.

[34] Majid Hussain, "Fuel Consumption Patterns in High Altitude Zones of Kashmir and Ladakh," *Energy Environment Monitor* (India), vol. 3, No. 2, September 1987, pp. 57-62.

[35] J.S. Singh, Uma Pandey, and A.K. Tiwari, op. cit., footnote 26; Kedar Lal Shrestha, *Energy Strategies in Nepal and Technological Options* (Nepal: Research Center for Applied Science and Technology, Tribhuvan University, for the End-Use Oriented Global Energy Workshop, Sao Paulo, Brazil, June 1984). The World Bank Energy Sector Assessment for Nepal estimated that 16 percent of all labor went for fuelwood and animal fodder collection.

[36] E. Ernest, "Fuel Consumption Among Rural Families in Upper Volta, West Africa," paper presented at Eighth World Forestry Conference, Jakarta, Indonesia, 1978.

[37] Varun Vidyarthi, "Energy and the Poor in an Indian Village," op. cit., footnote 33.

[38] Phil O'Keefe and Barry Munslow, "Resolving the Irresolvable: The Fuelwood Problem in Eastern and Southern Africa," paper presented at the ESMAP Eastern and Southern Africa Household Energy Planning Seminar, Harare, Zimbabwe, Feb. 1-5, 1988.

[39] Irene Tinker, "The Real Rural Energy Crisis: Women's Time," *Energy Journal*, vol. 8, special issue, 1987, pp. 125-146.

[40] Geoffrey Barnard and Lars Kristoferson, *Agricultural Residues as Fuel in the Third World* (Washington, DC, and London: Earthscan and International Institute for Environment and Development, Energy Information Program, Technical Report No. 4, 1985).

[41] G.C. Aggarwal and N.T. Singh, "Energy and Economic Returns From Cattle Dung as Manure and Fuel," *Energy*, vol. 9, No. 1, 1984, pp. 87-90; see also G.C. Aggarwal, "Judicious Use of Dung in the Third World," *Energy*, vol. 14, No. 6, 1989, pp. 349-352; Eric Eckholm et al., *Fuelwood: The Energy Crisis That Won't Go Away* (London: Earthscan, 1984), p. 105; Ken Newcombe, World Bank, Energy Department, *An Economic Justification for Rural Afforestation: The Case of Ethiopia* (Washington, DC: World Bank, 1984).

[42] Varun Vidyarthi, op. cit., footnote 33.

the subsistence agricultural labor.[43] As women's work often does not produce any cash revenue, they are limited in their ability to introduce improved technologies. Improving labor productivity and energy efficiency in rural areas will thus require special attention to the role of women.

The migration of men to look for urban work leaves women to fulfill traditional male roles as well as their own. In Uttar Pradesh, India, the male to female ratio in villages is 1:1.4 for the working age group of 15 to 50 years.[44] In Kenya, a quarter of rural households are headed by women—in Botswana, 40 percent.[45] Yet the remittances of the migrants can make an important contribution to rural household finances.

Children, too, play an important role in rural labor, freeing adults to perform more difficult tasks.[46] In Bangladesh, for example, children begin performing certain tasks as early as age 4. By age 12, boys become net producers—producing more than they consume--and are nearly as efficient in wage work as men. By age 15, boys have produced more than their cumulative consumption from birth, and by 22 they have compensated for their own and one sibling's cumulative consumption.[47] The important role of children in farming helps explain high fertility rates in rural areas.

The Role of Commercial Biomass in the Rural Economy

While much biomass is used locally, rural areas are also the source of substantial amounts of fuelwood (both firewood and charcoal) used in towns.[48] This trade pumps significant amounts of cash into the rural economy and provides much-needed employment to rural dwellers during non-agricultural seasons. Such marketing networks can be quite extensive and complex.[49]

In many countries, people in the poorest areas, where conditions do not permit expansion of crop or animal production and natural woody vegetation is the only resource, depend heavily on sales of firewood for their income.[50] In India, "headloading" (individuals carrying wood to urban markets on their heads) has become an important source of income for perhaps 2 to 3 million people.[51] Similarly, when crops fail, charcoal production[52] or the cutting of wood from farm hedgerows[53] provides alternatives for earning cash.

The response of rural peoples to fuel shortages varies widely. Some sell wood to urban markets and use the lower quality residues themselves. Others use dung for fuel rather than for fertilizer. In Malawi, to grow sufficient fuel for household use on the typical family farm would displace maize worth perhaps 30 times more; collecting "free" wood

---

[43] A 1928 survey of 140 Sub-Saharan ethnic groups found that women "carried a major responsibility for food farming" in 85 percent of the cases, and did all but the initial land clearing in 40 percent of the cases. In contrast, the Muslim custom of Purdah, for example, tends to keep women near their homes and away from the fields in Bangladesh. See: Jeanne Koopman Henn, op. cit., footnote 29; Mead T. Cain, "The Economic Activities of Children in a Village in Bangladesh," *Population and Development Review*, vol. 3, No. 3, September 1977, pp. 201-227; Gloria L. Scott and Marilyn Carr, World Bank, "The Impact of Technology Choice on Rural Women in Bangladesh," Staff Working Paper No. 731, Washington, DC, 1985.

[44] J.S. Singh, Uma Pandey, and A.K. Tiwari, "Man and Forests: A Central Himalayan Case Study," op. cit., footnote 26.

[45] World Bank, *Population Growth and Policies in Sub-Saharan Africa* (Washington, DC: World Bank, 1986), p. 39.

[46] Ingrid Palmer has noted: "Children's labor, especially daughters', is usually more significant than husbands' in easing a work bottleneck for women." Ingrid Palmer, "Seasonal Dimensions of Women's Roles," in Robert Chambers, Richard Longhurst, and Arnold Pacey (eds.), *Seasonal Dimensions to Rural Poverty*, op. cit., footnote 27.

[47] Mead T. Cain, op. cit., footnote 43.

[48] The value of commercialized fuelwood and charcoal exceeds 10 percent of the gross domestic product in countries such as Burkina Faso, Ethiopia, and Rwanda and exceeds 5 percent in Liberia, Indonesia, Zaire, Mali, and Haiti. Philip Wardle and Massimo Palmieri, "What Does Fuelwood Really Cost?," *UNASYLVA*, vol. 33, No. 131, 1981, pp. 20-23. George F. Taylor, II, and Moustafa Soumare, "Strategies for Forestry Development in the West African Sahel: An Overview," *Rural Africana*, Nos. 23 and 24, fall 1985 and winter 1986.

[49] Alain Bertrand, "Marketing Networks for Forest Fuels to Supply Urban Centers in the Sahel," *Rural Africana*, Nos. 23 and 24, fall 1985 and winter 1986.

[50] J.E.M. Arnold, "Wood Energy and Rural Communities," *Natural Resources Forum*, vol. 3, 1979, pp. 229-252; Centre for Science and Environment, *The State of India's Environment 1984-85: The Second Citizen's Report* (New Delhi, India: 1985).

[51] Centre For Science and Environment, Ibid., p. 189.

[52] D.O. Hall and P.J. de Groot, "Biomass For Fuel and Food—A Parallel Necessity," draft for *Advances in Solar Energy*, Karl W. Boer (ed.), vol. 3, Jan. 10, 1986; Rafiqul Huda Chaudhury, "The Seasonality of Prices and Wages in Bangladesh," Robert Chambers, Richard Longhurst, and Arnold Pacey (eds.), *Seasonal Dimensions to Rural Poverty*, op. cit., footnote 27.

[53] Rick J. Van Den Beldt, "Supplying Firewood for Household Energy," M. Nurul Islam, Richard Morse, and M. Hadi Soesastro (eds.) *Rural Energy to Meet Development Needs* (Boulder, CO: Westview Press, 1984).

proves much easier.[54] In contrast, aerial surveys of Kenya have shown that hedgerow planting increases with population density—demonstrating that villagers respond to the reduced opportunity of collecting free wood from communal lands by growing their own.[55]

These considerations mean that although people in rural areas may appear to use energy—as well as many other resources—in a technically inefficient manner compared with what is possible with modern commercial technologies, they use energy efficiently in the broader context given the difficult constraints of limited resources, technology, and capital that they face.[56] Rather than maximizing production as is done in modern industrial society, traditional peoples focus on minimizing risk in the face of the vagaries of drought and other natural disasters. In so far as traditional peoples are operating rationally within their decision framework, changes are required in that framework through the introduction of external inputs—financial, managerial, material, technical.

Energy and the Environment

Many developing countries are experiencing widespread environmental degradation in both rural and urban areas.[57] Rural areas are experiencing deforestation, desertification, soil erosion (with associated downstream flooding and siltation), and air pollution. In many urban areas of developing countries, levels of air pollution far exceed those in industrialized countries. Water supplies, too, are often heavily polluted.

The role of energy in environmental degradation is complex. On the one hand, energy, used wisely, can potentially provide several important environmental benefits in developing countries. For example, greater energy inputs into agriculture in the form of tractive power, fertilizer, and irrigation, for example, can substantially improve agricultural productivities where soils and climates are appropriate, and help slow the expansion of agricultural lands into tropical forests or environmentally fragile lands that would otherwise be needed to feed a burgeoning population. At the same time, however, modern agriculture can also cause environmental damage: by overuse of pesticides, herbicides, and fertilizers; by waterlogging and salinizing irrigated lands; and by use of these techniques under inappropriate soil and climatic conditions.[58]

In addition to providing environmental benefits, energy production, conversion, and use also contribute to environmental degradation. Coal mining disturbs surface lands and waters and may also contaminate underground or surface waters if excavated material is not properly managed. Dust and emissions from coal mining and preparation can contribute to local air pollution. Oil and gas production and transport can also lead to land disturbance and water contamination. The combustion of fossil fuels—in refineries, power stations, and by end users—contributes to air pollution through adding sulfur dioxide, particulates, carbon monoxide, nitrogen oxides, and carbon dioxide emissions, leading to acid rain, urban smog, and potentially global warming. The development of hydro resources can flood large tracts of land, uprooting people and leading to loss of forests and wildlife habitat; disrupt the natural flow of rivers; and contribute to the increased incidence of debilitating diseases such as schistosomiasis. Nuclear energy has the potential to release toxic and radioactive materials, and poses problems of weapons proliferation.

---

[54]D. French, "The Economics of BioEnergy in Developing Countries," H. Egneus et al. (eds.), *Bioenergy 84. Volume V. Bioenergy in Developing Countries* (Amsterdam: Elsevier, 1985). It is estimated that 90 percent of all rural households collect all their wood; 10 percent purchase some of their wood at $0.50/m^4 or $0.04/GJ. Urban households buy their wood at a cost of $0.12/GJ. In contrast, plantation-derived fuelwood can cost $1.50 to $2.00/GJ. A farmer could plant trees, but the loss of 0.4 hectare of farmland reduces maize production by a total of $125 and profit by $30. In contrast, trees produced on 0.4 hectare will be worth $6 in 7 years.

[55]P.N. Bradley, N. Chavangi, and A. Van Gelder, "Development Research and Energy Planning in Kenya," *AMBIO*, vol. 14, No. 4-5, 1985, pp. 228-236.

[56]Notable examples of such studies include: N.H. Ravindranath et al., An Indian Village Agricultural Ecosystem—Case Study of Ungra Village, Part I: Main Observations, *Biomass*, vol. 1, No. 1, September 1981, pp. 61-76; Amulya Kumar N. Reddy, "An Indian Village Agricultural Ecosystem—Case Study of Ungra Village, Part II: Discussion," *Biomass*, vol. 1, No. 1, September 1981, pp. 77-88; M. B. Coughenour et al., op. cit., footnote 26.

[57]The interactions between energy and the environment are analyzed in detail in ch.5 of the interim report of this project U.S. Congress, Office of Technology Assessment, *Energy in Developing Countries* OTA-E-486 (Washington DC: U.S. Government Printing Office, January 1991).

[58]U.S. Congress, Office of Technology Assessment, *Enhancing Agriculture in Africa: A Role for U.S. Development Assistance*, OTA-F-356 (Springfield, VA: National Technical Information Service, 1988). Some note, however, that even steep or acid-infertile lands can be productive over long periods as shown by the centuries of terraced rice farming in Asia or continuous sugar-cane cropping in the Dominican Republic. See Ricardo Radulovich, "A View on Tropical Deforestation," *Nature*, vol. 346, No., 6281, July 19, 1990, p. 214.

Table 2-7—Causes and Consequences of Environmental Degradation in Rural Areas

Consequence	Direct cause	Underlying cause
Deforestation Loss of biodiversity Soil erosion Flooding	Shifting agriculture Permanent agriculture Permanent pasture	Population growth Poverty Lack of land tenure Low-level agricultural inputs Mechanization of agriculture and/or the consolidation of agricultural lands
	Commercial logging	Destructive logging, lack of forest management and protection, poor reforestation Increased access to forests along logging roads for farmers and ranchers
	Agriculture, ranching, and logging	Production for export markets Fiscal policies and legislation, in part to promote exports of primary products due to need for foreign exchange to service debt. Inappropriate economic valuations of natural resources and biodiversity
	Use of biomass fuels	Inefficient use of fuelwood; overcutting of fuelwood resources
	Use of forest biomass for fodder	Shortages and lack of alternative sources of fodder
Desertification	Agricultural expansion onto fragile lands Overgrazing Burning of grasslands	Population growth Poverty Lack of land tenure Low-level agriculture and/or the consolidation of agricultural lands
	Use of biomass fuels	Inefficient use of fuelwood; overcutting of fuelwood sources Lack of access to higher quality fuels and stoves.
	Climate change	Various; not well understood
Air pollution	Slash and burn agriculture Burning of grasslands	Population growth Poverty Lack of land tenure Low-level agricultural inputs
Salinization and water-logging of irrigated lands	Poor planning and management Inadequate investment in infrastructure	Lack of access to high-quality or alternative sources of fodder Cheap or free water contributing to inefficiency

SOURCE: Office of Technology Assessment, 1992.

Energy efficient technologies can moderate these environmental impacts while providing the energy services needed for development. An increased role for renewable energy technologies and natural gas could also reduce these adverse environmental impacts.

It is important to note, however, that energy is not the sole contributor to environmental degradation in developing countries, especially in rural areas. Others include population growth, inequitable land tenure, unsustainable agricultural and forestry practices, industrialization, and government policy (see table 2-7).

Population pressure is a major cause of environmental degradation in rural areas. As rural populations grow, the demands on the land for food, fuel, and fodder increase accordingly while, in many developing countries, the low agricultural productivities of traditional cultivation techniques have difficulty keeping up. Farmers then face three basic choices: they can ''mine'' the land—taking more out of it than they put in—until the land is exhausted; they can migrate to new lands—often marginal and ecologically fragile (the best lands are often already in use)—or to poor urban areas; or they can increase the level of (capital-, energy-, and labor-intensive) agricultural inputs—mechanical traction, fertilizer, and irrigation—into the land in order to raise yields.

The latter strategy could also include higher inputs of information and management as might be the case for intercropping, agroforestry, integrated crop-livestock, or other sophisticated agricultural systems.[59]

When farmers migrate to new lands, woodlands are cleared for cropland and pastures. Woodlands are also commercially logged. The use of biomass for fuel or fodder places further demands on woodlands and grasslands, particularly in arid regions with high population densities. Farming, ranching, logging, and the use of biomass fuels are all necessary if the people dependent on these resources are to survive. But these various pressures can also have a variety of negative impacts: destruction of tropical forests and biodiversity; desertification; soil erosion and increased downstream flooding and siltation; and air pollution—local, regional, and global.

Rapid population growth, along with inadequate infrastructure and economic and industrial growth with minimal or inadequately enforced environmental controls, have also led to high levels of pollution in urban areas. Levels of sulfur dioxide, particulates, ground-level ozone, and nitrogen oxides often exceed those in industrialized countries. Major sources include electricity generation, transportation, and industrial production. Greater use of fossil fuels in the modern, primarily urban, sector can also lead to environmental degradation and pollution in the rural areas where fuels are extracted from the ground and transported to the cities, and where hydroelectric facilities are sited.

Many have viewed environmental costs—degradation and pollution of the natural resource base—as the price that must be paid in order to develop economically.[60] Increasingly, however, others argue that environmental protection and economic development are tightly interconnected and mutually supportive.[61] Energy efficiency may allow these polar positions on economic growth and environmental quality to be sidestepped altogether. As detailed in this report, energy efficient technologies usually reduce both initial capital and life cycle operating costs, contributing to economic growth. At the same time, energy efficient technologies, by using less energy to provide a given service, reduce the adverse impacts of energy production and use while still providing the energy services needed for development.

Greenhouse Gases and Global Climate Change

The environmental impacts described above are largely limited to the individual countries concerned. Some activities—notably, the production and use of fossil fuels, deforestation, the use of chlorofluorocarbons (CFCs), and others—can have a wider impact, including impacts on the global climate through the "enhanced" greenhouse effect. These issues have been explored in depth in several recent publications.[62]

The "natural" greenhouse effect is a well-established scientific fact. In the absence of the natural greenhouse effect, the average surface temperature of the Earth would be -18 °C instead of the actual $+15$ °C. This $+33$ °C increase in average surface temperature is due to the presence of naturally occurring greenhouse gases—principally carbon dioxide, methane, and water vapor. Today, increases in atmospheric concentrations of these and other greenhouse gases due to the burning of fossil fuels, deforestation, the use of CFCs, and other human-induced changes in the biosphere are leading to an enhancement of this naturally occurring greenhouse effect. Table 2-8 lists some of the leading sources of these greenhouse gases. A recent review by over 200 leading scientists from 25

---

[59] U.S. Congress, Office of Technology Assessment, Ibid.

[60] Clem Tisdell, "Sustainable Development: Differing Perspectives of Ecologists and Economists, and Relevance to LDCs," *World Development*, vol. 16, No. 3, 1988, pp. 373-384.

[61] World Commission on Environment and Development, *Our Common Future* (New York, NY: Oxford University Press, 1987).

[62] U.S. Congress, Office of Technology Assessment, *Changing By Degrees: Steps to Reduce Greenhouse Gases*, OTA-O-482 (Washington, DC: U.S. Government Printing Office, February 1991); Intergovernmental Panel on Climate Change (IPCC), Meteorological Organization/U.N. Environment Program, *Scientific Assessment of Climate Change, Summary and Report* (Cambridge, United Kingdom: Cambridge University Press, 1990); Michael Grubb, *Energy Policies and the Greenhouse Effect, Volume one: Policy Appraisal*, Royal Institute of International Affairs (Aldershot, Hants, UK: Dartmouth Publishing Company, 1990); National Academy of Sciences et. al., *Policy Implications of Greenhouse Warming* (Washington, DC: National Academy Press, 1991); World Resources Institute, *Greenhouse Warming: Negotiating a Global Regime* (Washington, DC: World Resources Institute, January 1991); William A. Nitze, *The Greenhouse Effect: Formulating a Convention* (London: Royal Institute of International Affairs and Washington DC: Environmental Law Institute, 1990); Allen L. Hammond, Eric Rodenburg, and William R. Moomaw, "Calculating National Accountability for Climate Change," *Environment*, vol. 33, No. 1, 1991, pp.11-15, 33-35; David G. Victor, "How to Slow Global Warming," *Nature*, vol. 349, No. 6309, Feb. 7, 1991, pp.451-456.

Table 2-8—Sources of Greenhouse Gases

Greenhouse gas	Principal sources
Carbon Dioxide	Fossil fuel combustion
	Deforestation, land use changes
	Cement production
Methane	Fossil fuel production (coal mines, oil and gas wells, gas pipelines)
	Fossil fuel combustion
	Landfills
	Rice cultivation
	Animal husbandry
	Biomass combustion and decay
Chlorofluorocarbons	Synthetics used in refrigerators and air conditioners
	Used in manufacturing processes as blowing agent, cleaning agent
Nitrous Oxide	Fertilizers
	Fossil fuel combustion
	Biomass combustion
	Deforestation and land use changes

Adapted from: Michael Grubb, *Royal Institute of International Affairs*, "Energy Policies and the Greenhouse Effect, Volume one: Policy Appraisal," (Aldershot, Hants, England: Dartmouth Publishing Co., 1990); and Dilip R. Ahuja, "Estimating Regional Anthropogenic Emissions of Greenhouse Gases," forthcoming, T.N. Khoshoo and M. Sharma (eds.); "The Indian Geosphere Biosphere" (New Delhi, Indian National Science Academy, Vikas Publishing House, 1991).

countries concluded that this increase in greenhouse gas concentrations will raise the average surface temperature of the Earth (see box 2-B).

Based on current models and under "business-as-usual" scenarios, the Intergovernmental Panel on Climate Change (IPCC) scientists predict that global mean temperature will increase at a rate of about 0.3 °C per decade during the next century, a rate higher than that seen over the past 10,000 years. This would mean a nearly 1 °C increase over present day global average temperatures by 2025 and a 3 °C increase by 2100. In addition to increases in mean global temperature, other effects expected to occur with global warming include increases in sea level[63] and shifts in regional temperature, wind, rainfall, and storm patterns. These, in turn, are expected to submerge low-lying coastal areas and wetlands, threaten buildings and other structures, and increase the salinity of coastal aquifers and estuaries. Such changes could disrupt human communities and aquatic and terrestrial ecosystems, and affect food production and water availability.[64] Many developing countries will be particularly vulnerable to these effects due to their high degree of dependence on subsistence or low-input agriculture. Some developing countries may also be heavily impacted by flooding of their low lying lands.

In 1985, according to estimates for the IPCC Working Group III, developing countries contributed slightly more than one-quarter (26 percent) of annual global **energy** sector CO_2 emissions; three-fourths came from the industrialized market countries and the centrally planned European countries (including the U.S.S.R.). By 2025, with expanding populations and rapidly increasing energy use, developing countries are projected by the IPCC to produce roughly 44 percent of global energy sector CO_2 emissions. Even so, per capita emissions of CO_2 will continue to be much lower in the developing countries compared with the industrial countries.

While the CO_2 emissions from the commercial energy sector are fairly well known, there are large uncertainties about the contribution of emissions from traditional fuels, and from deforestation and other land use changes. Estimates of the CO_2 emissions from tropical deforestation differ by a

---

[63] The IPCC working group predicted an average rate of global mean sea level rise of about 6 cm per decade over the next century, 20 cm by 2030 and 65 cm by the end of the century with significant regional variations. This increase is primarily due to thermal expansion of the oceans and melting of some land ice.

[64] Intergovernmental Panel on Climate Change (IPCC), World Meterological Organization/U.N. Environment Program, "Policymaker's Summary of the Potential Impacts of Climate Change: Report from Working Group II to the IPCC," May 1990, p. 8.

Box 2-B—Highlights of the Intergovernmental Panel on Climate Change 1990 Scientific Assessment

Several hundred scientists from 25 countries prepared and reviewed the scientific data on climate change under the auspices of the World Meteorological Organization and the United Nations Environment Program. This Intergovernmental Panel on Climate Change summarized their findings as follows:

The IPCC is certain that:
- there is a natural greenhouse effect which already keeps the Earth warmer than it would otherwise be.
- emissions resulting from human activities are substantially increasing the atmospheric concentrations of the greenhouse gases: carbon dioxide, methane, chlorofluorocarbons (CFCs), and nitrous oxide. These increases will enhance the greenhouse effect, resulting on average in an additional warming of the Earth's surface. The main greenhouse gas, water vapor, will increase in response to global warming and further enhance it.

The IPCC calculates with confidence that:
- atmospheric concentrations of the long-lived gases (CO_2, N_2O, and the CFCs) adjust only slowly to changes in emissions. Continued emissions of these gases at present rates would commit us to increased concentrations for centuries ahead. The longer emissions continue to increase at present-day rates, the greater reductions would have to be for concentrations to stabilize at a given level.
- the long-lived gases would require immediate reductions in emissions from human activities of over 60 percent to stabilize their concentrations at today's levels; methane would require a 15 to 20 percent reduction.

Based on current model results, the IPCC predicts that:
- under the IPCC Business-As-Usual Scenario, global mean temperature will increase about 0.3 °C per decade (with an uncertainty range of 0.2 to 0.5 °C per decade); this is greater than that seen over the past 10,000 years. This will result in a likely increase in global mean temperature reaching about 1 °C above the present value by 2025 and 3 °C before the end of the 21st century.
- land surfaces will warm more rapidly than the ocean, and high northern latitudes will warm more than the global mean in winter.
- regional climate changes will differ from the global mean, although confidence in the prediction of the detail of regional changes is low. Temperature increases in Southern Europe and central North America are predicted to be higher than the global mean, accompanied on average by reduced summer precipitation and soil moisture.
- global mean sea level will rise about 6 cm per decade over the next century, rising about 20 cm by 2030 and 65 cm by the end of the 21st century.

All predictions are subject to many uncertainties with regard to the timing, magnitude, and regional patterns of climate change due to incomplete understanding of:
- sources and sinks of greenhouse gases,
- clouds,
- oceans, and
- polar ice sheets.

These processes are already partially understood, and the IPCC is confident that the uncertainties can be reduced by further research. However, the complexity of the system means that surprises cannot be ruled out.

The IPCC judgment is that:
- Global mean surface air temperature has increased by 0.3 to 0.6 °C over the last 100 years, with the 5 global-average warmest years occurring in the 1980s. Over the same period global sea level has increased by 10-20 cm.
- The size of this warming is broadly consistent with predictions of climate models, but it is also of the same magnitude as natural climate variability. Thus, the observed temperature increase could be largely due to natural variability; alternatively, this variability and other human factors could have offset a still larger human-induced greenhouse warming. The unequivocal detection of the enhanced greenhouse effect from observations is not likely for a decade or more.

SOURCE: Intergovernmental Panel on Climate Change, *Scientific Assessment of Climate Change, Summary and Report*, World Meteorological Organization/U.N. Environment Program (Cambridge, United Kingdom: Cambridge University Press, 1990).

Table 2-9—Parameters for Key Greenhouse Gases

	CO_2	CH_4	CFC-11	CFC-12	N_2O
Atmospheric concentration					
Pre-industrial, 1750-1800	280 ppmv	0.8 ppmv	0 pptv	0 pptv	288 ppbv
Present day, 1990	353 ppmv	1.72 ppmv	280 ppmv	484 ppmv	310 ppbv
Current annual rate of change	1.8 ppmv (0.5%)	0.015 ppmv (0.9%)	9.5 pptv (4%)	17 pptv (4%)	0.8 ppbv (0.25%)
Atmospheric lifetime (years)	(50-200)[a]	10	65	130	150
Global warming potential relative to carbon dioxide for today's atmospheric composition:					
Instantaneous potential, per molecule	1	21	12,000		
20-year time horizon, per kg	1	63	4,500	7,100	270
100-year time horizon, per kg	1	21	3,500	7,300	290
500-year time horizon, per kg	1	9	1,500	4,500	190
Contribution to radiant forcing,					
1765-1990	61%	23%	2.5%	5.7%	4.1%
1980-1990	55%	15%	5%	12%	6%
Reduction required to stabilize concentrations at current levels	60-?%	15-20%	70-75%	75-85%	70-80%

KEY: ppm(b,t)v = parts per million (billion, trillion) by volume.
[a] Carbon dioxide absorption by the oceans, atmosphere, soils, and plants can not be described by a single overall atmospheric lifetime.
SOURCE: Adapted from Intergovernmental Panel on Climate Change, *Scientific Assessment of Climate Change, Summary and Report*, World Meteorological Organization/U.N. Environment Program (Cambridge, United Kingdom: Cambridge University Press, 1990).

factor of four.[65] By various estimates, deforestation could be the source of between roughly 7 to 35 percent of total annual CO_2 emissions. Overall, the available estimates suggest that developing countries currently contribute somewhere between 30 to 55 percent of total global annual CO_2 emissions. Developing countries also account for at least half of the global anthropogenic generation of two other important greenhouse gases, methane and nitrogen oxides. There similarly remain large uncertainties, however, about the sources and size of these emissions: for example, about the methane emissions from rice paddies and from animal husbandry.

Controlling emissions can slow potential global warming. Emission control strategies that countries could consider today include improved energy efficiency and cleaner energy sources—strategies that also often have economic benefits. The expansion of forested areas, improved livestock waste management, altered use and formulation of fertilizers, improved management of landfills and wastewater treatment, and the elimination of the most greenhouse active CFCs might also reduce or offset emissions. Reducing CO_2 emissions by 60 percent or more (see table 2-9) to stabilize atmospheric concentrations at current levels, however, is a formidable challenge with today's technologies. The United States, for example, would have to reduce per capita consumption of fossil fuels to less than 10 percent of current levels—a more than 90 percent reduction—if such an emissions rate were applied uniformly across today's global population. On the other hand, many developing countries would not currently need to cut fossil fuel use to meet such an emissions target, but might be constrained were they to expand fossil fuel use in the future.

Achieving meaningful reductions in emissions will require unprecedented levels of international cooperation and must include developing countries. In addition to the technological challenges for the energy, agriculture, and industrial sectors, governments of the industrial and developing countries face challenges in improving and expanding institutional mechanisms for transferring technologies that can provide vital energy services while limiting emissions.

---

[65] Intergovernmental Panel on Climate Change (IPCC), World Meterological Organization/U.N. Environment Program "Policymaker's Summary of the Formulation of Response Strategies: Report Prepared for IPCC by Working Group III," June 1990, p. 5. IPCC Working Group 1, "Scientific Assessment of Climate Change: Peer Reviewed Assessment for WG1 Plenary Meeting, May 1990," Apr. 30, 1990, p. 1-9.

CONCLUSION

The magnitude of these problems suggests the need for new approaches to providing the energy services needed for economic development. This report, the final report of this OTA assessment,[66] evaluates the role of technology in better providing energy services for development. By ''technology'' this assessment includes not only hardware, but also the knowledge, skills, spare parts, and other infrastructure that permit equipment to be used effectively. Further, any discussion of technology must recognize the key role of institutional and policy considerations, as they frequently combine to provide adverse incentives to improved energy technologies on both the demand and supply side. As shown in following chapters, however, there is an historic opportunity to use more efficient energy end use and supply technologies to meet the growing demand for energy services in developing countries while at the same time minimizing financial, environmental, and other costs.

---

[66] The first report was: U.S. Congress, Office of Technology Assessment, *Energy in Developing Countries,* OTA-E-486 (Washington, DC: U.S. Government Printing Office, January 1991).

Chapter 3
Energy Services: Residential and Commercial

Photo credit: Sam Baldwin

Contents

	Page
INTRODUCTION AND SUMMARY	47
COOKING	50
WATER HEATING	60
LIGHTING	63
REFRIGERATION	71
SPACE CONDITIONING	77
ELECTRONIC EQUIPMENT	82
BARRIERS TO CONSUMER PURCHASE OF ENERGY EFFICIENT APPLIANCES	83
POLICY RESPONSES	84
CONCLUSION	88

Boxes

Box	Page
3-A. Improved Stoves in Developing Countries	61
3-B. Research Needs for Improved Biomass Stoves in Developing Countries	64
3-C. Development of Electronic Ballasts in the United States and Brazil	71
3-D. Lighting Efficiency Programs in Europe	72
3-E. The Brazilian PROCEL Program	85
3-F. The Appliance Industry: Obstacles and Opportunities in Manufacturing Efficient Appliances	86
3-G. Integrated Resource Planning	87

Chapter 3
Energy Services: Residential and Commercial

INTRODUCTION AND SUMMARY

Energy use in the residential/commercial sector of developing countries typically accounts today for about one-third of commercial[1] (primarily fossil fuel-based) energy use and two-thirds or more of traditional biomass fuel use (see table 3-1). In traditional rural areas, the primary residential energy services demanded are cooking, water heating, lighting, water hauling, grain grinding, and in colder areas, space heating. Economic and social development foster dramatic changes in each of these energy services. This can be seen today by tracing how these energy services change from traditional rural areas to modern urban areas (see tables 3-2, 3-3).

Traditional rural areas around the world use biomass fuels—wood, crop residues, animal dung—in open "three-stone" fires or simple fireplaces to cook their food. As incomes grow and access to improved fuels becomes more reliable, people shift to purchased fuels such as charcoal or coal,[2] kerosene, LPG (or natural gas) and/or electricity for their cooking and water heating needs. Lighting technologies similarly shift with income and fuel access from open fires to kerosene lamps to electric lights. Water hauling and grain grinding—back breaking labor that typically requires several hours of household labor each day in traditional rural and many poor urban areas—is largely eliminated in modern urban areas: water is piped to the proximity or directly to the household and grain is ground in commercial mechanical mills before being sold in the market.

Completely new residential/commercial energy services are provided in the modern economy as well. Refrigeration allows long term storage of food. Electric fans or air conditioners improve personal comfort even in the hottest and most humid of climates. Electronic equipment such as televisions or radios at home and computers at modern offices provide entertainment and information services.

Together, six energy services account for most direct[3] residential and commercial energy use in traditional rural as well as modern urban areas: cooking, water heating, lighting, refrigeration, space conditioning, and electronic equipment (see table 3-2). Each of these is examined in detail below. A variety of other appliances also contribute modestly to energy use in industrial countries and are likely to become more important in developing countries in the future. These include dishwashers, clothes washers and dryers, electric irons, and others. For example, dish and clothes washers/dryers may account for as much as 10 percent of household electricity use in the United States.[4]

Of particular interest in this chapter are those services provided by electricity. This focus is based on the rapid growth in demand for these services and their high cost. Lighting, refrigeration, air conditioning, and information services are primarily powered by electricity; cooking and water heating often use electricity.

The residential/commercial sector in developing countries has a rapidly growing demand for energy services, particularly electric services. This demand is driven by a number of factors, including: rapid

---

[1]Note the two distinct meanings of "commercial" used here. Commercial, as in the residential/commercial sector, refers to commercial buildings—retail stores, offices, hotels, restaurants, etc. Depending on the data source, it also usually includes government offices, schools, hospitals, and other public buildings. Commercial, as in commercial fuels, refers to those fuels that are purchased for cash in the marketplace. These fuels include primarily coal, oil, gas, and electricity. Although biomass fuels are also sold for cash in public markets in many areas, they are nevertheless often still counted as a traditional fuel rather than a commercial fuel. This distinction for biomass fuels varies with the data source. Elsewhere in this report, commercial is also used to refer to those technologies which can be purchased in the market or near-commercial—will soon be available.

[2]The intermediate shift to charcoal occurs primarily in areas with a tradition of charcoal use. The shift to coal is primarily in China, parts of India, and a few other countries where there are large supplies of coal and limited alternatives. See: U.S. Congress, Office of Technology Assessment, *Energy In Developing Countries*, OTA-E-486 (Washington, DC: U.S. Government Printing Office, January 1991), and Willem Floor, World Bank, personal communication, 1991.

[3]Only "direct" energy use in the residential/commercial sector is considered here—that used to power stoves, lights, refrigerators, air conditioners, etc. Indirect energy use—that used to produce the steel and cement for constructing commercial buildings or to haul goods sold to the residential sector—are considered separately in the industrial and transport sectors.

[4]Leo Ranier, Steve Greenberg, and Alan Meier, "The Miscellaneous Electrical Energy Use in Homes," *American Council for an Energy Efficient Economy, 1990 Summer Study on Energy Efficiency in Buildings* (Washington, DC: 1990).

Table 3-1—Total Delivered Energy by Sector, in Selected Regions of the World, 1985 (exajoules)[a]

Region	Residential/commercial		Industry		Transport		Total		Total energy
	Commercial fuels	Traditional fuels[b]	Commercial fuels	Traditional fuels[b]	Commercial fuels	Traditional fuels[b]	Commercial fuels	Traditional fuels[b]	
Africa	1.0	4.0	2.0	0.2	1.5	NA	4.4	4.1	8.5
Latin America	2.3	2.6	4.1	0.8	3.8	NA	10.1	3.4	13.5
India and China	7.3	4.7	13.0	0.2	2.0	NA	22.2	4.8	27.1
Other Asia	1.9	3.2	4.0	0.4	1.9	NA	7.8	3.6	11.3
Developing countries	12.5	14.5	23.1	1.6	9.2	NA	44.5	15.9	60.4
United States	16.8	NA	16.4	NA	18.6	NA	51.8	NA	51.8

NA = Not available or not applicable.
NOTES: This is delivered energy and does not include conversion losses from fuel to electricity, in refineries, etc. The residential and commercial sector also includes others (e.g., public services, etc.) that do not fit in industry or transport. Traditional fuels such as wood are included under commercial fuels for the United States.
[a]Exajoule (10^{18} Joules) equals 0.9478 Quads. To convert to Quads, multiply the above values by 0.9478.
[b]These estimates of traditional fuels are lower than those generally observed in field studies. See references below.
SOURCE: U.S. Congress, Office of Technology Assessment, *Energy in Developing Countries*, OTA-E-486 (Washington, DC: U.S. Government Printing Office, January 1991) p. 49.

population growth, economic growth, and urbanization; the transition from traditional biomass to modern commercial fuels (made possible by growing access to these energy carriers); and the dramatic reductions in the real costs of consumer goods—refrigerators, air conditioners, televisions, etc.—made possible by modern materials and manufacturing techniques. Further, as women in developing countries increasingly enter the formal workforce, the demand for and the means to purchase labor-and time-saving household appliances such as dishwashers, clothes washers, or refrigerators (to store food and thus reduce the frequency of grocery shopping) can be expected to grow rapidly. Factors limiting appliance penetration include capital and operating costs, availability, and the lack of electric service, particularly in rural areas. In Brazil, for example, 90 percent of urban households but only 34 percent of rural households have electric service.[5]

The increase in consumer appliances is creating an explosive demand for energy both directly to power these goods and indirectly to manufacture and distribute them. A recent review of 21 of the largest developing countries in Asia, Latin America, and Africa found electricity use to be growing faster in the residential than in other sectors in all but four countries. Growth rates in residential electricity use averaged about 12 percent in Asian countries examined, 10 percent in African countries, and 5 percent in Latin American countries.[6] This rapid growth is further straining many electric power systems that are already having difficulty meeting demand and poses serious financial, institutional, and environmental problems for developing countries.

The costs and difficulties of meeting residential electricity needs are particularly high. Much of the residential load comes during early evening and contributes to system peak demand; costs of providing peak electricity demand are high as generating equipment to meet this peak is left largely idle during other hours and premium fuels are often used. Residential demand is also spread over many widely dispersed small users. The cost of serving this demand is high due to the extensive distribution system required and because transmission and distribution losses are proportionately larger due to both the long transmission distances and the low voltages the power is supplied at. Finally, electricity sold to residential consumers in many countries is subsidized for social and political reasons; this is compounded by poor billing and collection practices and by theft.

A European or U.S. level of electricity services, using today's most commonly adopted residential and commercial technologies, requires an annual systemwide—including the upstream costs of generating electricity supplies—capital investment of

---

[5]Gilberto De Martino Jannuzzi, "Residential Energy Demand in Brazil by Income Classes," *Energy Policy*, vol. 17, No. 3, p. 256, 1989.

[6]Stephen Meyers et al., "Energy Efficiency and Household Electric Appliances in Developing and Newly Industrialized Countries," report No. LBL-29678 (Berkeley, CA: Lawrence Berkeley Laboratory, December 1990).

Table 3-2—Per Capita Primary Energy Use by Service in Traditional and Modern Economies

	Traditional India, 1980		Modern U.S. 1988	
	Fuel GJ/cap	Electricity[a] GJ/cap	Fuel GJ/cap	Electricity[a] GJ/cap
Cooking	7.8	—	1.4	2.7
Water heating	1.9	—	6.0	6.1
Lighting	0.3	0.1	—	10.7
Refrigeration	—	—	—	5.7
Space conditioning	—	—	30.9	38.4
Information services	—	—	—	2.4[b c]
Subtotal	10.0	0.1	38.3	66.0
Other energy services	—	—	3.3	4.7

— = Not available or very small.
[a] Electricity converted to primary fuel equivalent using a conversion factor of 0.33 for generation, transmission, and distribution combined.
[b] Leo Ranier, Steve Greenberg, and Alan Meier, "The Miscellaneous Electrical Energy Use in Homes," in ACEEE 1990 Summer Study on Energy Efficiency in Buildings, American Council for an Energy Efficient Economy, Washington, DC 1990.
[c] Les Norford, Ari Rabl, Jeffrey Harris and Jacques Roturier, "Electronic Office Equipment: The Impact of Market Trends and Technology on End-Use Demand for Electricity," in Thomas B. Johansson, Birgit Bodlund, and Robert H. Williams, Electricity: Efficient End-Use and New Generation Technologies, and Their Planning Implications (Lund, Sweden: Lund University Press, 1989). Estimated information services electricity demand has been subtracted from the Holtberg et al "other" category.
SOURCES: 1. Data for India is adapted from: "Rural Energy Consumption Patterns: A Field Study," ASTRA, Indian Institute of Science, Bangalore, 1981; and N.H. Ravindranath, et al., "An Indian Village Agricultural Ecosystem—Case Study of Ungra Village, Part I: Main Observations," Biomass, vol. 1, 1981, pp. 61-76. 2. Primary source of residential and commercial data for the United States is: Paul D. Holtberg, Thomas J. Woods, Marie L. Lihn, and Nancy C. McCabe, "Baseline Projection Data Book: 1989 GRI Baseline Projection of U.S. Energy Supply and Demand to 2010," Gas Research Institute, Washington, DC.

Table 3-3—Per Capita Energy Use by Service in Selected Countries (gigajoules)

	Brazil	China	India	Kenya	Taiwan	U.S.
Residential	6.2	11.7	5.5	16.9	8.9	64.9
Cooking	5.3	8.5	5.0	16.4	4.7	3.5
Lighting	0.3	0.4	0.5	0.5	0.7	NA
Appliances	0.6	NA	0.05	NA	3.1	13.0
Space conditioning	0.05	2.8	NA	NA	1.0	38.2
Commercial	1.5	0.7	0.26	0.4	4.2	45.2
Space conditioning	0.4	NA	0.13	0.24	1.9	NA
Lighting	0.5	NA	0.05	0.16	0.8	7.2
Appliances	0.06	NA	0.07	NA	1.5	NA

NA= Not available or not applicable.
SOURCE: U.S. Congress, Office of Technology Assessment, Energy in Developing Countries, OTA-E-486 (Springfield, VA: National Technical Information Service, January 1991). Table 3-3 and app. 3-A.

over $200 per person (see figure 3-1).[7] For the five billion people that will be living in developing countries in the year 2000, an annual capital investment of over $1 trillion would be required. This level of investment is 10 times that currently projected for all electric services in developing countries, including industry and agriculture. To meet even a part of these huge capital needs for energy services as well as to save funds for all the other pressing needs of development, energy systems must be carefully optimized.

It is now recognized in the industrial countries that high efficiency technologies for lighting, refrigeration, space conditioning, and other needs are usually cost effective on a life cycle basis: the higher initial capital cost of these technologies to consumers is counterbalanced by their lower operating costs. Equally important for capital constrained developing countries, high efficiency technologies dramatically reduce the need for expensive utility generating plants. When the total systemwide costs—including both the generation equipment and the end use appliance—are summed, energy efficient equipment usually allows a substantial reduction in the total societal capital costs compared to the less efficient equipment now commonly in use (see figure 3-1).

[7] Note that this does not include all the costs, particularly in the end use sector. See app. A of this report for details.

If the most efficient commercial or near-commercial residential and commercial electricity service technologies were purchased, societies could realize systemwide capital savings of over 10 percent, operating savings of about 40 percent, and electricity savings of 60 percent (see figure 3-1 and app. A at the back of this report) compared to the conventional technology now in use in the industrial and developing countries.[8] A brief listing of a few of these improved technologies is given in table 3-4 and appendix A. These capital and operating savings could then be invested in other pressing development needs. The reduced energy use would also lessen a host of local, regional, and global environmental impacts. Further cost effective efficiency improvements are possible beyond the few technologies surveyed below. Developing countries have the opportunity to leapfrog past today's industrial countries in their selection of residential/commercial energy service technologies.

Figure 3-1 does not, however, include the environmental or other costs of energy use. Although these costs are difficult to quantify, they cannot be ignored. Rather than choosing the least-cost energy service on a purely financial basis as in figure 3-1, even more efficient energy service technologies could be justified on the assumption that their use will mitigate some of these as yet unmeasured external costs. Large scale production will usually dramatically reduce the premium for these higher efficiency technologies as well.

Yet consumers generally do not adequately invest in these technologies due to a variety of market barriers (see table 3-5). These include a resistance to the higher cost of efficient appliances. In the United States, some studies find that this resistance implies an effective discount rate 10 times actual market discount rates.[9] This extreme sensitivity to the first cost of a consumer good is likely to be an even more important constraint in the developing countries. The critical role of this first cost sensitivity can be seen in figure 3-1B for the dramatic shift in capital costs with more efficient end-use equipment from utilities to consumers.

Implementing high efficiency technologies in the residential/commercial sector will require institutional changes and reallocation of funds, but probably will not demand significantly higher levels of technical manpower to put into place. Since much of the technology is interchangeable—a compact fluorescent lightbulb for an incandescent, an efficient refrigerator for an inefficient one—a large amount of technical expertise is not required. There are exceptions, however, including daylighting design in commercial buildings, which requires training of architects and building engineers, or training of construction workers to properly incorporate building-shell improvements. Some additional work may also be needed to incorporate design features for handling the widely fluctuating voltages found in developing countries into electronic ballasts for fluorescent lights and into adjustable speed electronic drives in refrigerators, building ventilation systems, or other equipment, or for other adaptations to developing-country conditions. A variety of these possible policy responses are summarized in table 3-6.

Public interventions to redress inadequacies of the marketplace, however, also carry substantial risks. A few of these difficulties are listed in table 3-7.

There is no shortage of technical opportunities. The technologies examined here are all currently commercially available or nearly so. Many more technologies are at various stages of research, development, and demonstration, but this chapter does not attempt to enumerate them. Rather, it examines the potential of existing or near commercial energy efficient technologies.

COOKING

The most important energy service in many developing countries today is cooking food. In rural areas of developing countries, traditional fuels—wood, crop residues, and dung—are the primary fuels used for cooking; in many urban areas, charcoal is also used. More than half of the world's people depend on these crude polluting biomass fuels for their cooking and other energy needs.

---

[8] These figures differ somewhat from those shown in ch. 1 due to the exclusion of industrial motors here. This assumes the mix of energy services, technologies, and utilization rates as detailed in app. A.

[9] Henry Ruderman, Mark D. Levine, and James E. McMahon, "The Behavior of the Market for Energy Efficiency in Residential Appliances Including Heating and Cooling Equipment," *The Energy Journal*, vol. 8, No. 1, January 1987, pp. 101-124; U.S. Department of Energy, technical support document, "Energy Conservation Standards for Consumer Products: Refrigerators and Furnaces," DOE/CE-0277, November 1989; Malcolm Gladwell, "Consumers' Choices About Money Consistently Defy Common Sense," *The Washington Post*, Feb. 12, 1990, p.A3.

Figure 3-1A—Total System-Wide Capital and Lifecycle Operating Costs and Energy Consumption for Conventional and High Efficiency Residential/Commercial Energy Service Technologies

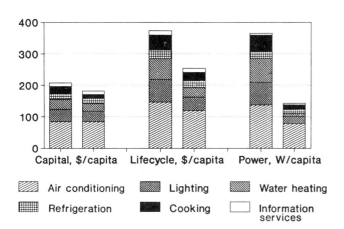

Figure 3-1B—Allocation of Capital Costs for Conventional and High Efficiency Residential/Commercial Energy Service Technologies

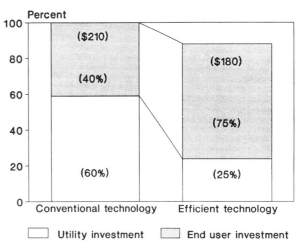

Over a wide range of conditions high efficiency technologies have a lower system wide capital cost than the conventional technologies because the increased capital cost to consumers is more than offset by the decreased capital investment required in upstream electricity generating plants. Lifecycle operating costs are lower because the increased capital costs to the consumer are more than offset by the lower electricity costs to operate the equipment.

SOURCE: U.S. Congress, Office of Technology Assessment, 1992. See app. A for details of this calculation, including the assumptions and sensitivity analysis.

Although the high efficiency technologies have lower system wide capital costs, they dramatically shift the capital costs from the utility to the consumer. The high capital cost of efficient equipment to consumers is the reason why it is not more heavily invested in—even though it provides net savings to society overall and, indirectly, to the consumer. Means of reducing this initial capital cost barrier to the consumer are critical and can be found in table 3-6.

Table 3-4—Selected Energy Efficient Technologies

Energy service	Technologies/remarks
Residential/commercial cooking	Improved biomass stoves, kerosene stoves, LPG and gas stoves, microwave ovens. Improved wood-and charcoal-burning stoves have demonstrated 30-percent savings in the field. Further development work is needed to make them inherently low emitters of pollutants while maintaining high efficiencies. As biomass will continue to be the primary fuel for the rural and urban poor for many years to come, this is a particularly important technology to develop.
Water heating	Increased insulation, flow restrictors, traps to prevent thermosyphoning. Solar water heaters can be very cost effective and perform well in many areas.
Lighting	Compact fluorescent bulbs, voltage stabilizing electronic ballasts, high performance reflectors, task lighting, high efficiency fluorescent lights, high pressure sodium lights, daylighting, advanced lighting controls, and spectrally selective window coatings.
Refrigeration	More and/or improved (aerogel, vacuum, etc.) insulants, improved motor/compressor systems for small scales, voltage stabilizing adjustable speed drives, load management techniques, adaptive defrost, two compressor systems.
Space conditioning	White roofs, planting trees, awnings, roof sprays, spectrally selective window coatings, increasing insulation and reducing infiltration (while using air-to-air heat exchangers if indoor air quality becomes a problem), air economizers, adjustable speed electronic drives, updating design rules for ventilation systems, improved motors/compressors/coils in air conditioners.
Information services	CMOS integrated circuits, power management techniques, flat panel displays.

NOTE: Technologies can be viewed as on a spectrum of: C) Commercially available; A) commercially available in industrial countries but needing Adaptation to the conditions of developing countries; N) Near commercial development, and R) requiring further Research and development. Since most technologies have variations at many points on this spectrum—for example, compact fluorescents are available, may need further adaptation in developing countries in some cases, have improved phosphors or other advances near commercial development, and may have more fundamental advances under research—the status of these technologies—C,A,N,R—will not be discussed; instead, particular opportunities will be presented.

SOURCE: Office of Technology Assessment, 1992.

Table 3-5—Barriers to Investment in Energy Efficient Technologies

Technical

Availability
: High efficiency technologies and their needed support infrastructure of skilled manpower and spare parts may not be locally available. Foreign exchange may not be available to purchase critical spare parts. For the residential and commercial sectors, in particular, high efficiency technologies need to be marketed in a complete package to allow "one stop" shopping.

Culture
: Culture is rarely an impediment to the use of energy efficient technologies, although it is frequently cited as a problem in disseminating technologies in rural areas. In most cases, the technology itself is found to have significant technical shortcomings or is unable to meet the multiple uses desired.

Design rules
: Conventional design rules often lead to excessive oversizing of equipment—raising capital cost and wasting energy.

Diagnostics
: Technologies for measuring the efficiency of equipment, as in energy audits, are often awkward and inaccurate. Some of them may require shutting down a commercial business or making intrusive measurements, such as cutting holes in pipes or ducts to make flow and pressure drop measurements.

Infrastructure
: The available infrastructure within a developing country may not be able to adequately support a particular high efficiency technology. This might include an electric power system with frequent brownouts or blackouts that the high efficiency technology is unable to handle well; dirty fuels that clog injectors; or poor water quality for high performance boilers. The developing country may also lack a reliable spare parts supply system and trained manpower to ensure adequate maintenance. Finally, the existing infrastructure might impede the implementation of a more efficient technology system. An extensive road system and/or little land use planning, for example, might slow or stop the development of an efficient mass transport system.

Reliability
: Innovative high efficiency equipment may not have a well proven history of reliability, particularly under developing country conditions, as for other equipment.

Research, development, demonstration
: Developing countries may lack the financial means and the technical manpower to do needed RD&D in energy efficient technologies, or to make the needed adaptations in existing energy efficient technologies in use in the industrial countries to meet the conditions—such as large fluctuations in power supply voltage and frequency—in developing countries. Technology development and adaptation is particularly needed in rural enterprises and other activities.

Scale
: Energy efficient technologies developed in the industrial countries are often too large in scale to be applicable in developing countries, given their smaller markets and lower quality transport infrastructure.

Scorekeeping methods
: Methods of "measuring" energy savings may not be sufficiently accurate yet for the purpose of paying utilities or energy service companies for the savings that they have achieved. This must be contrasted with the ease of measuring the power generated or used. It is a particularly important issue for utilities that usually earn revenues solely on the basis of energy sold and so have little incentive to assist efficiency efforts.

Technical/managerial manpower
: There is generally a shortage of skilled technical and managerial manpower in developing countries for installing, operating, and maintaining energy efficient equipment. This may not be a significant problem where turnkey equipment is used.

Financial/economic

Behavior
: Users may waste energy, for example, by leaving lights on. In some cases, however, this seeming waste may be done in order to meet other user needs, such as personal security.

Cost
: The high initial cost of energy efficient equipment to the end user and the high effective discount rate used by the end user discourage investment.

Currency exchange rate
: Fluctuations in the currency exchange rate raises the financial risk to firms who import high efficiency equipment with foreign exchange denominated loans.

Dispersed energy savings
: Energy efficiency improvements are scattered throughout the residential and commercial sectors and are difficult to identify and exploit. In contrast conventional energy supplies may be more expensive, but are readily and reliably identified and employed. This tends to give planners a supply side bias irrespective of the potential of efficiency improvements.

Table 3-5—Barriers to Investment in Energy Efficient Technologies—Continued

Financial accounting/budgeting methods
    Commercial accounts for paying energy bills may be separated from accounts for capital investment in more efficient equipment. Budgets for more efficient equipment may be rationed, forcing energy efficiency improvements to compete with each other for scarce budgeted funds even though the return on investment may be much higher than the overall cost of capital to the firm.
International energy prices
    Uncertainty of international energy prices, such as oil, raises risks that price drops will reverse the profitability of investments in efficiency.
Multiple needs
    The multiple roles and needs served by an existing technology may not be adequately met by a new energy efficient technology. Draft animals, for example, can provide meat, milk, leather, and dung in addition to traction power. They also reproduce. Mechanical drive only provides traction.
Risk
    Particularly in poor rural and urban areas, people are highly risk averse; they have to be if they are to survive through the vagaries of drought and other disasters. That villagers are risk averse should not, however, be construed to mean that they are technology averse. A variety of technologies have been adopted very rapidly in poor rural and urban areas.
Seasonality
    Rural life is dominated by the seasons, with sharp labor shortages during the agricultural season and serious underemployment during the rest of the year that rural enterprises can only partly support. Capital investment in efficient agricultural or rural commercial technologies is relatively more expensive as it must pay for itself during just the fraction of the year it is used.
Secondary interest
    Energy efficiency is often of secondary interest to potential users. In industry, for example, efficiency must compete with other equipment parameters—quality and quantity of product; timeliness, reliability, and flexibility; etc.—as well as other factors of production when investment choices are made and when the scarce time of skilled manpower is allocated. These are aspects of overall corporate strategy to improve profitability and competitiveness.
Secondhand markets
    Low efficiency equipment may be widely circulated in secondhand markets in developing equipments, either among industries within developing countries, or perhaps as "gifts" or "hand-me-downs" from industrial countries. Further, users who anticipate selling equipment into the secondhand market after only a few years may neither realize energy savings over a long enough period to cover the cost premium of the more efficient equipment nor, if secondhand markets provide no premium for high efficiency equipment, advantage in its sale.
Subsidized energy prices
    Energy prices in developing countries are often controlled at well below the long run marginal cost, reducing end-user incentive to invest in more efficient equipment. Energy prices may be subsidized for reasons of social equity, support for strategic economic sectors, or others, and with frequent adverse results. On the other hand, however, the low cost of power results in substantial financial costs to the utilities, providing them a potential incentive to invest in more efficient equipment on behalf of the user.
Threshold level of energy/cost savings
    Users may not find a moderate level of energy or cost savings, particularly if spread over many different pieces of equipment, sufficiently attractive to justify the investment of technical or managerial manpower needed to realize the savings.

Institutional
Bias
    There is often a bias towards a small number of large projects, usually for energy supply, than for small projects, usually energy efficiency, due to administrative simplicity and to minimize transaction costs.
Disconnect between purchaser/user
    In a rental/lease arrangement, the owner will avoid paying the higher capital cost of more efficient equipment while the rentor/lessor is stuck with the resulting higher energy bills. Similarly, women in some countries may not have a strong role in household purchase decisions and may not themselves earn a cash income for their labor, but must use inefficient appliances purchased for them.
Disconnect between user/utility
    Even though the total system capital cost is generally lower for energy efficient equipment, it is the user who pays for the more efficient equipment but only recoups the investment over the equipment lifetime while the utility sees an immediate capital savings.
Information
    Potential users of energy efficient equipment may lack information on the opportunities and savings.
Intellectual property rights
    Energy efficient technologies may be patented and the royalties for use may add to the initial costs for the equipment.
Political instability
    Political instability raises risks to those who would invest in more efficient equipment that would only pay off in the mid-to long-term.
Turnkey system
    Turnkey and other package systems are often directly adopted by commercial or industrial operations in developing countries. In many cases, however, the equipment within these systems is based on minimizing capital cost rather than minimizing life cycle operating costs.

SOURCE: Office of Technology Assessment, 1992.

Table 3-6—Policy Options

Alternative financial arrangements
> Currently, the capital costs of generation equipment are paid by the utility and the capital costs of enduse equipment are paid by the enduser. As shown above and in ch. 2, the high effective discount rate of the enduser as well as this separation between utility and user (or for leased equipment, the separation between owner and user) leads to much lower levels of investment in end-use equipment efficiency than is justified on the basis of either total system capital costs or life cycle operating costs. Alternative financial arrangements to redress this "disconnect" might range from the enduser choosing equipment according to the total life cycle cost and paying this cost in monthly installments on the utility bill; to the enduser paying a front-end deposit or posting a bond to the utility to cover the life cycle operating costs of the equipment, against which the utility would charge the capital cost of the equipment and the monthly electricity bills. Either of these approaches would force the enduser to directly face the total systemwide life cycle costs of the equipment when purchasing it. See also Integrated Resource Planning, below.

Data collection
> The range of opportunities for energy efficient equipment, enduser preferences, and operating conditions are not well known in many countries. Data collection, including detailed field studies, would help guide policy decisions.

Demonstrations
> Many potential users of energy efficient equipment or processes remain unaware of the potential savings or unconvinced of the reliability and practicability of these changes under local conditions. Demonstration programs can show the effectiveness of the equipment, pinpoint potential problems, and thereby convince potential users of the benefits of these changes.

Design tools
> Computer design tools can be developed, validated, demonstrated, and widely disseminated to potential users.

Direct installation
> In some cases, particularly where the cost of energy is subsidized by State operated utilities or where peak loads are reduced, the direct installation of energy efficient equipment or processes at low-or no-cost by the utility can reduce costs for both the utility and the user.

Energy audits
> Energy audits by a skilled team, either enterprise employees or from the outside—perhaps associated with an energy service company—can provide highly useful specific information on where energy can be saved. In new plants or in retrofits, submetering of equipment in order to maintain an ongoing record of energy use can also be a very useful means of monitoring performance.

Energy service companies
> Energy service companies (ESCos)—third parties that focus primarily on energy efficiency improvements within a factory and are paid according to how much energy they save—can play a valuable role in implementing energy efficiency gains. They can bring great expertise and experience to bear on the problem, and as their goal is saving energy rather than maintaining production, they are able to devote greater effort and focus to conservation activities. On the other hand, industry employees are sometimes reluctant to work with ESCos, believing that they could implement efficiency activities equally well if they had the time; and worrying that any changes to process or related equipment by the ESCo could disrupt the production line. Generic forms of contracts for ESCo services to industry need to be developed in order to adequately protect both parties, and pilot programs with ESCo's can demonstrate the potential savings by ESCos and their ability to avoid disruption to processes. Compensation for the work done by ESCos should be based on "measured" energy savings, not on the basis of listing the measures taken, irrespective of their effectiveness. Utility programs that provide for competitive bidding on energy savings risk paying the enduser and ESCo twice, once for the energy saved and once for the lower utility bill. This problem can be minimized by appropriately sharing the costs and benefits.

Extension efforts
> Extension efforts may be useful at several levels. The efficiency and productivity of traditional rural industries might be significantly increased in a cost-effective manner with the introduction of a limited set of modern technologies and management tools. To do this, however, is extremely difficult due to the small and scattered nature of traditional rural industries and the large extension effort needed to reach it. Large industry in developing countries has many of the same needs—technical, managerial, and financial assistance—but can be reached more readily.

Grants
> *See* Direct installation, above.

Information programs
> Lack of awareness about the potential of energy efficient equipment can be countered through a variety of information programs, including distribution of relevant literature directly to the industries concerned, presentation of competitions and awards for energy efficiency improvements; and others

Integrated resource planning (IRP)
> Currently, utilities base their investment budgets on a comparison of the costs of different sources of generating capacity—coal, oil, gas, hydro, etc.—and the supply option that has the lowest cost for its particular application is chosen. Integrated resource planning expands this "least cost" planning system to include end use efficiency as an alternative to supply expansion in providing energy services. If energy efficiency is shown to be the lowest cost way of providing energy services, then under IRP utilities would invest in energy efficiency rather than new generating capacity.

Labeling programs
> The efficiency of equipment can be clearly listed by labels. This provides purchasers a means of comparing alternatives. Measuring the efficiency of equipment, however, needs to be done in conjunction with standardized test procedures, perhaps established and monitored by regional test centers, rather than relying on disparate and perhaps misleading manufacturer claims.

Loans/rebates
> Loans or rebates from the utility to the purchaser of energy efficient equipment can lower the first cost barrier seen by the user, and if incorporated in the utility rate base, can also prove profitable for them. On the other hand, users that would have purchased efficient equipment anyway then effectively get the loan/rebate for free—the "free rider" problem. This reduces the effectiveness of the utility program by raising the cost per additional user involved. This problem can be minimized by restricting the loans/rebates to high efficiency equipment for which there is little market penetration, or by other means.

Table 3-6—Policy Options—Continued

Marketing programs
    A variety of marketing tools might be used to increase awareness of energy efficient technologies and increase their attractiveness. These might include radio, TV, and newspaper ads, billboards, public demonstrations, product endorsements, and many others.

Pricing policies
    Energy prices should reflect costs, an obviously highly politicized issue in many countries. Where prices are heavily subsidized, the introduction of energy efficient equipment might be done in conjunction with price rationalization in order to minimize the price shock to users. Prices alone, however, are often insufficient to ensure full utilization of cost effective energy efficient technologies. There are too many other market failures as discussed above. As evidence of this, even the United States has adopted efficiency standards for a variety of appliances.

Private power
    Opportunities to cogenerate or otherwise produce private power have frequently not been taken advantage of because State-owned or controlled utilities have refused to purchase privately generated power at reasonable costs—many State electricity boards simply refuse to take privately generated power; many States impose a sales tax on self-generated electricity; many states decrease the maximum power available to enterprises with onsite generation capabilities, and then are reluctant to provide back-up power when cogeneration systems are down. In other cases, well-intentioned self generation taxes intended to prevent use of inefficient generators by industry penalizes efficient cogeneration. Finally, power is subsidized in many areas, making it difficult for private power to compete. Changes in laws mandating utility purchase of private power—such as that established by the U.S. Public Utility Regulatory Policies Act (PURPA) laws—at reasonable rates would allow many of these opportunities to be seized. This should include establishing generic contracts that provide adequate protection to all concerned parties but that can be readily developed and implemented.

Protocols for equipment interfaces
    The use of power line carriers or other techniques of utility load management will require the development of common equipment interfaces and signalling techniques.

Rate incentives
    See loans/rebates.

R&D: equipment, processes, design rules
    Examples of R&D needs are listed in the text. R&D programs might be established at regional centers of excellence in developing countries, possibly in conjunction with sister research institutes in the industrial countries.

Regional test/R&D centers
    Regional centers of excellence are needed to help gather a crucial mass of highly skilled technical manpower at a single site. The technologies to be developed should focus on those amenable to mass production while maintaining quality control under field conditions. Researchers and field extension agents should, in many cases, make greater use of market mechanisms to guide technology development efforts and to ensure accountability.

Scorekeeping: savings/validation
    Technologies and software for "measuring" energy savings need to be further developed and their effectiveness validated under field conditions. This would be a particularly valuable activity at regional centers of excellence.

Secondhand markets—standards
    Efficiency labels or standards might be set for secondhand equipment. This might be particularly valuable for such things as secondhand factories sold to developing countries.

Standards for equipment/process efficiency
    Many industrial countries have chosen to largely accept the financial "disconnect" between the utility and the user. Instead of providing low-cost, easily available capital to the enduser and at the same time incorporating the full life cycle cost of the end-use equipment in the initial purchase, many industrial countries are attempting to overcome the economic and energy inefficiency of this disconnect by specifying minimum efficiency standards for appliances, buildings (residential and commercial), and, in some cases, industrial equipment.

Tax credits, accelerated depreciation
    A variety of tax incentives—tax credits, accelerated depreciation, etc.—to stimulate investment in energy efficient or other desirable energy technologies might be employed.

Training programs
    Training programs are needed in order to ensure adequate technical/managerial manpower. In addition, means of adequately compensating highly skilled and capable manpower are needed. Currently, skilled manpower—trained at government expense—is frequently attracted away from developing country governmental organizations by the higher salaries of the private sector. Similarly, a more clear career path is needed for skilled technical and managerial manpower in efficiency just as utility operations now provide a career path for those interested in energy supply.

Utility demand/supply planning
    Methodologies for integrated supply and demand least-cost planning have been developed by the industrial countries. These should now be adapted to the needs of developing countries and utility planners and regulators trained in their use.

Utility regulation
    Utility regulations that inhibit the generation of private power (see above) or limit the role of the utility in implementing energy efficiency improvements on the supply or demand side need to be reevaluated. Means of rewarding utilities for energy saved as well as energy generated need to be explored (see also scorekeeping). This might include incorporation of energy efficient equipment into the utility ratebase.

SOURCE: Office of Technology Assessment, 1992.

Table 3-7—Some Problems of State Intervention

- Individuals may know more about their own preferences and circumstances than the government.
- Government planning may increase risk by pointing everyone in the same direction—governments may make bigger mistakes than markets.
- Government planning may be more rigid and inflexible than private decisionmaking since complex decisionmaking machinery may be involved in government.
- Government may be incapable of administering detailed plans.
- Government controls may prevent private sector individual initiative if there are many bureaucratic obstacles.
- Organizations and individuals require incentives to work, innovate, control costs, and allocate efficiently, and the discipline and rewards of the market cannot easily be replicated within public enterprises and organizations.
- Different levels and parts of government may be poorly coordinated in the absence of the equilibrating signals provided by the market, particularly where groups or regions with different interests are involved.
- Markets place constraints on what can be achieved by government, for example, resale of commodities on black markets and activities in the informal sector can disrupt rationing or other nonlinear pricing or taxation schemes. This is the general problem of "incentive compatibility."
- Controls create resource-using activities to influence those controls through lobbying and corruption—often called rent-seeking or directly unproductive activities in the literature.
- Planning may be manipulated by privileged and powerful groups that act in their own interests and further, planning creates groups with a vested interest in planning, for example, bureaucrats or industrialists who obtain protected positions.
- Governments may be dominated by narrow interest groups interested in their own welfare and sometimes actively hostile to large sections of the population. Planning may intensify their power.

SOURCE: N. Stern, "The Economics of Development: A Survey," *Economic Journal,* vol. 99, No. 397, September, 1989.

Higher incomes and reliable fuel supplies enable people to switch to modern stoves and cleaner fuels such as kerosene,[10] LPG, and electricity (and, potentially, modern biomass[11])—a transition that is widely observed around the world largely irrespective of cultural traditions (see figure 3-2).[12] These technologies are preferred for their convenience, comfort, cleanliness, ease of operation, speed, efficiency, and other attributes.[13]

The efficiency, cost, and performance of stoves alone generally increase as consumers shift progressively from wood stoves to charcoal, kerosene, LPG or gas, and electric stoves (see figure 3-3).[14] Background data and supporting documentation for the stoves shown in the figure, as well as for other stoves, are summarized in appendix A and discussed elsewhere.[15]

---

[10] For example, kerosene is the predominant household cooking fuel in parts of Asia—35 percent, 42 percent, and 30 percent of urban households in India, Pakistan, and Sri Lanka, respectively. See: Willem Floor and Robert van der Plas, "Kerosene Stoves: Their Performance, Use, and Constraints," draft report prepared for the World Bank and U.N. Development Program, Energy Sector Management Assistance Program (Washington, DC: Mar. 7, 1991).

[11] Modern technologies can use biomass directly or convert it into liquid or gaseous fuels or into electricity with low emissions. See ch. 6 and Samuel F. Baldwin, U.S. Congress, Office of Technology Assessment "Cooking Technologies," staff working paper, Aug. 15, 1991.

[12] Cultural factors frequently have been cited as a barrier to the adoption of improved wood, charcoal, or other stoves/fuels. Cultural factors may play a role in stove/fuel choice; a wide variety of stoves and fuels have nevertheless been adopted across the full range of class, cultural, and income groups in developing countries, and a strong preference is displayed for superior stoves/fuels such as kerosene, LPG, or electricity. More typically, the reason why various improved biomass stoves have not been adopted by the targeted developing country group is that the proposed technology did not work well or did not meet the multiple needs of the user (see box 3-A).

[13] These issues, particularly the problems of smoke pollution and other environmental impacts from traditional biomass stoves/fuels, the effort required to forage for biomass, and the role of women are discussed in a previous OTA report—U.S. Congress, Office of Technology Assessment, *Energy in Developing Countries,* op. cit., footnote 2.

[14] Even among high efficiency stoves, such as those using LPG or gas, there can be further improvements in efficiency. In practice, the high efficiency of gas stoves can be largely negated when pilot lights are used. Gas stoves that are lit by electric ignition systems (or simply matches) typically use just half the energy that stoves with pilot lights consume. Higher efficiency natural gas stoves under development combine advanced ceramic materials and new designs to augment both infrared and convective heating in a burner with very low emissions. There are similarly many potential efficiency improvements in ovens (such as convection ovens), pots used to cook with (such as pressure cookers), and other cooking devices. See Howard S. Geller, American Council for an Energy Efficient Economy, "Residential Equipment Efficiency: A State of the Art Review," contractor report prepared for the Office of Technology Assessment, May, 1988; Jorgen S. Norgard, "Low Electricity Appliances—Options for the Future,' *Electricity: Efficient End Use and New Generation Technologies, and Their Planning Implications* (Lund, Sweden: Lund University Press, 1989); K.C. Shukla, J.R. Hurley, and M. Grimanis, *Development of an Efficient, Low NOx Domestic Gas Range Cooktop, Phase II,* Report No. TE4311-36-85, Thermo Electron Corp., Waltham, MA, for the Gas Research Institute, Chicago, IL, GRI-85/0080, 1985.

[15] Samuel F. Baldwin, U.S. Congress, Office of Technology Assessment, "Cooking Technologies," staff working paper, Aug. 15, 1991.

Figure 3-2—Choice of Cooking Fuel by Income for Five Medium-Sized Towns in Kenya

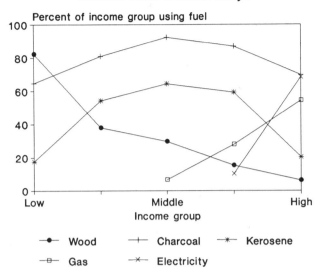

Many households use more than one fuel depending on the particular food cooked and the supply and cost of fuel. Note the shift in fuel choice from wood to charcoal and kerosene, and then from charcoal and kerosene to gas and electricity. This transition is very complex and not yet well understood. Factors that affect a household's shift to modern stoves and fuels include: household income and fuel producing assets (land, trees, animals, etc.); reliability of access to modern fuels; relative cost of traditional and modern fuels and stoves; level of education of the head of household; cooking habits; division of labor and control of finances within the household; and the relative performance of the stove/fuel.

SOURCE: John Soussan, "Fuel Transitions Within Households," Discussion paper No. 35, Walter Elkan et. al. (eds.), *Transitions Between Traditional and Commercial Energy in the Third World* (Guildford, Surrey, United Kingdom: Surrey Energy Economics Center, University of Surrey, January 1987).

Figure 3-3—Representative Efficiencies and Capital Costs for Various Stoves

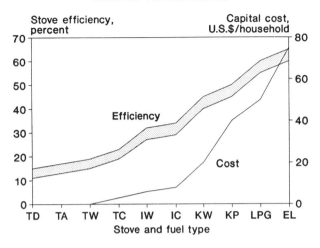

Stoves listed are: (TD), (TA), (TW), (TC)—traditional stoves using dried animal dung, agricultural residues, wood, and charcoal, respectively; (IW), (IC)—improved wood and charcoal stoves; (KW), (KP)—kerosene wick and kerosene pressure stoves; (LPG)—LPG or natural gas stoves; and (EL)—electric resistance stove. Efficiencies and capital costs are for the stove alone and do not include upstream capital costs for producing and delivering fuel. The range of performance both in the laboratory and in the field is much larger than that suggested by this figure and is affected by such factors as the size of the stove and pot, the climate (wind), the quality of the fuel used, the care with which the stove is operated, the type of cooking done, and many other factors. The type of material that the pot is made of is also a significant factor: aluminum pots are almost twice as efficient as traditional clay pots due to their better conduction of heat.

SOURCE: Samuel F. Baldwin, "Cooking Technologies," Office of Technology Assessment, staff working paper; and U.S. Congress, Office of Technology Assessment, *Energy in Developing Countries*, OTA-E-486 (Washington, DC: U.S. Government Printing Office, January 1991).

Efficiencies and costs tell a much different story, however, when examined from a system, rather than the individual purchaser's, perspective. When the energy losses of converting wood to charcoal, fuel to electricity, refining petroleum products, and transporting these fuels to consumers, etc. are included, the system efficiency of delivering cooking energy by charcoal and electric stoves, in particular, drops precipitously (see figure 3-4).[16] These are important attributes to consider when evaluating the environmental impacts and financial feasibility of different cooking systems.

As for the efficiency estimates, there are substantial variations between the capital costs for individual stoves and for the entire cooking system. This is particularly notable in the case of electricity where

---

[16] Actual capital and operating costs will vary widely from these nominal values according to local fuel and stove costs, taxes, and other factors. Actual stove and system efficiencies and other performance factors will vary widely according to household size, diet, income, fuel availability, cooking habits, activity level, season, and many other factors. Some of these factors also tend to change at the same time stove type is changed. Migrants to urban areas may simultaneously change their stove and fuel type, family size, diet, etc. Financial savings gained by moving up to more efficient stoves may also induce greater energy consumption as diet is changed, cooking habits relax, or more food is consumed. For example, estimates are that 15-25 percent of the savings with improved charcoal stoves will be offset through subsequent, income induced consumption. The systemwide fuel savings achieved by going from traditional wood stoves to kerosene or LPG stoves also tend to be less than that expected from simply comparing the efficiency of different stoves as measured in the laboratory. See Donald W. Jones, "Some Simple Economics of Improved Cookstove Programs in Developing Countries," *Resources and Energy*, vol. 10, 1988, pp. 247-264; Kevin B. Fitzgerald, Douglas Barnes, and Gordon McGranahan, World Bank, Industry and Energy Department, *Interfuel Substitution and Changes in the Way Households Use Energy: The Case of Cooking and Lighting Behavior in Urban Java* (Washington, DC: World Bank, June 13, 1990).

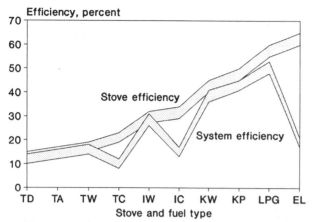

Figure 3-4—Stove and System Efficiencies

Stove efficiencies are nominal values for the stove alone; system efficiencies include the energy losses in producing, converting, and delivering fuel to the consumer. Note particularly the low system efficiencies for charcoal (TC and IC) and electric (EL) stoves due to the large energy losses in converting wood to charcoal and fuel to electricity.

SOURCE: Samuel F. Baldwin, "Cooking Technologies," Office of Technology Assessment, staff working paper.

the upstream costs of generation, transmission and distribution, and other facilities are much larger than the capital cost that the consumer sees for the stove itself (see figure 3-5).[17]

There can be, however, a substantial reduction (depending on relative fuel and stove prices) in both operating costs and energy use in going from traditional stoves using commercially purchased fuelwood to improved biomass, gas, or kerosene stoves (see figure 3-6). There may be opportunities to substitute high performance biomass stoves for traditional ones or to substitute liquid or gas stoves for biomass stoves.[18] Local variations in stove and fuel costs[19] and availability, and in consumer perceptions of stove performance, convenience, and other attributes will then determine consumer choice. Regardless, there are substantial differences in systemwide capital and operating costs for different stoves, many of which are not directly seen by the consumer.

Public policy can help shift consumers toward the more economically and environmentally promising cooking technologies as judged from the national perspective. In particular, improved biomass stoves may be the most cost effective option for the near- to mid-term but require significant additional work to improve their performance (see box 3-A, 3-B, figure 3-7). In rural areas, biomass is likely to be the fuel of necessity for cooking for many years to come.

Alternatively, particularly in urban areas, liquid or gas fueled stoves may offer the consumer greater convenience and performance at a reasonable cost. Foreign exchange requirements, however, will usually require that stove and system efficiencies be maximized and that as much as possible of the stove and other system equipment be manufactured in-country.

Finally, although past efforts in solar cooking have generally been disappointing,[20] recent work suggests that solar box ovens may yet offer an opportunity to meet a portion of cooking needs in areas with high levels of sunshine and cuisine that is adaptable to that style of cooking (i.e, boiling or baking, not frying).[21] The potential of solar cookers nevertheless remains highly controversial due to past failures. They remain expensive, bulky, and fragile; may require changes in cooking practice in many areas; and materials to repair them with are often difficult to obtain. Much more extensive field trials will be needed if the actual potential of newer designs is to be determined.

---

[17]The impact of electric cooking on the grid can be substantial. For example, more than one-third of electricity consumption in Costa Rica and Guatemala is for cooking, where about half of all electrified households have electric stoves. Annual electricity consumption for cooking in Guatemala is roughly 2,500 kilowatt hour (kWh)/year per household compared to 700 kWh/y in the United States. In parts of Asia, electric rice cookers are becoming a substantial electric load. These demands can cause significant local problems for the power grid where loads are low and there is little diversified demand. With larger, more diversified grids, however, such demands pose fewer difficulties. Further, in some cases such as the use of microwave ovens for some baking, electric cooking may lower total energy use compared to, for example, baking in a conventional gas oven. Costa Rican and Guatemalan data from: Andrea N. Ketoff and Omar R. Masera, "Household Electricity Demand in Latin America," *ACEEE 1990 Summer Study on Energy Efficiency in Buildings* (Washington, DC: 1990).

[18]Some argue that the biomass not used in cooking could instead be diverted to high efficiency electricity generation or process heat applications. Backing biomass out of the household sector in this manner poses significant difficulties in the collection and transport of the biomass to a central facility. This will be discussed in chs. 5 and 6.

[19]The values shown are for residential stoves. Commercial stoves exhibit similar trends, with a general shift upwards in efficiency and down in cost per quantity of food cooked due to scale effects. Further, this analysis does not separately consider foreign exchange costs.

[20]Richard Pinon, "About Solar Cookers," *Passive Solar Journal*, vol. 2, No. 2, 1983, pp. 133-146.

[21]Daniel M. Kammen and William F. Lankford, "Cooking in the Sunshine," *Nature*, vol. 348, Nov. 29, 1990, pp. 385-386.

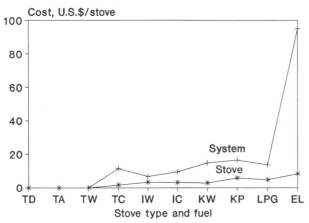

Figure 3-5—Stove and System Capital Costs

When system costs are included, electric stoves can be seen to be particularly expensive. There is a wide range of costs around these nominal values. Note the logarithmic scale.

SOURCE: Samuel F. Baldwin, "Cooking Technologies," Office of Technology Assessment, staff working paper.

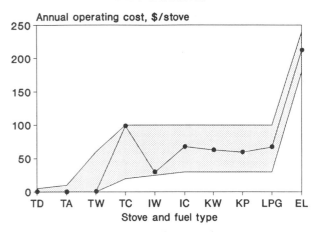

Figure 3-6—Annual Cost of Cooking for Different Stoves

Data points show the cost as estimated from the nominal values. The gray band suggests the wide variation in costs using any particular stove depending on local stove and fuel costs, diet, and a host of other factors.

SOURCE: Samuel F. Baldwin, "Cooking Technologies," Office of Technology Assessment, staff working paper.

In the long term (assuming income growth and the ability to finance imports), the transition to high quality liquid and gas fuels for cooking is inexorable.[22] With this transition, substantial amounts of labor now expended to gather biomass fuels in rural areas may be freed; the time and attention needed to cook when using crude biomass fuels may be substantially reduced; and household, local, and regional air pollution from smoky biomass (or coal) fires may be largely eliminated. On the other hand, high quality fuels will increase financial costs to the individual consumer and, if fuels or stove equipment are imported, could have significant impacts on national trade balances and foreign exchange holdings.

The transition to modern stoves and fuels thus offers users many benefits—reduced time, labor, and possibly fuel use for cooking, and reduced local air pollution[23]—but this transition is also often sharply constrained due to their frequently higher capital and operating costs and uncertain fuel supplies. Means of lowering capital and operating costs and ensuring the reliability of supply are needed if the poor are to gain access to these clean, high efficiency technologies. Further, this transition could impose a substantial financial burden on poor nations.

A large scale transition to LPG, for example, would require a significant investment in both capital equipment and ongoing fuel costs. Optimistically assuming that the capital cost of LPG systems would average $50 per household, including bottling, storage, and transport, the investment would be roughly 3.5 percent of current Gross National Product and 20 percent of the annual value-added in manufacturing for the three billion people in the lowest income countries.[24] The LPG used[25] would be equivalent to one-fourth total commercial energy consumption today by these countries and would be

---

[22] It may be possible, however, to use biomass in a domestic gasifying stove or in a central gasification plant from which it is then piped to the household to generate producer gas for cooking. Examples of this are discussed in Samuel F. Baldwin, U.S. Congress, Office of Technology Assessment, "Cooking Technologies," staff working paper, Aug. 15, 1991.

[23] It might, however, increase global carbon dioxide emissions.

[24] World Bank, *World Development Report 1989* (New York, NY: Oxford University Press, 1989), tables 1 and 6.

[25] Assuming a per-capita power rate for cooking with LPG systems of 100 watts. This is comparable to that seen in the United States and about twice that seen in European countries. It is likely that people in developing countries would continue to eat less processed foods, less restaurant food, and probably more grains and so would continue to use somewhat more fuel than that seen in households in the industrialized countries. Energy use rates for household cooking in different countries are given in K. Krishna Prasad, "Cooking Energy," paper presented at Workshop on End Use Focussed Global Energy Strategy, Princeton University, Princeton, New Jersey Apr. 21-29, 1982.

a significant fraction of their export earnings.[26] Significant economic growth and a gradual phase-in of these technologies are needed if these costs are to be absorbed.

WATER HEATING

The demand for hot water—for bathing, washing clothes and dishes, and other uses—is a significant energy end use in industrial countries and will become more important in developing countries as their economies grow. Hot water demand for a single family in the United States in 1985 was roughly 200 liters/day—40 percent for showers, 25 percent for clothes washers, 20 percent for dishwashers, and 15 percent for other uses—or about 50 liters per person each day.[27] On average, heating water is the second largest use of energy, after space conditioning (heating and cooling), in the U.S. residential and commercial sectors (see table 3-2).

The use of hot water is also important in developing countries, even with their generally warmer climates. For example, electric water heating accounts for 20 to 25 percent of residential electricity demand in Brazil, Colombia, Ecuador, and Guatemala.[28] In Brazil, instant (rather than storage) electric water heaters are widely used for showers. In use, they draw as much as 7 kilowatts and account for perhaps half of the residential peak electricity demand in Sao Paulo.[29] In a study of six villages in India, water heating used about one-fifth as much fuel as cooking.[30] A study of households that used fuelwood in the large urban area of Bangalore, India, found fuelwood consumption almost equally divided between cooking and water heating.[31]

In many rural and poor urban areas, use of hot water remains modest today but will generally increase in the future. There are many technical approaches to water heating that greatly reduce fuel consumption and, from the societal perspective, can reduce capital costs and life cycle operating costs. The most simple, and perhaps the most readily applicable in developing countries, is the use of solar water heaters. These can range from simple devices that simply hold the water in a container exposed to the sun, to complex devices that actively monitor water temperatures and move it into large storage tanks until it is needed. The fuel demands of these systems are generally quite low.[32]

A variety of solar water heating technologies are now well developed. The 7,400 square meter building in which the Office of Technology Assessment is housed has a solar system on its roof that provides all of the hot water and up to 70 percent of the space heat needed. Solar water heating systems for developing countries can, in principal, be lower in cost and more efficient than those in the industrial countries. Complicated and expensive protection mechanisms used in cold climates to keep the water from freezing and breaking the system are largely unnecessary in the warm developing countries. Solar water heater efficiencies are generally higher in the developing countries because of the warm outside temperatures.

Solar water heaters have been put into widespread use in some countries. There are more than two million solar water heaters in Japan and 600,000 in Israel. More than 50 firms manufacture or market solar water heaters in Turkey; a total area of 19,000 square meters of collectors was installed in 1982. China has an installed collector area of 150,000 square meters; Kenya has about 19,000 residential

---

[26]World Bank, *World Development Report 1989* (New York, NY: Oxford University Press, 1989), table 5, and converting kilogram oil equivalent to energy at 42 megajoules (MJ)/kg.

[27]American Council for an Energy Efficient Economy, ''Residential Conservation Power Plant Study: Phase I—Technical Potential,'' Pacific Gas and Electric Co.

[28]Andrea N. Ketoff and Omar R. Masera, ''Household Electricity Demand in Latin America,'' *ACEEE 1990 Summer Study on Energy Efficiency in Buildings* (Washington, DC: 1990).

[29]Stephen Meyers et al., *Energy Efficiency and Household Electric Appliances in Developing and Newly Industrialized Countries*, report No. LBL-29678 (Berkeley, CA: Lawrence Berkeley Laboratory, December 1990); Alfredo Behrens and Stefano Consonni, ''Hot Showers for Energy Rich Countries,''*Energy*, vol. 15, No. 9, pp. 821-829, 1990.

[30]ASTRA, ''Rural Energy Consumption Patterns: A Field Study,'' Centre for the Application of Science and Technology to Rural Areas, Indian Institute of Science, Bangalore, India, 1981.

[31]Amulya Kumar N. Reddy and B. Sudhakar Reddy, ''Energy in a Stratified Society,'' *Economic and Political Weekly*, vol. XVIII, No. 41, October 1983, p. 1762.

[32]Fuel may be required even in a warm climate to supplement solar heating when there is heavy cloud cover.

Box 3-A—Improved Stoves in Developing Countries

Traditional stoves, with their high levels of smoke and heat, awkward use, and heavy demand for fuel (and its attendant laborious collection), have long been a central focus of efforts to improve the lives of women in developing countries. Gandhian organizations developed stoves with chimneys to reduce indoor smoke pollution beginning in the 1930s. S.P. Raju worked on stove design at the Hyderabad Engineering Research Laboratory and published a pamphlet, "Smokeless Kitchens for the Millions" in 1953 in which he promised women five freedoms—from smoke, from soot, from heat, from waste, and from fire risk. Following their own energy crisis in 1973, the Western World began to perceive corresponding energy problems in developing countries, particularly that of fuelwood use in traditional stoves and a possible connection between fuelwood use in cooking and deforestation.[1]

Following the path pioneered in India and elsewhere, a variety of improved stove programs were begun by western donors and host countries in the mid- to late-1970s. Program design and technology choice were strongly influenced by the "appropriate technology" movement: local materials were used, designs were kept simple (and supposedly low-cost), and low-skill labor intensive construction was emphasized both for its own sake and to allow easy maintenance.

The first generation of woodstoves that resulted from these considerations were typically thick blocks of a sand and clay mixture; chambers and holes were carved in the block for the fire, for the pots to sit on, and for the smoke to exit to the chimney. A fire was built under the first pot; the second and subsequent pots were heated by the hot gases flowing toward the chimney. These were known as "massive multi-pot" stoves (see figure 3-7A).

In many cases, little or no testing was done of these first generation designs before dissemination programs began. Numerous competing projects were launched—some countries in West Africa had a dozen or more largely independent stove programs supported by different international nongovernmental organizations, bilateral and multilateral aid agencies, and domestic organizations. These countries never individually had a critical mass of technical manpower to do proper design work, perform detailed field evaluations or followup, or conduct careful economic or social analysis. Operating independently and with little interprogram communication, the same mistakes were repeated.

When field evaluations of these massive stoves began, serious problems began to surface. Field surveys found that most users quickly returned to using their traditional stoves; laboratory and field studies alike showed that these massive stoves often used more fuel and emitted as much smoke into the kitchen as traditional stoves; the stoves cost users a significant amount of time and effort (and sometimes money) to build; and the stoves were hard to maintain—they cracked and crumbled easily in the heat of the fire.

Contrary to then accepted wisdom, it appeared that traditional stoves were actually well optimized for the local materials, pots, and other conditions from many, many years of trial and error. To do better required sustained technical input in design, quality control in production, careful field testing and followup, and extensive input at every stage from women users. These factors were missing from nearly all the programs. Improved heat resistant materials such as metals or ceramics also proved important.

In response to the numerous independent stove programs, their lack of success, and the need for firm technological underpinnings, efforts were launched in the early 1980s to coordinate these programs and develop the technological foundations necessary. In West Africa, for example, IBM-Europe and the Club du Sahel asked Volunteers in Technical Assistance, a private U.S. volunteer organization, to work with CILSS (The Interstate Committee To Fight Drought in the Sahel—an eight country African organization) in a program funded by USAID and IBM-Europe, and later by the Netherlands to coordinate disparate technical development and dissemination efforts. The technical effort showed that, at least under West African conditions, massive stoves typically used substantially more wood than traditional stoves. Using the principles of engineering combustion and heat transfer, a "second generation" of simple lightweight stoves made of metal or ceramic was developed that achieved fuel savings in the field of about one-third compared with traditional stoves (see figure 3-7B). To realize these savings required that critical stove dimensions be maintained to within a fraction of an inch. In turn, this required that the

---

[1] It is now generally acknowledged that use of wood for fuel is not usually a strong contributor to deforestation. There are a few exceptions, particularly arid regions where biomass growth is low, or around urban or industrial areas with unusually high levels of fuel demand. See: Office of Technology Assessment, *Energy In Developing Countries*, (Washington, DC: U.S. Government Printing Office, January 1991).

> **Box 3-A—Improved Stoves in Developing Countries—Continued**
>
> stoves be "mass-produced" to match standard pot sizes. This was best done at central sites where tight quality control could be ensured. In comparison to massive stoves that were hand-crafted on site, centralized production also had the advantages of more rapid production, lower stove and program costs, and ready commercialization. For the user, the lightweight stoves had the advantage of somewhat more rapid cooking, protection from burns that are likely when cooking over an open fire, and portability: the stove could be moved across a courtyard to take advantage of afternoon shade or across town when the household moved.
>
> These stoves have proven popular in the field. A World Bank program to disseminate lightweight metal stoves in Niamey, Niger, set an ambitious target of 20,000 stoves in 2 years; 40,000 stoves were sold. Similar large-scale efforts are now underway in Mali, Burkina Faso, and elsewhere.
>
> Other countries have similarly enjoyed considerable success recently with improved stove programs. Some 130,000 improved jiko stoves (which use charcoal rather than wood, as above) with average savings of 15 to 40 percent are now sold annually in Kenya. In China, an estimated 80 million improved stoves had been disseminated by the end of 1987. In Karnataka State, India, over 100,000 massive stoves have been disseminated, and with careful technical input and field followup, have also realized savings. Ironically, the widespread failure of the first generation "massive" stoves discredited stove programs generally and caused many aid organizations to cut back financial supports just as significant improvements in stove performance were finally being realized with these second generation lightweight and (in the case of Karnataka State) massive stoves.
>
> Centralized mass production and commercial sale of lightweight stoves has, however, generally shifted the programmatic focus of remaining stove efforts from traditional rural areas largely outside the cash economy to poor urban areas. Effective means of addressing rural areas have not yet been developed. Considerable work also remains to further refine biomass stove designs so as to improve efficiencies and reduce noxious emissions.
>
> SOURCES: S.P. Raju, "Smokeless Kitchens for the Millions," Christian Literature Society, Madras, India, 1953, reprinted 1961. Samuel F. Baldwin, *Biomass Stoves: Engineering Design, Development, and Dissemination* (Rosslyn, VA and Princeton, NJ: Volunteers in Technical Assistance and Center for Energy and Environmental Studies, Princeton University, 1986); Kirk R. Smith, "Dialectics of Improved Stoves," *Economic and Political Weekly*, Mar. 11, 1989; Jas Gill, "Improved Stoves in Developing Countries," *Energy Policy*, April 1987, pp. 135-144, and Addendum, Energy Policy, June 1987, pp. 283-285; Margaret Crouch, "Expansion of Benefits: Fuel Efficient Cookstoves in the Sahel, The VITA Experience," *Volunteers in Technical Assistance*, Rosslyn, VA, July 1989; H. Mike Jones, "Energy Efficient Stoves In East Africa: An Assessment of the Kenya Ceramic Jiko (Stove) Program," Oak Ridge National Laboratory, report No. 89-01, Jan. 31, 1989. "China: County-Level Rural Energy Assessments: A Joint Study of ESMAP and Chinese Experts," Activity Completion report No. 101/89; World Bank/UNDP, May 1989.

solar hot water heaters; Papua New Guinea has about 8,000.[33]

The initial capital cost to the individual consumer is substantially higher for a solar water heater than for an electric heater.[34] When the upstream capital costs for the electric utility are included, however, the total capital cost to society is almost 30 percent greater for the electric water heater than for the solar water heater. When the total operating costs (capital plus fuel) to the user are considered, the solar heated water costs less than one-half that heated by electric heaters (see app. A). These comparisons are summarized in figure 3-8 for electric resistance water heaters, electric heat pump water heaters,[35] and solar water heaters of comparable performance. Gas or liquid fueled water heaters can also be compared with solar water heaters but are not considered here.

The efficiency of hot water systems can also be improved by: increasing the amount of insulation on the storage tanks and distribution pipes, and by using traps to prevent convective loops from carrying heat away through distribution lines. Low flow shower-

---

[33] Christopher Hurst, "Establishing New Markets for Mature Energy Equipment in Developing Countries: Experience with Windmills, Hydro Powered Mills and Solar Water Heaters," *World Development*, vol.18, No.4, 1990, pp. 605-615.

[34] The comparison here will be restricted to electric water heaters. Water heaters using natural gas or other fuels are also widely available and extensively used in many countries. The financial advantages of solar water heating may be largely offset or negated compared to gas or other types of water heaters.

[35] Note that in large commercial applications, heat pump water heaters can be coupled with air conditioning loads and can then be more economic under a variety of circumstances.

Figure 3-7A—A Massive Multipot Stove

A broken massive multipot stove is in the foreground. Only one of the two potholes is in use. A second pot to the side of the massive stove is supported by two stones and the stove wall. To the rear are women grinding grain with traditional mortars and pestles.
Site: near Niamey, Niger. **Photo credit:** Sam Baldwin.

Figure 3-7B—A Lightweight Metal Stove

Construction of an improved lightweight stove by a metalsmith.
Site: Ouagadougou, Burkina Faso. **Photo credit:** Sam Baldwin.

heads and cold water washing of clothes can reduce the quantity of hot water needed.

LIGHTING

Lighting accounts for only a small fraction of total national energy use in both developing and industrial countries, typically ranging from about 2 to 5 percent of the total (see table 3-8). Lighting's share of electricity use is higher, ranging from about 8 to 17 percent of total electricity use in industrial countries,[36] with similar shares in developing countries. In India, for example, lighting accounts for about 17 percent of total national electricity consumption and about 30 to 35 percent of the peak power demand.[37]

Lighting merits particular attention as it plays a very important social role in domestic life and in commerce and industry—enabling activities at night or where natural lighting is inadequate. As rural incomes increase, or as people move to urban areas and gain greater access to modern fuels and electricity, lighting services and the energy used to provide them increase dramatically.

As shown elsewhere,[38] lighting technologies follow a fairly clear technological progression in performance, efficiency, and cost—going from the simple open fire, to kerosene wick or pressure lamps, to the use of electric-powered incandescent or fluorescent lamps. Consumers' choices of lighting technologies largely follow the same progression as household incomes increase and as electricity becomes available.

The shift to electric lighting is observed everywhere electricity has been made available. In contrast to kerosene lamps or other nonelectric lighting technologies, electric lighting is clean (within the home), relatively safe, easy to operate, relatively efficient, and provides a high quality light.

---

[36]Terry McGowan, "Energy Efficient Lighting," *Electricity: Efficient End Use and New Generation Technologies, and Their Planning Implications* (Lund, Sweden: Lund University Press, 1989).

[37]Ashok Gadgil and Gilberto De Martino Jannuzzi, *Conservation Potential of Compact Fluorescent Lamps in India and Brazil*, report No. LBL-27210 (Berkeley, CA: Lawrence Berkeley Laboratory, July 1989).

[38]U.S. Congress, Office of Technology Assessment, *Energy in Developing Countries*, op. cit., footnote 2.

Box 3-B—Research Needs for Improved Biomass Stoves in Developing Countries

The use of biomass resources for fuel contributes to several problems: they generate air pollution—particularly in village households—and possibly greenhouse gases;[1] they demand large amounts of time and labor; and in some areas they can contribute to deforestation.[2] Nevertheless, biomass will continue to be the primary cooking fuel for rural and poor urban areas in developing countries for many years to come. Higher quality liquid or gas fuels are simply too expensive and too irregular in supply to supplant biomass anytime soon.

Although biomass will continue to be a primary fuel in the mid-term, it may be possible to improve the performance of biomass stoves significantly. Some successes have already been realized (Box 3-A) but much more could be done. Significant technical challenges remain to be overcome, however, if clean burning, fuel efficient biomass stoves are to be developed.

The performance of a biomass stove is the product of several difficult technical tradeoffs. Fuel efficiency is improved by narrowing the gap between the pot wall and the stove to increase convective heat transfer; by limiting the flow of air into the stove to increase the average temperature of the combustion gases; by lowering the pot closer to the fire to increase the fraction of radiant heat from the fire that is intercepted by the pot; and by other means. But too narrow a gap can choke the fire and greatly increase cooking times; too little airflow into and through the stove can increase emissions of carbon monoxide and hazardous smoke; and lowering the pot closer to the fire can prevent complete combustion and greatly increase smoke emissions—this can easily be seen by putting an object into the flame of a candle. Further, unlike commercial fuels such as kerosene or LPG, biomass fuels vary markedly in density and form—from grass to logs, in composition, moisture content, and a host of other factors important in determining combustion characteristics. Combustion of biomass is also extremely complex, involving many thousands of interacting chemical species. Achieving both high fuel efficiency and low smoke emissions in a cook stove thus remains a substantial technical challenge.

The technical complexity of this task has several important implications.

First, the technical complexity requires a high level of technical expertise that is often difficult to assemble given the often small size of existing improved stove projects in developing countries. An alternative approach might be to form regional centers of excellence in developing countries where a relatively large number of researchers can be brought together to form the critical mass of skilled manpower that is needed. Such a center would draw the best manpower and concentrate the research effort, but base selection and continued participation on peer-reviewed performance. When the technology was developed, these researchers could then return to their respective regions to direct further development and dissemination efforts.

Such regional centers of excellence could themselves have several important benefits. They would provide an opportunity to further develop institutional capacity in the developing countries through goal oriented research focussed on technologies of particular interest in developing countries.

Regional centers of excellence might also offer a means of enhancing the training of scientists and engineers. Too often, students returning to smaller developing countries with a Masters or Doctoral degree in science or engineering are expected to immediately play an important role in national research organizations. In contrast, in the industrial countries, scientists and engineers often spend many years—even after receipt of their doctorate—working under the tutelage of a more experienced researcher, both as a postdoctoral fellow and as a member of a research team. Such experience is important. Much of what is required to select a viable approach to solving a research problem, to direct the research, and to manage the research budget and related administrative matters is not taught in school; it is instead learned through the modern equivalent of an apprenticeship. A regional center of excellence would provide budding scientists and engineers more opportunity to learn such skills from mentors, rather than being expected to learn it all by trial and error on their own.

Regional centers of excellence in developing countries have been successfully developed for agricultural research, development, and field trials. An example is the International Rice Research Institute. The experience of these institutions may hold useful lessons for the development of energy technologies.

---

[1] If the biomass is used on a sustainable basis, then greenhouse gas emissions are essentially eliminated. In contrast, fossil fuels always generate greenhouse gases.

[2] See Office of Technology Assessment, *Energy in Developing Countries*, OTA-E-486 (Washington, DC: U.S. Government Printing Office, January 1991).

Second, appropriate technologists have often suggested that technologies must be adapted to a very large extent to the locality in which they are used and that failure to make such adaptations is a principal cause of failure. In this context, it is important to recognize that a number of different factors might be "adapted," including the underlying technology, product design, manufacturing process, or means of product dissemination.

The basic technology of a stove or of most other devices will remain largely the same globally. Indeed, the technical complexity of biomass combustion and the similarities of wood in Latin America, Africa, or Asia suggests that much of the basic research might be best undertaken at the global level, as long as it was closely coupled to nearby field trials.

Variations in the practical use of stoves between regions may require some adaptation of the product design—such as the means of holding pots down when vigorously stirring, modified shapes to hold different types of pots, etc. Nevertheless, gas and kerosene stoves have been used equally well in the United States, Africa, and Asia with little or no adaptation of their design.

The choice of fabrication method is more commonly an aspect that might be adapted to local conditions. Although the maintenance of precise dimensions in stoves will generally require mass production and quality control techniques, these can be implemented in ways as varied as village metalsmith artisanal production using standardized templates to automated metal stamping facilities. Similarly, the choice of materials can also be varied somewhat depending on local conditions, from low quality scrap metal recovered from barrels or wrecked cars to new high quality steel alloys or ceramics.

Finally, product dissemination methods must be adapted to local conditions if they are to be successful.

Third, past failures in improved stove programs indicate the need to link laboratory research activities with practical field experience. It is important to avoid the creation of laboratory curiosities with no practical field application. Means of rewarding researchers that successfully see their work through to full scale commercialization need to be explored.

Fourth, the complexity of biomass combustion and heat transfer requires close attention to design and quality control. This is particularly significant because of the importance of the informal artisanal sector in disseminating such technologies as biomass stoves. Greater effort needs to be made to work with this important sector in terms of upgrading their production technologies, improving their access to adequate finance, developing better means of technology transfer, providing training, and other issues.

Finally, the experience with improved biomass stoves has shown the market mechanism to be a particularly valuable tool in urban areas for weeding out poor technology designs or weighing alternatives through competition. The market mechanism is not as effective, however, in rural areas largely outside the cash economy. Methods of adequately meeting the needs in these areas are needed.

SOURCES: Samuel F. Baldwin, *Biomass Stoves: Engineering Design, Development, and Dissemination*, (Rosslyn, VA and Princeton, NJ: Volunteers In Technical Assistance and Center for Energy and Environmental Studies, Princeton University, 1986). Sam Baldwin et al. "Improved Woodburning Cookstoves: Signs of Success," *AMBIO*, vol. 14, No. 4-5, pp. 280-287, 1985. Dilip R. Ahuja, "Research Needs for Improving Biofuel Burning Cookstove Technologies," *Natural Resources Forum*, May 1990, pp.125-134; Kirk R. Smith, *Biomass Fuels, Air Pollution, and Health: A Global Review* (New York, NY: Plenum Publishing Co. 1987); K. Krishna Prasad, E. Sangen, and P. Visser, "Woodburning Cookstoves," in James P. Hartnett and Thomas F. Irvine, Jr. *Advances in Heat Transfer*, (New York, NY: Academic Press, 1985).

As incomes increase with economic development, households begin to buy other appliances—radios, TVs, fans, refrigerators, and air conditioners. Electricity use for lighting usually continues to increase, but is then only a small fraction of total residential electricity use (see figure 3-9). Electricity use for lighting in the commercial and service sectors also grows rapidly as the economy expands.

The demand for lighting has continued to increase in the industrialized countries over the past 30 years as incomes have increased. Today, lighting ranges from roughly 20 to 100 million lumen-hours per capita per year (Mlm-hr/cap-yr) in the industrial countries, with most in the range of 25 to 40 Mlm-hr/cap-yr.[39] In comparison, annual household light production in South Bombay, India, varies with

---

[39] Terry McGowan, "Energy Efficient Lighting," *Electricity: Efficient End Use and New Generation Technologies, and Their Planning Implications* (Lund, Sweden: Lund University Press, 1988).

Figure 3-8—Capital Costs, Operating Costs, and Electricity Consumption for Electric Resistance, Electric Heat Pump, and Solar Water Heaters

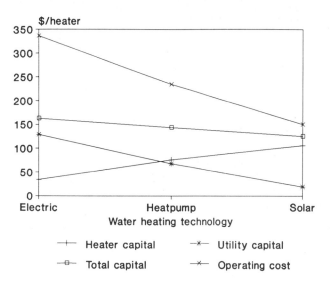

Capital costs to the individual consumer are substantially higher for solar water heaters than for electric resistance heaters. When upstream utility investment is included, however, the capital costs for solar water heaters are modestly lower than the systemwide capital costs for electric resistance and heat pump water heaters. Lifecycle operating costs to the consumer are dramatically lower for solar heaters than for electric heaters.

SOURCES: See app. A. Cost data for electric resistance water heater and solar water heater, installed costs, are from Sunpower Solar Hot Water Systems, Barbados, West Indies, 1990, provided by Robert van der Plas, World Bank, Washington, DC. Heat pump water heater data is from Howard S. Geller, "Residential Equipment Efficiency: A State-of-the-Art Review," American Council for an Energy-Efficient Economy, contractor report for the Office of Technology Assessment, May, 1988. See also, Electric Power Research Institute, "Heat Pump Water Heaters: An Efficient Alternative for Commercial Use," Palo Alto, CA.

household income from about 1 to 3 Mlm-hr/cap-yr;[40] light production in the commercial sector might double these numbers. This is equivalent to a light output level 10 to 30 percent that of the lowest industrialized countries.

Figure 3-9—Household Electricity Use for Lighting v. Household Income, Brazil

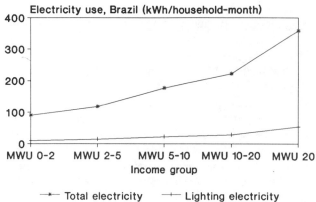

This graph shows that electricity use for lighting continues to grow with income even in a relatively prosperous developing country such as Brazil. Lighting electricity is, however, only a small fraction of total household electricity use in this case. MWU are minimum wage units.

SOURCE: Ashok Gadgil and Gilberto De Martino Jannuzzi, *Conservation Potential of Compact Fluorescent Lamps in India and Brazil* (Lawrence Berkeley Laboratory, Berkeley, CA and Universidade Estadual de Campinas, Brazil: June 23, 1989).

The choice of electric lighting technology has a strong impact on upstream utility investment, and this impact is accentuated by lighting's large contribution to the peak utility load. Incandescent lights are the least efficient electric lighting technology and have the highest capital and operating costs (see figure 3-10A). Currently, roughly 60 percent of total lighting electricity used in India is consumed by incandescent lightbulbs, and a higher fraction of residential lighting is used in incandescent bulbs.[41] In Brazil, some 95 percent of residential lighting is with incandescents,[42] about the same as in the United States.[43]

Much more efficient, cost effective lighting technologies are available.[44] For example, compact fluorescent lights with the same light output as an

---

[40] Calculated from Gadgil, op. cit., footnote 37.

[41] Ashok Gadgil and Gilberto De Martino Jannuzzi, op. cit., footnote 37.

[42] Gilberto De Martino Jannuzzi, et al., "Energy Efficient Lighting in Brazil and India: Potential and Issues of Technology Diffusion," Apr. 28, 1991, draft.

[43] Howard S. Geller, American Council for an Energy Efficient Economy, "Residential Equipment Efficiency: A State-Of-The Art Review," contractor report prepared for the Office of Technology Assessment, May, 1988. Note, however, that most lighting electricity in the United States is used in the commercial sector.

[44] Much more than just the energy efficiency and light output of the lighting hardware should be considered in practical applications. Lighting hardware is also characterized by its chromaticity (the color of the light), correlated color temperature (the temperature of a perfect emitter that has the same chromaticity as the lighting hardware), and color rendering index (how realistic the colors of objects illuminated by the light appear to be). Sources: Robert van der Plas and A.B. de Graaff, "A Comparison of Lamps for Domestic Lighting in Developing Countries," (Washington, DC: World Bank, Industry and Energy Department, June 1988); General Electric, "Lighting Application Bulletin: Specifying Light and Color."

Table 3-8—Lighting as a Share of National Energy and Electricity Use

	Energy use for lighting, GJ/cap				Percent of national total	
	Residential	Commercial	Industrial	Total	Energy	Electricity
Brazil	0.3	0.5	0.1	0.9	2.1	14.6
China	0.4	NA	NA	NA	NA	NA
India	0.5	0.05	0.05	0.6	5.1	14.2
Kenya	0.5	0.16	NA	0.66	2.6	27.2
Taiwan	0.7	0.8	NA	1.5	2.2	6.9
U.S.	NA	7.2	NA	NA	NA	NA

NA = Not available or not applicable.
SOURCE: U.S. Congress, Office of Technology Assessment, *Energy in Developing Countries*, OTA-E-486 (Springfield, VA: National Technical Information Service, Washington, DC: U.S. Government Printing Office, January 1991). Table 3-3 and app. 3-A.

incandescent have an initial capital cost for the lamp alone that is 20 times that of the incandescent but use just one-fourth as much electricity and last 10 times as long.[45] When the longer lifetime of these lamps and the required capital investment in utility equipment to power these different lights are taken into account, the total capital cost of compact fluorescent lighting is half that of incandescent lighting. Further, the total operating cost—including the cost of the lamp and the electricity to power it—of the compact fluorescent is one-third the cost of the incandescent. These results are summarized in figure 3-10 and are tabulated in appendix A.[46]

The high first cost of a compact fluorescent is a significant deterrent. (Of course, in places where a light is rarely used—such as a closet—it makes good economic sense to use a low cost, lower efficiency light, see figure 3-10.) The highly subsidized price of electricity in many developing countries reduces the consumers perceived benefit of higher efficiency lights. Making such an investment also poses a significant risk for the poor if such an expensive lamp is accidentally broken or stolen.

Utilities in developing countries may directly benefit by reducing the amount of subsidized power that they sell. It will often be less costly for the utility to subsidize the sale of efficient lightbulbs than to continue to subsidize the cost of electricity, particularly when utilities must finance expensive new generation capacity.[47]

If compact fluorescents are to be used in developing countries, special conditions should be taken into account. For example, conventional core-coil ballasts may have difficulties where power line voltage or frequency fluctuations are excessively large; large overvoltages, for example, could cause the lamp to burn out prematurely. Electronic ballasts can be used, however, particularly if modified with special voltage regulators.[48] At the same time, the electronic ballast raises the lamp efficiency (from 50) to about 60 lumens/watt. In contrast, to make standard incandescents more robust to the voltage fluctuations found in developing countries, the filament is made heavier at the cost of lowering the efficiency from 12 to typically 10 lumens per watt.[49] As many of the applications in developing countries will be new installations, properly sized and designed fixtures to hold compact fluorescents can be installed from the beginning, avoiding the difficulty and expense of retrofitting fixtures that industrial countries are now encountering.

There are similar opportunities in commercial buildings, even though they often now use fluorescent lighting. A standard fluorescent tube, not

---

[45] Prices are highly uncertain due to a variety of factors. Prices to original equipment manufacturers for the PL-13—the equivalent of a 60 W incandescent—are estimated at $7: $3.50 for the 10,000 hour glass element and $3.50 for the 20,000 hour base. Ashok Gadgil and Gilberto De Martino Januzzi, op. cit., footnote 37.

[46] See also: Howard S. Geller, "Electricity Conservation in Brazil: Status Report and Analysis," contractor report prepared for the Office of Technology Assessment, November 1990; Howard S. Geller, *Efficient Electricity Use: A Development Strategy for Brazil*, (Washington, DC: American Council for an Energy Efficient Economy, 1991).

[47] A detailed discussion of these various perspectives can be found in: Ashok Gadgil and Gilberto De Martino Januzzi, op. cit., footnote 37.

[48] These voltage regulators might be as simple as a zener diode or a neon lamp and could add about $1 to the factory cost of the lamp. This could allow the lamp to operate while voltages fluctuated around 220 V, from 80 V to 650 V. In contrast, core-coil ballasts can only withstand a 2 percent fluctuation in frequency and about +/-10 percent fluctuation in voltage. Total factory cost of the lamp with a modified electronic ballast might then be about $12. Ashok Gadgil, Lawrence Berkeley Laboratory, personal communication, Jan. 23, 1991.

[49] Ashok Gadgil, ibid.

Figure 3-10A—First (Capital) Cost, Annualized Capital Cost, and Efficiency of a 60 Watt Incandescent Bulb and Its Equivalent Compact Fluorescent Bulb

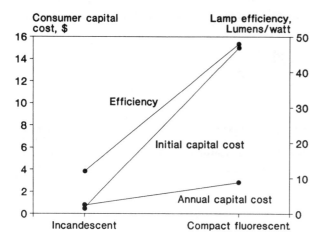

This figure indicates the barrier to buying the compact fluorescent bulb faced by the first-cost sensitive consumer. It also shows that the longer lifetime of the compact fluorescent bulb, shown on the basis of annualized capital cost, largely offsets its higher initial capital cost.

SOURCE: See app. A for details.

Figure 3-10B—Annualized Capital Cost for Incandescent and Compact Fluorescent Lamps, Upstream Utility Generation, Transmission, and Distribution Equipment to Power the Lamps, and the Two Combined to Give the Total Annual Capital Investment

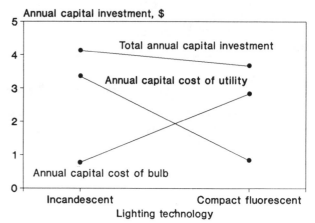

The total annualized capital investment declines substantially for the high efficiency compact fluorescent light bulbs.

SOURCE: See app. A for parameters in the case of 4 hours of operation per day. Note that the capital costs do not include switches, interior house wiring, etc.

Figure 3-10C—Annual Capital and Operating Costs for Different Lamps

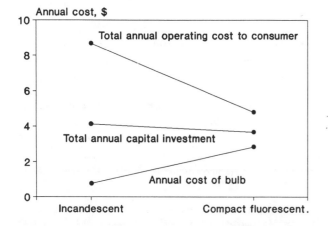

The total cost to the consumer decreases substantially for the high efficiency compact fluorescent.

SOURCE: See app. A.

Figure 3-10D—Cost Effectiveness of Compact Fluorescent Bulbs Versus Incandescents for Different Daily Operating Times

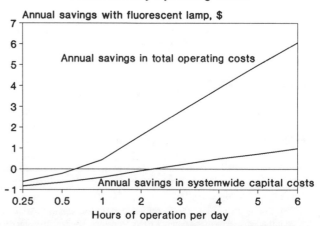

As the daily use of a compact fluorescent bulb increases, it is increasingly cost effective both in terms of systemwide capital costs and total annual operating costs (capital plus fuel). Only for daily operating times of less than about 40 minutes per day does the higher initial capital cost of the compact fluorescent outweigh savings in electricity costs over the lamps lifetime.

SOURCE: See app. A for parameters.

including the ballast, produces 3,150 lumens with a 40-watt (W) input for an efficiency of roughly 79 lumens per watt.[50] Through a variety of improvements in design and materials, tubes with efficiencies of more than 95 lumens per watt are now available.

To operate these lamps, ballasts must be used that lower their overall efficiency.[51] Conventional core-coil ballasts are typically made of aluminum wire and low quality laminated iron cores and consume typically 8 to 18 W per lamp.[52] This lowers the lamp efficiency from about 79 lm/W to about 55-65 lm/W—an efficiency loss of as much as one-third.

High efficiency core-coil ballasts use high quality materials—iron and copper—to cut ballast losses in half, to 3 to 4 W per lamp. Electronic ballasts can cut losses in half a second time, to 2 W, and at the same time increase the efficiency of the fluorescent tube itself, improve the quality of the light generated, reduce sensitivity to voltage and frequency fluctuations, and—for advanced designs—allow the lamp to be dimmed when, for example, daylighting is available (see boxes 3-C, and 3-D). The overall efficiency of a high efficiency tube and an electronic ballast is over 90 lm/W—about 40 percent higher than the standard lamp and ballast. If poorly designed, however, electronic ballasts can generate power line harmonics that may interfere with other electronic equipment or increase transmission and distribution losses somewhat.[53] This technology is now well established in the U.S. market, with annual sales approaching the two million mark in 1988.[54]

Reducing the temperature of the lamp 20 °C by raising efficiency and providing air cooling can also raise the operating efficiencies of lamps by as much as 10 percent.[55]

Efficiency improvements in the lighting systems of commercial buildings are particularly important in the newly industrializing countries due to the rapid growth of this sector. Commercial buildings in the ASEAN countries—Indonesia, Malaysia, Philippines, Singapore, and Thailand—now consume about one-third of the total electricity generated in the region, and of the electricity used, roughly one-third goes to lighting.[56] Lighting's share of commercial building electricity use is even higher in many other countries.

Further efficiency improvements in the lamp itself are possible.[57] In addition, better reflectors and diffusers can improve direction of the light generated and can allow the number of lamps per fixture to be reduced. Fixtures in many developing countries are particularly low quality due to the lack of highly reflective coatings. In Brazil, the combination of high efficiency lamps, ballasts, and fixtures was found to raise the useful light output of a fluorescent lamp system by a factor of three (see table 3-9).[58]

Better controls, such as occupancy sensors or photocell controlled dimmers used with advanced

---

[50]Based on a standard 4-foot long cool white fluorescent tube, with an output of 3,150 lumens and an input power of 40 W for the lamp and 15-20 W for the ballast. see: Terry McGowan, op. cit., footnote 36; and Gautam S. Dutt, "End Use Oriented Energy Strategies for Member Developing Countries (With Special Reference to India)," draft report to the Asian Development Bank, Manila, Philippines, January 1991.

[51]Core-coil ballasts are current limiting inductors used to prevent high currents from developing in the lamp and burning it out. Without the ballast, the plasma discharge of the fluorescent lamp—with its negative coefficient of resistance—would support an increasing current and basically turn the lamp into a miniature arc welder.

[52]The higher figure is found for low quality ballasts operating a single tube. See: Howard S. Geller, op. cit., footnote 46; and Gautam S. Dutt, "End Use Oriented Energy Strategies for Member Developing Countries (With Special Reference to India), draft report to the Asian Development Bank, Manila, Phillipines, January 1991.

[53]Rudolph R. Verderber, Oliver C. Morse, Francis M. Rubinstein, "Performance of Electronic Ballast and Controls With 34-and 40-Watt F40 Fluorescent Lamps," *Transactions on Industrial Applications*, vol. 25, No. 6, 1989, pp. 1049-1059; Amory B. Lovins and Robert Sardinsky, *The State of the Art: Lighting* (Snowmass, CO: Rocky Mountain Institute, 1988).

[54]Verderber, ibid.

[55]M. Sminovitch et al., *The Energy Conservation Potential Associated With Thermally Efficient Fluorescent Fixtures*, report No. LBL-27315 (Berkeley, CA: Lawrence Berkeley Laboratory, June 1989).

[56]M.D. Levine, J.F. Busch, and J.J. Deringer, "Overview of Building Energy Conservation Activities in ASEAN," paper presented at the ASEAN Special Sessions of the ASHRAE Far East Conference on Air Conditioning in Hot Climates, Kuala Lumpur, Malaysia, Oct. 26-28, 1989, Lawrence Berkeley Laboratory, report No. LBL-28639, Berkeley, California.

[57]The combination of isotopic enrichment, two photon phosphors, and gighertz operation, among others, may be able to push efficiencies to as high as 230 lumens/watt. See: Samuel Berman, "Energy and Lighting," *Energy Sources: Conservation and Renewables*, (New York, NY: American Physical Society, 1985).

[58]H. Geller, op. cit., footnote 46.

electronic ballasts, can help ensure that the light generated is neither excessive nor in unoccupied spaces. Finally, the controlled use of daylight, particularly with filters to keep out unwanted heat, can dramatically cut lighting needs in buildings. Combined, the use of a photocell or otherwise controlled dimmer that adjusts for daylighting or other lighting can reduce electricity use in many buildings by perhaps 35 percent or more without degrading lighting quality or quantity.[59] Retrofits of the lamps and light fixtures at the U.S. Environmental Protection Agency building in Washington, DC recently allowed it to cut its lighting electricity use by 57 percent.[60]

Hardware is, however, only a part of the total consideration in developing an efficient, attractive lighting system. Lighting design must also include such factors as producing attractive displays for retailers and maintaining high productivity among office workers—avoiding glare, flicker and hum, and strobing on computer display screens (for the time when computers are as prevalent in developing-country offices as they are today in industrial countries) while providing sufficient light for employees to work efficiently. For a typical office in which wages are 90 percent and lighting electricity is 3 percent of operating costs, it makes little sense to reduce lighting by one-third—a 1-percent savings—if it lowers worker productivity by 10 percent—a 9-percent cost. On the other hand, overlighting is also found widely and offers a substantial opportunity for savings. In Brazil, for example, overlighting of one-third or more has been found in a number of office buildings surveyed.[61]

Therefore, along with improved lighting hardware, careful attention is also needed to lighting design. Considerable savings, for example, may be possible by carefully and appropriately lighting the task—i.e., the desk—and reducing background lighting levels. Good lighting is design intensive; the importance of design should not be minimized. Design tools need to be further developed and widely disseminated so that developing countries can also take advantage of these opportunities.

As in other cases, even the most careful technical assessment of potential can be blunted when users fail to clean lamps, leave them running for longer hours than needed, or—in response to lower monthly electricity bills—add more lights. In some applications, consumers may also prefer to use incandescent lights for their warmer color or because the very short periods of use—such as in a closet—do not warrant the larger expense of a high efficiency light. Market costs for all lighting hardware that reflect the total capital and operating costs—including upstream capital investment in generation, transmission, and distribution facilities—will help to ensure that all choices of lighting equipment are more economically rational.

Improvements in lighting efficiency have a synergistic impact on other important uses of electricity such as air conditioning. In large buildings or residences cooled by air conditioning, every kW of lighting power saved cuts the needed air conditioning by as much as one-third kW.[62] Together, saving this 1.3 kW of electrical power reduces, for example, coal consumption at a rate of 4.5 kW or 0.63 kilograms (kg) per hour.[63] Reducing the lighting load will also reduce the size of air conditioner and related equipment needed, saving capital costs. Of course, in the winter the heat generated by lights can help warm the building and reduce space heating requirements. This is likely to be less important in most developing countries.

---

[59] Terry McGowan, op. cit. footnote 36; *See also*: Y.J. Huang, B. Thom, B. Ramadan, "A Daylighting Design Tool For Singapore Based on DOE-2.1C Simulations," in *Proceedings of the ASEAN Special Sessions of the ASHRAE Far East Conference on Air Conditioning in Hot Climates*, Kuala Lumpur, Malaysia, Oct. 26-28, 1989, Lawrence Berkeley Laboratory, Report. LBL-28639. Of particular importance in implementing effective daylighting systems is the development of high performance algorithms for controlling the daylighting system and of computer design tools for designing the office workspace. See: R.R. Verderber, F.M. Rubinstein, and G. Ward, "Photoelectric control of Daylight-Following Lighting Systems," report No. CU-6243 (Palo Alto, CA: Electric Power Research Institute, 1989).

[60] Matthew L. Wald, "E.P.A. Urging Electricity Efficiency," *The New York Times*, Jan. 16, 1991.

[61] Howard S. Geller, op. cit., footnote 46.

[62] One kW of heat input into a building from lights adds a load of 3,412 BTU/hour to the cooling load. For an air conditioning unit with an energy efficiency ratio—Btu per hour of cooling capacity divided by watts of electrical input—of 10.0, this demands an additional 3412/10 = 341 W or one-third kW of power for cooling. Because the power required to move air through ducts increases as the cube of the flow velocity, there may be further energy savings by reducing the flow rates necessary to cool the building when using efficient lighting.

[63] Assuming technical losses in transmission and distribution of 15 percent, a 1.34 kW load requires 1.58 kW of power to be generated. At an average generation efficiency of 35 percent, this requires burning 4.5 kW of coal, or 16 MJ/hour. At an average calorific value of 25.58 MJ/kg, this saves 0.63 kg of coal per hour.

> **Box 3-C—Development of Electronic Ballasts in the United States and Brazil**
>
> The Department of Energy established a Lighting Research Program at the Lawrence Berkeley Laboratory (LBL) in 1976. This program was to accelerate the introduction of new energy efficient lighting technologies in the marketplace. One of the first technologies targeted was high-frequency electronic ballasts. Data showed that lamps operated at frequencies above 10 kHz were 15 to 30 percent more efficient; advances in transistor and switching power supply technologies and reductions in their costs during the 1960s and 1970s made such high frequency operation both possible and potentially cost-effective.
>
> LBL looked for companies that would share development costs and who, if the project was successful, could manufacture the electronic ballasts for the marketplace. None of the major manufacturers of conventional core coil ballasts indicated any interest, but fourteen small entrepreneurial firms responded. LBL selected two firms in 1977. IOTA engineering proposed a low-cost nondimmable design; Stevens Electronics proposed a sophisticated, higher efficiency design that could adjustably reduce lamp light output by 90 percent. The first prototypes showed the 15-percent efficiency gain due to higher frequency lamp operation, and saved 10 percent more by reducing losses in the ballast compared to conventional core coil ballasts—a total efficiency gain of 25 percent. These savings were then confirmed in field tests in PG&Es main office building in San Francisco. These field trials also provided data on reliability that was then incorporated into the prototype design.
>
> Despite the success of the trials, major U.S. ballast manufacturers showed little interest. Meanwhile, a number of small U.S. firms and, beginning in 1980, several foreign firms—Toshiba (Japan) and O.Y. Helver (Finland)—picked up on the LBL work and began manufacturing solid-state ballasts during 1980-84. In 1984 to 1986, major U.S. manufacturers such as General Electric, GTE, Advanced Transformer, and Universal Manufacturing entered the market. By 1986, one million solid state ballasts were sold annually in the United States and sales were increasing 60 percent per year.
>
> The year 1985 witnessed the beginnings of the Brazilian effort to develop and produce solid-state ballasts. Researchers at the University of Sao Paulo tested potential designs and, with funding from the Sao Paulo State electrical utility, produced about 50 prototype units in 1986-87 that demonstrated performance comparable to electronic ballasts produced in industrial countries. The technology was transferred to a private company (Begli) in 1988 with subsequent scaleup of production and tests to 100 demonstration units, then 350 units, and finally mass production in 1990.
>
> Several features of this case are important. The United States took the lead in initially developing and proving the technology. In part based on this work, Brazilian researchers could then move more quickly and with some confidence that their efforts would result in a workable design. Key elements in Brazil were a capable staff of research and development scientists and engineers that could effectively adapt a foreign technology to locally produced components and conditions; critical (and farsighted) early and long-term support from the Brazilian Government and electric utility; gradual and careful scaleup of production with detailed demonstrations and field testing; and effective technology transfer—at an early stage—to a private company for production and commercialization.
>
> SOURCE: Howard Geller, Jeffrey P. Harris, Mark D. Levine, and Arthur H. Rosenfeld, "The Role of Federal Research and Development In Advancing Energy Efficiency: A $50 Billion Contribution to the U.S. Economy," *Annual Review of Energy* Vol. 12, 1987, pp. 357-395; and Howard S. Geller, "Electricity Conservation in Brazil: Status Report and Analysis," contractor report for the Office of Technology Assessment, November 1990, and to be published as "Efficient Electricity Use: A Development Strategy for Brazil," American Council for An Energy Efficient Economy, Washington, DC, 1991.

Finally, energy, systemwide capital cost, and operating cost savings are also possible in street lighting and other lighting applications. High pressure sodium lamps, for example, have much higher efficiencies than other types of street lighting and, because of their longer lifetime, have lower maintenance costs as well.

Cost effective lighting technologies are available, but under the present institutional structure dividing responsibility for the supply and the use of electricity between the utility and the consumer, substantial inefficiencies arise in the consumer choice of end-use technology and in the allocation of capital between energy supply and end-use efficiency.

REFRIGERATION

Refrigerator ownership is at present quite low in most developing countries, but it is increasing rapidly. This rapid penetration is also occurring at much lower income levels than was the case in the

Box 3-D—Lighting Efficiency Programs in Europe

Electric utilities in Sweden, Denmark, the Netherlands, and West Germany operated 40 residential lighting efficiency programs in late 1988/early 1989. These programs provided compact fluorescent lamps (CFLs) to consumers through a variety of means, including give-aways, rebates, wholesale discounts, tax breaks by governments, and by adding small monthly payments for the CFL to the consumer's electric bill.

These programs provided more than two million CFLs to the 4.9 million eligible households, accounting for 80 to 95 percent of all CFLs placed in use in the residential sector during that time. The average cost of these programs—including both the CFL and administrative costs—was equivalent to about $0.02 per kWh of electricity saved, much less than the cost of generating electricity. The most cost-effective programs were those that simply gave the CFL away for free and thus minimized administrative costs. By increasing the demand for CFLs, these programs also lowered the retail price for CFLs by 20 to 50 percent even after the programs ended.

Among program participants, two-thirds were satisfied with the brightness and color of the lamps, but more than half were dissatisfied with the lamp's appearance. Many had trouble fitting the CFLs to their existing lighting fixtures. Nonparticipants lacked interest or knowledge of the program, considered CFL prices excessive, or found the CFL too large or heavy for their needs. Many of these technical shortcomings are being addressed by manufacturers with the development of smaller CFLs, electronic ballasts, and other improvements.

SOURCES: Evan Mills, Agneta Persson, Joseph Strahl, "The Inception and Proliferation of European Residential Lighting Efficiency Programs," in ACEEE 1990 Summer Study on Energy Efficiency in Buildings, American Council for an Energy Efficient Economy, Washington, DC 1990; Evan Mills, "Evaluation of European Lighting Programmes: Utilities Finance Energy Efficiency," *Energy Policy*, April 1991, pp. 266-278.

Table 3-9—Cost and Performance of Commercial Lighting Improvements, Brazil

	Standard[a]	Efficient[b]
Performance		
Power input	192 W	60 W
Rated light output	10800 lm	5000 lm
Useful light output	2260 lm	2100 lm
Capital costs		
Lamps	$ 9.70	$ 5.80
Ballasts	$ 33.30	$ 16.65
Reflectors	NA	$ 33.95
Subtotal	$ 43.00	$ 56.40
Annualized capital costs	$ 12.66	$ 16.60
Annual energy use		
Direct electricity use	507 kWh	203 kWh
Air conditioning energy[c]	142 kWh	57 kWh
Total electricity use	649 kWh	260 kWh
Utility costs[d]		
Capital investment	$296.00	$118.00
Annualized capital cost	$ 23.85	$ 9.51
Annual electricity costs	$ 35.70	$ 14.30
System wide costs		
Total annual capital cost	$ 36.50	$ 27.60
Total annual operating	$ 48.36	$ 30.90

NA = Not applicable.
[a] Based on four 40 W tubes with conventional core-coil ballast in a standard fixture with completely exposed lamps.
[b] Based on two 32 W high efficiency tubes with a mirrored glass reflector. Useful output is so high because of: 1) the narrow 32 W tubes trap less light in the fixture; 2) the mirrored reflector increases useful light output.
[c] This is the amount of air conditioning power needed to remove the heat generated by the lights.
[d] Utility capital costs are set at estimated marginal prices as calculated in app. A. Electricity prices are set at prevailing Brazilian rates for large commercial users of $0.055/kWh.

SOURCE: Adapted from Howard S. Geller, "'Electricity Conservation in Brazil: Status Report and Analysis," contractor report to the Office of Technology Assessment, November 1990, tables 15 and 16; published as "Efficient Electricity Use: A Development Strategy for Brazil," American Council for an Energy Efficient Economy, Washington, DC 1991.

United States or other industrial countries due to the declining real cost of refrigerators as materials and manufacturing techniques have improved (see figure 3-11).

Refrigerators are typically not the first appliance acquired by a household when it gets electric service; lights are first, followed by refrigerators or other equipment depending on household income, region, and other factors (see figure 3-12). In India, fans are typically among the first appliances acquired, followed by televisions and refrigerators. In Brazil, even relatively poorer, newly electrified households often have televisions and refrigerators, as these appliances are comparatively inexpensive and are available secondhand.[64]

The refrigerators used in developing countries are typically half the size of American refrigerators or less. They are also much less efficient than the best refrigerators now commercially available. (The average refrigerator used in the United States is similarly much less efficient than the best available).

---

[64] Gadgil and Jannuzzi, op. cit., footnote 37.

Figure 3-11A—Reduction in the Real Cost of Refrigerators Over Time in the United States

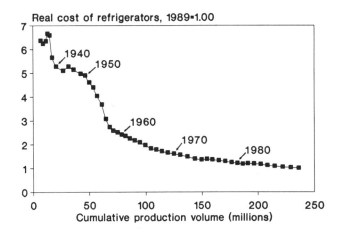

Over the past 40 years, the real price of refrigerators has dropped by almost a factor of 5 (measured by the consumer price index for refrigerators divided by either the overall change in consumer price index or by the GNP deflator). For developing countries, such price reductions are allowing households to invest in refrigerators at a much earlier point in time than was the case for the United States and other industrialized countries at a similar level of development.

SOURCE: U.S. Congress, Office of Technology Assessment, *Energy in Developing Countries*.

Figure 3-11B—Nationwide Refrigerator Penetration Versus Per Capita GDP Corrected for Purchasing Power Parity

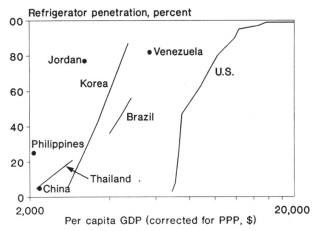

This figure shows that a much higher percentage of the developing-country households shown own refrigerators than did households in the United States at a similar level of economic development as measured by PPP corrected per-capita incomes. Note that the per-capita GDPs have been smoothed—otherwise the curves shown would reverse themselves during serious recessions—so that the curves are monotonic along the GDP axis.

SOURCE: GDP data is from: Angus Maddison, "The World Economy in the 20th Century," (Paris, France: Organization for Economic Co-Operation and Development, 1989); and Robert Summers and Alan Heston, "A New Set of International Comparisons of Real Product and Price Levels Estimates for 130 Countries, 1950-1985," *The Review of Income and Wealth*, vol. 34, pp. 1-25, 1988. See especially the diskettes accompanying the article. Refrigerator penetration data for the United States is from: Donald W. Jones, "Energy Use and Fuel Substitution in Economic Development: What Happened in Developed Countries and What Might be Expected in Developing Countries?" Oak Ridge National Laboratory, ORNL-6433, August 1988; Refrigerator penetration data for the various developing countries shown is from: S. Meyers et al., "Energy Efficiency and Household Electric Appliances in Developing and Newly Industrialized Countries," Lawrence Berkeley Laboratory, LBL-29678, December 1990.

In Indonesia, for example, most refrigerators are assembled locally from imported components and, in general, do not take advantage of proven energy efficiency features such as rotary compressors and increased insulation.[65] In many countries, popular models still use fiberglass insulation rather than better polyurethane foam.

The efficiency of refrigerators has improved in many countries in recent years. In the United States, energy consumption by the average new refrigerator dropped from about 1,700 kWh per year in 1972 to about 975 kWh by 1990, an efficiency improvement of about 3 percent per year.[66] Factors contributing to this efficiency improvement included:

- The technological push of low cost efficiency improvements such as shifting from fiberglass to polyurethane foam insulation, and the use of more efficient motors, compressors, and heat exchangers;
- the market pull generated by a more well informed public—mandatory efficiency labeling was instituted in 1980—faced with a 40 percent real increase in residential electricity prices between 1973 and 1984; and
- the regulatory shove of minimum efficiency standards first adopted in California in 1977 and nationally in 1987.[67]

The energy consumption of the average new refrigerator in Brazil similarly dropped by almost 13 percent between 1986 and 1990, a rate of improve-

---

[65]Lee Schipper, "Efficient Household Electricity Use in Indonesia," Lawrence Berkeley Laboratory, Draft, January 1989.

[66]Association of Home Appliance Manufacturers, "Major Home Appliance Industry Fact Book, 1990/91," Chicago, IL, 1991.

[67]Howard S. Geller, *Energy Efficient Appliances* (Washington, DC: American Council for an Energy Efficient Economy, June 1983); Howard S. Geller, American Council for an Energy Efficient Economy, "Residential Equipment Efficiency," contractor report prepared for the Office of Technology Assessment, May 1988.

Figure 3-12—Electric Appliance Ownership in Urban Java, 1988

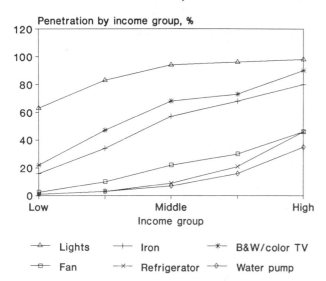

This figure shows the rapid penetration and relative importance within household purchasing patterns of lights, TVs, irons, fans, refrigerators, and water pumps. Income groups (share of households) in ascending order are: less than 75,000 Rp/month (24 percent), 75-120 (22 percent), 121-185 (21 percent), 186-295 (14 percent), and greater than 295 (9 percent).

SOURCE: Lee Schipper and Stephen Meyers, "Improving Appliance Efficiency in Indonesia," *Energy Policy*, forthcoming.

ment of about 2.5 percent per year (see table 3-10).[68] Brazilian manufacturers, however, are unable to make use of the very efficient motor-compressors in domestic refrigerators that Brazil manufactures and exports, as these units cannot readily tolerate the voltage fluctuations found in Brazil (but substantial improvements in refrigerator performance are never-

Table 3-10—Progress in Improving the Efficiency of Refrigerators in Brazil

Year	Electricity consumption (kWh/year)		
	Best	Worst	Average
1986	440	570	490
1987	440	490	460
1988	440	490	460
1989	335	490	435
1990	335	490	435

For single-door 250 to 300 liter refrigerators

SOURCE: Howard S. Geller, "Electricity Conservation in Brazil: Status Report and Analysis," contractor report prepared for the Office of Technology Assessment, August 1990; published as "Efficient Electricity Use: A Development Strategy for Brazil," American Council for an Energy Efficient Economy, Washington, DC, 1991.

theless possible).[69] In South Korea, energy consumption by the average new refrigerator dropped by a remarkable 65 percent between 1980 and 1987, a rate of improvement of about 12 percent per year (see table 3-11).[70] The initial costs of achieving these efficiency gains have generally been small, and have been more than offset by savings in electricity bills.

In the United States, the failure of market forces alone to push the energy efficiency of refrigerators and other products sufficiently rapidly has led to enactment of laws mandating efficiency standards for 13 types of consumer products.[71] Different levels of technology for implementing these standards in an 18 cubic foot (510 liter)[72] top-mount automatic defrost refrigerator/freezer—the most popular design in the U.S. market with 73 percent of annual

---

[68]Howard S. Geller, op. cit. footnote 46.

[69]Howard Geller, Ibid.

[70]Stephen Meyers et al., op. cit., footnote 29. Note that these appliances are not strictly comparable in terms of their actual power consumption. The Brazilian and South Korean refrigerators are just half the size of the average American refrigerator, and the features offered on these refrigerators differ significantly, with varying impacts on their energy consumption.

[71]These laws include: the Energy Policy and Conservation Act (Public Law 94-163) as amended by the National Energy Conservation Policy Act (P.L. 95-619), the National Appliance Energy Conservation Act of 1987 (P.L. 100-12), and the Appliance Energy Conservation Amendments of 1988 (P.L. 100-357). Note that although Law 95-619 required standards, the U.S. Department of Energy never issued them. Action has only begun with Laws 100-12 and 100-357. The 13 products covered are: 1) refrigerators and freezers, 2) room air conditioners, 3) central air conditioners and heat pumps, 4) water heaters, 5) furnaces, 6) dishwashers, 7) clothes washers, 8) clothes dryers, 9) direct space heating equipment, 10) kitchen ranges and ovens, 11) pool heaters, 12) television sets, and 13) fluorescent lamp ballasts. Other products can be included at the discretion of the Secretary of Energy. *See*: U.S. Department of Energy, *Technical Support Document: Energy Conservation Standards for Consumer Products: Refrigerators, Furnaces, and Television Sets*, report No. DOE/CE-0239 (Washington, R : U.S. Department of Energy, November 1988); and U.S. Department of Energy, *Technical Support Document: Energy Conservation Standards for Consumer Products: Refrigerators and Furnaces*, report No. DOE/CE-0277 (Washington, DC: U.S. Department of Energy, November 1989).

[72]Efficiency ratings for refrigerator/freezers are normally cited in the United States in terms of the refrigerators adjusted volume, given by the volume of interior refrigerated space plus the 1.63 times the interior volume of the freezer compartment. The refrigerator/freezer cited here has an adjusted volume of 20.8 cubic feet (590 liters).

[73]U.S. Department of Energy, *Technical Support Document: Energy Conservation Standards for Consumer Products: Refrigerators and Furnaces*, op. cit. footnote #71.

Table 3-11—Progress in Improving the Efficiency of Appliances Produced in Korea

Year	Refrigerator 200 liter (kWh/year)	Room AC 7,100 Btu/hr (Btu/hr-W)	Color TV 14 inch (W)
1980	672	7.6	82
1981	456	7.8	69
1982	336	8.4	55
1983	312	9.0	57
1984	288	11.0	54
1985	264	11.3	56
1986	240	11.3	62
1987	240	11.3	60

SOURCE: Stephen Meyers et al., "Energy Efficiency and Household Electric Appliances in Developing and Newly Industrialized Countries," Lawrence Berkeley Laboratory, report No. LBL-29678, December 1990.

Table 3-12—Description of Refrigerator/Freezer Technology Levels

Level	Description
A	Baseline—18 ft^3 (20.8 ft^3 adjusted volume) refrigerator/freezer, side wall insulated with 2.2 inches foam in freezer, 1.9 inches foam in refrigerator; door insulated with 1.5 inches foam in freezer and 1.5 inches fiberglass in refrigerator; back insulated with 2.2 inches foam; Features include improved thermal seal gasket, antisweat switch, 4.5 EER[a] compressor, bottom-mounted condenser, auto-defrost timer, 10 W evaporator, and 13.5 W condenser fans.
B	Baseline—Level A plus enhanced evaporator
C	Level B plus Door Foam Insulation
D	Level C plus 5.05 EER Compressor
E	Level D plus 2 inches door insulation
F	Level E plus more efficient evaporator and condenser fans
G	Level F plus 2.6 inches/2.3 inches side insulation and 2.6 inches back insulation
H	Level F plus 3.0 inches/2.7 inches side insulation and 3.0 inches back insulation
I	Level F plus evacuated panel (K=0.055)
J	Level I plus two compressor system
K	Level J plus adaptive defrost

[a]EER is the Energy Efficiency Ratio measured in terms of BTU/hr cooling output divided by watts of electrical power input.

SOURCE: *Technical Support Document: Energy Conservation Standards for Consumer Products: Refrigerators and Furnaces* (Washington, DC: U.S. Department of Energy, November 1989) publication DOE/CE-0277.

sales[73]—that does not use ozone damaging (CFCs)[74] (CFC-11, CFC-12) are listed in table 3-12 and shown in figure 3-13.[75] Although this refrigerator is larger and has more features than those typically found in developing countries today, it will nevertheless be used here to demonstrate the potential of energy efficiency improvements because: the size and variety of features in refrigerators used in developing countries is increasing and will likely continue to increase in the future; and the relative impact of different technical options to improve refrigerator performance is similar.

The cost of this model refrigerator increases slowly as its efficiency is improved. Between the baseline—Technology A—and Technology H, for example, the energy consumption of the refrigerator decreases by one-third while its U.S. retail cost increases by just 12 percent. For consumers, this modest increase in cost nevertheless appears to be a very large barrier in practice. Studies of appliance purchases in the United States have shown that consumers behave as if there was a discount rate on refrigerators (and other consumer goods) of more than 60 percent.[76] That is, consumers do not buy more efficient refrigerators unless the energy saved pays for its higher first cost in less than about one and a half years, providing a net savings to the consumer for the rest of its typical 20-year life.[77] Indeed, consumers may often not even consider energy savings and may rarely actually compute the potential payback of more efficient models. In developing countries, higher first costs may prove to be an even larger barrier in practice, due to the lack of cash or access to credit.

Every time a consumer purchases a refrigerator, however, he or she commits the nation as a whole to a large investment in upstream power generation, transmission, and distribution equipment.[78] Pur-

---

[74]L.E. Manzer, "The CFC-Ozone Issue: Progress on the Development of Alternatives to CFCs," *Science*, vol. 249, July 6, 1990, pp. 31-35.

[75]This analysis is based on Department of Energy standard refrigeration testing procedures and simulations. There are indications that these testing procedures overestimate the use of electricity in actual practice by 20 to 25 percent. This suggests that allocations of electricity use in residences may give too much weight to refrigerators and seriously underestimate certain other uses. See, for example: Michael Shepard, et al., *The State of the Art: Appliances*, (Snowmass, CO: Rocky Mountain Institute, August 1990).

[76]Henry Ruderman, Mark D. Levine, James E. McMahon, "The Behavior of the Market for Energy Efficiency in Residential Appliances Including Heating and Cooling Equipment," *Energy Journal*, vol. 8, No. 1, January 1987, pp. 101-124; Harry Chernoff, "Individual Purchase Criteria for Energy Related Durables: The Misuse of Life Cycle Cost," *Energy Journal*, vol. 4, No. 4, October 1983, pp. 81-86; Malcolm Gladwell, "Consumers' Choices About Money Consistently Defy Common Sense," *The Washington Post*, Feb. 12, 1990, p. A3.

[77]Average lifetimes for refrigerators in the United States are 19 years. *See*: U.S. Department of Energy, *Technical Support Document: Energy conservation Standards for Consumer Products: Refrigerators and Furnaces*, op. cit., footnote #71.

[78]Note that this applies primarily to where new investments must be made or existing equipment is being retired and must be replaced.

chasing an inefficient refrigerator at a slightly lower cost to the consumer commits the nation to a larger investment in utility equipment. The total capital investment required per refrigerator is shown in figure 3-13. Although the cost to the consumer for the refrigerator increases, the total capital cost including the cost of electricity generating equipment decreases through technology "I" and remains slightly lower for the most efficient technology considered than for the baseline technology—which is still better than the technology found in most developing countries.

Similarly, the total operating cost to consumers (including the discounted first cost of the refrigerator and the cost of electricity) decreases steadily as the efficiency of the refrigerator is increased (see figure 3-13). Thus, purchasing the most efficient refrigerators examined here saves the consumer money over the lifetime of the appliance and saves the Nation both initial capital investment and fuel costs to power the refrigerator. This can free capital for investment in other critical development needs.

Yet further improvements are possible. The improvements listed in table 3-12 were chosen based in large part on whether or not U.S. appliance manufacturers could implement them in a 3 year period—by 1992/93. Over a longer time span, additional cost effective improvements may be possible.[79] In fact, more efficient commercial designs have already been developed. One U.S. company[80] now makes and commercially sells in small lots a refrigerator/freezer (18.5 ft^3 adjusted volume) that uses just 280 kWh/yr[81]—just over half the energy used by the most efficient design listed in table 3-12 and figure 3-13 (costs are high, however). Further improve-

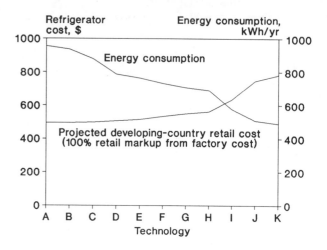

Figure 3-13A—Retail Costs and Energy Consumption for Technology Improvements in Refrigerators

Projected developing country retail costs are assumed to be marked up by 100-percent from the factory cost for each technology. This 100 percent markup is somewhat lower than the retail markups assumed by Lawrence Berkeley for the United States, but was chosen to be more representative of retail overheads in developing countries. Retail costs do not include any additional markup for taxes or tariffs.
SOURCE: See table 3-12 and app. A.

ments are also possible using better insulation—including various types of vacuum insulation,[82] electronic adjustable speed drives,[83] and other changes.

Much of the potential improvement in refrigerator performance can be achieved without resorting to high efficiency motor/compressor systems.[84] This is of interest in developing countries where large voltage fluctuations may limit the use of high efficiency motors.

---

[79] See, for example: David B. Goldstein, Peter M. Miller, and Robert K. Watson, "Developing Cost Curves for Conserved Energy in New Refrigerators and Freezers: Demonstration of Methodology and Detailed Engineering Results," Natural Resources Defense Council, San Francisco, CA, and American Council for an Energy Efficient Economy, Washington, DC, Jan. 15, 1987.

[80] Sun Frost Co., Arcata, CA, cited in Michael Shephard et al., *The State of the Art: Appliances* (Snowmass, CO: Rocky Mountain Institute, August 1990).

[81] Note that this is the company test procedure—not the standard Department of Energy test procedure—for 70 °F ambient temperatures; at a high 90 °F ambient temperature, the energy consumption is 365 kWh/yr. The refrigerator is also not strictly comparable to the other ones listed because it is manual rather than automatic defrost.

[82] Extensive research is now being done on soft vacuum panels containing powder, aerogels, and hard vacuum panels, among others. See, for example, Michael Shepard et al. *The State of the Art: Appliances* (Snowmass, CO: Rocky Mountain Institute, August 1990).

[83] S. Zubair, V. Bahel, and M. Arshad, "Capacity Control of Air Conditioning Systems by Power Inverters," *Energy*, vol. 14, No. 3, 1989, pp. 141-151.

[84] *See*: U.S. Department of Energy, technical support document, "Energy Conservation Standards for Consumer Products: Refrigerators, Furnaces, and Television Sets," report No. DOE/CE-0239 (Washington, DC: U.S. Department of Energy, November 1988); and U.S. Department of Energy, technical support document, "Energy Conservation Standards for Consumer Products: Refrigerators and Furnaces," Report No. DOE/CE-0277 (Washington, DC: U.S. Department of Energy, November 1989); and David B. Goldstein, Peter M. Miller, and Robert K. Watson, "Developing Cost Curves for Conserved Energy in New Refrigerators and Freezers: Demonstration of Methodology and Detailed Engineering Results," Natural Resources Defense Council, San Francisco, CA and American Council for an Energy Efficient Economy, Washington, DC.

Figure 3-13B—Refrigerator Retail Cost, Utility Capital Investment To Power the Refrigerator, Total System Capital Investment, and Total Annual Operating Costs for Different Refrigerator Technologies

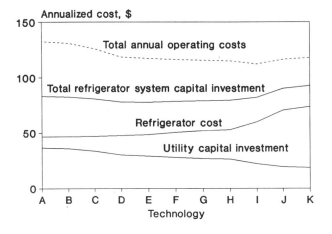

This diagram shows that the total capital cost—including both the retail cost of the refrigerator in the developing country and the capital cost of utility investment in generation, transmission, and distribution equipment—decreases slightly with more efficient refrigerators until technology "E" is reached, whereupon it increases slightly. Technology "A" represents a more efficient refrigerator than most of those now sold in the U.S. market. Total annual operating costs, including the annualized capital cost of the refrigerator and the cost of electricity to power it, decrease with more efficient technologies until Technology "I" is reached. The total annual operating cost even with the most efficient and expensive refrigerator, technology "K", however, is substantially lower than that for the baseline technology "A".

SOURCE: See app. A.

Table 3-13—Commercial Refrigeration Efficiency Improvements

Technology	Energy savings
Glass doors	40-50%
Strip curtains	10-20% and more
Parallel unequal compressors	13-27%
Variable speed compressor control	NA
Evaporative pre-coolers	8-11% depending on climate
Floating head pressure control	2-15%

SOURCE: A. Usibelli et. al., "Commercial-Sector Conservation Technologies," Lawrence Berkeley Laboratory Report No. LBL-18543.

In this situation, electronic adjustable speed drives (ASDs) (see ch. 4) offer several additional opportunities, particularly as advances increase their reliability and reduce their cost. First, the ASD could be used to buffer fluctuations in line frequency and voltage and allow higher efficiency motors/compressors to be used. Second, if standard protocols can be agreed on, these ASDs could, at little additional cost, be programmed to be controlled by high frequency signals sent by the utility over the power lines or by other means. The utility, for example, might use this technique to cycle off a certain fraction of the refrigerators for short periods in order to prevent blackouts or brownouts when large utility generating plants failed or peak demand was excessive. Such techniques are already in use in the United States with air conditioners and have proven cost effective even with retrofits. If such systems could be built into new refrigerators at little or no cost, this might be a useful means of improving power reliability in developing countries.[85]

Large cost-effective reductions in energy consumption are also possible with commercial refrigerators, particularly in the retail food industry. The use of glass doored rather than open refrigerator cases; improved glass doors and door seals (and the subsequent elimination of antisweat heaters); improved compressors; improved display lighting; and other improvements can significantly reduce electricity consumption (see table 3-13). Improvements primarily in the compressors of grocery store refrigeration systems alone have demonstrated overall electricity savings of about 23 percent and reduced peak demands by 30 percent.[86]

SPACE CONDITIONING

Space conditioning includes heating, cooling, and ventilating residential and commercial buildings in order to create more comfortable conditions. Space heating is important only in a few colder or mountainous areas in developing countries. An example is Northern China: nearly one-fifth of China's total annual coal and 5 percent of China's annual biomass consumptions is used for space heating.[87] Residences rarely have any insulation and

---

[85]Samuel F. Baldwin, "Energy Efficient Electric Motor Drive Systems," *Electricity: Efficient End Use and New Generation Technologies, and Their Planning Implications* (Lund, Sweden: Lund University Press, 1989).

[86]D.H. Walker and G.I. Deming, *Supermarket Refrigeration Modeling and Field Demonstration*, report No. CU-6268, (Palo Alto, CA: Electric Power Research Institute, 1989).

[87]Vaclav Smil, "China's Energy," contractor report prepared for the Office of Technology Assessment, 1990.

often have large gaps around doors and windows.[88] Indoor temperatures in these homes are controlled not by a thermostat or by comfort requirements, but by fuel supply—and fuel, though cheap, is scarce. In Kezuo county, Northeast China, for example, average indoor temperatures are near the freezing point during the winter, compared to average outdoor temperatures of –3 °C to –5 °C with lows of –25 °C. Additions to coal supply, more efficient stoves, or better wall insulation would thus result mainly in comfort improvements and not in energy savings.

Many developing countries have temperate climates year around that require little space heating or cooling. In Latin America, examples include Mexico City, Sao Paulo, Caracas, and Buenos Aires.[89] Nevertheless, large amounts of electricity may still be used for ventilation in the commercial buildings that use mechanical ventilation systems. Even with Sao Paulo's temperate climate, for example, roughly 20 percent of the total electricity use in all commercial buildings is for air conditioning; when buildings with central air conditioning are considered alone, air conditioning accounts for half of energy use.[90] Proper design of the air handling system—fans, ducts, controls, etc.—and careful choice of components can substantially lower both capital investment and/or energy consumption in these systems.[91] These issues are examined in chapter 4 within the broader context of electric motor drive systems.

In hotter climates, air conditioning systems are desirable but are now found only in the highest income households in developing countries.[92] In contrast, most people in the United States who need it have air conditionering and, on a national average (including cool regions and buildings without air conditioning), use about 1,400 kWh per person per year to cool residential and commercial buildings.[93]

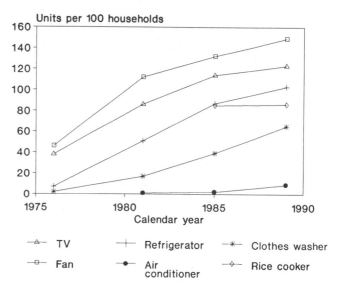

Figure 3-14—Appliance Ownership in South Korea

This figure shows the rapid increase in ownership of various appliances in Korea.

SOURCE: S. Meyers et al., "Energy Efficiency and Household Electric Appliances in Developing and Newly Industrialized Countries," Lawrence Berkeley Laboratory, LBL-29678, December 1990.

The use of air conditioning or other cooling techniques is likely to grow rapidly in importance in most developing countries,[94] and may eventually dominate electrical energy use in the residential/commercial sector in the hottest and most humid countries. Already, active space ventilation by electric fans has become popular in many areas where there is reliable electric service and costs are affordable. In Korea, household fans have increased rapidly in number to the current level of 1.5 per household (see figure 3-14). Electric fan ownership in Beijing, China jumped from 47 percent of

---

[88] Robert M. Wirtshafter, "Energy Conservation Standards for Buildings in China," *Energy*, vol. 13, No. 3, 1988, pp. 265-274; Robert M. Wirtshafter and Chang Song-ying, "Energy Conservation in Chinese Housing," *Energy Policy*, April 1987, pp. 158-168.

[89] Andrea N. Ketoff and Omar R. Masera, "Household Electricity Demand in Latin America," *American Council for an Energy Efficient Economy, 1990 Summer Study on Energy Efficiency in Buildings* (Washington, DC: American Council for an Energy Efficient Economy, 1990).

[90] Howard S. Geller, op. cit., footnote 46.

[91] J. Barrie Graham, "Air Handling," *Technology Menu for Efficient End Use of Energy: Volume I. Movement of Material*, Environmental and Energy Systems Studies, Lund University, Lund Sweden, 1989.

[92] Jayant Sathaye and Stephen Meyers, "Energy Use in Cities of the Developing Countries," *Annual Review of Energy* (Palo Alto, CA: Annual Reviews, Inc., 1985).

[93] Adapted from Paul D. Holtberg, et al., *Baseline Projection Data Book: 1991 Edition of the GRI Baseline Projection of U.S. Energy Supply and Demand to 2010*, Gas Research Institute, Washington, DC 1991.

[94] All 50 of the hottest cities in the world are in the developing world—the hottest is Djibouti, with an average annual high temperature of 113 F°. None of the 50 coldest cities in the world are in the developing world (V. Showers, *World Facts and Figures* [New York, NY: John Wiley and Sons, 1979]).

households in 1981 to 77 percent in 1984.[95] Households are effectively restrained from using electric air conditioning, however; electricity tariffs increase sharply for usage sufficient to support an air conditioner.[96] On the other hand, one-in-five households in Rio de Janeiro, Brazil, now has an air conditioner and typically uses 600-800 kWh/yr. In coastal Mexico, those with air conditioners typically use about 1,800 kWh/yr due to the long cooling season and the low efficiency of their systems.[97] In Thailand, air conditioning is projected to become the dominant demand over the next decade (see table 3-14).

In hot/humid climates with low quality construction of homes, electricity use for air conditioning can be much greater than these estimates if the desire for cooling is to be fully satisfied. A study of uninsulated concrete block homes in southern Florida found an average of nearly 8,200 kWh used per year for cooling, or 4.14 kWh per square foot of living space in the house.[98] Uninsulated concrete block construction is common in much of the developing world.

There are a variety of ways that ventilation/cooling needs can be met. First, external heat gain by the building can be minimized. Shade trees;[99] awnings that allow windows to receive indirect light but minimize the entry of direct sunlight that would heat the room;[100] exterior or interior shades; reflective or tinted coatings[101] on windows;[102] insulated windows;[103] light colored roofs; roof sprays; and wall and roof insulation[104] can each cut building heat gain. Natural ventilation and use of the ground for cooling can also be effective.

Many of these techniques are used in traditional building styles and have proven highly effective.[105] Increasing urbanization and the use of commercial building materials have made some of these traditional practices less practical and less popular, however. Cramped urban areas often have fewer shade trees and less opportunity for natural ventilation, while suffering higher temperatures due to the urban "heat island" effect. Sheet metal has often

Table 3-14—Estimates of Electricity Consumption, Bangkok

Appliance	Power (W)	Usage (hours)	Annual consumption (kWh)
Color TV	79	2,014	159
Refrigerator	109	5,760	628
Rice cooker	1,149	230	264
Clothes washer	1,567	91	143
Air conditioner:			
Window	1,815	1,442	2,617
Central	2,257	1,564	3,530
Ceiling fan	77	2,061	159
Water heater	4,418	54	239

SOURCE: Load Forecast Working Group, 1989, Thailand, as cited in: Stephen Meyers et al., "Energy Efficiency and Household Electric Appliances in Developing and Newly Industrialized Countries," Lawrence Berkeley Laboratory, report No. LBL-29678, December 1990.

---

[95] J. Sathaye, A. Ghirardi, and L. Schipper, "Energy Demand in Developing Countries: A Sectoral Analysis of Recent Trends," *Annual Review of Energy* (Palo Alto, CA: Annual Reviews, Inc., 1987), pp. 253-281.

[96] J. Sathaye et al., "An End Use Approach to Development of Long Term Energy Demand Scenarios for Developing Countries," report No. LBL-25611 (Berkeley, CA: Lawrence Berkeley Laboratory, February 1989). Prices increase several times for usage above 80-100 kWh/month.

[97] Andrea N. Ketoff and Omar R. Masera, "Household Electricity Demand in Latin America," *ACEEE 1990 Summer Study on Energy Efficiency in Buildings* (Washington, DC: 1990).

[98] Danny S. Parker, "Monitored Residential Space Cooling Electricity Consumption in a Hot Humid Climate: Magnitude, Variation and Reduction From Retrofits," *ACEEE 1990 Summer Study on Energy Efficiency in Buildings* (Washington, DC: 1990).

[99] U.S. Department of Interior, National Park Service, *Plants/People/and Environmental Quality*, (Washington, DC: U.S. Government Printing Office, 1972).

[100] Aladar Olgyay and Victor Olgyay, *Solar Control and Shading Devices*, (Princeton, NJ: Princeton University Press, 1957).

[101] Other coatings of particular interest are spectrally selective coatings that allow visible light to enter but keep infrared light out; photochromic coatings (like sunglasses) that become darker as light intensity increases; thermochromic coatings that become darker as temperatures increase; and electrochromic coatings whose transmissivity can be adjusted using an applied voltage.

[102] Claes Goran Granqvist, "Energy Efficient Windows: Options with Present and Forthcoming Technology," *Electricity: Efficient End Use and New Generation Technologies, and Their Planning Implications* (Lund, Sweden: Lund University Press, 1989).

[103] Ashok Gadgil et al., "Advanced Lighting and Window Technologies for Reducing Electricity Consumption and Peak Demand: Overseas Manufacturing and Marketing Opportunities," report LBL-30389 (Berkeley, CA: Lawrence Berkeley Laboratory, Mar. 29, 1991).

[104] Insulation must be sized so that it optimizes the tradeoff between gaining heat from outside versus losing internal heat to the outside.

[105] Lim Jee Yuan, "Traditional Housing: A Solution to Homelessness in the Third World: The Malaysian Example," *The Ecologist*, vol. 18, No. 1, 1988, pp. 16-23; Mehdi N. Bahadori, "Passive Cooling Systems in Iranian Architecture," *Scientific American*, vol. 238, 1978, p. 144-154; R.K. Hill, "Utilization of Solar Energy for an Improved Environment Within Housing for the Humid Tropics," Division of Building Research, CSIRO, Victoria Australia, 1974.

replaced thatch in urban as well as many rural areas—it is more durable, but also leads to higher interior temperatures. Good design in residential as well as in commercial construction can capture the cooling benefits of the above techniques while providing the durability and performance of modern construction materials. Simulation studies of good commercial building design in Brazil found that air conditioning electricity use could be reduced 60 to 75 percent by the use of these and other design features compared to conventional buildings.[106]

Even retrofits can be highly cost effective in some cases. Putting reflective plastic film on windows to cut heat gain, for example, can pay for itself in some climates in less than 2 years.[107]

Second, internal heat gains could likewise be kept low. This is accomplished in part by using the most energy efficient appliances—lights, refrigerators, ventilation fans, electronic equipment. Each unit of energy saved by using more efficient appliances can also reduce cooling energy requirements—where air conditioning is used—by 0.2 to 0.4 units of energy.

Third, high efficiency mechanical cooling equipment can be used. Variable-pitch fans and variable speed motor drives can increase the efficiency of ventilation equipment by one-third.[108] Direct and indirect evaporative coolers and absorption chillers can also be effective in some climates. Air-to-air heat exchangers can reduce heat loss/gain of ventilation air brought in from outside.[109] Gas fired absorption chillers and engine driven chillers can be useful on large buildings.

Conventional electric powered air conditioning equipment is also increasing in efficiency and becoming more widely available. Between the late 1970s and mid-1980s, air-and water-cooled centrifugal chillers in the United States improved their energy efficiency ratings from averages of 7.5 to a best of 10 and from 13 to a best of 17-19, respectively.[110] Electronic adjustable speed drive systems have proven very effective in air conditioning units, reducing energy use by 25 percent over conventional fixed speed systems. Adjustable speed AC systems now account for over half the Japanese air conditioning market with sales of more than one million units annually.[111] In addition to large energy savings, advantages of these adjustable speed systems include: better capacity control; better temperature and humidity control; longer lifetime; reduced maintenance; and others. As discussed above for refrigerators, the use of an electronic adjustable speed drive in an air conditioner may also allow the use of higher efficiency motors where voltage and

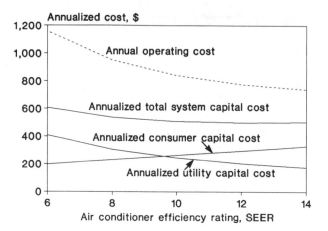

Figure 3-15A—Annualized Consumer, Utility, Total, and Consumer Operating Costs for Different Levels of Air Conditioner Efficiency Rating (SEER)

This figure shows that the annualized cost of more efficient air conditioners increases for consumers but the corresponding annualized cost of utility generation, transmission, and distribution equipment to power these air conditioners falls a little faster for the assumed intensity of use. The total annual operating costs for the consumer decrease substantially with the more efficient air conditioner.

SOURCE: See app. A for the 8,000 kWh annual cooling power case.

---

[106] Howard S. Geller, op. cit., footnote 46.

[107] A west window can gain up to 200 Btu/hr per square foot in the late afternoon. A reflective plastic film for windows can reduce this heat gain by 80 percent at a cost of $2.00 per square foot. This saves 23 W of air conditioning power for a system with an SEER of 7.0 (0.8x200/7=23 W). If there are 3 hours of sun on this window for 6 months of cooling season per year, then it saves 3x182x0.023kWx$0.09/kWh=$1.13/year. Window gain and costs for plastic films from: American Council for an Energy Efficient Economy, "Residential Conservation Power Plant Study: Phase I—Technical Potential." op. cit. footnote #27.

[108] Samuel F. Baldwin, "Energy Efficient Electric Motor Drive Systems," *Electricity: Efficient End Use and New Generation Technologies, and Their Planning Implications* (Lund, Sweden: Lund University Press, 1989).

[109] Edward Vine, "Air-To-Air Heat Exchangers and the Indoor Environment," *Energy* vol. 12, No. 12, 1987, pp. 1209-1215.

[110] Howard S. Geller, "Commercial Building Equipment Efficiency: A State-Of-The-Art Review," contractor report prepared for the Office of Technology Assessment, May 1988.

[111] S. Zubair, V. Bahel, M. Arshad, "Capacity Control of Air Conditioning Systems by Power Inverters," *Energy*, vol. 14, No. 3, 1989, pp. 141-151.

Figure 3-15B—Cost Effectiveness of an Efficient Air Conditioner for Different Levels of Annual Usage

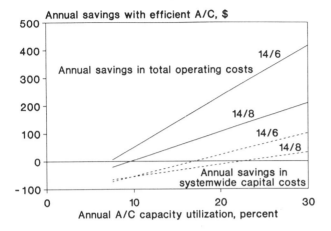

This figure shows that the cost effectiveness—both systemside capital costs and total (capital plus power) operating costs—of an efficient air conditioner increases as annual usage increases. Shown is the comparison of a high efficiency SEER 14 air conditioner with one that has a low SEER of 6 and one that has an average SEER of 8. For the high efficiency air conditioner to cease being cost effective on a lifecycle basis, it must be used at less than 10 percent of its annual capacity (this does not directly translate into annual days of cooling due to air conditioner cycling). Air conditioner costs do not include taxes or tariffs.
SOURCE: See app. A for parameters.

frequency fluctuations in the power line would otherwise cause the high performance system to stall and burn out. Other possible improvements include larger heat exchangers, improved control systems, occupancy sensors, and others.[112]

The range of efficiencies of air conditioners now available in the United States is quite large. For room air conditioners, the average energy efficiency rating (EER) is 8 with a best of 12; for central air conditioners, the average sold in 1988 had an SEER (seasonal EER) of 9 with the best on the market at 16.9—nearly twice as efficient as the average.[113]

High efficiency systems are, or could be, available in most developing countries. The most efficient room air conditioner sold in the United States in 1989 was assembled in Brazil using an imported rotary compressor. If such rotary compressors were similarly used in air conditioners sold in Brazil, they could reduce electricity consumption by 20 to 40 percent.

As shown in figure 3-15, the capital cost to consumers for more efficient air conditioners rises significantly. From a societal perspective, however, this increased cost to consumers is usually (depending on the amount of cooling time, etc.) offset by the decrease in the cost of utility generation, transmission, and distribution equipment needed to power it. For systems in hot and/or humid areas that operate more of the year, the total system capital cost significantly decreases for more efficient systems.[114] The total annual operating cost to the consumer also decreases for the more efficient system.

Similarly, in the commercial sector, numerous technologies have been developed to reduce the energy consumption of space cooling; some of these are listed in table 3-15.[115] Although they do not save much energy directly,[116] thermal storage systems can store ''cold'' in building concrete walls, in water, in ice, or in other media during the night for use during the heat of the day. This reduces the peak load on electric utilities, reducing the need for expensive peaking capacity.

Although improvements in the air conditioner itself usually reduce capital costs and life cycle operating costs, building insulation, shading devices, insulating windows, and numerous other improvements described above can often be even more cost effective. For example, high efficiency windows[117] can reduce annual heat gain by a building in, for example, Thailand by an average of some 180 kWh per square meter of window at an

---

[112] Stephen Meyers et al., ''Energy Efficiency and Household Electric Appliances in Developing and Newly Industrialized Countries,'' draft report No. LBL-29678 (Berkeley, CA: Lawrence Berkeley Laboratory, October 1990).

[113] American Council for an Energy Efficient Economy, *The Most Energy Efficient Appliances: 1989 to 1990 Edition* (Washington, DC: American Council for an Energy Efficient Economy, 1990).

[114] Note that this cited decrease in capital cost with operating time is based on the highly conservative assumption that the capital cost of electricity generating equipment is considered only when it is being used to power the air conditioner, while the capital cost of the air conditioner is fixed irrespective of operating time. See app. A at the back of this report for details.

[115] A. Usibelli et al., ''Commercial Sector Conservation Technologies,'' report No. DE-AC03-76SF00098 (Berkeley, CA: Lawrence Berkeley Laboratory, February 1985); American Gas Association, ''1988 Commercial Gas Cooling Fact Sheet and Market Assessment Summary,'' issue brief 1988-15 (Arlington, VA: Nov. 4, 1988).

[116] In fact, they could increase total energy consumption.

[117] Argon filled, spectrally selective double pane window.

Table 3-15—Commercial Space Cooling Equipment Technologies

Technology	Percent reduction in cooling energy demand	
High efficiency mechanical cooling		
Small, air cooled	20-50	
Water cooled	10-25	
Absorption chiller	10-50	
Part load COP improvement	15-30	
Gas fired absorption chillers	100	of electricity
Adjust evaporator/condenser temperatures	3-10	
Outside air economizers	15-80	depending on climate
Direct and indirect evaporative cooling	5-50	depending on climate
Cooling tower	0-10	
Off peak ice/chilled water storage	0	but reduces peak load
Mass storage and night venting	NA	
Dessicant cooling systems	NA	
Roof insulation	NA	
Light-colored roofs	NA	
Roof-spray cooling	NA	
Heat removing light fixtures	NA	
Air-to-air heat exchangers	NA	

Technology	Percent reduction in ventilation energy demand
Variable air volume systems	18-80
Fan shutoff during unoccupied hours	60
Motion/sensor control of ventilation	17-40
Energy efficient motors	3-11
Variable speed electronic motor drives	10-40

NA = Not available.
SOURCE: A. Usibelli et al., "Commercial-Sector Conservation Technologies," Lawrence Berkeley Laboratory, LBL-18543, February 1985.

incremental cost of $7 per square meter, compared to conventional uncoated single pane glass. For typical air conditioner efficiencies, this avoids the use of 60 kWh of electricity per year for cooling.[118] Corresponding annualized capital costs are an incremental $0.56 for the high efficiency window, which is offset by a four times greater capital savings of $2.20 in reduced investment in utility generation, transmission, and distribution equipment, not including the smaller air conditioner this makes possible. Similarly, building insulation can be extremely cost effective.[119]

ELECTRONIC EQUIPMENT

Electronic equipment has become a significant consumer of power in the commercial sector of industrial countries in recent years as personal TUs, computers, photocopiers, facsimile machines, and others have gained importance. Studies in the United States have found miscellaneous loads on receptacles within offices—primarily due to the use of various types of electronic equipment—to range from 9 to 24 W per square meter of office floor space (W/m^2) with an average of 17 W/m^2. The corresponding power demand for lighting was 19 W/m^2. A second case study in the United States found lighting loads in an office building to average 16 W/m^2 from 8am to 10pm during weekdays and drop to near zero at night and on weekends, while office equipment averaged 12 W/m^2 during the day, but only dropped to 6 W/m^2 at night and on weekends as much of it was left on—usually unnecessarily. A study in Australia similarly found a range going as high as 66 W/m^2 with an average of 15 W/m^2.[120] Electronic equipment such as televisions and stereos are also important end uses for electricity in the home. Although the corresponding demands for power by electronic equipment are currently far lower than this in developing countries, these loads

---
[118] Ashok Gadgil et al., "Advanced Lighting and Window Technologies for Reducing Electricity Consumption and Peak Demand: Overseas Manufacturing and Marketing Opportunities," report No. LBL-30389 (Berkeley, CA: Lawrence Berkeley Laboratory, Mar. 29, 1991).

[119] See, for example: Kuwait Institute for Scientific Research, *Economics of Thermal Insulation in Hot Climates* (Kuwait, August 1982).

[120] L. Norford et al., "Electricity Use In Information Technologies," *Annual Review of Energy*, vol. 15, 1990, pp. 423-453.

Table 3-16—Energy Use by Display Technologies

Display	Approximate power demand watts/cm^2
Monochrome CRTs	0.06
Color CRTs	0.13
LCD, nonbacklit	0.004
LCD, backlit	0.03
Electroluminescent	0.02
Plasma	0.02

SOURCE: L. Norford et. al., "Electricity Use In Information Technologies," *Annual Review of Energy*, vol. 15, pp. 423-453, 1990; Michael Shepard, "The State of the Art: Appliances," Rocky Mountain Institute, Snowmass, CO, 1990.

can be expected to markedly increase in the future. Appropriate planning in developing countries for efficient electronic equipment now could save substantial capital investment in the electric supply sector in the future.

Electronic equipment is usually chosen on the basis of performance and cost rather than on its electric power requirements. In recent years, however, the proliferation of high performance, low power equipment such as laptop computers makes it possible to also consider energy consumption when choosing electronic equipment. The power requirements of this equipment is substantially reduced through a variety of means—the use of CMOS[121] integrated circuits, liquid crystal displays (see table 3-16), and various power management techniques. Measured electric power consumption for some of the most popular types of electronic equipment are given elsewhere.[122]

A variety of energy efficiency improvements in electronic equipment may quickly pay for themselves—both from the society perspective of systemwide capital investment and from the consumer perspective of life cycle cost. Improvements in the energy efficiency of TVs in South Korea is shown in table 3-11; cost and efficiency data for modest efficiency improvements in color TVs in the United States are shown in tables 3-17 and 3-18. Table 3-18, for example, shows that a 2-percent increase in factory unit costs of more efficient color TVs can yield energy savings of up to 17 percent. Corresponding calculations of systemwide capital and life cycle costs are shown in appendix A.

Much larger improvements may be possible. For example, laptop computers use less than one-tenth the power of desktop machines. Although the current premium of as much as $500 or more for a laptop compared to a desktop could only be justified on the basis of systemwide capital costs or life cycle costs if the machine were left on virtually 24 hours per day[123] (as some offices do), many of these energy saving design features could be incorporated into a desktop machine at much lower costs. Timers or occupancy sensors could also reduce this energy consumption significantly.

BARRIERS TO CONSUMER PURCHASE OF ENERGY EFFICIENT APPLIANCES

Energy efficient appliances are often highly cost effective: their higher initial cost is more than offset by lower electricity bills over their lifetime. Further, the higher initial cost of efficient appliances to consumers ignores the upstream cost savings in capital equipment to generate the power needed to operate them. From a system and societal perspective, energy efficient appliances often cost less in capital and less to operate. Yet consumers frequently fail to take advantage of these opportunities. A variety of reasons for this have been summarized in table 3-5; a few of these issues are presented below:[124]

- Consumers may not have access to information about the costs and benefits of energy efficiency.
- Consumers often do not have market access to high efficiency appliances.
- In many cases, consumers may effectively require savings of greater than a certain threshold value before they will make the effort to

---

[121] CMOS means Complementary Metal Oxide Semiconductor

[122] L. Norford et al., "Electricity Use In Information Technologies," *Annual Review of Energy*, vol. 15, 1990 pp. 423-453.

[123] For example, comparing a laptop consuming 16 W to a desktop AT computer using 166 W, the difference in power consumption of 150 W has an annual upstream capital cost of about $50 and an annual electricity consumption of $120. Over a 5 year period at a 7-percent real discount rate, these costs have a present value of about $700, or $200 more than the premium on the typical namebrand laptop today.

[124] Sources: U.S. Department of Energy, *Technical Support Document: Energy Conservation Standards for Consumer Products: Refrigerators and Furnaces*, report No. DOE/CE-0277 (Washington, DC: U.S. Government Printing Office, November 1989). Henry Ruderman, Mark D. Levine, and James E. McMahon, "The Behavior of the Market for Energy Efficiency in Residential Appliances Including Heating and Cooling Equipment," *The Energy Journal*, vol. 8, No. 1, January 1987, pp. 101-124; Malcolm Gladwell, "*The Washington Post*, Feb. 12, 1990, p. A3.

Table 3-17—Power Demand by Color TVs

Size (inches)	Power (watts)			Annual energy consumption
	White	Black	Standby	
13"-14"	69	42	0.0	122 Mechanical off/on
13"-14"	69	48	4.9	161 kWh electronic off/on
19"-20"	100	60	4.4	205
26"-27"	134	87	6.2	284

SOURCE: U.S. Department of Energy, "Technical Support Document: Energy Conservation Standards for Consumer Products: Refrigerators, Furnaces, and Television Sets," National Technical Information Service, Springfield, VA, November 1988, see p. 3-60, table 3-28.

Table 3-18—Cost and Efficiency Data for 19-20 inch Color TV Sets

Level	Design option	Factory unit cost	Energy consumption
0	Baseline (100/60W)	$158.00	205 kWhr/yr
1	Reduce standby power to 2W	$160.15	184
2	Reduce screen power by 5 (93/55W)	$161.45	176
3	Increase display efficiency (91/53W)	$161.75	171

Baseline: Electronic tuning with standby power of 4.4 W; white picture/black picture of 100 W/60 W.
SOURCE: U.S. Department of Energy, "Technical Support Document: Energy Conservation Standards for Consumer Products: Refrigerators, Furnaces, and Television Sets," National Technical Information Service, Springfield, VA, November 1988, see p. 3-60, table 3-29.

locate and purchase a more efficient appliance. Individually these savings may be small, but to society overall they may sum to a very large benefit. Higher efficiency appliances may also have an excessively large premium—they might be loaded with unnecessary extras or be used to subsidize the cost of less efficient models.

- Consumers—even when they are aware of the advantages of higher efficiency appliances—tend to be extremely sensitive to the first cost of an appliance. In developing countries, this sensitivity to first cost may be even greater: consumers may simply not have access to the additional capital needed for a more efficient appliance.
- Consumers may have their electricity costs heavily subsidized. On average, the cost of electricity to consumers in developing countries is just 60 percent of the cost of producing it.[125]
- Consumers are often not the ones who purchase the appliances that they use. The building contractor or landlord often purchase the appliances used in the building and base their choice on lowest first cost rather than life cycle operating costs. Their tenants, not they, must pay the cost of operating this inefficient equipment.
- Consumers do not directly see the high upstream cost of capital equipment to power their inefficient appliances. While consumers face high interest rates, utilities can borrow at commercial (or better) interest rates over typically a 30-year period.
- Many consumers purchase secondhand goods where energy efficiency information is unavailable, and efficiency is not usually considered. Correspondingly, consumers who plan to sell their used appliances, knowing that they cannot get a premium for a more efficient appliance, may initially choose a less efficient design.

Together, these market failures or inefficiencies pose a powerful barrier to the rapid adoption of more efficient residential and commercial energy technologies.

POLICY RESPONSES

Numerous policy responses have been used in both industrial and developing countries to deal with the market failures listed above and in table 3-5. A variety of these policy responses are summarized in

---

[125] A. Mashayekhi, World Bank, Industry and Energy Department, "Review of Electricity Tariffs in Developing Countries During the 1980s," Industry and Energy Department working paper, *Energy Working Paper No. 32*, November 1990.

> ### Box 3-E—The Brazilian PROCEL Program
>
> The Brazilian Government established PROCEL—a nationwide electricity conservation program—within Eletrobras—the national utility holding company—in 1985. PROCEL funds or otherwise supports activities at utilities, universities, private manufacturers, and elsewhere within Brazil. These activities include:
>
> - Research and Development—more energy-efficient technologies—refrigerators, lights, motors, controls, etc.;
> - Education and information programs—testing the efficiency of equipment and labeling it in the marketplace, conducting energy audits of industries and commercial buildings, and promoting energy efficiency through publications and public events;
> - Financial Assistance and Direct Installation programs—providing low-interest loans or directly installing more efficient equipment, such as street lights; and
> - Setting standards for equipment efficiency—such as refrigerators (a 5-percent annual increase in average efficiency is now required for new models produced between 1994 to 1998 on top of the best performance currently achieved), and lights (standard incandescents are to be phased out in favor of more efficient incandescents and compact fluorescents, among other improvements).
>
> Total cumulative funding of PROCEL reached $20 million by 1990 with perhaps an equal amount from State and local utilities and other organizations. The more than 150 projects funded had resulted in direct savings of 1,000 gigawatt/hour/yr (GWh) by 1989, allowing utilities to defer at least $600 million in new generating capacity. This is a return some 15 times greater than total investment.
>
> As a result of these successes, the PROCEL program was planned to expand to nearly $35 million in funding during 1990. Even this large sum is less than 1 percent of current annual utility investment. Long term goals are to save 10 percent of national electricity use by 2000 and 14 percent by 2010, equal to 88 Terawatt/hour (TWh)—the equivalent of nearly half of total Brazilian electricity consumption in 1988.
>
> SOURCES: Howard S. Geller and Jose R. Moreira, "Brazil's National Electricity Conservation Program (Procel): Progress and Lessons," paper presented at the Conference, "DSM and the Global Environment," Apr. 22-23, 1991, Arlington, VA.; and Howard S. Geller, "Efficient Electricity Use: A Development Strategy for Brazil," American Council for An Energy Efficient Economy, Washington, DC, 1991.

table 3-6, and a few of them are discussed in more detail below. One should carefully note, however, some of the potential problems of State intervention in the marketplace (see table 3-7). Examples of policies that have been implemented in Brazil are discussed in box 3-E.

The United States and other industrial countries could take the lead in developing efficient technologies, offering large scale markets within which manufacturing costs can be brought down, and in generally proving both the concept and the potential for savings (see box 3-F).

If these technologies are first developed and proven in industrial countries, widespread distribution of information to potential developing country users is then more readily possible and networks for distributing the technologies can be more easily established. In many cases, however, some local adaptation of the technology may be necessary—such as making the technology more robust in the presence of large voltage fluctuations. National or regional centers of excellence might play an important role in adapting these technologies to local conditions or in developing new technologies (see box 3-B). Such centers have played a central role in improving agricultural productivities in many developing countries and are particularly noted for their role in the "green revolution."

Regional or national centers of excellence might also perform such tasks as establishing standard methodologies for measuring and comparing equipment efficiency; developing "scorekeeping" techniques for determining energy savings in the field; collecting data; conducting field energy audits or extension; and other activities, in addition to technology adaptation or more basic research, development, and demonstration.

These activities may be made easier by following the lead of the industrial countries, but they can also be done independently of the industrial countries. Brazil, Korea, Taiwan, China, and a number of other countries have shown the potential for independent energy efficiency activities.

> ### Box 3-F—The Appliance Industry: Obstacles and Opportunities in Manufacturing Efficient Appliances
>
> The appliances sold in developing countries vary from finished goods imported from overseas, to "kits" assembled from mostly imported components, to largely locally manufactured components and finished products. The terms under which these appliances are made vary as widely, from wholly owned foreign subsidiaries, to products made under license to foreign firms, to small local entrepreneurs with few foreign components or technological inputs. Limits on the efficiency of locally assembled or manufactured appliances may range from inability to get license agreements at sufficiently attractive terms to high import duties.
>
> One of the most effective means of improving appliance efficiency is to work with manufacturers directly. In many developing countries, manufacturers produce appliances based on technology many years behind the state-of-the-art found in the industrial countries: they are often able to get this obsolete technology at relatively low cost from leading-edge manufacturers who, in turn, could not earn much additional return on the technology except by selling or licensing it to developing countries where markets may be less discriminating. In some cases, import tariffs may prevent the use of high efficiency components or equipment at the (unrecognized) cost of even greater imports of utility generating equipment.
>
> By assisting licensing of high-efficiency technologies, by joint ventures, or by other means including adjustment of import duties on high efficiency components, developing countries might gain better access to these technologies and substantially reduce total systemwide capital requirements when utility investment is considered. Joint ventures between international and Thai manufacturers have enabled them to improve the energy efficiency of their refrigerators, but import tariffs largely prevented the use of high efficiency rotary compressors until 1990. The drive to manufacture for export markets has also played an important role in the Thai effort to improve the quality and efficiency of the refrigerators and other appliances that they manufacture.
>
> Governments and research institutes can also work directly with domestic manufacturers. Funding from the State utility enabled researchers at the University of Sao Paulo to develop high efficiency solid state ballasts for fluorescent lights and transfer the technology to a private company for manufacturing (box 3-C). The Brazilian Procel program has worked with private manufacturers to establish voluntary energy efficiency protocols for refrigerators, lamps, ballasts, and motors.
>
> SOURCES: S. Meyers et al., "Energy Efficiency and Household Electric Appliances in Developing and Newly Industrialized Countries," Lawrence Berkeley Laboratory, LBL-29678, December 1990; Howard S. Geller and Jose R. Moreira, "Brazil's National Electricity Conservation Program (Procel): Progress and Lessons," paper presented at the Conference, "DSM and the Global Environment," Apr. 22-23, 1991, Arlington, VA.

Many, if not most, energy efficient technology development activities are best done by the private sector. These can be done either independently—spurred by national initiatives—or as joint ventures or in other forms of partnerships between manufacturers in the developing and industrialized countries. Means of encouraging these partnerships need to be explored.

In many developing countries, electricity prices paid by residential and commercial users are far below supply costs, with the difference made up by government subsidies. Raising prices would encourage the adoption of energy efficient appliances, as well as improving the government budgetary situation. Raising electricity prices, however, is a difficult political issue. High efficiency equipment may offer a way out of this dilemma. If more efficient equipment could be introduced at the same time as higher prices, the cost of energy services to consumers need not increase nearly as much, if at all.

Higher prices alone are often insufficient to ensure full utilization of cost effective energy efficient technologies because of the market failures described in table 3-5. Even the United States, where electricity prices are much closer to long run supply costs, has reinforced the price effect with mandatory efficiency standards for a variety of appliances.

Finally, and perhaps most importantly, means of better reflecting total societal costs in consumer investment decisionmaking could be explored. Currently, the capital costs of generation equipment are paid by the utility and the capital costs of end-use equipment are paid by the end user. As shown above, the high implicit discount rate of the end user as well as this separation between utility and user (or for leased equipment, the separation between owner and

Box 3-G—Integrated Resource Planning[1]

Conceptually, Integrated Resource Planning[2] (IRP) is straightforward. Planners rank by cost all the different energy supply and energy end use technologies that might be used to provide an energy service, and implement them beginning with the lowest cost opportunities. Thus, various electricity supply technologies such as conventional coal plants, steam-injected gas turbines, and combined-cycle plants are compared with each other and with end-use technologies such as compact fluorescent lights, adjustable speed electronic drives for motors, and increased insulation in buildings to reduce air conditioning loads. Of all the different possibilities, the lowest cost options are chosen for investment.

The manner in which energy institutions are organized, however, has not encouraged the implementation of integrated resource planning. Under the traditional regulatory framework found in most countries, utilities are in the business of selling energy supplies, not energy services. Each kiloWatthour (kWh) sold by an electric utility increases gross earnings, no matter how much it costs to generate; conversely, each kWh saved by using an energy efficient technology decreases earnings, no matter how little it cost to implement.[3] Similarly, displacing utility generated power with purchases of power from nonutility sources such as industrial cogeneration usually reduces utility earnings. These considerations often hold even where electricity costs are heavily subsidized—the State simply replenishes utility funds while utility managers and workers are rewarded in terms of job security, increased salaries or staffs, etc. for the amount of electricity generated, irrespective of its cost and usefulness.

In contrast, Integrated Resource Planning changes the regulatory framework in order to encourage utilities and others to implement the least-cost demand and supply options. Among other changes, regulators allow utilities to earn income based on the net benefits from investments in energy efficiency improvements. This focuses the financial, managerial, and technical skills of the utility on some of the market failures on the demand side (table 3-E) and helps realize some of the most important policy responses (table 3-6), especially the capital cost-related ones.

Factors that should be considered in IRP programs include: providing appropriate financial rewards for utilities to support efficiency improvements as well as supply—decoupling utility profits from the number of kWh sold—in order to minimize the overall cost of supplying energy services; ensuring that the startup costs of the IRP program and the administrative complexity and overheads are kept to a minimum; developing adequate methods for "measuring" savings (also known as scorekeeping); avoiding the "free rider" problem and others.

Numerous utilities in the United States have begun to implement Integrated Resource Planning Programs. In developing countries, the efforts in IRP have generally been much more limited to date. One exception, however, is the PROCEL program in Brazil (see box 3-E). Shown in figure 3-16 is a supply curve of the equivalent cost of energy of different end-use technologies in Brazil. Based on such a supply curve, the utility planner can tailor programs to maximize the utility return-on-investment in energy efficiency improvements.

---

[1]Sources and further reading: David Moskovitz, "Profits and Progress Through Least-Cost Planning," National Association of Regulatory Utility Commissioners, Washington, DC, November, 1989; Jonathan Koomey, Arthur H. Rosenfeld, and Ashok Gadgil, "Conservation Screening Curves to Compare Efficiency Investments to Power Plants," *Energy Policy*, October 1990, pp. 774-782; Thomas B. Johansson, Birgit Bodlund, and Robert H. Williams, *Electricity: Efficient End-Use and New Generation Technologies and Their Planning Implications*, (Lund, Sweden: Lund University Press, 1989); Howard S. Geller, "Efficient Electricity Use: A Development Strategy for Brazil," contractor report for the Office of Technology Assessment, (Washington, DC: American Council for an Energy Efficient Economy, 1991); *Proceedings: 5th National Demand-Side Management Conference*, Electric Power Research Institute, Palo Alto, CA. report CU-7394; 1991; P. Herman, et. al., "End-Use Technical Assessment Guide, volume 4: Fundamentals and Methods," Electric Power Research Institute EPRI CU-7222, vol. 4, April 1991, Palo Alto, CA; Linda Berry and Eric Hirst, "The U.S. DOE Least-Cost Utility Planning Program," *Energy*, vol. 15, No. 12, pp. 1107-1117, 1990; Glenn Zorpette, "Utilities Get Serious About Efficiency," *IEEE Spectrum*, May 1991, pp. 42-43.

[2]Other names associated with Integrated Resource Planning include Least Cost Planning and Demand Side Management. Least Cost Planning has sometimes been taken to mean only comparisons of energy supply options, with no comparisons with end use options. Demand side management commonly examines end-use options, with no comparisons with energy supply options.

[3]Adapted from David Moskovitz, op. cit., footnote 1.

Figure 3-16—Electricity Efficiency Supply Curve, Brazil, 2010

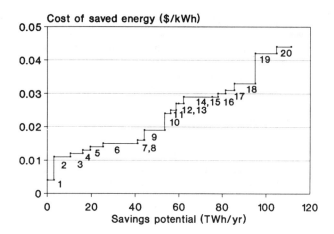

This curve shows the equivalent cost of saving energy for various improvements in the energy efficiency of end-use technologies. (1,14,15)—improvements to residential and commercial (residential/commercial) refrigerators and freezers; (2)—more efficient industrial furnaces and boilers; (3,13) more efficient res/com air conditioning; (4,17) more efficient residential electric water heating; (5,19) more efficient industrial motors and adjustable speed drives; (6)—miscellaneous industrial improvements; (7) more efficient industrial electrochemical processes; (8,9,10,11,12,16,18,20)—various improvements to residential, commercial, industrial, and public lighting.

SOURCE: Howard S. Geller, *Efficient Electricity Use: A Development Strategy for Brazil*, contractor report for the Office of Technology Assessment (Washington, DC: American Council for An Energy Efficient Economy, 1991).

user) leads to much lower levels of investment in end-use equipment efficiency than is justified on the basis of either total system capital costs or life cycle operating costs.

A powerful tool to redress this "disconnect" is Integrated Resource Planning (see box 3-G). If utilities planned on a systemwide energy services basis, they could use resources that would otherwise have been devoted to expanding capacity to financing efficient appliances. Examples of such innovative financing approaches might range from the end user choosing equipment according to the total life cycle cost and paying this cost in monthly installments on the utility bill; to the end user paying a front-end deposit or posting a bond to the utility to cover the life cycle operating costs of the equipment, against which the utility would charge the capital cost of the equipment and the monthly electricity bills. Either of these approaches would force the end user to directly face the total life cycle costs of the equipment when purchasing it.

CONCLUSION

This review of energy efficient and/or alternative technologies for the residential and commercial sectors shows that substantial reductions in society wide capital costs, life cycle operating costs, and energy consumption are possible. Achieving these savings will require, however, significant longterm efforts and institutional changes to overcome a variety of market and institutional failures (see table 3-5). Many approaches to these failures are possible (see table 3-6). The United States can help in this effort (see ch. 8) by accelerating programs of research, development, and demonstration of energy efficient technologies, by providing technical assistance and training in both technology and institutional change, and by setting an example.

Chapter 4
Energy Services: Industry and Agriculture

Photo credit: Appropriate Technologies, International

Contents

	Page
INTRODUCTION AND SUMMARY	91
Small Scale Industry	91
Modern Industry	94
Agriculture	103
MOTOR DRIVE SYSTEMS	103
Motor Drive System Design	104
Motors	107
Pumps and Fans	108
Adjustable Speed Drives	112
Pipes and Ducts	115
Systems	117
MODERN INDUSTRIAL PROCESSES	117
Steel	117
Cement	123
Pulp and Paper	127
Chemicals	129
IMPROVING THE EFFECTIVENESS OF MATERIAL USAGE	130
BARRIERS TO ENERGY EFFICIENCY IMPROVEMENTS IN INDUSTRY	132
Small Scale	132
National Infrastructures	133
Available Technology	133
Financial Constraints	133
POLICY RESPONSES	134
Information	134
Availability	135
Capital Cost	136
AGRICULTURE	136
Irrigation	137
Traction	140
CONCLUSION	141

Box

Box	Page
4-A. Agricultural Energy Use in Context	139

Chapter 4
Energy Services: Industry and Agriculture

INTRODUCTION AND SUMMARY

The industrial sector is growing rapidly in many developing countries and is a key element in their drive for economic development and modernization. Between 1980 and 1989, annual growth of the industrial sector averaged 8.7 percent for the low income countries and 3.8 percent for the middle income countries. There was wide variation within these averages, however, ranging from a 5 percent or worse annual decline in production for Bolivia, Liberia, and Trinidad and Tobago to over 10 percent annual growth for China, South Korea, Indonesia, and several others.[1]

The industrial sector typically consumes 40 to 60 percent of total commercial fossil energy used in developing countries (see tables 4-1 to 4-3);[2] it also makes heavy use of traditional biomass fuels, often traded in commercial markets. For example, biomass fuels supply up to 40 percent of the industrial energy used in Indonesia, 28 percent in Thailand, 17 percent in Brazil, and similarly large fractions in many other countries.[3] Per capita industrial energy consumption in the developing world is 5 to 10 percent of the U.S. level (see table 4-2).

A significant portion of industrial energy is used technically inefficiently, with serious economic and environmental impacts. Improving overall performance of the industrial sector, not just the efficient use of energy, will be necessary if these countries are to compete in world markets and to provide a high standard of living for their citizens.

The industrial sector of developing countries includes a broad range of firms in size and sophistication. At one end of the spectrum are small traditional firms,[4] largely located in rural areas,[5] which use relatively energy inefficient and low productivity manufacturing technologies. These small manufacturing enterprises may, however, be operating efficiently in the broader economic context, given the available inputs of capital, labor, and materials.[6] High transport and marketing costs and small market size may prevent larger, more technically efficient firms from competing effectively with these traditional cottage industries. Over time, a few smaller companies grow into medium, and/or large ones as the transport infrastructure improves and incomes rise—increasing the size of markets and providing economies of scale that turn the advantage to larger firms.[7] At this end of the spectrum are large modern firms, often with multinational parent companies, that have globally competitive manufacturing capabilities.

Small Scale Industry

In many developing countries today, one-half to three-quarters of manufacturing employment is in small scale establishments; the remainder is divided more-or-less evenly between medium and large operations.[8] Small scale industry supplies one-fourth to one-half or more of manufacturing gross domestic product (GDP).[9] Much of the employment in the small traditional and primarily rural industries is based on seasonal labor available during the nonagriculturally active times of year; typically a fourth to a third of rural nonfarm employment is for

---

[1] World Bank, *World Development Report, 1990* (New York, NY: Oxford University Press, 1990), pp. 180-181.

[2] Jayant Sathaye, Andre Ghirardi, and Lee Schipper, "Energy Demand in Developing Countries: A Sectoral Analysis of Recent Trends," *Annual Review of Energy*, vol. 12, 1987, pp. 253-281, table 5.

[3] Joy Dunkerley et al., *Energy Strategies for Developing Countries* (Baltimore, MD: Johns Hopkins University Press, 1981), p. 265.

[4] A variety of definitions and terms are used for small scale industry, including: household, cottage, micro, tiny, small, and others. Some of these terms are used with distinct meanings according to the number employed, the location of the enterprise, and its assets. See Carl Liedholm and Donald Mead, "Small Scale Industries in Developing Countries: Empirical Evidence and Policy Implications," Michigan State University International Development Paper No. 9, Department of Agricultural Economics, East Lansing, Michigan, 1987.

[5] Here, rural means localities with 20,000 people or less. See Carl Liedholm and Donald Mead, Ibid.

[6] This is based on a social benefit-cost analysis rather than total factor productivity. Ibid. See p. 68, ff.

[7] Dennis Anderson, "Small Industry in Developing Countries: Some Issues," World Bank Staff Working Paper No. 518, 1982.

[8] Ibid. See tables 1 and 2.

[9] Carl Liedholm and Donald Mead, op. cit., footnote 4.

Table 4-1—Total Delivered Energy by Sector, in Selected Regions of the World, 1985 (exajoules)[a]

Region	Residential/commercial		Industry		Transport		Total		Total energy
	Commercial fuels	Traditional fuels[b]	Commercial fuels	Traditional fuels[b]	Commercial fuels	Traditional fuels[b]	Commercial fuels	Traditional fuels[b]	
Africa	1.0	4.0	2.0	0.2	1.5	NA	4.4	4.1	8.5
Latin America[c]	2.3	2.6	4.1	0.8	3.8	NA	10.1	3.4	13.5
India and China	7.3	4.7	13.0	0.2	2.0	NA	22.2	4.8	27.1
Other Asia[d]	1.9	3.2	4.0	0.4	1.9	NA	7.8	3.6	11.3
Developing countries	12.5	14.5	23.1	1.6	9.2	NA	44.5	15.9	60.4
United States	16.8	NA	16.4	NA	18.6	NA	51.8	NA	51.8

NA = Not available or not applicable.
NOTES: This is delivered energy and does not include conversion losses from fuel to electricity, in refineries, etc. The residential and commercial sector also includes others (e.g., public services, etc.) that do not fit in industry or transport. Traditional fuels such as wood are included under commercial fuels for the United States.
[a] 1 exajoule (10^{18} Joules) equals 0.9478 Quads. To convert to Quads, multiply the above values by 0.9478.
[b] These estimates of traditional fuels are lower than those generally observed in field studies. See references below.
[c] Olade estimates these values, left to right, as 1.79 EJ, 1.50 EJ; 4.81 EJ, 0.5 EJ; 3.94 EJ,—; 10.54 EJ, 2.0 EJ; 12.54 EJ (Gabriel Sanchez-Sierra, Executive Secretary, Organization LatinoAmerica de Energia, Quito, Ecuador, personal communication, July 15, 1991.)
[d] Does not include Japan.
SOURCE: U.S. Congress, Office of Technology Assessment, *Energy in Developing Countries*, OTA-E-486 (Washington, DC: U.S. Government Printing Office, January 1991) p. 49.

Table 4-2—Delivered Energy Per Capita by Sector in Selected Regions, 1985 (includes traditional fuels, in gigajoules)

Region	Residential and Commercial	Industry	Transport	Total
Africa	11.8	5.2	3.5	20.5
Latin America	12.7	12.5	9.7	34.9
India and China	6.7	7.3	1.1	15.1
Other Asia	7.2	6.2	2.7	16.1
United States	69.8	68.5	77.5	215.8

NOTE: These estimates do not include conversion losses in the energy sector, and underestimate the quantity of traditional fuels used compared to that observed in field studies. In Latin America, an alternative set of estimates are, left to right, 8.2 GJ, 13.3 GJ, 9.9 GJ, 31.4 GJ (Gabriel Sanchez-Sierra, Executive Secretary, Organization Latino America de Energia, Quito, Ecuador, personal communication, July 15, 1991.)
SOURCE: U.S. Congress, Office of Technology Assessment, *Energy in Developing Countries*, OTA-E-486 (Washington, DC: U.S. Government Printing Office, January 1991).

Table 4-3—Per Capita Energy Use by Service in Selected Countries (gigajoules)

	Brazil	China	India	Kenya	Taiwan	U.S.
Residential	6.2	11.7	5.5	16.9	8.9	64.9
cooking	5.3	8.5	5.0	16.4	4.7	3.5
lighting	0.3	0.4	0.5	0.5	0.7	NA
appliances	0.6	NA	0.05	NA	3.1	13.0[a]
Commercial	1.5	0.7	0.26	0.4	4.2	45.2
cooking	0.4	NA	0.13	0.24	1.9	NA
lighting	0.5	NA	0.05	0.16	0.8	7.2
appliances	0.6	NA	0.07	NA	1.5	NA
Industrial	19.4	13.8	4.1	4.8	39.2	94.1
process heat	17.5	10.2	2.7	NA	NA	55.8
motor drive	1.6	3.6	1.3	NA	NA	20.4
lighting	0.1	NA	0.05	NA	NA	NA
Transport	13.3	1.2	1.3	2.7	11.5	80.8
road	12.0	0.2	0.8	1.8	10.1	66.7
rail	0.2	0.7	0.4	0.2	0.1	2.0
air	0.7	NA	0.1	0.7	0.7	11.3
Agriculture	2.1	1.8	0.6	0.5	2.6	2.5
Total	43.4	27.0	11.7	25.6	67.7	288.0

NA = Not available or not applicable.
[a] This is the combined total for appliances and lighting.
SOURCE: U.S. Congress, Office of Technology Assessment, *Energy In Developing Countries*, OTA-E-486 (Washington, DC, U.S. Government Printing Office, January 1991).

Photo credit: Appropriate Technologies, International

Some small scale rural industries, such as pottery-making, are frequently owned or staffed by women.

manufacturing.[10] This is an important source of income and employment for the rural and poor urban sectors.[11] Small scale industry also provides important inputs into larger scale industries—particularly in Asia—and into other key sectors such as agriculture.

Traditional rural industries include: crop processing activities; beer brewing; textiles and garment production; carpentry, masonry, and other construction; leatherworking and shoemaking; brick and pottery production; blacksmithing; and many others. These activities are often divided along tribal, class, or family lines and skills are usually passed along through the equivalent of apprenticeships.

Small industry in developing countries does not, however, mean exclusive use of traditional fuels. Many small shops use electric welding equipment, for example, and others would follow suit if electricity were available. The provision of modern fuels and power, such as electricity, offers significant opportunities to improve the productivity and quality of small manufacturing operations. On the other hand, many shops that work on modern equipment, ranging from autos to electric motors, often use the most primitive means and fuels available to perform the work. Shops that rewind electric motors, for example, often simply burn the windings off in an open fire rather than in a temperature controlled oven. This has potentially serious impacts on the performance of the motor after it is rewound.[12]

A key concern of the commercial establishment is fuel supply reliability: this has often led establishments, particularly those further from urban centers, to prefer firewood over more modern fuels.[13] Establishments that use wood in large volumes may also realize cost savings over modern fuels such as kerosene and liquified petroleum gases (LPG), particularly when they are imported.

The energy efficiency of traditional industry can be low (see table 4-4). The introduction of modern engineering analysis, design, and technology—including modern diagnostic instrumentation and analysis tools—into the traditional sector offers significant opportunities for improving the efficiency of traditional industry and improving product quality while minimizing capital investment. There are numerous examples. Principles of engineering combustion and heat transfer have been used to improve the energy efficiency of traditional stoves used for brewing beer, heating water for dyeing cloth, or other process heat needs.[14] Modern downdraft kiln designs have been introduced in West Africa and other regions to improve the energy efficiency of firing traditional clay pots and other goods. At the same time, these kilns can substan-

---

[10]Dennis Anderson and Mark Leiserson, "Rural Nonfarm Employment in Developing countries," *Economic Development and Cultural Change* 28, No. 2, 1980, table A2, p. 245; cited in Donald W. Jones, Oak Ridge National Laboratory, "Energy Requirements for Rural Development," Report No. ORNL-6468, June 1988. Measurements of this employment are very sensitive to the timing of the survey, how employment is defined, and the responsiveness of those interviewed. Under-reporting of nonfarm household employment is common; in some African countries it is reportedly as high as 40%. Dennis Anderson, "Small Industry in Developing Countries: Some Issues," World Bank Staff Working Paper No. 518, 1982.

[11]"Rural Small-Scale Industries and Employment in Africa and Asia," Ed. Enyinna Chuta and S.V. Sethuraman, International Labor Office, Geneva, 1984; Harold Lubell and Charbel Zarour, "Resilience Amidst Crisis: The Informal Sector of Dakar," *International Labour Review*, vol. 129, No. 3, 1990, pp. 387-396.

[12]Samuel F. Baldwin and Emile Finlay, "Energy-Efficient Electric Motor Drive Systems: A Field Study of the Jamaican Sugar Industry," Center for Energy and Environmental Studies, Princeton University, Working Paper No. 94. February 1988. Note that the oven must be carefully temperature controlled as well if damage to the windings is to be avoided.

[13]M. Macauley, M. Naimuddin, P.C. Agarwal, and J. Dunkerley, "Fuelwood Use In Urban Areas: A Case Study of Raipur, India," *The Energy Journal*, vol. 10, No. 3, July 1989, pp. 157-180.

[14]Samuel F. Baldwin, *Biomass Stoves: Engineering Design, Development, and Dissemination* (Arlington, VA and Princeton, NJ: Volunteers in Technical Assistance and Center for Energy and Environmental Studies, 1986).

Table 4-4—Efficiency of Fuel Use In Traditional (Developing Countries) and Modern (Industrial Countries) Commercial and Industrial Operations

Activity	Location	Estimated efficiency of traditional technology	Estimated efficiency of modern technology in U.S.
Cooking	West Africa	15-19%	50-60%
Beer brewing	Burkina Faso	15-17	79[a]
	Burkina Faso	0.3-0.7	6[a]
Tobacco drying	Tanzania	0.5	(36%)[b]
Tea drying	Tanzania	2.9	NA
Baking	Sudan	12-19	43
	India	16.0	NA
	Guatemala	3.0	NA
Fish smoking	Tanzania	2-3	NA
Brick firing	Sudan	8-16	6-11
	India	6.4	NA
	Uganda	5-10	NA
Foundry work	Indonesia	3.0	40

NA = Not available or not applicable.
[a]These are two different measures of the energy efficiency of the process.
[b]A proposal for a high efficiency tobacco curing barn with this efficiency can be found in H. Kadete, "Energy Conservation in Tobacco Curing," *Energy*, vol. 14, No. 7, pp. 415-420, 1989.
SOURCE: U.S. Congress, Office of Technology Assessment, *Energy in Developing Countries*, OTA-E-486 (Washington, DC: U.S. Government Printing Office, January 1991).

tially improve the quality of the firing and reduce losses due to breakage.[15] Air-to-air heat exchangers for traditional foundry processes—such as melting scrap aluminum to cast pots—could recuperate perhaps a hundred times as much waste heat as would be required to power the small hand- or electric-driven blower powering the heat exchanger.[16] A variety of technologies for both traditional and modern industry are listed in table 4-5.

Small firms often face substantial obstacles to improving the efficiency of their operations, including energy use. These include: inadequate access to credit; internal lack of technical and managerial skills; inadequate infrastructure—roads, water, electricity; poor access to raw materials; poor access to markets; and sometimes systematic opposition by larger, better established, and better politically connected formal industry. Some of these barriers are summarized for both small and large industry in table 4-6 and possible policy responses are listed in table 4-7.[17]

There are numerous technical opportunities for improving the efficiency of energy use in small scale industry in developing countries. Financial, technical, and managerial extension efforts will be needed to realize these opportunities, a difficult task given the highly dispersed, small scale, and informal nature of this sector.

Modern Industry

Energy use by the small scale sector, though significant in many developing countries today, is likely to decrease in the future as a percentage of total industrial energy use. Much of the growth in industrial energy use will instead be in large scale energy intensive materials, such as steel and cement, needed to develop a modern economy. Such industries will be the primary focus here.

---

[15]In West Africa, for example, these products have traditionally been fired on open bonfires with correspondingly large energy losses, high breakage rates and low quality.

[16]Assuming a fan efficiency of 40 percent; and including the energy losses in converting fuel to electricity or food to muscle drive. See: Samuel F. Baldwin, op. cit., footnote 14.

[17]See Carl Liedholm and Donald Mead, op. cit., footnote 4. Hubert Schmitz, "Growth Constraints on Small Scale Manufacturing in Developing Countries: A Critical Review," *World Development*, vol. 10, No. 6, pp. 429-450, 1982; Hernando de Soto, *The Other Path: The Invisible Revolution in the Third World* (New York, NY: Harper & Row, 1989); Robert N. Gwynne, *New Horizons? Third World Industrialization in an International Framework* (New York, NY: John Wiley & Sons, 1990); Dennis Anderson, op. cit., footnote 7; Harold Lubell and Charbel Zarour, op. cit., footnote 11; G. Norcliffe, D. Freeman, and N. Miles, "Rural Industrialisation in Kenya," Enyinna Chuta and S.V. Sethuraman (eds.), *Rural Small-Scale Industries in Africa and Asia* (Geneva: International Labour Organisation, 1984).

Table 4-5—Selected Energy Efficient Technologies for the Industrial Sector

Energy service	Technologies and remarks[a]
Traditional technologies	The application of modern technologies and techniques to traditional technologies is a key area. So-called appropriate technology, however, has generally failed to accomplish much due to the general lack of highly skilled scientists and engineers in the efforts, the excessive emphasis on using traditional or local materials, and other factors.
Motor driven systems	High efficiency motors, pumps, fans, etc., electronic adjustable speed drives, optimized pipe, duct, etc. dimensions. Standardized testing is needed to compare performance on a uniform basis. Motors, etc. must also be properly maintained: with proper lubrication, adjustments to gears, belts, etc., maintaining phase balance of input electric power, etc. Improved design tools for sizing and controlling equipment needed. Standard protocols needed to incorporate load management techniques into adjustable speed drive.
Efficient process heat systems	Using waste heat recovery systems, including heat exchangers and vapor recompression systems; steam system improvements, including increased insulation, steam traps, desuperheating plant steam as needed, and plugging leaks; monitoring heat exchanger fouling; maintaining and upgrading furnaces--adjusting burners and excess air, preheating air intake, etc.; insulating steam lines, furnaces, etc.; improved combustion controls; and many other technologies and techniques. Particularly important is scaling down these technologies for use in smaller scale developing-country plants; and selectively adapting these technologies for developing-country conditions--where labor is lower cost, but may be less well trained for handling advanced equipment.
Processes	High efficiency industrial processes are at all stages of development. Particular attention needs to be given to directing and adapting this research to developing-country needs, taking into account: the lower labor costs and the scarcity of capital; the less well developed infrastructure (i.e., the frequent voltage fluctuations); the lower availability of highly-skilled technical and managerial manpower; and in some cases, the relatively less reliable maintenance infrastructure (i.e., making particular types of automatic controls desirable where they can prevent damage due to irregular maintenance).
Cogeneration	High pressure steam turbines, engines, gas turbines. Improved cogeneration technologies coming available include steam injected gas turbines, and others.
Equipment testing procedures, standards, and diagnostic equipment	Regional test centers for establishing uniform standards and testing equipment to it could be established. Diagnostic equipment and procedures need to be adapted to developing-country conditions.
Efficient design rules and design software	Research, development, and field verification needed for design rules in sizing and controlling plant and equipment.
Advanced materials	High performance materials can dramatically reduce the volume of energy-intensive materials required. Particularly important for developing countries are high performance structural materials such as cements, steel, and plastics.
Quality control and just-in-time inventory control	Reduce defects and rework, material handling, and inventory costs; and improve productivity.
Recycling	Well established in developing countries. Development work needed to recycle complex composite materials and systems, e.g., as electronic equipment.

[a]Technologies can be viewed as on a spectrum of: (C) Commercially available; (A) commercially available in industrial countries but needing Adaptation to the conditions of developing countries; (N) Near commercial development; and (R) requiring further Research and development. Since most technologies have variations at many points on this spectrum—e.g., compact fluorescents are available, may need further adaptation in developing countries in some cases, have improved phosphors or other advances near commercial development, and may have more fundamental advances under research--the status of these technologies—C,A,N,R—will not be discussed; instead, particular opportunities will be presented.

SOURCES: Office of Technology Assessment; K.E. Nelson, "Use These Ideas to Cut Waste," *Hydrocarbon Processing,* March 1990; Julio R. Gamba, David A. Caplin, and John J. Mulckhuyse, "Industrial Energy Rationalization in Developing Countries," World Bank, Johns Hopkins University Press, Baltimore, MD, 1986; U.S. Department of Commerce, National Bureau of Standards Handbook 115, "Energy Conservation Program Guide for Industry and Commerce," September 1974.

Modern large scale industries in developing countries are modeled after their counterparts in industrialized countries but are often operated at significantly lower efficiencies.[18] A few energy intensive materials—steel, cement, chemicals (especially fertilizer), and paper—account for much of the energy used by industry (see tables 4-8 to 4-10 and figure 4-1) and the total energy used to produce these materials will increase rapidly as developing countries build their national infrastructures of roads, buildings, industry, and power. For example, steel production in developing countries increased at an

---

[18]Detailed reviews of energy conservation in industrial plants can be found in: Julio R. Gamba, David A. Caplin, and John J. Mulckhuyse, *Industrial Energy Rationalization in Developing Countries* (Baltimore, MD: Johns Hopkins University Press, 1986).

Table 4-6—Barriers to Investment in Energy Efficient Technologies

Technical

Availability
: High efficiency technologies and their needed support infrastructure of skilled manpower and spare parts may not be locally available. Foreign exchange may not be available to purchase critical spare parts. For the residential and commercial sectors, in particular, high efficiency technologies need to be marketed in a complete package to allow "one stop" shopping.

Culture
: Culture is rarely an impediment to the use of energy efficient technologies, although it is frequently cited as a problem in disseminating technologies in rural areas. In most cases, the technology itself is found to have significant technical shortcomings or is unable to meet the multiple uses desired.

Design rules
: Conventional design rules often lead to excessive oversizing of equipment—raising capital cost and wasting energy.

Diagnostics
: Technologies for measuring the efficiency of equipment, as in industrial energy audits, are often awkward and inaccurate. Some of them may require shutting down a production line or making intrusive measurements, such as cutting holes in pipes or ducts to make flow and pressure drop measurements.

Infrastructure
: The available infrastructure within a developing country may not be able to adequately support a particular high efficiency technology. This might include an electric power system with frequent brownouts or blackouts that the high efficiency technology is unable to handle well; dirty fuels that clog injectors; or poor water quality for high performance boilers. The developing country may also lack a reliable spare parts supply system and trained manpower to ensure adequate maintenance. Finally, the existing infrastructure might impede the implementation of a more efficient technology system. An extensive road system and/or little land use planning, for example, might slow or stop the development of an efficient mass transport system.

Reliability
: Innovative high efficiency equipment may not have a well proven history of reliability, particularly under developing-country conditions, as for other equipment.

Research, development, demonstration
: Developing countries may lack the financial means and the technical manpower to do needed RD&D in energy efficient technologies, or to make the needed adaptations in existing energy efficient technologies in use in the industrial countries to meet the conditions—e.g., large fluctuations in power supply voltage and frequency—in developing countries. Technology development and adaptation are particularly needed in rural industry and other activities.

Scale
: Energy efficient technologies developed in the industrial countries are often too large in scale to be applicable in developing countries, given their smaller markets and lower quality transport infrastructure.

Scorekeeping methods
: Methods of "measuring" energy savings may not be sufficiently accurate yet for the purpose of paying utilities or energy service companies for the savings that they have achieved. This must be contrasted with the ease of measuring the power generated or used. It is a particularly important issue for utilities, which usually earn revenues solely on the basis of energy sold and so have little incentive to assist efficiency efforts.

Technical and managerial manpower
: There is generally a shortage of skilled technical and managerial manpower in developing countries for installing, operating, and maintaining energy efficient equipment. This may not be a significant problem where turnkey equipment is used.

Financial and economic

Behavior
: Users may waste energy, for example, by leaving lights on. In some cases, such seeming waste may be done for important reasons. Bus drivers in developing countries often leave their engines on for long periods, at a significant cost in fuel, in order to avoid jumpstarting their vehicle if the starter is broken, or to prevent customers from thinking (if the engine is off) that their vehicle is broken and going to a competitor whose engine is running.

Cost
: The high initial cost of energy efficient equipment to the end user and the high effective discount rate used by the end user discourage investment.

Currency exchange rate
: Fluctuations in the currency exchange rate raises the financial risk to firms who import high efficiency equipment with foreign exchange denominated loans.

Dispersed energy savings
: Energy-efficiency improvements are scattered throughout the industrial and other sectors and are difficult to identify and exploit. In contrast conventional energy supplies may be more expensive, but are readily and reliably identified and employed. This tends to give planners a supply side bias irrespective of the potential of efficiency improvements.

Financial accounting and budgeting methods
: Factory accounts for paying energy bills may be separated from accounts for capital investment in more efficient equipment. Budgets for more efficient equipment may be rationed, forcing energy efficiency improvements to compete with each other for scarce budgeted funds even though the return on investment in efficiency may be much higher than the overall cost of capital to the firm.

International energy prices
: Uncertainty of international energy prices, such as oil, raises risks that price drops will reverse the profitability of investments in efficiency.

Multiple needs
: The multiple roles and needs served by an existing technology may not be adequately met by a new energy efficient technology. Draft animals, for example, can provide meat, milk, leather, and dung in addition to traction power. They also reproduce. Mechanical drive only provides traction.

Risk
: Particularly in poor rural and urban areas, people are highly risk averse; they have to be if they are to survive the vagaries of drought and other disasters. That villagers are risk averse should not, however, be construed to mean that they are technology averse. A variety of technologies have been adopted very rapidly in poor rural and urban areas.

Seasonality
: Rural life is dominated by the seasons, with sharp labor shortages during the agricultural season and serious underemployment during the rest of the year that rural industry can only partly support. Capital investment in efficient agricultural or rural industrial technologies is relatively more expensive as it must pay for itself during just the fraction of the year it is used.

Secondary interest
: Energy efficiency is often of secondary interest to potential users. In industry, for example, efficiency must compete with other equipment parameters—quality and quantity of product; timeliness, reliability, and flexibility; etc.—as well as other factors of production when investment choices are made and when the scarce time of skilled manpower is allocated. These are aspects of overall corporate strategy to improve profitability and competitiveness.

Secondhand markets
: Low efficiency equipment may be widely circulated in secondhand markets in developing equipments, either among industries within developing countries, or perhaps as gifts or hand-me-downs from industrial countries. Further, users who anticipate selling equipment into the second hand market after only a few years may neither realize energy savings over a long enough period to cover the cost premium of the more efficient equipment nor, if secondhand markets provide no premium for high efficiency equipment, gain advantage in its sale.

Subsidized energy prices
: Energy prices in developing countries are often controlled at well below the long run marginal cost, reducing end user incentive to invest in more efficient equipment. Energy prices may be subsidized for reasons of social equity, support for strategic economic sectors, or others, and with frequent adverse results. On the other hand, however, the low cost of power results in substantial financial costs to the utilities, providing them a potential incentive to invest in more efficient equipment on behalf of the user.

Threshold level of energy and cost savings
: Users may not find a moderate level of energy or cost savings, particularly if spread over many different pieces of equipment, sufficiently attractive to justify the investment of technical or managerial manpower needed to realize the savings.

Unstable and/or low energy prices
: Oil prices, in particular, have been volatile in recent years. This poses the risk that investments in other energy supplies will become uneconomic if the price of oil drops.

Institutional

Bias
: There is often a bias towards a small number of large projects, usually for energy supply, than for small projects, usually energy efficiency, due to administrative simplicity and to minimize transaction costs.

Disconnect between purchaser and user
: In a rental or lease arrangement, the owner will avoid paying the higher capital cost of more efficient equipment while the rentor or lessor is stuck with the resulting higher energy bills. Similarly, women in some countries may not have a strong role in household purchase decisions and may not themselves earn a cash income for their labor, but must use inefficient appliances purchased for them.

Disconnect between user and utility
: Even though the total system capital cost is generally lower for energy efficient equipment, it is the user who pays for the more efficient equipment but only recoups the investment over the equipment lifetime while the utility sees an immediate capital savings.

Information
: Potential users of energy efficient equipment may lack information on the opportunities and savings.

Intellectual property rights
: Energy efficient technologies may be patented and the royalties for use may add to the initial costs for the equipment.

Political instability
: Political instability raises risks to those who would invest in more efficient equipment that would only pay off in the mid-to long-term.

Turnkey systems
: Turnkey and other package systems are often directly adopted by commercial or industrial operations in developing countries. In many cases, however, the equipment within these systems is based on minimizing capital cost rather than minimizing lifecycle operating costs.

SOURCE: Office of Technology Assessment, 1992.

Table 4-7—Policy Options

Alternative financial arrangements
Currently, the capital costs of generation equipment are payed by the utility and the capital costs of end use equipment are payed by the end user. The high effective discount rate of the end user as well as this separation between utility and user (or for leased equipment, the separation between owner and user) leads to much lower levels of investment in end use equipment efficiency than is justified on the basis of either total system capital costs or lifecycle operating costs. Alternative financial arrangements to redress this "disconnect" might range from the end user choosing equipment according to the total lifecycle cost and paying this cost in monthly installments on the utility bill; to the end user paying a front-end deposit or posting a bond to the utility to cover the lifecycle operating costs of the equipment, against which the utility would charge the capital cost of the equipment and the monthly electricity bills. Either of these approaches would force the end user to directly face the total lifecycle costs of the equipment when purchasing it. See also Integrated Resource Planning.

Data collection
The range of opportunities for energy efficient equipment, end user preferences, and operating conditions are not well known in many countries. Data collection, including detailed field studies, would help guide policy decisions.

Demonstrations
Many potential users of energy efficient equipment or processes remain unaware of the potential savings or unconvinced of the reliability and practicability of these changes under local conditions. Demonstration programs can show the effectiveness of the equipment, pinpoint potential problems, and in so doing convince potential users of the benefits of these changes.

Design tools
Computer design tools can be developed, validated, demonstrated, and widely disseminated to potential users.

Direct installation
In some cases, particularly where the cost of energy is subsidized by State operated utilities or where peak loads are reduced, the direct installation of energy efficient equipment or processes at low- or no-cost by the utility can reduce costs for both the utility and the user.

Energy audits
Energy audits by a skilled team, either factory employees or from the outside—perhaps associated with an energy service Co.—can provide highly useful specific information on where energy can be saved. In new plants or in retrofits, submetering of equipment in order to maintain an ongoing record of energy use can also be a very useful means of monitoring performance.

Energy service companies
Energy service companies—third parties that focus primarily on energy efficiency improvements within a factory and are paid according to how much energy they save—can play a valuable role in implementing energy efficiency gains. They can bring great expertise and experience to bear on the problem, and as their goal is saving energy rather than maintaining production, they are able to devote greater effort and focus to conservation activities. On the other hand, industry employees are sometimes reluctant to work with ESCos, believing that they could implement efficiency activities equally well if they had the time; and worrying that any changes to process or related equipment by the ESCo could disrupt the production line. Generic forms of contracts for ESCo services to industry need to be developed in order to adequately protect both parties, and pilot programs with ESCos can demonstrate the potential savings by ESCos and their ability to avoid disruption to processes. Compensation for the work done by ESCos should be based on "measured" energy savings, not on the basis of listing the measures taken, irrespective of their effectiveness. Utility programs that provide for competitive bidding on energy savings risk paying the end-user and ESCo twice, once for the energy saved and once for the lower utility bill. This problem can be minimized by appropriately sharing the costs and benefits.

Extension efforts
Extension efforts may be useful at several levels. The efficiency and productivity of traditional rural industries might be significantly increased in a cost-effective manner with the introduction of a limited set of modern technologies and management tools. To do this, however, is extremely difficult due to the small and scattered nature of traditional rural industries and the large extension effort needed to reach it. Large industry in developing countries has many of the same needs—technical, managerial, and financial assistance—but can be reached more readily.

Grants
See "Direct installation," above.

Information programs
Lack of awareness about the potential of energy efficient equipment can be countered through a variety of information programs, including distribution of relevant literature directly to the industries concerned; presentation of competitions and awards for energy efficiency improvements.

Integrated resource planning (IRP)
Currently, utilities base their investment budgets on a comparison of the costs of different sources of generating capacity—coal, oil, gas, hydro, etc.—and the supply option that has the lowest cost for its particular application is chosen. Integrated Resource Planning expands this "least cost" planning system to include end use efficiency as an alternative to supply expansion in providing energy services. If energy efficiency is shown to be the lowest cost way of providing energy services, then under IRP utilities would invest in energy efficiency rather than new generating capacity.

Labeling programs
The efficiency of equipment can be clearly listed by labels. This provides purchasers a means of comparing alternatives. Measuring the efficiency of equipment, however, needs to be done in conjunction with standardized test procedures, perhaps established and monitored by regional test centers, rather than relying on disparate and perhaps misleading manufacturer claims.

Loans or rebates
Loans or rebates from the utility to the purchaser of energy efficient equipment can lower the first cost barrier seen by the user, and if incorporated in the utility rate base, can also prove profitable for them. On the other hand, users that would have purchased efficient equipment anyway then effectively get the loan or rebate for free—the "free rider" problem. This reduces the effectiveness of the utility program by raising the cost per additional user involved. This problem can be minimized by restricting the loans or rebates to the highest efficiency equipment for which there is little market penetration.

Marketing programs
  A variety of marketing tools might be used to increase awareness of energy efficient technologies and increase their attractiveness. These might include radio, TV, and newspaper ads, billboards, public demonstrations, product endorsements, and many others.
Pricing policies
  Energy prices should reflect costs, an obviously highly politicized issue in many countries. Where prices are heavily subsidized, the introduction of energy efficient equipment might be done in conjunction with price rationalization in order to minimize the price shock to users. Prices alone, however, are often insufficient to ensure full utilization of cost effective energy efficient technologies. There are too many other market failures as discussed above. As evidence of this, even the United States has adopted efficiency standards for a variety of appliances.
Private power
  Opportunities to cogenerate or otherwise produce private power have frequently not been taken advantage of because State-owned or controlled utilities have refused to purchase privately generated power at reasonable costs—many State electricity boards simply refuse to take privately generated power; many States impose a sales tax on self-generated electricity; many states decrease the maximum power available to industries with onsite generation capabilities, and then are reluctant to provide back-up power when cogeneration systems are down. In other cases, well-intentioned self generation taxes intended to prevent use of inefficient generators by industry penalizes efficient cogeneration. Finally, power is subsidized in many areas, making it difficult for private power to compete. Changes in laws mandating utility purchase of private power—such as that established by the U.S. Public Utilities Regulatory Policy Act laws—at reasonable rates would allow many of these opportunities to be seized. This should include establishing generic contracts that provide adequate protection to all concerned parties but that can be readily developed and implemented.
Protocols for equipment interfaces
  The use of power line carriers or other techniques of utility load management will require common equipment interfaces and signalling techniques.
Rate incentives
  See "Loans or rebates."
R&D: equipment processes, design rules
  Examples of R&D needs are listed in the text. R&D programs might be established at regional centers of excellence in developing countries, possibly in conjunction with sister research institutes in the industrial countries.
Regional test and R&D centers
  Regional centers of excellence are needed to help gather a critical mass of highly skilled technical manpower at a single site. The technologies to be developed should focus on those amenable to mass production while maintaining quality control under field conditions. Researchers and field extension agents should, in many cases, make greater use of market mechanisms to guide technology development efforts and to ensure accountability.
Scorekeeping: savings and validation
  Technologies and software for "measuring" energy savings need to be further developed and their effectiveness validated under field conditions. This would be a particularly valuable activity at regional centers of excellence.
Secondhand markets—standards
  Efficiency labels or standards might be set for secondhand equipment. This might be particularly valuable for such things as secondhand factories sold to developing countries.
Standards for equipment and process efficiency
  Many industrial countries have chosen to largely accept the financial "disconnect" between the utility and the user. Instead of providing low-cost, easily available capital to the end user and at the same time incorporating the full lifecycle cost of the end use equipment in the initial purchase, many industrial countries are attempting to overcome the economic and energy inefficiency of this disconnect by specifying minimum efficiency standards for appliances, buildings (residential and commercial), and, in some cases, industrial equipment.
Tax credits, accelerated depreciation
  A variety of tax incentives—tax credits, accelerated depreciation, etc.—to stimulate investment in energy efficient or other desirable energy technologies might be employed.
Training programs
  Training programs are needed in order to ensure adequate technical or managerial manpower. In addition, means of adequately compensating highly skilled and capable manpower are needed. Currently, skilled manpower—trained at government expense—is frequently attracted away from developing-country governmental organizations by the higher salaries of the private sector. Similarly, a more clear career path is needed for skilled technical and managerial manpower in energy efficiency just as utility operations now provide a career path for those interested in energy supply.
Utility demand and supply planning
  Methodologies for integrated supply and demand least-cost planning have been developed by the industrial countries. These should now be adapted to the needs of developing countries and utility planners and regulators trained in their use.
Utility regulation
  Utility regulations that inhibit the generation of private power (see above) or limit the role of the utility in implementing energy efficiency improvements on the supply or demand side need to be reevaluated. Means of rewarding utilities for energy saved as well as energy generated need to be explored (see also scorekeeping). This might include incorporation of energy-efficient equipment into the utility ratebase.

SOURCE: Office of Technology Assessment, 1992.

Table 4-8—Energy Consumption in Industry, China, 1980

Sector	Final energy use	
	Exajoules	Percent
Basic metals (iron and steel)	2.38	25.7
Chemicals (fertilizer)	2.23	24.1
Building materials (cement, brick, tile)	1.44	15.6
Pulp and paper	0.25	2.7
Machine building	0.82	8.8
Textiles	0.64	6.9
Food, beverages, tobacco	0.38	4.1
Other	1.12	12.1
Total	9.26	100.0

SOURCE: *China: The Energy Sector,* World Bank Country Study, Washington, DC, 1985.

Table 4-9—Energy Consumption in industry, India, 1985/86

Sector	Coal PJ[a]	Oil PJ	Electricity[b] PJ	Total PJ
Paper	26.0	NA	4.1	30.1
Chemicals[c]	5.3	87.2	21.9	114.4
Cement	106.0	5.7	6.1	117.8
Primary metals	232.0	19.3	25.2	276.5
Textiles	23.6	22.1	13.2	58.9
Food	NA	12.2	2.8	15.0
Subtotal	393.0	147.0	73.0	613.0
Total industry	683.0	207.0	194.0	1,084.0

Coal-derived and natural gas not included.
[a]One Petajoule (PJ) = 1 million Gigagoules (GJ) = 10^{15} Joules.
[b]Electricity is given in units of primary energy, converted from delivered kWh at 33 percent conversion efficiency fuel to electricity and 85 percent transmission & distribution efficiency, for overall efficiency of 28 percent.
[c]Includes plastics, rubber, petrochemicals, fertilizer, paint, other, etc.
SOURCE: Ashok Desai, "Energy, Technology and Environment in India," contractor report to the Office of Technology

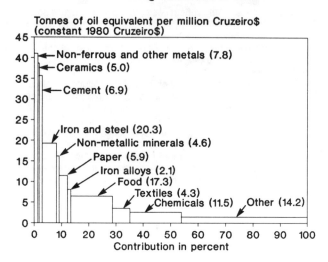

Figure 4-1—Final Energy Intensity Versus Manufacturing Value Added for Brazilian Manufacturing Industries in 1980

The number displayed for each bar is the sector's contribution to total final energy use in Brazilian manufacturing, in percent.
SOURCE: Jose Goldemberg et al., *Energy for Development* (Washington, DC: World Resources Institute, 1987).

overall average rate of 7.3 percent between 1974 and 1982—significantly faster than the 5 percent rate of overall gross national product (GNP) growth.[19] At higher incomes, however, consumption of steel levels off as the market saturates (see figure 4-2). A similar trend has been found for a wide variety of materials.[20] A more detailed analysis has shown that at about U.S.$2,000 per capita GDP, both the production and the consumption of energy intensive materials largely saturates—the elasticity of production and consumption with GDP is near zero.[21] This point of saturation is, however, still a long way off for most developing countries.

Although consumer goods such as electric lights, refrigerators, televisions, and automobiles are less energy intensive to assemble than basic materials, large amounts of energy are nevertheless used in their manufacture. Most of this energy is for low temperature process heat and electricity to power motors driving pumps, fans, compressors, conveyors, and machine tools.[22] The consumption of consumer goods is also increasing rapidly in developing countries, aided by their declining real cost

---

[19]Maurice Y. Meunier and Oscar de Bruyn Kops, World Bank, "Energy Efficiency in the Steel Industry with Emphasis on Developing Countries," World Bank Technical Paper No. 22, 1984; and World Bank, *World Development Report 1990* (New York, NY: Oxford University Press, 1990), indicator tables 1 and 26. Note that steel production, however, is concentrated in a few countries.

[20]Robert H. Williams, Eric D. Larson, and Marc H. Ross, "Materials, Affluence, and Industrial Energy Use," *Annual Review of Energy*, vol. 12, 1987, pp. 99-144. Simply put, there is a limit to the number of steel/cement intensive cars, refrigerators, buildings, roads, bridges, pipelines, etc. that a person needs. Eventually, consumption levels tend to plateau at replacement levels. When these wants for basic materials are fulfilled, people tend to spend incremental income on higher value-added materials—such as those with a high-quality finish—or on less material intensive but higher value added consumer goods. It is important to recognize these trends so as to not overestimate future demands for these energy-intensive materials.

[21]Alan M. Strout, "Energy-Intensive Materials," Ashok V. Desai (ed.), *Patterns of Energy Use in Developing Countries* (New Delhi, India: Wiley Eastern Limited, 1990), pp. 106-107.

[22]The transition from an economy focussed on heavy industry producing energy intensive materials such as steel and cement to an economy that concentrates on the manufacture of consumer goods can have a profound impact on the overall energy intensity of the economy.

Table 4-10—Per Capita Primary Industrial Energy Use in The United States

	U.S., 1980s	
	Fuel	Electricity[a]
Industry	(GJ/cap)	
By subsector:		
Primary metals	10.1	6.1
Chemicals	11.1	6.3
Refining	23.5	1.9
Stone, clay, glass, cement . . .	3.3	1.5
Pulp and paper	9.2	3.9
Food	3.4	2.0
Textiles	0.9	1.2
Machinery[b]	3.0	5.1
Subtotal	64.5	28.0
Other industry	2.5	3.1
By service:		
Motor drive	NA	21.7
Lighting and other	NA	2.9
Electrolytic	NA	4.4
Boilers	28.6	NA
Process heat	17.0	2.2
Feedstocks	21.4	NA

NA = Not available or very small.
[a]Electricity converted to primary fuel equivalent using a conversion factor of 0.33 for generation, transmission, and distribution combined.
[b]Includes the categories fabricated metal, machinery, electrical equipment, and transportation equipment.
SOURCE: Primary source of industrial and agricultural data is: OTA, Energy Efficiency in the Industrial Sector, forthcoming; data is for 1985.

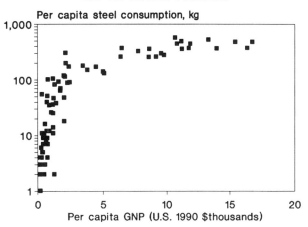

Figure 4-2—Per Capita Steel Consumption Versus GNP for Various Countries

The saturation of the steel market at higher income levels as national infrastructures are developed is readily seen in linear or logarithmic plots. It is shown here in a semilog plot so as to better display both low-end and high-end data. Each data point represents a country.

SOURCE: U.S. Congress, Office of Technology Assessment, *Energy in Developing Countries*, OTA-E-486 (Washington, DC: U.S. Government Printing Office, January 1991).

with new materials and improved manufacturing techniques (see ch. 2).

There are numerous opportunities to reduce the energy intensity of delivering industrial goods and their corresponding services (see table 4-5). Further, at least for electricity using technologies, these efficiency improvements generally lower both the total installed capital costs viewed from a societal perspective—factory investment plus upstream utility or other energy supply investment—as well as the life cycle operating cost for the user. These society wide financial advantages are shown in figures 4-3a and b together with the sharp shift in capital cost from the utility to the end user. Means of easing these capital costs to the user are needed if these high efficiency technologies are to be adopted and society to realize their financial and environmental advantages. These efficient technologies can also provide significant improvements in plant productivity, quality, and competitiveness.

Among the many opportunities for improving industrial energy efficiency, three are singled out here. It is important to view these opportunities, however, not as simple retrofits or equipment upgrades, but rather as one facet of the larger drive to modernize the industrial sector.

First, there are numerous opportunities to improve the energy efficiency of existing and new electric motor drive systems in industry. These include the use of motors designed for energy efficiency; the use of mechanical or electronic adjustable speed drives in order to match the speed of the motor to the load it is driving; improved pumps, fans, and other driven equipment; and improved design methods for properly sizing and interconnecting the many complicated components of a complete motor drive system. Motor drive will be a key focus here due to the importance of the electric sector in developing countries.

Second, there are a variety of ways in which specific manufacturing processes can be made more energy efficient. These are examined below for four industries—steel, cement, chemicals (fertilizer), and pulp and paper. These measures include housekeeping, retrofits of existing plants, and the establishment of state-of-the-art or advanced processes in new plants. Efficiency opportunities include the use of waste heat. Low temperature waste heat can be used in-plant to preheat materials; it can be used

Figure 4-3A—Total Systemwide Capital and Life Cycle Operating Costs and Energy Consumption for Conventional and High Efficiency Industrial Motor Drive Technologies

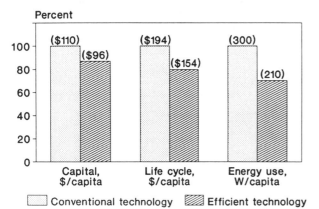

Over a wide range of conditions high efficiency technologies have a lower systemwide capital cost than the conventional technologies because the increased capital cost to consumers is more than offset by the decreased capital investment required in upstream electricity generating plants. Life cycle operating costs are lower because the increased capital costs to the consumer are more than offset by the lower electricity costs to operate the equipment.

SOURCE: U.S. Congress, Office of Technology Assessment, 1992. See app. A for details of this calculation, including the assumptions and a sensitivity analysis.

Figure 4-3B—Allocation of Capital Costs for Conventional and High Efficiency Industrial Motor Drive Technologies

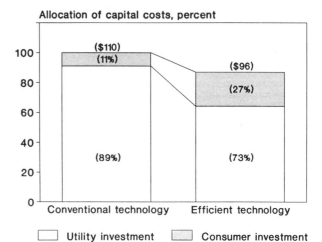

Although the high efficiency technologies have lower systemwide capital costs, they dramatically shift the capital costs from the utility to the industrial or agricultural user. The high capital cost of efficient equipment to consumers is the reason why it is not more heavily invested in—even though it provides net savings to society overall and, indirectly, to the consumer. Means of reducing this initial capital cost barrier to the consumer are critical.

SOURCE: U.S. Congress, Office of Technology Assessment, 1992. See app. A for details of this calculation, including the assumptions and a sensitivity analysis.

outside of plant in neighboring industry or perhaps for district heating schemes in regions with cold weather such as northern China. High temperature waste heat can also be used in (or generated by) cogeneration systems. The focus here will be on the equipment and processes themselves.

Third, energy intensive materials can be used more effectively than at present. Smaller quantities of higher performance materials, such as high strength steel alloys, can often be substituted for larger quantities of lower performance materials. Materials can be more extensively recycled and products reused or remanufactured. Quality control as well as such techniques as near-net shape processing in producing basic materials or finished consumer goods can play an important role in saving energy by reducing the amount of scrap and reworking that is necessary.

Although the basic technologies remain the same, other factors—raw materials, capital, labor, techni- cal and managerial manpower, political, trade regimes, and many others—vary dramatically between countries. These cases range from informal rural cottage industries, to protected nationalized industries with little external technical input, to subsidiaries of multinationals that have access to the best technologies available.[23] This wide range of conditions and capabilities requires a similarly wide range of policies in order to respond appropriately.

When considering various efficiency options, it is important to consider the time scales involved. At a sustained rate of growth in the industrial sector of 8 percent annually,[24] the manufacturing plant that exists now in a country will constitute less than half of the total manufacturing plant of the country in 10 years and less than one-fourth in 20 years. Housekeeping and retrofits of existing plants may be important over the short term for energy savings and for instilling a consciousness about the importance of energy savings. With such rapid growth, however,

---

[23] Depending on local conditions and corporate strategic plans, however, the corporation may not make use of the best technology it has available.

[24] The average annual growth rate of industry in the low income economies between 1965 and 1988 was about 8.8 percent. See: World Bank, *World Development Report 1990* (New York, NY: Oxford University Press, 1990), indicator table 2.

housekeeping and retrofits will play a much less significant role in saving energy over the longer term than ensuring that new investments are targeted towards high efficiency state-of-the-art plants or even the adoption of advanced manufacturing processes. The choice of targeting existing plant and equipment or new investments for energy efficiency will then depend on the complex tradeoffs of the potential gains versus the limited manpower and capital that can be invested.

On the other hand, the industrial sector is growing at a much slower rate in many other countries, particularly Africa and Latin America. In these areas, retrofits of existing plant and equipment must be emphasized—especially where capital constraints inhibit modernization generally.[25]

There are a variety of barriers to improving the energy efficiency in the industrial sector of developing countries. Some of these barriers as well as potential policy responses are summarized in tables 4-6 and 4-7.

Agriculture

Relatively little commercial energy is used directly in the agricultural sector, ranging from less than 1 percent of the national total in the United States to perhaps 5 to 8 percent of the national totals in developing countries such as Brazil, China, India, and Kenya. The developing countries, however, will need to increase energy intensive inputs—fertilizer, irrigation, improved crop varieties, and animal or mechanical traction—into agriculture if they are to keep up with their rapid population growth.[26] The agricultural sector is particularly important for the role it plays in improving the living standards of the rural poor.

The efficiency of many agricultural operations can be improved. These include improvements in pumping systems, mechanical traction, and in the production and application of chemicals such as fertilizers (discussed under industry). There are also many opportunities for decentralized power production through the use of renewables (see ch. 6) or for improving or changing the task itself—such as using drip irrigation or even going to advanced agroforestry or other agricultural techniques. A considerable effort to provide extension services will be necessary, however, if these opportunities are to be realized by the highly dispersed agricultural sector.

MOTOR DRIVE SYSTEMS[27]

Traditional industry, agriculture, transport, and household activities rely primarily on human and animal muscle for mechanical power. When only muscle power is available, many hours can be spent on "enabling" activities, such as hauling water or grinding grain, rather than on more directly economically productive activities. Productive industrial or agricultural activities themselves are sharply limited by the low efficiency and output of muscle power. If the productivity of people in developing countries is to be increased, modern motor drive technologies and supporting infrastructures must be made available at affordable costs. As these technologies are adopted, energy use—especially electricity—will increase rapidly.

The efficiency, convenience, and high degree of control of electric motors provide dramatic efficiency and productivity improvements in industry, agriculture, and other sectors.[28] This led to a rapid transition in the industrialized countries from water and steam powered drive to electric drive in the early 1900s (see figure 4-4); the electricity intensity of industry continues to increase today in industrialized as well as developing countries.

Electric motor drive today consumes an estimated 58 to 68 percent of the electricity used in the United States and even more in the industrial sector alone. Motor drive is similarly important in developing countries (see tables 4-11 to 4-14). Electric motors

---

[25]Gabriel Sanchez-Sierra, Organizacion LatinoAmericana de Energia, personal communication, July 15, 1991.

[26]The increase in energy use will be somewhat less for integrated agroforestry or other advanced sustainable agricultural approaches.

[27]Principal sources for this section are: Samuel F. Baldwin, "Energy-Efficient Electric Motor Drive Systems," Thomas B. Johansson, Birgit Bodlund, and Robert H. Williams (eds.), *Electricity: Efficient End-Use and New Generation Technologies, and Their Planning Implications* (Lund, Sweden: Lund University Press, 1989, pp. 21-58), used with permission; Samuel F. Baldwin, "The Materials Revolution and Energy Efficient Electric Motor Drive Systems," *Annual Review of Energy*, vol. 13, 1988, pp. 67-94, used with permission from the *Annual Review of Energy*, vol. 13, copyright 1988 by Annual Reviews, Inc.; and Samuel F. Baldwin, "Energy-Efficient Electric Motor Drive Systems," Princeton University, Center for Energy and Environmental Studies, Working Papers No. 91, 92, 93, and 94, February 1988.

[28]Samuel F. Baldwin, "The Materials Revolution and Energy Efficient Electric Motor Drive Systems," *Annual Review of Energy*, vol. 13, 1988, pp. 67-94; W.D. Devine, Jr., "Historical Perspective on Electrification in Manufacturing," S. Schurr and S. Sonenblum (eds.), *Electricity Use: Productive Efficiency and Economic Growth* (Palo Alto, CA: Electric Power Research Institute, 1986).

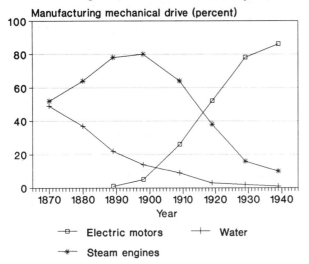

Figure 4-4—Percentage of Manufacturing Mechanical Drive From Water Power, Steam Engines, and Electric Motors by Year

SOURCE: Samuel F. Baldwin, "The Materials Revolution and Energy-Efficient Electric Motor Drive Systems," *Annual Review of Energy*, vol. 13, 1988, pp. 67-94.

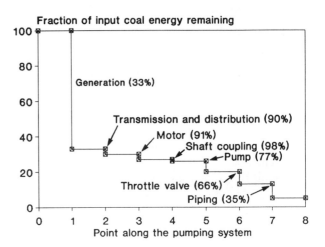

Figure 4-5—Energy Losses in an Example Electric Motor-Driven Pumping System in the United States

This figure shows the useful energy remaining at each stage of a pumping system. The values in parentheses are the efficiencies of the particular device at each stage.

SOURCE: Samuel F. Baldwin, "Energy-Efficient Electric Motor Drive Systems," Thomas B. Johansson, Birgit Bodlund, and Robert H. Williams (eds.), *Electricity: Efficient End-Use and New Generation Technologies and Their Planning Implications* (Lund, Sweden: Lund University Press, 1989) pp. 21-58.

are the workhorses of modern industrial society. They run home refrigerators; drive office air conditioners; power industries' pumps, fans, and compressors; and keep cities' water supplies flowing.

Significant efficiency improvements are possible in electric motors and the systems that they drive. These gains in energy efficiency usually reduce the total systemwide—including both the end user and the upstream utility—capital investment required as well as the life cycle operating costs for the user. These improvements will be discussed here in terms of the overall design and performance of motor drive systems and the performance of individual components.

Motor Drive System Design

Electric motor drive systems are often large and complex, involving numerous interacting components. The most common types of motor drive systems in industry include pumps, fans, compressors, conveyers, machine tools, and various rollers, crushers, and other direct-drive systems (see table 4-14). A pumping system is shown in figure 4-5 together with the efficiencies and net useful energy remaining at each point along the system.

Motor driven pumps, fans, and other system components are usually deliberately designed to be oversized, that is, to have excess capacity.[29] These components are often oversized partly because it is difficult to predict system flow rates and friction factors accurately in advance, and to allow for the effect of the buildup of deposits on duct and pipe walls over time. Motors are oversized to handle starting electrical, mechanical, and thermal stresses, particularly with high-inertia loads; to provide a safety margin for the worst-case load over their lifetimes; and occasionally to handle plant expansion or perhaps to be widely interchangeable within the plant. In developing countries, oversizing motors is often important for preventing motor stall and possible burnout when the line voltage drops. More generally, these system components are oversized because the increased energy and capital costs to the end user are perceived to be less than the risk of equipment failure. In manufacturing, for example, average electricity costs in the United States are just 2.8 percent of the value-added;[30] the cost of motor

---

[29] On the other hand, pipes and ducts are often undersized to reduce capital costs.

[30] M. Ross, "Trends in the Use of Electricity in Manufacturing," *IEEE Technology and Society*, vol. 5, No. 1, March 1986, pp. 18-22

Table 4-11—Industrial Electricity End Use in Brazil, 1984

Industry	Percent of total industrial electricity consumption	Fraction of subsector total for each end use (percent)					
		Motor	Process heat	Direct heat	Electro-chemical	Light	Other
Nonferrous metals	20.9	32	1	35	32	1	NA
Iron and steel	12.4	1	NA	98	NA	1	NA
Chemicals	11.9	79	5	4	9	3	NA
Food and beverage	9.0	6	78	16	NA	1	3
Paper and pulp	6.5	87	8	2	NA	3	NA
Mining and pelletization	5.6	50	NA	49	NA	1	NA
Textiles	5.3	89	4	1	NA	5	1
Steel alloys	4.8	7	NA	92	NA	1	NA
Ceramics	3.9	65	NA	34	NA	1	NA
Cement	2.7	91	NA	6	NA	3	1
Other	17.0	76	2	16	NA	5	1
Total[a]	100.0	49	10	32	NA	2	NA

NA = Not applicable or not available.
[a]Total industrial electricity use was 10 terawatthours.

SOURCE: Howard S. Geller, "Electricity Conservation in Brazil: Status Report and Analysis," contractor report, prepared for the Office of Technology Assessment, March 1990; published as *Efficient Electricity Use: A Development Strategy for Brazil* (Washington, DC: American Council for an Energy Efficient Economy, 1991).

Table 4-12—Projected Electricity Consumption in India by Sector and End Use, 1990 (percent of total national electricity use)

Sector	Total[a]	Industrial process			Lighting	Space conditioning		Appliances		Other miscellaneous
		Motor drive	Electrolysis	Process heat		Cooling/ ventilation	Heating	Refrigeration	Other	
Residential	13.0	NA	NA	NA	4.2	3.5	NA	1.5	1.0	2.9
Urban	10.4	NA	NA	NA	2.9	2.9	NA	1.2	1.0	2.4
Rural	2.6	NA	NA	NA	1.3	0.5	NA	0.3	NA	0.5
Commercial	11.2	NA	NA	NA	4.8	1.6	1.5	0.4	0.8	2.1
Agriculture	18.4	18.4	NA	NA	NA	NA	NA	NA	NA	NA
Industrial	54.8	33.4	10.8	5.5	5.1	NA	NA	NA	NA	NA
Primary metals[b]	17.2	6.4	6.9	3.0	0.9	NA	NA	NA	NA	NA
Chemicals	13.8	8.8	3.6	0.1	1.3	NA	NA	NA	NA	NA
Textiles	10.2	7.8	NA	0.4	2.1	NA	NA	NA	NA	NA
Coal, cement	6.8	5.8	0.2	0.5	0.4	NA	NA	NA	NA	NA
Secondary metals[c]	3.4	1.5	NA	1.4	0.2	NA	NA	NA	NA	NA
Paper	3.4	3.0	NA	0.1	0.3	NA	NA	NA	NA	NA
Railway traction	2.6	2.6	NA	NA	NA	NA	NA	NA	NA	NA
Motor drive	61.4	54.4	NA	NA	NA	5.1	NA	1.9	NA	NA

NA = Not available or not applicable.
[a]Total national consumption is projected to be 249.1 terawatthours in 1990.
[b]Aluminum, nonferrous, iron, and steel.
[c]Iron and steel.

SOURCE: Ahmad Faruqui, Greg Wikler, and Susan Shaffer, "Application of Demand-Side Management (DSM) To Relieve Electricity Shortages in India," contractor report prepared for the Office of Technology Assessment, April 1990.

Table 4-13—Electricity Consumption (GWh/year) by Service for 24 Industries in Karnataka, India (1984-85)

Industry	Motors	Process heat	Electrolysis	Lights cooling	Total	Share (percent)
Aluminum	43.4	0.0	343.0	1.57	387.9	22.4
Primary steel	61.6	233.6	0.0	4.83	300.0	17.3
Fertilizer	206.9	0.0	0.0	1.17	208.1	12.0
Paper	193.4	0.0	0.0	1.05	194.5	11.2
Cement	185.6	0.0	0.0	1.18	186.8	10.8
Secondary steel	68.6	113.2	0.0	1.42	183.2	10.6
Ferro alloys	3.4	118.8	0.0	0.22	122.5	7.1
Caustic soda	13.3	0.0	95.6	0.48	109.4	6.3
Graphite	7.3	33.9	0.0	0.99	42.1	2.4
Total	783.6	499.5	438.6	12.90	1,735.0	100.0
Share (percent)	45.2	28.8	25.3	0.70	100.0	

SOURCE: Amulya Kumar N. Reddy, et al., "A Development-Focussed End-Use-Oriented Energy Scenario for Karnataka: Part 2—Electricity," Department of Management Studies, Indian Institute of Science, Bangalore, India.

Table 4-14—Disaggregation of Electricity Consumption (GWh/yr) by Motors for 24 Industries in Karnataka, India (1984-85)

Industry	Shaft power	Compressors	Pumps	Fans	Refining	Material handling	Agitating	Total	Shares (percent)
Aluminum	1.4	9.9	21.5	8.3	0.0	1.1	1.2	43.4	5.5
Primary steel	34.9	13.6	6.0	0.0	0.0	6.3	0.0	61.6	7.9
Fertilizer	2.3	164.2	39.0	0.0	0.0	1.4	0.0	206.9	26.4
Paper	48.5	4.7	66.5	9.2	53.2	3.5	7.7	193.4	24.7
Cement	97.6	13.7	10.5	58.3	0.0	5.6	0.0	185.6	23.7
Secondry steel	44.4	6.4	11.4	0.0	0.0	6.5	0.0	68.6	8.8
Ferro alloys	0.8	0.8	0.2	1.7	0.0	0.0	0.0	3.4	0.4
Caustic soda	0.4	3.0	6.6	2.5	0.0	0.3	0.4	13.3	1.7
Graphite	4.1	0.3	0.7	0.0	0.0	2.2	0.0	7.3	0.9

SOURCE: Amulya Kumar N. Reddy, et al., "A Development-Focussed End-Use-Oriented Energy Scenario for Karnataka: Part 2—Electricity," Department of Management Studies, Indian Institute of Science, Bangalore, India.

failure and unplanned shutdown of the entire process line can be a much more severe penalty.[31]

The intent of this design process is understandable and reasonable. Manufacturers, design engineers, and users all want to ensure that the system can meet the demands placed on it at every stage of the process. Although oversizing can provide direct safety margins for equipment, it can have unintended negative side effects. For example, when each successive element of a drive system is sized to handle the load presented by the previous component plus a safety margin, oversizing can quickly become excessive and require throttling valves on pumps or throttling vanes on fans to limit flow. In addition, many systems need variable outputs. Space heating and cooling, manufacturing, municipal water pumping, and most other motor drive loads vary with the time of day, the season, and even the health of the economy—such variations can be quite large.

Traditionally, throttling valves or vanes have been the principal means by which flow is controlled. This is, however, an extremely inefficient means of limiting flow. It is analogous to driving a car with the gas pedal floored, and then controlling the car's speed with the brake. Other systems of control, however, have generally been too expensive, less reliable, frequently difficult to control, or are themselves inefficient. Even simple approaches, such as turning a motor on and off to limit output—as with a home refrigerator—results in significant (if unseen) energy losses.

The direct and indirect energy losses due to such control strategies include part load operation, poor

---

[31] A.D. Little, Inc., *Energy Efficiency and Electric Motors*, U.S. Department of Commerce Report No. NTIS PB-259 129 (Springfield, VA: National Technical Information Service, 1976).

power factor, throttling losses, excess duct or pipe friction, and pump or fan operation off the design point, among others. Standard engineering design rules, together with manufacturing safety margins, sometimes automatically lead to significant inefficiencies of this type, of which the user may be unaware. Industrial and commercial pumps, fans, and compressors, for example, have estimated average losses of 20 to 25 percent or more due to throttling or other inefficient control strategies alone.[32] The losses of complete systems can be much greater than this, as shown in figure 4-5. The largest single loss is in the process of electricity generation, but additional substantial losses take place at every stage throughout the system.

Motors

The efficiency of standard electric motors is often significantly lower in developing countries than in industrialized countries due to the use of lower quality materials in motor construction and improper techniques in maintenance, repair, and rewind (see figure 4-6).[33]

High efficiency motors are readily available in industrial countries and are sometimes available in developing countries, but in some cases cannot be used because of the poor quality of the electric power available. In Brazil, for example, the largest manufacturer of smaller motors exports more efficient models than those sold at home. These motors, comparable in efficiency to standard motors in the United States, often cannot be used in Brazil due to the excessive variation in the power line voltage. If the line voltage drops too much, the motor can stall and possibly burn out. Electronic Adjustable Speed Drives (ASDs) could buffer such voltage fluctuations, allowing use of the higher efficiency motor and providing significant energy savings (see below).

Typical costs for one class of standard and high efficiency electric motors in the industrial countries are shown in figure 4-7.[34] An industrial motor,

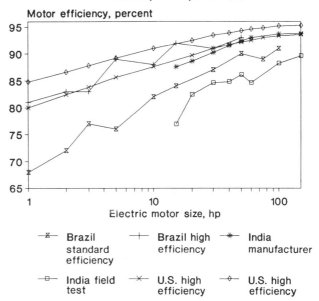

Figure 4-6—Efficiency of Electric Motors in the United States, Brazil, and India

SOURCE: U.S. Congress, Office of Technology Assessment, *Energy In Developing Countries*, OTA-E-486 (Washington, DC: U.S. Government Printing Office, January 1991).

however, can use electricity worth perhaps four times its capital cost annually.[35] Thus, from the user perspective even the relatively small efficiency gains of high efficiency motors can quickly pay for their increased capital costs, particularly in a new installation or in replacing a damaged motor.

The benefits of improved motor efficiency are even more pronounced from the system perspective. In this case, the total capital costs of the system— including the upstream capital investment in utility generation, transmission, and distribution equipment along with the cost of the electric motor— decrease with improved motor efficiency (see figures 4-8a and b). Not only are total system capital costs reduced by using more efficient motors, but the substantial utility fuel use and operating costs are avoided. Total annual operating costs to users are also lower.

---

[32] Samuel F. Baldwin, "Energy-Efficient Electric Motor Drive Systems," Thomas B. Johansson, Birgit Bodlund, and Robert H. Williams (eds.), op. cit., footnote 27; William J. McDonald and Herbert N. Hickok, "Energy Losses in Electrical Power Systems," *IEEE Transactions on Industry Applications*, vol. IA-21, No. 4, 1985, pp. 124-136.

[33] Samuel F. Baldwin and Emile Finlay, op. cit., footnote 12. In particular, when motors are rewound they are sometimes simply put on an open fire to burn the insulation off the windings rather than putting them in temperature controlled ovens. This can damage the insulation between the core laminations and leads to greater losses.

[34] The unit horsepower (hp) is used here rather than kW in order to distinguish between kW of input electric power and hp of output shaft power. One horsepower is equal to 0.7457 kW.

[35] Operating 5,000 hours per year at 70-percent load, a 90-percent efficient motor annually consumes 2,900 kWh per hp. At $0.07/kWh this costs $200. In comparison, motors cost roughly $50/hp.

Figure 4-7—Efficiency and Cost of Standard and Energy Efficient Motors in an Industrial Country

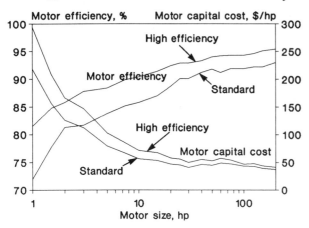

This figure shows that for all sizes of motors, high efficiency models have a higher capital cost than standard models. This additional capital cost generally decreases as motor size increases. Note that these values are given per horsepower of motor capacity. Large motors will cost more in total, but less per unit capacity.

SOURCE: Marbek Resource Consultants, Ltd., "Energy Efficient Motors in Canada: Technologies, Market Factors and Penetration Rates," Energy Conservation Branch, Energy, Mines and Resources, Canada, November 1987. See app. A of this report for details.

Small but highly cost effective efficiency gains are also possible in many cases by increasing the size of the cables running to the motor and making other improvements in power system equipment.[36] For example, national standards for electrical cable sizes are often based on minimizing fire hazard rather than energy use. Larger cables are even safer because they have lower electrical resistance and generate less heat. These same characteristics can also enable larger cables to often quickly pay for themselves in reduced electricity costs to the user alone, not including the substantial upstream benefits of reduced system capital costs.[37] Other improvements in motors[38] and motor operating conditions[39] are possible that can provide small but highly cost effective energy savings. New technologies are also becoming available, such as high performance permanent magnet motors, that allow further improvements in efficiency.[40]

Plant engineers in many countries, however, may have a difficult time choosing between high efficiency motors offered on the market as measured efficiencies vary substantially according to the test standard[41] used (see table 4-15). In general, the Japanese test standard suggests higher efficiencies than the European standards, and the European standard gives higher values than the North American. Thus, a motor purchased from Japan could have a higher reported test efficiency but a lower actual efficiency than a similar motor purchased from the United States due to the difference in testing methodologies. This variation might correspondingly play a role in market competitiveness for different firms. It suggests a need for international efficiency testing protocols, probably best administered in regional test centers in both developed and developing countries.

Pumps and Fans[42]

Pumps and fans are, overall, the most common motor driven equipment.[43] In principle, equipment is chosen so that the optimal efficiency is matched to the operating conditions. In practice, there are numerous design complications due to oversizing,

---

[36]William J. McDonald and Herbert N. Hickok, "Energy Losses in Electrical Power Systems," *IEEE Transactions on Industry Applications*, vol. IA-21, No. 4, 1985, pp. 803-819.

[37]Amory B. Lovins et. al., *The State of the Art: Drivepower* (Snowmass, CO: Rocky Mountain Institute, April 1989).

[38]Of particular interest is minimizing core damage due to excessive temperatures when motors are rewound; and improving operating motor maintenance such as lubrication, etc.

[39]Of particular interest may be improving the power factor, reducing voltage fluctuations on the line, balancing three-phase power, and controlling line harmonics.

[40]Samuel F. Baldwin, "Energy-Efficient Electric Motor Drive Systems," Thomas B. Johansson, Birgit Bodlund, and Robert H. Williams (eds.), op. cit., footnote 27; Amory B. Lovins et. al., *The State of the Art: Drivepower* (Snowmass, CO: Rocky Mountain Institute, April 1989).

[41]Note that the word "standard" here refers to the testing methodology used to measure motor performance. It does not mean a particular level of performance required by law or other codes.

[42]The use of the term "fans" here is intended to include industrial blowers.

[43]Pumps and fans are the primary focus here. The considerations for compressors are similar, with the exception that when using adjustable speed drives the larger static pressure that compressors work against make it more difficult to optimally match the performance map of the compressor to the system load curve across a wide range of operating conditions. Conveyer systems for solid materials also use substantial amounts of electricity in industry. A brief review of potential efficiency improvements in their design can be found in: William E. Biles, "Solids Conveying," *Technology Menu for Efficient End-Use of Energy: Volume 1: Movement of Material* (Lund, Sweden: Environmental and Energy Systems Studies, Lund University Press, 1989).

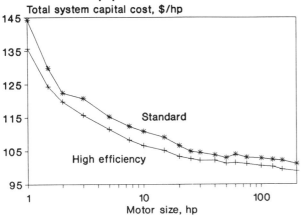

Figure 4-8A—Standard and High Efficiency Motor Capital Costs Including Motor and Upstream Utility Equipment To Power It

This figure shows that for all sizes of motors considered, total system capital costs are lower for high efficiency motors. Note that these values are given per horsepower of motor capacity. Large motors will cost more in total, but less per unit capacity.

SOURCE: U.S. Congress, Office of Technology Assessment, 1992. See app. A for details.

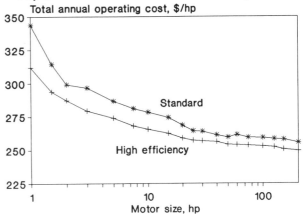

Figure 4-8B—Standard and High Efficiency Motor Operating Costs Including Both Annualized Motor Capital Investment and Annual Electricity Costs

This figure shows that for all sizes of motor considered, total annual operating costs are lower for high efficiency motors. Note that these values are given per horsepower of motor capacity.

SOURCE: U.S. Congress, Office of Technology Assessment, 1992. See app. A for details.

Table 4-15—Comparison of Efficiency Testing Standards for Motors

Standard	Motor size	Full-load efficiency (percent)			
		7.5 hp[a]		20 hp[a]	
CSA C390		80.3%		86.9%	
NEMA MG-1		80.3		86.9	
IEC 34-2		82.3		89.4	
JEC-37		85.0		90.4	
	Motor size	5 hp[b]	10 hp[b]	20 hp[b]	75 hp[b]
IEEE 112B		86.2%	86.9%	90.4%	90.0%
IEC 34-2		88.3	89.2	91.4	92.7
JEC 37		88.8	89.7	91.9	93.1

NOTE: CSA is the Canadian Standards Association; NEMA is the National Electrical Manufacturers' Association (U.S.); IEC is the International Electrotechnical Commission (Europe); JEC is the Japanese Electrotechnical Commission; and IEEE is the Institute of Electrical and Electronic Engineers (U.S.).

SOURCES: [a]Steven Nadel et al., "Energy-Efficient Motor Systems: A Handbook on Technology, Program, and Policy Opportunities," (Washington, DC: American Council for an Energy Efficient Economy, 1991); [b] John C. Andreas, "Energy-Efficient Electric Motors," (New York, NY: Marcel Dekker, Inc., 1982).

throttling, and other practices that move the system operating point away from the optimal efficiency.

Many pumps are also poorly designed and built, resulting in far lower efficiencies than technically and economically possible. Figure 4-9 shows the wide scatter in measured performance of new pumps in the United States. Factors that lower intrinsic pump efficiency include: excessive friction due to rough surfaces, poorly finished edges, and poorly shaped contours of the pump surfaces; internal leakage of fluid; and friction in the bearings and seals. Corresponding means of improving efficiencies include smoother and more carefully contoured internal surfaces; tighter tolerances;[44] and higher quality bearings. Further efficiency gains are often possible by operating the pump at a higher speed. Many of these efficiency improvements are possible at little cost.

Similar considerations apply to fans where large efficiency gains are likewise possible by choosing

---

[44]When pumping dirty water, etc. or mixed phase materials, larger tolerances may be preferable despite the efficiency penalty.

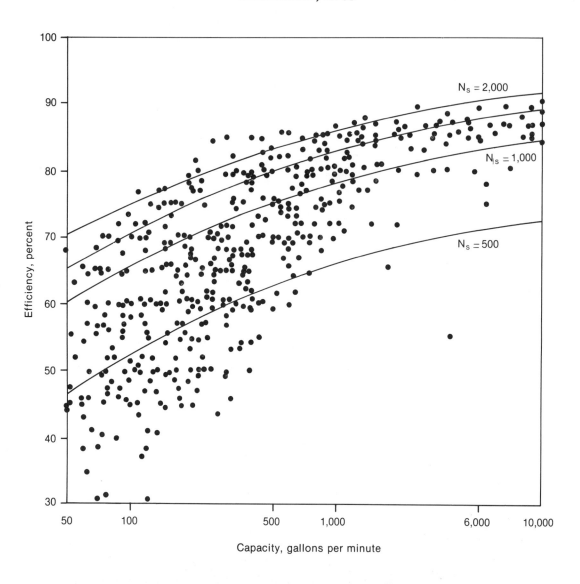

Figure 4-9—Actual New Pump Efficiencies at the Design Point, United States, 1970s

Note the wide scatter in efficiencies below what is expected in good design practice, as indicated by the solid lines for different pump speeds.

SOURCE: U.S. Department of Energy and Arthur D. Little, Inc., *Classification and Evaluation of Electric Motors and Pumps,* Report No. DOE-CS-0147 (Springfield, VA: National Technical Information Service, February 1980).

more efficient designs, operating them close to their design point, and by other means. Ranges for peak fan efficiencies are shown in table 4-16; many fans do not come close to these values. In some applications, of course, the highest efficiency designs cannot be applied.[45] Of particular note is that backwardly curved fans are essentially interchangeable with forwardly curved fans, but are typically 10 to 20 percentage points more efficient at a roughly 15 percent cost premium. Fan and air handling efficiencies are particularly important in commercial buildings, where up to half of the electricity use

---

[45]Lower efficiency fans might be used, for example, to handle corrosive gases or gases with erosive particles such as those coming off a furnace.

Table 4-16—Estimated Maximum Total Efficiencies for Assorted Fan Designs

Type	Peak efficiency range (percent)
Centrifugal fans:	
Airfoil, backwardly-curved	79-83
Modified radial	72-79
Radial	69-75
Pressure blower	58-68
Forwardly curved	60-65
Axial fans:	
Vaneaxial	78-85
Tubeaxial	67-72
Propeller	45-50

SOURCE: J. Barrie Graham, "Fans," *Technology Menu For Efficient End-Use of Energy, Volume 1: Movement of Material*, Environmental and Energy Systems Studies, Lund University, Lund Sweden, 1989.

occurs in air handling systems for building heating, cooling, and ventilation.[46]

Many pumps and fans are also poorly maintained and efficiencies can decrease markedly over time due to wear. One study of 84 large pumping systems, primarily in pulp and paper mills in Sweden and Finland, found that wear alone had reduced average pump efficiencies by 14 percentage points compared to their original performance.[47]

Improvements in a pump or fan have a large upstream multiplicative impact due to the losses in transmission and distribution (typically 15 percent in developing countries), motors, couplings, and other equipment between the pump/fan and the utility (see figure 4-5). An example of the very large potential upstream capital and operating cost savings that are possible by improving pump or fan performance are shown in table 4-17. On an annualized basis, upstream utility capital cost savings are nearly 10 times greater than the incremental cost of the improved motor and pump/fan. Operating costs of the less efficient system increase this differential even more. Two factors give rise to this huge leverage: the multiplicative effect of the energy losses noted above; and the much greater capital cost per unit energy supply/demand for utility equipment than for the motor/pump system.

Despite these potential savings, users of pumps/fans have often not paid particular attention to pump or fan efficiency. There are many reasons for this: Pump/fan performance is difficult to measure in the field and requires special equipment and effort,[48] and it can require cutting holes in pipes or ducts as well as even shutting down a process line. Many users are unaware of the potential savings. Users are not directly exposed to and do not consider the utility investment requirements that result from their choice of pump or fan. In some cases, industrial users' electricity rates are subsidized (although industrial users often pay full price or even cross subsidize the residential, commercial, or agricultural sectors). Finally, industrial investment decisions are often shifted away from energy efficiency improvements due to tax structures; tariffs on imported high efficiency equipment (even though that may result in greater imports of expensive utility generating equipment); and internal financial accounting controls. For example, accounting procedures may use a limited capital improvements budget wherein projects compete against each other for a limited pool of funds rather than against an external market interest rate determined rate-of-return criteria.

In addition, many users, unaware of the large potential energy savings, justifiably place very high premiums on the proven reliability of a certain type of pump/fan and its manufacturer; on minimizing spare parts inventories and simplifying maintenance; and on timely delivery of spares. For example, users are often unwilling to switch to a different manufacturer to get a few percent higher efficiency pump/fan for a specific application when their regular manufacturer's offerings do not provide a high efficiency pump/fan in that flow and pressure range. Many pumps/fans are incorporated in other equipment by Original Equipment Manufacturers

---

[46] J. Barrie Graham, "Fans," *Technology Menu For Efficient End-Use of Energy, Volume 1: Movement of Material* (Lund, Sweden: Environmental and Energy Systems Studies, Lund University Press, 1989); "Fans: A Special Report," *Power*, September 1983; Scott L. Englander, "Ventilation Control for Energy Conservation: Digitally Controlled Terminal Boxes and Variable Speed Drives," Princeton University, Center for Energy and Environmental Studies, Report No. 248, March 1990.

[47] Eric D. Larson and Lars J. Nilsson, "Electricity Use and Efficiency in Pumping and Air Handling Systems," paper presented at ASHRAE Transactions, June 1991 ASHRAE Meeting, Indianapolis, IN.

[48] For example, to measure pump efficiency requires that the motor power input to the pump and the pump power output (flow of liquid and pressure drop) be measured. To measure the motor power input to the pump in the field requires the use of power meters and tachometers. To measure the flow and pressure drop across a pump requires flow meters and pressure gauges and is usually an invasive procedure. For more information, see: Samuel F. Baldwin, "Energy-Efficient Electric Motor Drive Systems," Princeton University, Center for Energy and Environmental Studies, Working Papers No. 91, 92, 93, and 94, February 1988.

Table 4-17—Capital and Energy Savings With Improved Pumps

	Standard pump system			Efficient pump system		
	Efficiency percent	Power output kW	Annual capital cost dollars	Efficiency percent	Power output kW	Annual capital cost dollars
Generator	—	—	—	—	—	—
		17.4			15.6	
T&D	85%	—	—	85%	—	—
		14.8	$3,250		13.3	$2,936
Motor	90%	—	—	94%	—	—
		13.3	$ 107		12.5	$ 121
Pump	75%	—	—	80%	—	—
		10.0	$ 110		10.0	$ 132
Total capital			$3,467			$3,189
Investment in motor/pump			—			$ 36
Annual capital savings			—			$ 278
Annual electricity savings			—			$ 540

How to read this table: This table illustrates the impact of improved equipment efficiencies on capital expenditures for a hypothetical pumping system. In the standard pump case, the delivery of 10 kW in pumping power output requires a motor output of 13.3 kW, a transmission & distribution (T&D) system output to the motor of 14.8 kW, and a generator output of 17.4 kW due to the losses at each step of the system (efficiencies of 75%, 90%, and 85% respectively). The corresponding annual capital costs are $3,250 for generation plus T&D, $107 for the motor, and $110 for the pump. In the system with improved pump and motor efficiencies, to provide 10 kW in pumping power output, just 15.6 kW are needed at the generator busbar. The improved motor and pump have an annual capital cost $36 greater than in the standard case, but reduce the annual capital expenditure for generation and T&D equipment by $314, for a net capital savings of $278. Electricity (capital plus fuel) savings total $540 annually.

NOTES: This assumes a capital cost of $4,000/kW for delivered power from the generator through the T&D system as detailed in Appendix A at the back of this report. It assumes standard motors and pumps/fans cost $100/kW output, and efficient motors and pumps/fans cost 20 percent more. The pump is assumed to run 6000 hours per year corresponding to typical two shift operation at an industrial plant. Total capital costs are annualized using a 30-year 7-percent discount rate to get a capital recovery factor of 0.0806. The same 30 year lifetime and discount rate is assumed for the motor and a 15-year life is assumed for the pump (CRF=0.11). Note that this calculation assumes exact sizing of the motor and fan rather than sizing to the nearest available size. Finally, it assumes electricity costs $0.06/kWh corresponding to preferential industrial rates. The listed annual electricity savings include the annualized capital savings.

It has been estimated that for a 20-percent price premium on pumps, an average efficiency improvement of 10 percentage points is possible for pumps smaller than 5 hp (3.7 kW) decreasing to 2 percentage points for pumps larger than 125 hp (93 kW).[a] Similarly, backward-curved (blade) fans have been estimated to cost about 15 percent more than the equivalent forward-curved fans, but give a 10-20 percentage point improvement in efficiency—corresponding to savings of 2-4 times that listed above.[b]

[a]U.S. Department of Energy and Arthur D. Little, Inc, "Classification and Evaluation of Electric Motors and Pumps," DOE/CS-0147, 1980.
[b]Eric D. Larson and Lars J. Nilsson, "Electricity Use and Efficiency in Pumping and Air-Handling Systems," to be published in ASHRAE Transactions, presented at the June 1991 ASHRAE Meeting, Indianapolis, IN.
SOURCE: Office of Technology Assessment, 1992.

(OEMs) and sold as a package; lowest first cost, not energy efficiency, is usually the primary concern of an OEM.[49]

As for the case of electric motors, these considerations suggest the need for careful, standardized testing of new pump/fan efficiencies; ongoing monitoring of pump/fan performance in the field; wide dissemination of both new and field measured performance of pumps/fans; and finding a means of incorporating potential utility capital savings in the considerations of the industrial end user.

Adjustable Speed Drives

Adjustable speed drive (ASD) technology allows a significant change in system design and operation. In contrast to the conventional practice of oversizing systems, using constant-speed motors to drive them, and throttling excess flow with vanes or valves, ASDs and their associated sensors and controllers allow precise matching of motor speed to load. This can significantly improve overall system efficiencies both directly and indirectly. Further, this flexibility may allow design rules that lead to extreme oversizing to be relaxed, potentially reducing capital investment in some system components by the end user.

ASDs offer a number of other benefits as well. They often increase equipment lifetimes:

- by avoiding the back-pressures generated by conventional throttle valves or vanes used to limit the output of pumps or fans;

---

[49]U.S. Department of Energy and Arthur D. Little, Inc., "Classification and Evaluation of Electric Motors and Pumps," Report No. DOE/CS-0147 (Springfield, VA: National Technical Information Service, 1980).

- by permitting constant lubrication of bearings—in contrast to equipment operated in an on-off mode;
- by allowing operation at reduced speeds; and
- by permitting slow, controlled starts to reduce electrical stresses on motors, transformers, and switchgear, and to reduce mechanical stresses on motors, gears, and associated equipment.

Other ASD advantages include:

- isolation of the motor from the power line, which can reduce problems caused by varying or unbalanced line voltage;
- a ''ride through'' capability if there is a power failure for a few cycles;
- operation at higher speeds than the 60-Hz line frequency allows; and
- easy retrofits to existing equipment.

Finally, and perhaps most importantly:

- ASDs often provide better control over manufacturing processes or in other equipment than conventional systems can achieve, improving process and product quality.

This description may suggest that ASDs are hi-tech devices applicable only in advanced manufacturing plants in industrial countries. In fact, ASDs are likely to become ubiquitous and offer significant opportunities in developing countries—by buffering line voltage fluctuations, reducing systemwide capital costs, reducing user operating costs, reducing energy use, and improving the manufacturing process, among others.

Adjustable-speed drives are being successfully used in numerous applications today in OECD countries and their use could be extended to developing countries. In industry they are used in boiler fans in chemical plants, utility plants, packaging equipment, glass-blowing machines, cement factories, and many others. In commerce they are used in commercial refrigeration systems, heating, ventilation, and air conditioning (HVAC) systems for buildings, and other applications. In residences, they are used in air conditioning and heat pump systems.[50] Several hundred thousand ASDs have now been sold in the United States.

Individual case studies have documented energy savings of 30 to 50 percent in fan and boiler feedpumps, 20 to 25 percent in compressors, 30 to 35 percent in blowers and fans, 20 to 25 percent in pumps, 25 to 35 percent in central refrigeration systems, and 20 percent in air conditioning and heat pumps.[51] Of course, ASDs do not provide savings in constant-speed full load applications, so national energy savings will be less than might be calculated by assuming universal adoption at the above efficiencies.

The capital costs of ASDs have declined some 7 to 12 percent in real terms over the past 4 years[52] and are likely to continue slowly declining with improvements in ASD technology, manufacturing processes, and as further economies of scale and learning in manufacturing are achieved.[53] Figures 4-10a, b, and c show current capital costs for ASDs, and the corresponding total system capital costs (including those of the utility) and life cycle operating costs.[54] At the low end of the size scale (a fraction of a horsepower (hp)—not shown), the cost of mass produced ASDs used with residential heat

---

[50]Detailed listings of references for each of these cases are given in Samuel F. Baldwin, ''Energy-Efficient Electric Motor Drive Systems,'' Thomas B. Johansson, Birgit Bodlund, and Robert H. Williams (eds.), op. cit., footnote 27.

[51]Detailed listings of these case studies can be found in Baldwin, Ibid.

[52]Eric D. Larson and Lars J. Nilsson, ''Electricity Use and Efficiency in Pumping and Air Handling Systems,'' paper presented at ASHRAE Transactions, June 1991 ASHRAE Meeting, Indianapolis, IN.

[53]Costs will not decline as rapidly as they have for other electronics technologies, however. For example, much of the dramatic decline in cost and performance increases in the computer industry are possible because individual transistors on integrated circuits can be scaled down in size with no loss of function (actually, there is a gain in performance as higher system operating speeds are then possible). Computer circuits handle information, and there is no inherent lower limit on the amount of energy needed to carry a bit of information, at least down to inherent circuit noise. In contrast, ASDs handle large amounts of power—up to thousands of kilowatts—in order to drive a motor. To handle these powers requires large quantities of very high quality electronic materials, primarily silicon, and the quantities of material required cannot be scaled down as they can for a computer circuit. This greatly slows the pace at which costs can be reduced. For smaller loads such as refrigerators and air conditioners, however, power mosfets may allow the entire ASD to be fabricated on a single chip. This would allow significant cost reductions in this size range as fabrication yields improved.

[54]Additional details on ASD capital costs can be found in: Power Electronics Applications Center, *Directory Adjustable Speed Drives*, second edition (Palo Alto, CA: Electric Power Research Institute, 1987); and Everett B. Turner and Charles LeMone, ''Adjustable-Speed Drive Applications in the Oil and Gas Pipeline Industry,'' *IEEE Transactions on Industry Applications*, vol. 25, No. 1, pp. 30-35, 1989. In large sizes, e.g., 750 hp and above, ASD capital costs can be less than $100/hp at the low-cost end; at sizes of 5,000 hp and above, ASD capital costs approach $50-60/hp at the low-cost end and $100/hp at the high-cost end.

pumps and other appliances is reportedly down to $25/hp ($33/kilowatt).[55]

Systemwide financial considerations—including the cost of the ASD and the avoided utility capital investment—were recently applied in Canada. In that case, an ASD was chosen to drive a coal slurry pipeline rather than relying on a conventional throttling valve for control. The reason for this choice was simple. The cost to the company of the 3,500 hp (2.6 megawatts) ASD equipment and installation was Can$800,000; to raise the capacity of the local supply system to power the less efficient throttle controlled design would have cost the company Can$1,000,000 even including a capital cost sharing arrangement with the local utility and would have delayed the project 4 months. The company chose the ASD over the less efficient throttle valve to avoid those upstream costs and delays.[56]

ASDs will also be used increasingly in household and commercial refrigeration and air conditioning systems due to their efficiency and other advantages. By modifying the design of these ASDs and establishing standard protocols, direct implementation of load management techniques via power-line carriers or other means may then be possible at very low marginal costs. For example, circuitry to detect a power-line carrier signal and then to turn down or turn off a refrigerator or air conditioner could be incorporated directly into its ASD. Time-of-day electricity cost controls could be similarly implemented. Simple on-off load management techniques are already used in large scale systems in the United States and Europe. Southern California Edison, for example, has 100,000 air conditioners in a load management program networked via a VHF-FM radio system. Despite the high cost of establishing the system—approximately $120 per participating household just for the control devices, their installation, and marketing—benefits have been almost 4 times greater. Further, the program has been very

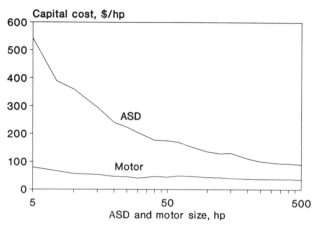

Figure 4-10A—Capital Costs for Electric Motors and ASDs Alone

This does not include utility investment nor additional engineering costs for designing and emplacing the ASD beyond those for the motor alone. These engineering costs can greatly increase the cost of an ASD system; with extensive experience, however, these additional engineering costs should be readily reducible. Note that these costs are given per horsepower of capacity. Large units will cost more in total, but less per unit capacity.

SOURCE: Capital costs for ASDs are from Steven Nadel et. al., *Energy Efficient Motor Systems: A Handbook on Technology, Programs, and Policy Opportunities* (Washington, DC: American Council for an Energy Efficient Economy, 1991).

well received; just 1.8 percent of the participants have withdrawn voluntarily. These techniques allow a substantial reduction in peak utility loads with corresponding savings.

ASDs do have some drawbacks. In particular, ASDs can distort the shape of the normal voltage waveform in the power grid.[57] This distortion can reduce the efficiency of motors, transformers, and other equipment. It can also interfere with computers, communications, and other equipment. Techniques are available to control this problem and further work to lower the costs of control is ongoing. With proper design up front—as opposed to onsite remediation after installation—harmonic control is relatively low cost and straightforward. Many OECD countries have established legislation limiting the distortion allowable from ASDs and related devices

---

[55] Steve Greenberg et. al., *Technology Assessment: Adjustable-Speed Motors and Motor Drives (Residential and Commercial Sectors)* (Berkeley, CA: Lawrence Berkeley Laboratory, 1988).

[56] Frank A. Dewinter and Brian J. Kedrosky, "The Application of a 3,500-hp Variable Frequency Drive for Pipeline Pump Control," *IEEE Transactions on Industry Applications*, vol. 25, No. 6, 1989, pp. 1019-1024.

[57] In more technical terms, ASDs are nonlinear electronic devices that can inject harmonics into the power line. Converters that use thyristors cause line notching—a brief short circuit that sends the line voltage abruptly to zero at that point. Furthermore, the current drawn is closer to a square wave than a sine wave. Converters that use diode bridges draw current in a periodic pulse and also cause distortion of the line voltage. A six-step inverter will generate characteristic harmonics at $nK+/-1 = 5,7,11,13,17, \ldots$ times the line frequency, each with maximum theoretical amplitudes equal to (1/harmonic) of the fundamental. Twelve-step inverters can, in principle, significantly reduce the generation of harmonics, but in practice system reactance and converter phase shifts somewhat reduce their advantage.

Figure 4-10B—System Capital Costs for a Motor (Plus Utility Investment to Power It) and for an ASD/Motor (Plus Utility) System

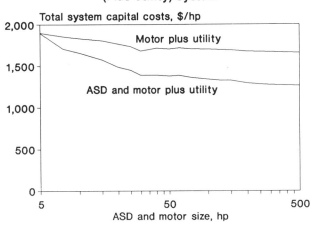

When upstream utility capital costs are included, the more efficient ASD/Motor system has lower capital costs than less efficient motors (plus utility) alone.
SOURCE: Office of Technology Assessment, 1992. See app. A for details.

Figure 4-10C—Annualized Operating Costs for Motor and ASD/Motor Systems

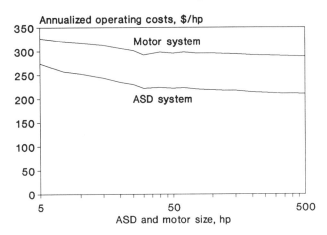

More efficient ASD/motor systems have lower annualized operating costs than less efficient motors alone.
SOURCE: Office of Technology Assessment, 1992. See app. A for details.

and much of this analysis should be readily transferable to developing countries.[58]

The total electricity savings possible by the use of ASDs is not yet well understood and more research is needed. Total savings will depend on a variety of factors including the rate of cost reduction of ASDs and corresponding market penetration, and the types of part- or variable-loads driven. Further, the individual savings achieved may depend on the development of new engineering design rules that fully exploit the opportunities presented by ASDs.

Pipes and Ducts

Pipes, ducts, and related fittings make up the final part of pump/fan systems, channeling the liquid or gas to where it is to be used. Appropriate design of a pipe/duct system can provide substantial energy savings in many applications.

In a pipe/duct system, energy is used to lift the liquid—water, petroleum, chemicals—or gas and to overcome friction in the pipe/duct and its fittings (including the throttling valve, if any). The energy required to lift the liquid/gas cannot be changed, but energy losses due to friction and related effects can be dramatically reduced by increasing the diameter of the pipe/duct, by using smoother pipe,[59] and by careful choice and spacing[60] of the fittings used. The energy savings achievable must be balanced against the various costs associated with a smoother or larger pipe or duct. In a building, increased pipe and particularly duct size can have substantial costs associated with it—by increasing the space needed between floors or reducing usable space. In industry, the costs are more often limited to the increased capital costs of the pipe/duct and related components alone. In theory, many cost related factors could be included in the analysis to determine the optimum pipe/duct diameter—capital costs for the pipe/duct, fittings, support structure, pump/fan, and motor; taxes and insurance; and energy savings. In practice, the analysis usually considers only the increased capital and O&M costs of the larger pipe/duct versus the energy savings.[61]

---

[58] For a brief review of this problem and various national standards for harmonic distortion from ASDs, see Samuel F. Baldwin, "Energy-Efficient Electric Motor Drive Systems," Thomas B. Johansson, Birgit Bodlund, and Robert H. Williams (eds.), op. cit., footnote 27.

[59] For example, the friction factor for commercial steel pipe is 30 times larger than that for the same diameter PVC plastic pipe. The impact of a larger friction factor on energy use varies, however, with pipe diameter, flow rates, and other factors.

[60] Proper spacing can reduce losses by as much as 30 to 40 percent.

[61] Eric D. Larson and Lars J. Nilsson, "Electricity Use and Efficiency In Pumping and Air-Handling Systems," paper presented at the American Society of Heating, Refrigerating, and Air Conditioning Engineers Meeting, June 23-25, 1991; Nicholas P. Cheremisinoff, "Piping," *Technology Menu for Efficient End-Use of Energy* (Lund, Sweden: Environmental and Energy Systems Studies, Lund University Press, 1989).

Figure 4-11A—Reduction in Energy Required To Pump a Liquid Through a Pipe as a Function of Pipe Diameter

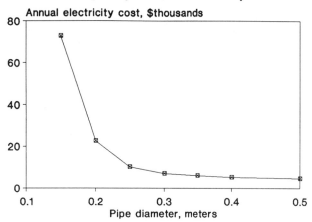

Assumptions: 6-meter static head; 30-meter friction head for 200 mm pipe diameter; 500-meter pipe; 90 l/s flow.

SOURCE: Nicholas P. Cheremisinoff,"Piping," *Technology Menu For Efficient End-Use of Energy, Volume I: Movement of Material* (Lund, Sweden: Environmental and Energy Systems Studies, Lund University Press, 1989).

Figure 4-11B—Annualized Capital Cost of Pipe, Electricity Cost, and Sum of the Pipe Capital Cost and Electricity Cost

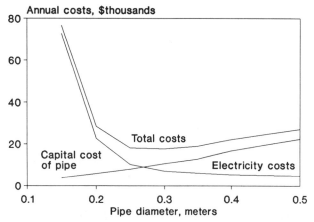

SOURCE: Nicholas P. Cheremisinoff,"Piping," *Technology Menu For Efficient End-Use of Energy, Volume I: Movement of Material* (Lund, Sweden: Environmental and Energy Systems Studies, Lund University Press, 1989); and app. A of this report.

Figure 4-11C—Annualized Total System Costs

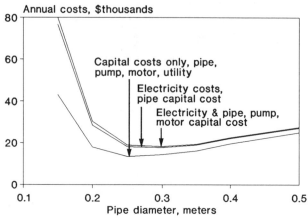

This figure shows annualized total system costs for three different types of system accounting: 1) capital costs only, including the pipe, pump, motor, and utility generation, transmission, and distribution equipment; 2) capital costs of the pipe only plus the cost of electricity; 3) capital costs of the pipe, pump, and motor plus the electricity costs to operate the system. Arrows indicate the minimum total cost for the three cases indicated.

Different discount rates and electricity prices can shift these various optimum points to higher or lower diameters. In general, however, the optima are relatively insensitive to changes in these parameters. For example, varying discount rates from 7 to 21 percent and electricity costs from $0.03 to 0.09 moved the optima from a low of 0.20 meter to a high of 0.30 meter. Over this same interval, electricity use varied by a factor of more than 3.

SOURCE: U.S. Congress, Office of Technology Assessment, 1992; see also app. A.

If, instead, the total system capital costs are considered—including the motor, pump, and upstream utility investment, then the optimum pipe/duct size will normally be increased compared to the case where only pipe capital and operating costs are considered. This is shown in figures 4-11a, b, and c. The change in the optimum diameter shown in figure 4-11c may seem small, but it reduces electricity use by one-third. It is also worth noting that, on all three accounting systems, the total cost is relatively flat from 0.2 meters to 0.5 meters. Over this same range, however, electricity use drops by a factor of 5 at the factory, or nearly 6 at the utility when transmission and distribution (T&D) losses are included.

Similar considerations apply to ducts. For building ventilation systems, new computerized design tools are being developed that minimize life cycle costs—including duct, fan, motor, and other capital costs, along with electricity costs. Compared to duct systems designed with today's conventional methodologies, case studies with these new design tools find total life cycle cost savings of roughly 20 to 50 percent and electricity savings of 25 to 55 percent at an increase in capital costs of 5 to 10 percent, depending on the particular parameters. Alternatively, for the same energy use, capital costs could be reduced by 20 percent or more compared to conventional designs.[62] Adjustable speed drives, high efficiency motors, and improved fans can provide additional savings, as outlined above.

Systems

These various opportunities for reducing electricity use are summarized in table 4-18. Just 4 percent of the input coal is converted to useful energy (to overcome static pressure drop) at the end of the piping, and losses are spread throughout the system. Thus, in this particular case, saving 1 unit of energy in piping system losses can save up to 25 units of energy worth of coal. ASDs achieve the greatest individual savings, and do so by eliminating shaft coupling and throttle valve losses, and by raising pump efficiencies by operating closer to the design point. Increasing the pipe diameter by 25 percent also offers large savings due to the high friction losses of this particular system and due to the great leverage such reductions have upstream in the system.[63]

Repeating the calculations for two-thirds flow gives similar savings: the increased throttling losses and reduced pump efficiency are roughly counterbalanced by decreased piping losses. For this particular system, combining all of the various improvements listed leads to electricity savings of some 65 percent (see table 4-18). Capital savings, however, would depend on the tradeoff between increased pipe size and decreased utility, motor, and pump size (see figure 4-11). The number of industrial or other systems in which such large savings can be found, however, is unknown but there are nevertheless many cases where reduced capital costs, reduced life cycle costs, and substantial energy savings are possible.

Many other motor drive system improvements are possible,[64] including the use of power factor correction capacitors or other devices, amorphous magnetic materials in distribution transformers and motor cores, and permanent magnet motors. For example, in India it has been estimated that power factor correction capacitors could reduce electricity demand for one-fifth the cost of building new supply.[65]

MODERN INDUSTRIAL PROCESSES

This section examines four large industries: steel, cement, pulp and paper, and chemicals (specifically fertilizer production). These industries are particularly important due to their role—especially steel and cement—in building national infrastructures of roads, buildings, and factories, and due to their very high energy intensities (see figure 4-1).[66] Other important industry-related energy issues, particularly the cogeneration of heat and electricity, are examined in chapter 6.[67]

Steel

Developing countries such as China, India, and Brazil devote about 20 percent—about the same as the OECD countries[68]—of industrial commercial[69] energy consumption to steel production. The top 10 producing countries account for about 90 percent of the crude steel made in the developing world; many other developing countries produce little or no steel.

Overall steel production has been increasing by a little over 7 percent per year in the developing countries (with a wide variation among individual countries), while remaining relatively constant in the industrialized countries (see figure 4-2). At current

---

[62]Robert J. Tsal, "Ducting," *Technology Menu for Efficient End-Use of Energy* (Lund, Sweden: Environmental and Energy Systems Studies, Lund University Press, 1989).

[63]This does not, however, take into account changes in pump or motor efficiency, etc. It also does not provide for the complex tradeoffs in capital costs for larger pipe (less the smaller motor and pump) versus the energy savings.

[64]Samuel F. Baldwin, "Energy-Efficient Electric Motor Drive Systems," Thomas B. Johansson, Birgit Bodlund, and Robert H. Williams (eds.), op. cit., footnote 27.

[65]Inter-Ministerial Working Group, subcommittee reports, "Report on Utilisation and Conservation of Energy," India, 1983.

[66]Due to the complexity of these industries and the wide variety of technical approaches (depending on the particular national circumstances of access to technology, capital, raw materials, trade policy, and a host of other factors), however, this will necessarily be a brief survey of these industries rather than an indepth examination.

[67]See also, U.S. Congress, Office of Technology Assessment, *Industrial and Commercial Cogeneration*, OTA-E-192 (Washington, DC: U.S. Government Printing Office, February 1983).

[68]Maurice Y. Meunier and Oscar de Bruyn Kops, op. cit., footnote 19.

[69]Commercial—primarily fossil or hydroelectric—energy as opposed to traditional biomass fuels used in many rural industries.

Table 4-18—Hypothetical Savings for Various Pumping System Configurations

	Baseline		Efficient system	
	eff[a]	rmn[b]	eff	rmn
Coal	NA	100%	NA	100%
Generation	33%	33	33%	33
Transmission and distribution	85	28	90	30
ASD	NA	NA	95	28
Electrical motor	90	25	93	26
Shaft coupling	98	25	98	26
Pump	75	19	80	21
Throttle valve	66	12	100	21
Piping system	35	4	56	12
Coal required	—	100 units	—	35 units
Savings	—	NA	—	65%

NA = not applicable.
[a] "eff" is the device efficiency in percent.
[b] "rmn" is the remaining energy as a percent of the starting total.
[c] "Savings" are the percentage point reductions in coal use made possible for the different component and system improvements while still providing the same useable energy at the end of the piping system.

Improvements include: an upgraded transmission & distribution system; an Adjustable Speed Drive (ASD); a high efficiency electric motor (note that the electric motor will lose up to a percentage point or so in efficiency due to harmonic generation by the ASD); an improved pump and operation of the pump at a higher intrinsic speed; the elimination of the throttle valve allowed by use of the ASD; and an increase in pipe diameter of 25 percent. Note that many baseline systems will have lower piping system losses than this example. For further details and additional potential improvements, see the reference.

SOURCE: Adapted from Samuel F. Baldwin, "Energy-Efficient Electric Motor Drive Systems," in *Electricity: Efficient End-Use and New Generation Technologies and Their Planning Implications* (Lund, Sweden: Lund University Press, 1989).

rates, steel production by developing countries will overtake that in the industrialized countries early in the next century. Developing countries are unlikely to use as much steel per capita as industrial countries did at their peak, however, as the performance of steel has increased substantially and there are now other lower cost and less energy intensive materials that can substitute for steel in many applications.

The energy efficiency of steel production in the developing countries varies widely. In some cases, it has significantly lagged that of the industrialized countries. Integrated steel plants in India and China currently use, on average, 45 to 53 gigajoules (GJ) (43 to 50 million Btus) per tonne of crude steel produced;[70] integrated steel plants in the United States and Japan use half as much energy. Some developing countries have made significant strides to reduce overall energy use in steel production. The Brazilians, for example, cut energy consumption from 34 GJ to 27 GJ per tonne of crude steel between 1975 and 1979 and the South Korean steel industry is among the most efficient in the world.[71]

There are seven major processes in conventional steel production:[72] preparation of the ore, production of coke, ironmaking, steelmaking, casting (or primary finishing), forming (or secondary finishing), and heat treating.[73] The corresponding energy use of each process is listed in table 4-19 for a state-of-the-

---

[70] These figures do not provide for differences in product mix, etc. For example, China produces much more cast iron than does the United States. Key plants in India, China, and elsewhere are often more efficient than these averages would suggest. In China, for example, energy use in key plants has been estimated at about 20 percent higher than that for the United States, if differences in the product mix are accounted for and the much more complex shaping and heat treating work done in the United States is ignored. If these differences in shaping and heat treatment are included, key Chinese plants use about one-third more energy than U.S. plants. Marc Ross and Liu Feng, "The Energy Efficiency of the Steel Industry of China," *Energy*, vol. 16, No. 5, 1991, pp. 833-848.

[71] Maurice Y. Meunier and Oscar de Bruyn Kops, op. cit., footnote 19; Sven Eketorp, "Energy Considerations of Classical and New Iron- and Steel-Making Technology," *Energy*, vol. 12, No. 10/11, 1987, pp. 1153-1168.

[72] U.S. Congress, Office of Technology Assessment, *Industrial Energy Use*, OTA-E-198 (Springfield, VA: National Technical Information Service, June 1983).

[73] This list is, of course, oversimplified. Among other simplifications, it does not include such important developments as secondary metallurgy, etc. and it artificially divides the steelmaking and primary finishing steps which, with the use of continuous casting, might more logically be listed as one process.

art steel mill using cost effective, commercially proven technologies. At each stage, different processes may be employed—depending on local resources and capabilities, and the type of steel being produced—with different impacts on capital expenditures and energy consumption. These alternative processes, particularly the direct use of coal for producing iron and advanced casting and forming techniques for steel, may dominate the industry in the future. The discussion here will nevertheless follow the conventional process flow chart.

Ore Preparation

Ore is prepared by crushing and grinding it, and then agglomerating it by pelletizing/sintering processes into marble sized pieces that can be fed into a blast furnace. More energy efficient agglomeration processes have been developed, some of which also improve pellet quality.[74]

Coking

Heating coal to temperatures of 1,650 to 2,000 °F for 12 to 18 hours boils off its volatiles and leaves coke, a fairly pure (80 to 90 percent) carbon. The coke serves three purposes within the blast furnace: as fuel to heat and melt the ore/iron; to physically support the ore; and as a chemical reducing agent. To meet these needs requires coke that has a low impurity/ash content and that is physically strong.

The efficiency of converting coal to coke has improved substantially in recent years. In the United States, for example, the energy used to produce a metric tonne of coke declined from 7.0 gigajoules (6.7 million Btus) in 1980 to 4.1 GJ (3.9 million Btus) by 1989.[75] More efficient processes are becoming available. For example, the traditional process for stopping the chemical reactions in the coke ovens was to quench the coke with water and vent the steam into the atmosphere. A much more efficient technique is dry coke quenching, in which the hot coke in the ovens is cooled by circulating a nonoxidizing gas through it to stop the chemical reactions and at the same time to capture the heat

Table 4-19—Energy Consumption in IISI Reference Steel Plants

	BF/BOF[a] GJ/tcs[c]	DRI/EAF[b] GJ/tcs	Scrap/EAF GJ/tcs
Ironmaking	13.8	11.7	NA
Coke oven	1.6	NA	NA
Sintering	1.4	NA	NA
BF	10.6	NA	NA
Steelmaking	–0.01	5.5	4.8
Casting[d]	0.4	0.3	0.3
Forming	2.8	2.3	2.3
Other	1.1	0.2	0.2
Total	18.1	20.0	7.6

NA = Not available or not applicable.
[a]This is for case A, commercially proven technologies, rather than for the IISI base case because, for example, dry quench coke plants and EAF scrap preheaters are in operation. More generally, these reference plants incorporate "proven, energy efficient technologies and operating practices which should be considered economically viable in the majority of steelmaking countries."
[b]75 percent directly reduced iron (DRI) and 25 percent scrap.
[c]GJ/tcs is Gigajoules per metric tonne of carbon steel. Electricity is valued at 9.63 MJ/kWh or a conversion efficiency, fuel to electricity, of 37 percent.
[d]Casting includes continuous casting and primary rolling mills.

SOURCE: International Iron and Steel Institute, "Energy and the Steel Industry," Brussels, 1982.

from the coke to generate steam or electric power. This process also improves coke quality and reduces environmental emissions.[76] Dryquenching is in use in India and other countries.

Charcoal is used in place of coke in roughly one-third of Brazilian ironmaking due to a lack of local coking coal and the lower cost of charcoal. In 1985, about 25 million tonnes (t) of wood with an energy content of 0.45 exajoules (EJ) (or 0.43 Quad) were converted to charcoal in Brazil, primarily for use in metals production. Between 1974 and 1986, improved reforestation techniques and improved forestry productivity improved forestry yields by 3.5 times, while improvements in the blast furnaces lowered the specific consumption of charcoal from 3.4 m^3/t to 2.6 m^3/t of hot metal.[77] A large portion of the wood used for charcoal, however, is obtained by clearing natural forests, rather than fuelwood plantations, with significant environmental impacts. This

---

[74]Examples of improved pelletizing/sintering processes include the MTU (Michigan Technological University), COBO (cold bonding), and Peridur processes for agglomeration, and various waste heat recovery schemes in the sintering process. See Sayed A. Azimi and Howard E. Lowitt, U.S. Department of Energy, "The U.S. Steel Industry: An Energy Perspective," Report No. DOE/RL/01830-T55 (Springfield, VA: National Technical Information Service, January 1988).

[75]Energetics, Inc., report for the U.S. Department of Energy, "Industrial Profiles: Steel," Report No. DE-AC01-87CE40762 (Springfield, VA: National Technical Information Service, December 1990).

[76]Jonathon P. Hicks, "The Search for a Cleaner Way to Make Steel," *The New York Times*, Mar. 21, 1990, p. D7.

[77]Francisco Lanna Leal and Benoni Torres, "The Iron and Steel Industry in Brazil," *Ironmaking and Steelmaking*, vol. 17, No. 1, 1990, pp. 1-7.

has caused considerable controversy over the use of charcoal for ironmaking in recent years.[78]

Ironmaking

Blast furnaces are most often used to convert iron ore to metallic iron. In this case, iron-bearing materials are fed into the top of the blast furnace together with limestone/dolomite and coke for fuel. Heated air (and sometimes additional fuel) is blown into the blast furnace from the bottom. The coke (and possibly other fuel) burns in the hot air blast to heat, chemically reduce, and melt the iron as it descends through the furnace. The limestone/dolomite combines with impurities in the iron to form a slag that floats on the molten iron and can be removed. Blast furnace efficiencies have been improved by a variety of means, including the use of "top pressure recovery turbines" to generate typically 8 to 15 megawatts of power using the pressure and heat from the blast furnace (see table 4-20).

A number of efforts are underway to directly produce iron with coal, rather than with coke. These processes would lower capital costs—by one-half to two-thirds compared to a similar sized 0.5 million metric tonnes per year plant, and by perhaps a third compared to a conventional plant six times larger—by avoiding coke ovens and, in some cases, agglomeration facilities. They would be more flexible, allow smaller scale operations, and would reduce environmental impacts. Energy requirements for these systems, however, range from about 13.3 GJ/t (12.6 million Btu/t) of hot metal for three-stage systems to as much as 30 GJ/t (28.4 million Btu/t) for single-stage systems. The low end of these energy requirements is about a 10-percent improvement over typical conventional blast furnaces and about 5 percent better than can be achieved with the best cost effective coke-based blast furnace technology available today. The high end has, however, much higher energy intensities than conventional best practice technologies today. Several of the systems make extensive use of electricity and thus suffer the conversion losses of fuel to electricity in generation and also carry a substantial additional capital cost for the generating facilities.[79] Coal based process plants are in operation in South Africa producing 300,000 tons per year.[80] These capital savings and other advantages may push direct coal-based processes into a dominant role in the steel industry of the future.

High quality iron ore can also be directly reduced to metallic iron (DRI) using natural gas. Direct reduction accounted for 85 percent of production in Venezuela[81] and 26 percent of Mexican production in 1986;[82] and for all of Indonesia's production.[83] Although direct reduction only accounts for 5 percent of Indian production, current construction will triple DRI capacity over the next several years.[84] There are also several DRI facilities in Africa.[85] Following direct reduction, the sponge iron is then converted to steel using electric arc furnaces.

Steelmaking

Steelmaking refines the pig iron produced in the ironmaking process. In steelmaking, chemical elements such as phosphorus, sulfur, and silicon are reduced and removed from the melt, and elements such as nickel, chromium, and other alloying agents are added to give the steel its desired properties.

Three types of furnaces are used to produce steel—the open hearth furnace, the basic oxygen furnace, and the electric arc furnace. The open hearth furnace is now outmoded—with higher capital costs, greater energy use, and lower productivity than the basic oxygen furnace—and is being phased out in most countries. By 1986, Korea and Taiwan had no production from open hearth furnaces, Brazil produced just 2.4 percent of its steel with open hearth

---

[78]Frank Ackerman and Paulo Eduardo Fernandes de Almeida, "Iron and Charcoal: The Industrial Fuelwood Crisis in Minas Gerais," *Energy Policy*, vol. 18, No. 7, September 1990, pp. 661-668; Anthony B. Anderson, "Smokestacks in the Rainforest: Industrial Development and Deforestation in the Amazon Basin," *World Development*, vol. 18, No. 9, 1990, pp. 1191-1205.

[79]R.B. Smith and M.J. Corbett, "Coal-Based Ironmaking," *Ironmaking and Steelmaking*, vol. 14, No. 2, 1987, pp. 49-75.

[80]"COREX Comes Onstream," *Iron Age*, March 1990, p. 35.

[81]C. Bodsworth, "The Iron and Steel Industry in Venezuela," *Ironmaking and Steelmaking*, vol. 16, No. 1, 1989, pp. 1-3.

[82]Organisation for Economic Cooperation and Development, *The Role of Technology in Iron and Steel Developments* (Paris, France: Organisation for Economic Cooperation and Development, 1989).

[83]H.K. Lloyd, "Steelmaking and Iron Ore Smelting in Indonesia," *Ironmaking and Steelmaking*, vol. 15, No. 2, 1988, pp. 53-55; T. Ariwibowo and Souren Ray, "Indonesia's Only Integrated Steel Plant—Krakatau Steel," *Iron and Steel Engineer*, January 1988, pp. 56-59.

[84]Amit Chatterjee, "The Steel Industry in India," *Ironmaking and Steelmaking*, vol. 17, No. 3, 1990, pp. 149-156.

[85]"The Emergence of an African Iron and Steel Industry," *Ironmaking and Steelmaking*, vol. 16, No. 6, 1989, pp. 357-361.

Table 4-20—Energy Conservation Opportunities and Costs in the Steel Industry[a]

Technology	Capital cost U.S. $millions	Energy saved MJ/ts	Type	Cost of saved energy $/GJ primary energy
Automatic ignition of coke oven flare	0.3-0.6	13-30	Fuel	$0.22
Top gas recovery[b] turbine	15-22	250-315	Electricity	$0.66[c]
Hot stove waste heat recovery	3-4.5	60-100	Fuel	$0.47
BOF gas recovery	125-150	750-850	Fuel	$1.70
Double insulation of skids	0.9-1.5	40-55	Fuel	$0.27
Coke dry quench[d]	120-200	420-630	Steam/electricity	$3.10
Coke oven automobile combustion control	4.5-7.5	33-50	Fuel	$1.45
Sinter heat recovery	37-67	150-290	Steam/electricity	$2.38
BF Bled gas recovery	3-4.5	29-42	Fuel	$1.06
BOF gas waste heat recovery	22-37	80-120	Steam	$3.00

[a]Converted from approximate 1980 to 1990$; converted from Mcal to MJ at 4.1868 J/cal; Marginal cost calculated based on midpoint of cost and energy savings ranges, assuming a 30-year plant life and 7-percent discount rate for a CRF of 0.080586, and an annual plant production of 8 million tons of carbon steel per year.
[b]Note the cost of top gas recovery turbine seems very low, even for capital cost alone.
[c]$0.66/GJ for electricity valued in terms of primary energy; or $0.007/kWh for electricity itself.
[d]Note that Coke dry quench improves quality of coke.
SOURCE: International Iron and Steel Institute, "Energy and the Steel Industry," Brussels, 1982.

furnaces, Mexico 12 percent, and Venezuela 18 percent. The share of steel produced in open hearth furnaces in India dropped to 68 percent by 1982/83 and to 47 percent in 1986/87, and is scheduled to be rapidly phased out after 1995. In China, open hearth furnaces accounted for only about one-quarter of production in 1988.[86]

Today's modern integrated plants—taking iron ore to finished steel—generally use basic oxygen furnaces (BOF). This process refines steel by blowing oxygen into the furnace to produce an intense chemical reaction.

Electric arc furnaces (EAF), which are used primarily to recycle scrap steel and to a lesser extent for DRI—have become increasingly important in recent years and are particularly attractive to developing countries because EAF-based minimills are economic at smaller scales than integrated plants using basic oxygen furnaces, and are less capital and (if the energy embodied in scrap is not included) less energy intensive per unit output. The drawbacks of EAF minimills include a limited set of products that can be produced—although this is changing with the development of thin slab casting;[87] the relatively high cost of electricity to power the arc furnaces (there is also a high upstream capital cost for generating plants); and the dependence on scrap (which may be in short supply in developing countries in the future) or on DRI.[88] Numerous minimills (including EAF and other operations) using steel scrap[89] have been established in developing countries. Korea and Taiwan each have about 50 minimills; Mexico has about 15;[90] and India and Brazil have over 150 each.[91]

The efficiency of EAFs can be improved by preheating scrap before it is fed into the furnace using the waste heat from the furnace,[92] by the use of ultra high power (UHP) electric arc furnaces, and

---

[86]Energy and Environmental Analysis, Inc., "Conserving Process Heat in Primary Industries of India and China," contractor report prepared for the Office of Technology Assessment, August, 1990; Amit Chatterjee, op. cit., footnote 84; Organisation for Economic Cooperation and Development, op. cit., footnote 82.

[87]Michael Schroeder and Walecia Konrad, "Nucor: Rolling Right Into Steel's Big Time," *Business Week*, No. 3188, Nov. 19, 1990, pp. 76-79.

[88]Maurice Y. Meunier and Oscar de Bruyn Kops, op. cit., footnote 19.

[89]R. Berlekamp, "Scrap—A Raw Material in Worldwide Demand for Steelmaking," *Ironmaking and Steelmaking*, vol. 17, No. 2, 1990, pp. 83-88. An interesting case study of scrap recovery can be found in: Mary Anne Weaver, "Great Ships Go To the Boneyard on a Lonely Beach in Pakistan," *Smithsonian Magazine*, vol. 21, No. 3, June 1990, pp. 30-40.

[90]Organisation for Economic Cooperation and Development, op. cit., footnote 82.

[91]Amit Chatterjee, op. cit., footnote 84; Francisco Lanna Leal and Benoni Torres, op. cit., footnote 77.

[92]Sayed A. Azimi and Howard E. Lowitt, U.S. Department of Energy, "The U.S. Steel Industry: An Energy Perspective," Report No. DOE/RL/01830-T55 (Springfield, VA: National Technical Information Service, January 1988).

by other means.[93] UHP furnaces shorten the cycle time and correspondingly improve furnace productivity and reduce energy use due to the shorter period at high temperature.[94] At the same time, however, UHP furnaces have higher power demand. This can put added strain on utilities that are already often short of capacity. Indeed, from 1976 to 1986, steel minimills in India were supplied with just 38 percent of the power they would need to be at full production.[95]

Energy use in many steel minimills in developing countries is much higher than that in the industrial nations. For example, the energy consumption in Indian minimills is typically 600 to 900 kilowatt-hours (kWh)/t versus as low as 400 kWh/t elsewhere. Many of the steel operations in developing countries, however, are unable to take advantage of proven energy conserving and productivity enhancing technologies due to their small scale and inadequate infrastructure. The average size of minimills in India is typically below 15 tonnes capacity, compared to 80 to 300 tonnes elsewhere. At such small scales, the capital costs for scrap preheaters and other technologies increase rapidly per unit output.

Casting

Casting or primary finishing includes casting and initial rolling of the steel into slabs for flat sheets, and blooms (rectangular) or billets (square) for structural shapes and bars. Ingot casting is done by pouring the liquid steel into molds to form ingots before reheating and initial rolling. Continuous casting pours the liquid steel directly into its semifinished shape, eliminating the intermediate steps of ingot casting and reheating.

Continuous casting has rapidly come to dominate the steel industry because it significantly lowers/eliminates capital investment for the stripper, reheating (soaking pit), and primary rolling mill; substantially reduces energy use (by 400 to 1,400 million joules (MJ)/tonne or 380 to 1,325 thousand Btus); gives a typically 5 to 10 percent or greater increase in yield—less metal must be scrapped as waste with corresponding energy savings—at often higher quality; and reduces emissions associated with ingot casting and heating.[96] As of 1986, several developing countries had higher shares of continuous casting than the United States (55 percent)—Korea 71 percent, Taiwan 88 percent, and Venezuela 71 percent; was comparable—Mexico 54 percent; or was slightly lower—Brazil 46 percent.[97] This capacity was largely developed in just a 10-year time period, from the mid-1970s to the mid-1980s. In contrast, all crude steel in integrated plants in India was ingot cast until 1982-83; only 10 percent of steel production was continuous cast as of 1986-87. Similarly, in China about 15 percent of steel was continuously cast by 1988.[98]

Other advances (at various stages of development) in casting offer even greater potential capital and energy savings, and productivity improvements. These include: thin slab casting;[99] thin strip casting;[100] net shape casting; and spray steel.[101]

Forming

Forming or secondary finishing transforms the steel into its final shape through a series of reheating, hot rolling, and cold rolling steps. Overall, hot rolling of strip, cold reduction, and finishing operations each consume about 125 kWh/t. Hot rolling thin cast strip to cold-rolled gages in an inert atmosphere could eliminate a series of scale re-

---

[93] L.L. Teoh, "Electric Arc Furnace Technology: Recent Developments and Future Trends," *Ironmaking and Steelmaking*, vol. 16, No. 5, 1989, pp. 303-313; Dick Hurd and John Kollar, "A Growing Technology in Electric Steelmaking," Electric Power Research Institute Center for Materials Production, Carnegie Mellon Research Institute newsletter, Pittsburgh, PA, January 1991.

[94] Sven Eketorpe, "Electrotechnologies and Steelmaking," Thomas B. Johansson, Birgit Bodlund, and Robert H. Williams (eds.), *Electricity: Efficient End-Use and New Generation Technologies and Their Planning Implications* (Lund, Sweden: Lund University Press, 1989), pp. 261-296.

[95] Amit Chatterjee, op. cit., footnote 84.

[96] Sayed A. Azimi and Howard E. Lowitt, op. cit., footnote 92.

[97] Organisation for Economic Cooperation and Development, op. cit., footnote 82.

[98] Energy and Environmental Analysis, Inc., op. cit., footnote 86.

[99] Michael Schroeder and Walecia Konrad, "Nucor: Rolling Right Into Steel's Big Time," *Business Week*, No. 3188, Nov. 19, 1990, pp. 76-79.

[100] Sayed A. Azimi and Howard E. Lowitt, op. cit., footnote 92.

[101] Gary McWilliams, "A Revolution in Steelmaking?" *Business Week*, No. 3179, Sept. 24, 1990, pp. 132-134.

moval, cleaning, and annealing steps and ultimately save 50 to 100 kWh/t.[102]

Heat Treating

After secondary finishing, the steel is again reheated and then allowed to cool slowly in order to relieve the stresses built up in the steel in the above rolling processes. This improves the strength and ductility of the final product. Heat treatment has also seen considerable efficiency gains in recent years. In the United States, energy consumption of heating and annealing furnaces dropped by nearly one-third between 1980 and 1989.[103] Continuous annealing and processing systems can further reduce energy use by about 460 to 700 MJ/tonne (440 to 660 thousand Btu) by combining five cold rolling batch processes into one.

By employing all currently cost-effective technologies, energy consumption in integrated steel mills can be reduced to about 18 GJ/tonne (17 million Btu), and to about 7.6 GJ/tonne (7.2 million Btu) in scrap fed electric arc furnaces (see table 4-19). The best steel mills in the world approach these levels of efficiency, including Japanese, U.S., South Korean, and others. Countries such as Brazil, Venezuela, and other industrializing countries are not far behind; while India and China have a considerable ways to yet go to reach these levels of efficiency.[104]

In the future, the steel industry could change dramatically from the above, particularly with the widespread introduction of processes for the direct production of iron from coal, greater use of electric arc furnaces, and extensive use of thin slab casting or other such techniques.

Cement

The cement industry typically consumes 2 to 6 percent, and sometimes more, of the commercial energy used in developing countries.[105] The use of cement is expected to increase rapidly as national infrastructures of roads, bridges, and buildings are

Table 4-21—Average Energy Intensities of Building Materials

Material	Energy intensity (MJ/kg)
Concrete aggregate	0.18
Concrete	0.8
Brick and tile	3.7
Cement	5.9
Plate glass	25.0
Steel	28.0

SOURCE: Mogens H. Fog and Kishore L. Nadkarni, "Energy Efficiency and Fuel Substitution in the Cement Industry with Emphasis on Developing Countries," World Bank Technical Paper No. 17, Washington, DC, 1983.

developed. Considerable effort is now being devoted to developing higher performance cements to ensure that this huge investment in infrastructure lasts a long time.[106] Despite the energy intensity of cement production, it is one of the least energy intensive construction materials when in its final form of concrete/aggregate (see tables 4-21, 4-22).

The value of cement is quite low compared to its weight. Because of this and because the raw materials for cement—limestone, various clay minerals, and silica sand—are widely available, cement is usually produced relatively near its point of use. In the United States, the maximum range for truck shipments of cement is about 300 km. In developing countries, where the transport infrastructure is less well developed, economical transport distances are often less. In China, for example, 150 to 200 km is the typical limit of transport; if transport over longer distance is needed, the construction of a new cement plant in the local area will be considered.[107] Thus, as a result of inadequate transport infrastructures, cement plants are often small and relatively inefficient.

The energy required to produce cement varies widely with the type of production process (see figure 4-12), quality of raw materials, plant management, operating conditions, and other factors. The performance of cement plants in developing coun-

---

[102] W.L. Roberts, Electric Power Research Institute, *Power Utilization in Flat Processing of Steel*, Report No. EM-5996 (Palo Alto, CA: Electric Power Research Institute, January 1989).

[103] Energetics, Inc., op. cit., footnote 75.

[104] Energy and Environmental Analysis, Inc., op. cit., footnote 86.

[105] Mogens H. Fog and Kishore L. Nadkarni, "Energy Efficiency and Fuel Substitution in the Cement Industry With Emphasis on Developing Countries," World Bank Technical Paper No. 17, 1983.

[106] For example, high performance cements are being developed at the Center for the Science and Technology of Advanced Cement-Based Materials, Northwestern University, Evanston, IL.

[107] Li Taoping, "Cement Industry in China," *Rock Products*, February 1985, p. 32.

Table 4-22—Energy Intensities of End Products Using Alternative Building Materials, MJ/m²

Structure	Concrete	Steel	Asphalt	Brick
Building wall	400	NA	NA	600
Bridge	4,000	8,000	NA	NA
Roadway	800	NA	3,000	NA

NA = Not available or not applicable.
SOURCE: Mogens H. Fog and Kishore L. Nadkarni, "Energy Efficiency and Fuel Substitution in the Cement Industry with Emphasis on Developing Countries," World Bank Technical Paper No. 17, Washington, DC, 1983.

tries also varies widely. Many plants approach the efficiency of the best in the industrialized countries, depending on when they were built and the conditions under which they are operated. Others show significant inefficiencies—using 25- to 50-percent more energy than efficient plants of the same type and with the same quality of raw materials input.[108]

The manufacture of cement involves three basic processes: the mining and preparation of the raw materials; clinker production; and finish grinding. Portland cement typically accounts for most of the cement in use. It consists of limestone (calcium carbonate), silica sand, alumina, iron ore, and small quantities of other materials. These materials are quarried, crushed, and mixed together, and are then processed at high temperatures—1,500 °C—in a large rotary kiln to produce marble sized pellets known as clinker. These pellets are then ground into a fine powder and mixed with gypsum for use as cement.

The two primary manufacturing processes are known as the wet and the dry processes. In the wet process, water is added to the raw materials so that the material fed into the kiln is a slurry. The wet process has been largely abandoned in recent years in favor of the dry process. In the dry process, the material fed into the kiln is a dry powder. Refinements in the dry process in recent years have focussed on making better use of the waste heat from the kiln to dry and preheat the material fed into the kiln—the preheater and precalciner processes. These have led to significant improvements in the overall energy efficiency of cement production. The relative fuel use of the wet and dry processes, however, differ considerably, with the dry process using somewhat more electrical energy and substantially less thermal energy.

Electricity use in cement production has increased in recent years due to the shift to the dry process, the use of more environmental controls, and more extensive use of preheaters, which require large fan systems working at high pressure drops. The move towards higher strength cements in recent years has also required additional electricity use in order to more finely grind the clinker.[109] On the other hand, less cement may be needed if higher strength types are used, reducing total energy consumption.

Regardless of the process, typically 80 to 85 percent of the total energy used in cement production is for high temperature processing. The rest is largely for grinding or related steps. A variety of different fuels are used in cement production, including, oil, coal, gas—together accounting for about 70 percent of the total energy used, and electricity—accounting for about 30 percent of the total energy used (on a primary energy basis).[110] In addition, Brazil has made extensive use of charcoal, and biomass gasifiers have been used in Finland (see ch. 6).[111]

The energy used in producing cement varies widely by country, depending primarily on what fraction of total cement production is done by the more energy efficient dry process. In 1980, the U.S. cement industry used about 5.86 GJ (5.6 million Btus) per tonne of cement, compared to 3.45 GJ (3.3 million Btus) per tonne for Japan and West Germany. The U.S. cement industry was then half wet process compared with 5 percent wet process in West Germany. Many developing countries are

---

[108] Mogens H. Fog and Kishore L. Nadkarni, op. cit., footnote 105. See figure 5-1, p. 39.

[109] Stewart W. Tresouthick and Alex Mishulovich, "Energy and Environment Considerations for the Cement Industry," paper presented at the Energy and the Environment in the 21st Century conference, Mar. 26-28, 1990, Massachusetts Institute of Technology, Cambridge, MA.

[110] Stewart W. Tresouthick and Alex Mishulovich, Ibid.

[111] Hannu Lyytinen, "Biomass Gasification As a Fuel Supply for Lime Kilns: Description of Recent Installations," *Tappi Journal*, vol. 70, No. 7, July 1987, pp. 77-80.

making rapid strides to improve the efficiency of their cement plants. Turkey, Tunisia, and the Philippines use the wet process for less than one-quarter of their cement production, with corresponding energy savings.[112] In Brazil, use of the dry process increased from 20 to 80 percent of production between 1970 and 1987.[113]

Similarly, most recent capacity additions in India have been the dry process. This resulted in the wet process share of the total dropping from about 70 percent in 1970 to 37 percent in 1984. The average kiln size in India has also increased; and large kilns tend to be more efficient than small ones. In the 1970s, the average wet process kiln size was around 400 tonnes per day and new kiln capacity for the dry process averaged about 750 tonnes per day. In the 1980s, new dry process capacity has averaged 1,800 tonnes per day, comparable to industrial countries.[114] Most Indian cement capacity is in large plants, totaling some 55.4 million tonnes per year capacity in 1988/89 spread over 96 plants operated by 49 companies. About 84 percent of this capacity is privately held. During the 1980s, mini-cement plants have also been established to make use of small and scattered limestone deposits and to meet local demands.

Energy costs are a large part of the total production cost for cement; consequently there is a significant push to save energy. In India during 1983/84, energy costs were about 44 percent of total production costs for wet process plants and 39 percent of the total for dry process plants. In response, 10 large scale plants have been or are being converted from the wet process to the dry process in recent years, and precalcinators have been installed on 35 kilns in 23 plants. A variety of other improvements are also being made, including the extensive installation of instrumentation to monitor processes and automatic controls to improve control and product quality; improved refractory materials

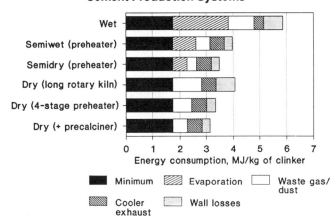

Figure 4-12—Energy Consumption for Alternative Cement Production Systems

SOURCE: Mogens H. Fog and Kishore L. Nadkarni, "Energy Efficiency and Fuel Substitution in the Cement Industry with Emphasis on Developing Countries," World Bank Technical Paper No. 17, 1983.

on the kilns; waste heat recovery systems; improved grinding equipment; and cogeneration equipment.[115]

Factors that contribute to the low efficiency of Indian cement production include low quality coal and limited, poor quality (variable voltage and frequency) electricity supply from the State utility grid. To compensate, the cement industry had installed 545 megawatts (MW) of captive generation capacity by 1988/89 with another 100 MW under construction.[116]

Cement production in China reached 145 million tonnes in 1985. Of the 366 rotary kilns in use, 235 used the dry process. Of these, 33 had preheaters and 11 had precalcinators. Cogeneration systems were used on about 50 of these kilns. There were also more than 4,000 vertical shaft kilns in use in 1985, providing about two-thirds of the total output of cement. Vertical shaft kilns are used primarily by smaller cement plants and can have lower efficiencies.[117] Average capacity of Chinese vertical shaft kilns is about 30,000 tonnes/year.[118]

---

[112]Mogens H. Fog and Kishore L. Nadkarni, op. cit., footnote 105, see figure 5-1, p. 39.

[113]Howard S. Geller and David Zylbersztajn, "Energy Intensity Trends in Brazil," *Annual Review of Energy*, vol. 16, 1991, pp. 179-203.

[114]Energy and Environmental Analysis, Inc., op. cit., footnote 86.

[115]Energy and Environmental Analysis, Inc., op. cit., footnote 86. Reported overall energy efficiencies of the Indian cement industry, however, have not improved much over the time period 1978-79 to 1983-84. Ministry of Industry, Bureau of Industrial Costs and Prices, Government of India, "Report on Energy Audit of Cement Industry," *Studies on the Structure of the Industrial Economy*, vol. 14 (New Delhi, March 1987).

[116]Energy and Environmental Analysis, Inc., op. cit., footnote 86.

[117]Energy and Environmental Analysis, Inc., op. cit., footnote 86. The data listed are, however, somewhat contradicted by that of the earlier report: Li Taoping, op. cit., footnote 107. The data in the report by Taoping are also inconsistent internally, so have not been included here.

[118]D.F. Stewart and B. Muhegi, "Strategies for Meeting Tanzania's Future Cement Needs," *Natural Resources Forum*, November 1989, pp. 294-302.

Although they have been largely abandoned in the industrial countries, vertical shaft kilns are used widely in China and might be practical in other countries. These plants offer lower installed capital costs and thus can serve smaller, dispersed markets.[119] On the other hand, the energy efficiency of vertical shaft kilns is typically low but potentially could be improved; and the product quality is often uneven due to poor distribution of raw materials within the kiln and interactions with the walls of the kiln. These problems might be redressed with further development and would allow more widespread use of this system.[120]

The Tanzanian cement industry was producing at just 22 percent of rated capacity by 1984, compared to 100 percent or better just 10 years earlier. Factors contributing to this decline in performance included:

- the lack of foreign exchange resulting in the frequent nonavailability of critical spare parts and other equipment when needed, resulting in delays and even complete shutdowns;
- inadequate industrial infrastructure resulting in long delays in making repairs at the few local workshops capable of doing the work;
- transport shortages and bottlenecks in providing raw materials; and
- shortages of skilled manpower in designing, constructing, and maintaining the plant and equipment.[121]

Substantial efficiency improvements above that of today's dry process are still possible. The grinding process, for example, is very inefficient, variously estimated at 2- to 5-percent efficient in breaking clinker apart into fine powder, the remainder goes into heat and vibration. A variety of improvements, including better separation and grinding equipment, and improved controls have reportedly raised grinding efficiencies by as much as 40 percent in some cases (see table 4-23).[122]

Similarly, there are numerous potential improvements in the high temperature processing of cement

Table 4-23—Technology Improvements in Grinding

Technology	Average potential energy savings (percent)
High efficiency classifiers	15
Roller mills	20
Controlled particle size distribution	27
Advanced mill internals	5
Separate grinding of components	5-10
Computer control	15
High performance sensors, including on kiln exhaust	NA
Nonmechanical comminution	NA
Total	40

NA = not available or not applicable.
SOURCE: Stewart W. Tresouthick and Alex Mishulovich, "Energy and Environment Considerations for the Cement Industry," Energy and the Environment in the 21st Century, March 26-28, 1990, Massachusetts Institute of Technology, Cambridge, MA.

(see table 4-24). These efficiency improvements can reduce total system capital costs as well as operating costs. Other energy conservation opportunities include waste heat recovery, cogeneration, and the use of ASDs on kiln fan motors and other motor driven equipment.[123]

Many cement plants operate with minimal environmental controls and poor quality raw material inputs can further impede cleanup. For example, the use of coals with high ash contents leads to poor combustion in some cases and consequently generates a high carbon monoxide content in the exhaust gases coming off the kiln. These gases will trip safety mechanisms in electrostatic precipitators to prevent an explosion, resulting in the venting of large amounts of dust. Properly functioning controls can, however, provide substantial benefits. For example, a typical rotary kiln without any dust controls can lose 110 to 125 kg of dust per tonne of clinker processed, or equivalently, an efficiency loss of 11 to 12 percent of the kilns output. Capturing this dust translates directly into a productivity increase

---

[119]Sanjay Sinha, "Small Versus Large in the Indian Cement Industry—David and Goliath Hand-In-Hand," Marilyn Carr (ed.), *Sustainable Industrial Development* (London: Intermediate Technology Publications Ltd., 1988), pp. 128-150.

[120]D.F. Stewart, "Options for Cement Production in Papua New Guinea: A Study in Choice of Technology," *World Development*, vol. 13, No. 4, 1985, pp. 639-651.

[121]D.F. Stewart and B. Muhegi, op. cit., footnote 118.

[122]Stewart W. Tresouthick and Alex Mishulovich, op. cit., footnote 109.

[123]Tore R. Nilsson, Bengt Sinner, and Ola V. Volden, "Optimized Production and Energy Conservation," *IEEE Transactions on Industry Application*, vol. IA-22, No. 3, 1986, pp. 442-445.

Table 4-24—Technology Improvements
in Pyroprocessing

Technology	Average potential energy savings (percent)
Computer control	3-10
Fluidized-bed reactor	10-30
Relaxed alkali specification	2-4
Low pressure-drop preheaters	5
Advanced sensors	2-5
Advanced preheater/precalciner	5-10
New mineralogical content of clinker	5-30
Total	NA

NA = Not available or not applicable.

SOURCE: Stewart W. Tresouthick and Alex Mishulovich, "Energy and Environment Considerations for the Cement Industry," Energy and the Environment in the 21st Century, Mar. 26-28, 1990, Massachusetts Institute of Technology, Cambridge, MA.

for the plant.[124] In modest quantities cement dust does not, however, cause particular damage to soils around the plant.

Pulp and Paper

Between 1961 and 1986, world paper production increased from about 78 million to 203 million tons annually. Developing countries produced about 16 percent of 1986 world paper output, Eastern Europe and the Soviet Union produced about 8 percent, and the remainder was produced in the OECD countries. Annual production in the developing countries, however, increased at an annual rate of about 7.3 percent between 1961 and 1980; in contrast, annual production in the industrial OECD countries increased at just 2.8 percent between 1971 and 1980.[125] At these rates, paper production by developing countries will overtake that of the industrialized countries by roughly 2020.

In the OECD countries in 1981, paper was the fourth largest industrial energy user, ranking behind iron and steel, chemicals, and petrochemicals, but was just ahead of cement.[126] Pulp and paper production is relatively less important in developing countries today due to the low rate of usage, but it is likely to become increasingly important in the future. In addition, the pulp and paper industry has the potential to become a significant generator of electricity for developing countries through the use of cogeneration systems supplied by captive waste products, especially bark and black liquor.

Papermaking typically consists of five process steps: wood preparation; pulping; bleaching; chemical recovery; and papermaking. The products produced include high quality writing and magazine paper, newsprint, tissue paper, cardboard, and various types of building construction papers.

Wood preparation consists of removing the tree bark and chipping the wood into small pieces. Pulping breaks apart the fibrous components of the wood into a form useful for making paper. This can be done through a variety of means. Most often chemical processes are used, but mechanical or other means are also employed perhaps 20 percent of the time. Each pulping process has certain advantages and disadvantages in terms of the quality, strength, cost, yield, and other attributes of the resulting paper produced. The residue from the chemical pulping process, i.e., the components of the wood that are not converted to pulp, is known as black liquor and is a large source of energy. Bleaching is often used to whiten the pulp before making paper (but is not necessary for paperboard and some other products). The papermaking process deposits the bleached pulp fibers in a sheet on a screen, drains and presses the water from it, and dries the sheet.[127] Table 4-25 lists energy consumption in pulp and paper mills in a variety of countries.

Table 4-26 compares energy use by paper mills using today's current mix of technologies for the United States, state-of-the-art paper mills, and advanced technologies. Substantial reductions in energy use are possible. Many of the technical improvements also lead to improved product quality. For example, improved presses (e.g., the extended nip press) used to squeeze the water out of the

---

[124]Sanjay Sinha, op. cit., footnote 119.

[125]Andrew J. Ewing, "Energy Efficiency in the Pulp and Paper Industry With Emphasis on Developing Countries," World Bank Technical Paper No. 34, 1985; and "1986-87 The World's Pulp, Paper, and Board Industry: Production and Trade," *Pulp and Paper International*, July 1988, pp. 50-51.

[126]In the OECD countries in 1981, the industrial energy use rankings were: steel—174 million tons oil equivalent (mtoe); chemicals—123 mtoe; petrochemicals—122 mtoe; pulp and paper—58 mtoe of commercial energy purchased, but probably twice this amount was consumed by this sector, the remainder coming from the use of waste products generated by the mills—such as bark and other materials; cement—49 mtoe. Source: Andrew J. Ewing, Ibid.

[127]For a more detailed description, see: U.S. Congress, Office of Technology Assessment, *Industrial Energy Use*, OTA-E-198 (Springfield, VA: National Technical Information Service, June 1983).

Table 4-25—Energy Consumption in Selected Pulp and Paper Mills

Country	Coverage		Average energy consumed	
	No. of mills	Percent of production	Purchased GJ/tp	Total GJ/tp
Colombia	4	80	34.6	50.3
Turkey	7	89	45.7	NA
India	4	13	79.4	112.0
Pakistan	5	90	56.3	59.7
Indonesia	15	38	33.2	50.3
Thailand	1	25	20.3	25.4
Average	NA	NA	38.3	54.0
OECD average	NA	NA	19.8	30.0

NOTE: these values of GJ/tonne of paper are below those listed in table 4-26 due to differences in definition, accounting conventions, assumed product mix, and other factors. The important feature to note is the relative performance—that the OECD average energy consumption is just over half that of the developing countries listed, and in table 4-26, that using advanced technologies can nearly halve energy consumption again, compared to the U.S. average.

SOURCE: Andrew J. Ewing, "Energy Efficiency in the Pulp and Paper Industry With Emphasis on Developing Countries," World Bank Technical Paper No. 34, 1985.

Table 4-26—Energy Use by Alternative Paper Production Technologies, United States

Process	Specific energy use, GJ/ton[a]		
	1988 average	State-of-the-art	Advanced technology
Wood preparation	0.48	0.40	0.40
Pulping	10.53	7.29	6.90
Bleaching	3.27	2.44	2.20
Chemical recovery	7.93	5.04	3.75
Papermaking	15.18	10.30	8.40
Auxiliary	3.41	3.12	3.12
Total	40.80	28.60	24.80

[a]These are averages for the net plant output. Energy requirements of individual processes can vary significantly from these averages depending on the type of paper product, reuse of waste streams within the plant, etc.

SOURCE: A. Elaahi and H.E. Lowitt, "The U.S. Pulp and Paper Industry: An Energy Perspective," April 1988, U.S. Department of Energy, Office of Industrial Programs, Report No. DOE/RL/01830-T57; Energetics, Inc. "Industry Profiles: Paper," U.S. Department of Energy, Office of Industrial Technologies, December 1990.

paper before drying, both reduce energy use by reducing the amount of water that must be evaporated and also improve fiber-to-fiber bonding, resulting in a sheet with higher strength. Alternatively, the same strength could be maintained while using lower grade—and less energy intensive—pulp. A large amount of electrical energy can be saved through the use of improved motor drive systems. Similarly, much of the process heat can be recovered through the use of various types of heat exchangers and recuperators, and through vapor recompression systems. The use of enzymes derived from wood rot funguses rather than traditional chemical or mechanical pulping might save a large amount of energy as well. Energy savings in the pulping process of 28 percent by using such enzymes compared to mechanical pulping have been demonstrated in Sweden.[128]

Of equal importance to conservation efforts within the plant is that the pulp and paper industry can greatly increase the amount of energy produced from its waste products. The typical pulp and paper operation has three principal waste streams that can provide energy: hog fuel, black liquor, and forest residues. Hog fuel is the bark, sawdust, and other scrap produced in reducing logs to feedstock for the pulping process. Hog fuels could supply about 3 GJ

---

[128] A. Elaahi and H.E. Lowitt, U.S. Department of Energy, Office of Industrial Programs, "The U.S. Pulp and Paper Industry: An Energy Perspective," Report No. DOE/RL/01830-T57, April 1988.

[129] As measured at one pulp mill in Sweden.

(2.8 million Btu) per tonne of pulp produced (GJ/tp).[129] Black liquor, from the chemical pulping process, averages an energy content of about 13 GJ (12.3 million Btu) per tonne of pulp. Other residues are currently left in the forest when harvesting the trees. A portion of these forest residues might be collected, but the long term impact this would have on forest soils would need to be examined closely. If fully recovered, the estimated energy content of forest residues would be about 25 GJ/tp (23.7 million Btu/tp). Combined, these energy resources total some 41 GJ/tp (38.9 million Btu/tp).[130]

Most kraft pulp mills (which use the chemical process resulting in the production of black liquor) currently use black liquor for cogenerating steam and electricity onsite. High efficiency steam injected gas turbine or combined cycle technology (see ch. 6) might be able to generate as much as 4,000 kWh of electricity per ton of pulp produced if all of the hog fuel, black liquor, and recoverable forest residues were used. Onsite needs today are typically about 740 kWh/tp of electricity plus some 4,300 kg/tp of steam, with the potential for significant reductions as discussed above. This would leave a substantial amount of power that could be sold to the grid. At $0.09/kWh, the saleable electricity (in excess of plant requirements) would have a value of some $230, equal to half the value of the pulp sold.[131]

Current and future research opportunities include the utilization of black liquor (physical, chemical and combustion properties), biological pulping, process monitoring and control, impulse drying of paper, advanced waste heat recovery, and others.[132]

Chemicals

The chemicals industry is extremely complex, involving the production of thousands of products in numerous competing processes. Moreover, the feedstocks and energy inputs into the chemical industry are often the same. Because of this complexity, the following discussion will be limited solely to the production of nitrogen based fertilizers, and will also serve as a transition to the discussion on the use of energy in the agricultural sector. More detailed reviews of energy use in the chemicals industry can be found elsewhere.[133] Finally, the use of electricity in the chemicals industry is dominated by pump, fan, compressor, and related mechanical drive needs. The above discussion of motor drive is directly applicable to these uses.

As of 1980, the energy used to produce fertilizers was about equal to all the commercial energy uses in the agricultural sector of the developing countries (not including China). Nitrogen fertilizers account for almost three-fourths of the energy consumption in the fertilizer sector, and nearly one-fifth is for packaging, transportation, and field application. The remainder is for phosphates (5 percent) and potash (3 percent). The production of nitrogen fertilizers is particularly energy intensive, and so will be the focus here.[134]

Nitrogen fertilizers are made primarily from ammonia. Energy use to produce ammonia has dropped from about 80 GJ/tonne (76 million Btu/t) of ammonia in the early 1940s to about 40 GJ/t (38 million Btu/t) for the average plant in the mid-1970s, to as low as 33 GJ/t (31 million Btu/t) today. The theoretical limit is about 21 GJ/t (20 million Btu/t); the practical limit has been estimated at about 28.5 GJ/t (27 million Btu/t). The fuel component of this total energy consumption is usually provided by natural gas, which serves both as a feedstock and as a fuel. Currently, about 80 percent of world ammonia production is done by chemically reforming natural gas. The (externally purchased) electric power component of this total energy consumption dropped from about 800 kWh/tonne in the early 1960s to about 20 kWh per tonne in the early 1980s. This large drop in electricity usage was made possible by the transition from reciprocating com-

---

[130] Eric D. Larson, "Prospects for Biomass-Gasifier Gas Turbine Cogeneration in the Forest Products Industry: A Scoping Study," Princeton University, Center for Energy and Environmental Studies Working Paper No. 113.

[131] Eric D. Larson, "Biomass-Gasifier/Gas-Turbine Applications in the Pulp and Paper Industry: An Initial Strategy for Reducing Electric Utility CO_2 Emissions," paper presented at the Conference on Biomass For Utility Applications, Electric Power Research Institute, Tampa, FL, Oct. 23-25, 1990; Eric D. Larson, "Prospects for Biomass-Gasifier Gas Turbine Cogeneration in the Forest Products Industry: A Scoping Study," Princeton University, Center for Energy and Environmental Studies Working Paper No. 113.

[132] A. Elaahi and H.E. Lowitt, op. cit., footnote 128.

[133] U.S. Congress, Office of Technology Assessment, op. cit., footnote 127; Energetics, Inc. for the U.S. Department of Energy, Office of Industrial Technologies, "Industry Profiles: Chemicals," Contract No. DE-AC01-87CE40762, December 1990.

[134] Roger Heath, John Mulckhuyse, and Subrahmanyan Venkataraman, "The Potential for Energy Efficiency in the Fertilizer Industry," World Bank Technical Paper No. 35, 1985.

Table 4-27—Efficiency Improvements in Nitrogen Fertilizer Production

	Energy savings	Capital cost ($millions)
Purge gas recovery	1.8 GJ/mt	$1.4
Synthesis convertor	1.12 GJ/mt	NA
Molecular sieves	0.6 GJ/mt	$1.8
Power recovery turbines	10 kWh/mt	$0.35
Combustion air preheaters	1 GJ/mt	$3.5
Surface condenser cleaners	NA	NA
Carbon dioxide removal	NA	NA
Use of flare gas	NA	$0.15
Feedstock saturation	0.6 GJ/mt	NA

NA = not available or not applicable; mt = metric tonne.
SOURCE: Roger Heath, John Mulckhuyse, and Subrahmanyan Venkataraman, "The Potential for Energy Efficiency in the Fertilizer Industry," World Bank Technical Paper No. 35, 1985.

pressors to centrifugal compressors and the extensive use of cogeneration from the high pressure steam.[135] Many more efficiency improvements are still possible as shown in table 4-27.

In many cases, pollution control measures in fertilizer plants have simultaneously allowed increases in production and reductions in energy use. Purge gas recovery and recycling, for example, captures hydrogen lost from the synthesis process, reducing fuel use and the need for compression. Typical savings are 1.8 GJ/t (1.7 million Btu/t) of ammonia at an installed cost of $1.4 million. These units are now standard in new plants. For a typical 1,000 tonne per day plant, some 600,000 GJ (570 billion Btu) would be saved annually. Over a 20-year plant lifetime, savings would total 12.6 million GJ (12 trillion Btu), at a discounted cost of about $0.25/GJ ($0.26/million Btu) in US$1990. New design methods, such as pinch technology for heat exchanger networks in chemical plants, also offer both energy and capital savings in new plant designs. Pinch technology, for example, realizes average energy savings in new plant designs of 40 percent.[136]

IMPROVING THE EFFECTIVENESS OF MATERIAL USAGE

The energy required to deliver industrial goods and services can often be lessened by using existing material more effectively or by changing the types of materials used.

Smaller quantities of higher performance materials can often be substituted for larger quantities of lower performance materials. The tensile strength of steel was increased by four times between 1910 and 1980, for example, allowing large reductions in the quantity of steel required in any particular application.[137] Plastics are being substituted for metal in many auto body parts—reducing weight and improving fuel efficiency, reducing industrial energy use, and improving corrosion resistance and durability.[138]

Materials can be more extensively recycled. Significant amounts of energy can be saved by recycling steel, aluminum, glass, paper, and other materials (see table 4-28). Even greater savings may be possible if, rather than melting down and recasting the material, the material can be used in exactly the same form as before—for example, if glass bottles are of a standard size and shape and can be simply washed out and reused. On the other hand, the increasing mixture of different materials, such as the use of plastics in automobile bodies, complicates recycling efforts if recyclability is not designed in.[139]

Extensive recycling is already done in developing countries through both informal and formal markets. In West Africa, for example, artisans routinely melt

---

[135] Ibid.

[136] Canan Ozgen et. al., "Designing Heat-Exchanger Networks for Energy Savings in Chemical Plants," *Energy*, vol. 14, No. 12, 1989, pp. 853-861.

[137] Economic Commission for Europe, *Evolution of the Specific Consumption of Steel* (New York, NY: United Nations, 1984).

[138] A detailed review of advanced materials is given in: U.S. Congress, Office of Technology Assessment, *Advanced Materials by Design*, OTA-E-351 (Washington, DC: U.S. Government Printing Office, June 1988).

[139] These issues are being examined in a forthcoming report from the U.S. Congress, Office of Technology Assessment, *Materials Technology: Integrating Environmental Goals With Product Design*.

Table 4-28—Energy Intensity of Primary and Recycled Materials

	Primary GJ/mt	Recycled GJ/mt	Savings (percent)
Aluminum	242-277	9.9-18.7	92-96
Glass	17.8	12.3	31
Paper			
Newsprint	51.6	40.4	22
Printing paper	78.8	50.5	36
Tissue paper	79.7	34.3	57
Liner board	16.8	42.2	(−151)
Plastic	NA	NA	92-98
Solvents	27.9	4.7	83
Steel[a]	18.1	7.6	58

NA = not available or not applicable; mt = metric tonne.
[a]See table 4-19.
SOURCE: U.S. Congress, Office of Technology Assessment, "Facing America's Trash: What Next For Municipal Solid Waste?" OTA-O-424 (Washington, DC: U.S. Government Printing Office), October 1989; Energetics, Inc., "Industry Profiles: Waste Utilization," U.S. Department of Energy, Office of Industrial Technologies, DE-AC01-87CE40762, December 1990.

down scrap aluminum and cast it into pots. In turn, aluminum pots perform much better than traditional fired clay pots for cooking. They provide significant energy savings due to their higher thermal conductivity and they rarely break—costing someone their supper. Improved (higher efficiency) biomass stoves (see ch. 3) are commonly made of recycled metal, culled from wrecked cars or oil drums. As the economies of developing countries grow, these informal recycling efforts may need additional incentives and/or capitalization to continue at a high level.

The quantity of energy intensive materials consumed can also be reduced by the use of energy efficient technologies. If energy is used efficiently, fewer power plants are needed. A typical coal-fired power plant might require 2,000 GJ (1,900 million Btu) of energy intensive construction materials per MW of capacity[140]—the equivalent of its entire power output for a month.[141] Similarly, if buses, light rail, or other systems substitute for automobiles, then both the cement needed to build the road and the steel needed to build the cars can be reduced per passenger-kilometer or freight-tonne-kilometer. Lightweight fuel efficient automobiles use less materials to construct and require fewer steel- and concrete-intensive refineries.

Quality control[142] can play an important role in saving energy by reducing the amount of scrap and reworking that is necessary. For example, in the early 1980s Ford Motor Co. rejected as much as 8 to 10 percent of the flat-rolled steel it obtained from U.S. producers such as Bethlehem Steel. This has since been lowered to 1 percent.[143] In India, modernization of the Rourkela steel plant is expected to raise the yield of liquid steel to steel slabs from 79 percent currently to 92.5 percent.[144] This alone will reduce energy consumption per unit of steel output by 17 percent.

Quality control is also important in assembling consumer goods. Factories that use outmoded methods of mass production (as opposed to lean production)—in which products are primarily inspected for defects after they are built—typically expend a quarter of

---

[140]Jyoti K. Parikh, "Capital Goods for Energy Development: Power Equipment for Developing Countries," *Annual Review of Energy*, vol. 11, 1986, pp. 417-450.

[141]Note, for example, that the additional material in a refrigerator will not primarily be energy intensive steel—except perhaps for a larger heat exchanger—but low density insulation with a correspondingly low energy-intensity. Thus the savings in materials by building fewer power plants are generally not offset by increases in material demands by the end-user.

[142]Genichi Taguchi and Don Clausing, "Robust Quality," *Harvard Business Review*, vol. 68, No. 1, January-February 1990, pp. 65-75; Raghu N. Kackar, "Taguchi's Quality Philosophy: Analysis and Commentary," *Quality Progress*, December 1986, pp. 21-28; Thomas B. Barker, "Quality Engineering by Design: Taguchi's Philosophy," *Quality Progress*, December 1986, pp. 32-42; Daniel E. Whitney, "Manufacturing By Design," *Harvard Business Review*, vol. 66, No. 4, July-August 1988, pp. 83-91.

[143]Energetics, Inc., op. cit., footnote 75; "Quality," *Business Week*, No. 3002, June 8, 1987, p. 131

[144]Amit Chatterjee, op. cit., footnote 84.

their effort on finding and fixing mistakes in the assembled products. Thus, a quarter of the time and labor, and a similar share of the energy expended is not used to produce goods, but rather to rework goods that were not built correctly the first time.[145] Statistical process control and just-in-time (JIT) inventory control can dramatically reduce these losses.

JIT inventory control contributes to reducing inventories and the amount of material handling required in a plant. In a conventional mass production factory, large inventories are kept on hand to be used as needed by the assembly line. This requires large amounts of handling and storage space in order to store these components when they first arrive at the assembly plant and to then retrieve them when needed by the assembly line. JIT eliminates this extra handling as well as reduces the need for expensive (and material intensive) storage areas. European automobile assembly plants, for example, keep 10 times as large an inventory of spare parts, more than 3 times the area for repairing defects in assembled cars, and one-third more total space per output than Japanese auto assembly plants that have fully adopted just-in-time assembly and other elements of "lean" production.[146]

Similar savings from lean production have been achieved in many other industries, from air conditioners to microwave ovens. The resulting savings are substantial in terms of reduced inventory costs, plant capital costs, improved labor productivity, improved product quality, and as a bonus, reduced energy consumption. Such savings are particularly important where there is a shortage of working capital and interest rates are high.[147] Finally, quality in the form of improved product lifetimes can reduce the frequency with which the product must be replaced.

These improvements in manufacturing—statistical process control, just-in-time inventory control, etc.—are necessary if a country is to be competitive in world markets. Such techniques can also provide substantial energy savings.

BARRIERS TO ENERGY EFFICIENCY IMPROVEMENTS IN INDUSTRY

A number of factors limit the efficiency, productivity, and performance of industrial operations in developing countries. These are summarized in table 4-6 and a few selected cases are discussed below. Plant managers and others are often making substantial efforts to improve energy efficiency, but are laboring under highly adverse conditions. As examples, manufacturing plants are often too small to be optimally efficient; available technologies are often low quality or obsolete; national infrastructures are often inadequate; the plant may lack foreign exchange to purchase critical components not available locally; and there is generally a lack of skilled technicians, engineers, and managers.

Small Scale

Iron and steel plants, cement plants, paper mills, and other industrial operations in developing countries are often much smaller than those in the industrialized countries due to:

- smaller markets;
- poor transportation infrastructures that limit the cost effective distance for transport;
- reduced capital costs and associated risks;
- the benefits (in labor-rich developing countries) of often greater employment per unit output.

For example, the average U.S. paper mill has an annual capacity of approximately 100,000 tons; the average Latin American mill has a capacity of 18,000 tons; in Africa 9,000 tons, and in Asia (except Japan) 5,000 tons.[148] These small plants require significantly more energy per unit output because of scale effects. In addition, the small size of these industries increases the cost of installing more energy efficient equipment per unit energy saved. These small mills must also often meet the full range of products demanded in a developing country, while using relatively less production

---

[145]Otis Port, "Quality," *Business Week*, No. 3002, June 8, 1987, p. 131.

[146]James P. Womack, Daniel T. Jones, and Daniel Roos, *The Machine That Changed The World* (New York, NY: MacMillan Publishing, 1990), p. 92.

[147]Note, however, that for JIT to work, reliable transportation and communications infrastructures are required.

[148]Andrew J. Ewing, op. cit., footnote 125, p. 45.

equipment with limited flexibility and with lower production runs per type of product.

National Infrastructures

Inadequate national infrastructures also reduce efficiency and productivity. Poor transport infrastructure, for example, reduces cost-effective transport distances and market sizes. Frequent electric power brownouts or cutoffs can seriously disrupt operations. The Indian steel minimill industry, for example, received just 38 percent of the power needed to run at full capacity between 1976 to 1986. In response, the steel, cement, and other industries have installed—at considerable expense—large amounts of onsite generation capacity. The Indian cement industry had installed some 545 MW of onsite capacity as of 1988-89. Overall, some 40 percent of electric power needs in India's cement industry are met with onsite generation, usually by low efficiency steam plants.[149]

Available Technology

Paper plants, for example, are often sold in developing countries as turnkey operations. Under these circumstances, one of the principal considerations is the initial capital cost. Oil fired boilers and generators are cheaper than waste wood or dual fuel boilers and/or cogeneration systems and so will usually be specified even though these oil-based systems will require very expensive fuel. Similarly, process control equipment may often be omitted in order to reduce capital costs, but then prevents monitoring and fine tuning plant processes.[150]

Financial Constraints

The additional capital cost to the end user of energy efficient equipment can be a substantial barrier to investment. Investment costs raise several different types of problems. The first is the "disconnect" between the user and the utility: the total system capital cost is often lower for energy efficient equipment, but the capital savings accrue to the utility while the capital costs are incurred by the user. Similarly, there is often a "disconnect" between the purchaser and the user. Where equipment is leased, the capital costs go to the purchaser—who will minimize expenditures by purchasing less efficient equipment, while the higher energy bills go to the user.

Second, potential users of energy efficient equipment—even large industrial organizations—will often have an effective discount rate that is much higher than that justified by their cost of capital, let alone social discount rates. This can dramatically shorten the amount of time they are willing to wait for energy savings to pay for the higher initial capital cost of energy efficient equipment. There are several reasons for these high effective discount rates. Foremost among these is that energy is just one component of overall corporate strategy to improve profitability and competitiveness, and often a rather minor component at that. Energy must compete with these other factors of production when investment choices are made and when the scarce time of skilled manpower is allocated. In particular, capacity expansion provides a much more visible investment alternative and is usually the preferred choice.

In existing plants, a pool of capital may be budgeted for different types of plant improvements, such as better process equipment or for improved energy efficiency. Different energy efficiency improvements then compete against each other for this limited budget rather than against an overall level of profitability. Energy efficiency projects with very high potential returns—but not the highest—may then be deferred.[151] In other cases, energy/cost savings above a certain—very high—threshold level may be required before management considers it worthwhile diverting the attention of their limited technical manpower from the business of keeping the factory running.

Investments in new plant and equipment may also focus on less efficient equipment than is desirable from a systemwide life cycle cost perspective. Factors that constrain these investments in developing countries include:

- aversion to the risk of relatively new, untested equipment, and the lack of an adequate infrastructure to repair the equipment should it fail;
- lack of technical manpower to install, operate, or maintain state-of-the-art equipment;
- the risk of currency exchange rate fluctuations when purchasing equipment denominated by

---

[149] Energy and Environmental Analysis, Inc., op. cit., footnote 86.

[150] Andrew J. Ewing, op. cit., footnote 125.

[151] Marc Ross, "Capital Budgeting Practices of Twelve Large Manufacturers," *Financial Management*, winter 1986, pp. 15-21.

foreign exchange loans; and
- the risk of political instability and consequent uncertainty in mid- to long-range loans.

Third, in many developing countries, fuel and electricity costs are highly subsidized so that investments in energy efficient equipment by the industrial user have very long payback times. For example, a recent review of 60 developing countries found the average cost of electricity to be $0.043/kWh (US$1990) while the average cost (including industrial, commercial, and residential users) of new supply was about $0.08/kWh.[152]

In addition, some countries use cost-plus pricing for key public sector dominated industries—allowing the industry to directly pass on the cost of energy and providing no incentive to conserve. India, for example, long used cost-plus pricing in their public sector industries. They have now largely eliminated this pricing system in cement, aluminum, and pulp and paper, and have modified this system in the iron and steel and in the fertilizer sectors in order to spur energy efficiency improvements. Cost-plus pricing remains in effect, however, in the Indian refinery sector.[153]

Fourth, the lack of foreign exchange to buy spare parts can also be a serious handicap. This has been an important factor in the decline of the Tanzanian cement industry, which operated at just 22 percent of rated capacity in 1984.[154]

Fifth, secondhand markets for equipment may provide no premium for efficient equipment, while at the same time keeping large amounts of inefficient equipment in operation, sometimes as "hand-me-downs" from industrial countries.

POLICY RESPONSES

There are a variety of possible policy responses to these barriers to energy efficiency improvements. These policy responses are summarized in table 4-7 and a few of these are discussed below.

In making their choice of industrial equipment, manufacturers must consider much more than energy efficiency. Decision criteria include: the financial return, the quality and quantity of product produced, the timeliness and reliability of the production equipment, and the flexibility of the equipment, among others.

Simply investing in high technology mills, as currently configured, does not necessarily meet these criteria. High technology mills can achieve higher efficiencies, but also tend to have high capacity—making them less suited to developing countries with their lower volume markets. High technology mills are expensive to maintain, require scarce technical manpower, and spares are often unavailable due to the lack of foreign exchange, lengthy licensing procedures, and high import duties. High technology mills may also not provide, for a variety of reasons, the savings desired. Continuous casting in the Chinese steel industry, for example, has so far provided energy savings at a cost several times greater than the price of the energy supply due to the mismatch between the product the continuous casters were designed to provide and that which was required.[155] Finally, high technology mills provide less employment—widely seen as a liability in developing countries with their large labor pools.[156]

Information

Even large industries may lack information on the opportunities for and potential savings of investing in energy efficient equipment. Policy responses might include information programs—particularly in conjunction with regional energy efficiency testing centers, labeling equipment with its energy consumption, training programs, and energy audits of industrial operations by groups established in-house or by outside experts—possibly supported by the government, utilities, or even private Energy Service Companies.

---

[152] A. Mashayekhi, World Bank, Industry and Energy Department, *Review of Electricity Tariffs in Developing Countries During the 1980s*, Industry and Energy Department Working Paper, Energy Series Paper No. 32 (Washington, DC: World Bank, November 1990).

[153] Ahmad Faruqui et. al., "Application of Demand-Side Management (DSM) to Relieve Electricity Shortages in India," contractor report prepared for the Office of Technology Assessment, April 1990.

[154] D.F. Stewart and B. Muhegi, op. cit., foonote 118.

[155] Mark D. Levine and Liu Xueyi, U.S. Department of Energy, Lawrence Berkeley Laboratory, *Energy Conservation Programs in the Peoples Republic of China*, Report No. LBL-29211 (Berkeley, CA: Lawrence Berkeley Laboratory, August, 1990).

[156] Vinod Bihari, "Problems and Solutions in Adopting Modern Technology At Steel Plants in Developing Countries," *Iron and Steel Engineer*, February 1988, pp. 26-31.

Energy audit services, when combined with a variety of supporting conditions including information, training, financial assistance, appropriate price incentives, and others, can be highly successful. In a USAID program in Kenya, for example, just 24 audits and 30 site visits, together with other supporting activities, resulted in annual savings of about US$1.1 million, at a cost of $136,000 annually in the pilot program.[157] Nonetheless, the types of efficiency improvements adopted in this program were limited to short to medium term pay-back measures, were primarily implemented with in-house staff, and were usually realized with very low cost used equipment or equipment made onsite or otherwise locally available. Very seldom were firms willing to borrow for these investments; most were financed out of maintenance budgets and internal funds.

Numerous countries now have audit programs,[158] but the degree of their effectiveness depends on the effort invested, the extensiveness of related support programs such as training and financial assistance, and other factors.

Numerous countries also have training or other informational programs. South Korea, for example, provided training sessions to some 89,000 people between 1974 and 1980 through their national energy management association.

Additionally, it might also be useful in many cases to establish uniform testing methodologies (test standards) to measure the performance of motors, ASDs, pumps, fans, and other equipment on a uniform basis. Such tests might best be done at regional centers with close institutional relationships to industrial country institutions.[159] Such a regional center might also play a role in concentrating a critical mass of manpower on key RD&D issues; developing computer design or diagnostic tools; establishing methodologies for field evaluation and extension teams; developing protocols for interfacing different types of efficient equipment—such as common forms of communication to control equipment via power line carriers or other means to allow utility load management; and others.

Availability

Where potential users are aware of the advantages of energy efficient equipment, they may still not be able to obtain it. Obstacles may range from an insufficient local market to be worth the expense for the vendor to develop it; lack of sufficient maintenance infrastructure to support use of the equipment; taxes and tariffs that prevent the import of the equipment even when its use would provide substantial capital and/or foreign exchange savings; and other factors.

Policy responses include local development of technologies, organizing local or regional buyers markets to develop sufficient demand to allow development of the market, providing special incentives to make the vendor's effort to enter a small market more attractive, relaxing taxes and tariffs on energy efficient equipment, and others. Local development efforts include the PROCEL (see ch. 3) program of Brazil, which has supported the development of more efficient refrigerators, water heaters, air conditioners, motors and controls, and lighting technologies.

Minimum efficiency standards have been set, for example, by Taiwan for numerous consumer appliances. Window air conditioners have improved in efficiency by over 40 percent since the standards were established. Building efficiency standards have been established in Singapore, South Korea, China, and on a voluntary basis are being established for Indonesia, the Philippines, and Thailand.[160] On the other hand, there are sometimes sound technical reasons for not mandating minimum efficiency standards for a product. Electric motors, for example, are more susceptible to stalling and burning out as their efficiency is increased if there are voltage fluctuations or if they are driving loads with very large starting torques. Mandating minimum efficiency for all (as opposed to certain types) industrial

---

[157] H. Mike Jones, "Kenya Renewable Energy Development Project: Energy Conservation and Planning, Final Report," contractor report prepared for the U.S. Agency for International Development, June 28, 1985.

[158] These include: Argentina, Bangladesh, Brazil, Costa Rica, Ecuador, Egypt, El Salvador, Ghana, Guatemala, Honduras, India, Indonesia, Jordan, Korea, Morocco, Pakistan, Panama, Philippines, Sri Lanka, Thailand, and Tunisia. Steven Nadel, Howard Geller, and Marc Ledbetter, "A Review of Electricity Conservation Programs for Developing Countries," draft report for the American Council for an Energy Efficiency Economy, Washington, DC, Jan. 1991; Amory Lovins, Director, Rocky Mountain Institute, personal communication, July 2, 1991.

[159] These might include the National Institute for Standards and Technology, Environmental Protection Agency (for environment-related technologies), the American National Standards Institute, the Institute for Electrical and Electronics Engineers, and others.

[160] Steven Nadel, Howard Geller, and Marc Ledbetter, op. cit., footnote 158.

Table 4-29—Commercial Energy Use in Agricultural Production, 1972/73

	Share				
	Fertilizers	Traction	Irrigation	Pesticides	Total
Industrial countries	35%	62%	1%	2%	107 Mtoe[a]
Developing countries	68%	22%	8%	2%	31 Mtoe

[a]Mtoe = million tons of oil equivalent.

SOURCE: "Agricultural Mechanization in Development: Guidelines for Strategy Formulation" (Rome, Italy: Food and Agriculture Organization of the United Nations, Agricultural Services Bulletin, No. 45, 1984).

motors might then sometimes lead to undesirable results if the motor were incorrectly specified.

In some cases, more efficient equipment may be available, but will not be considered for developing country applications due to its greater complexity and more difficult operation and maintenance. For example, high pressure steam boilers are widely available, but require more sophisticated water treatment—without which they can fail, and require substantial skills if they are to be operated safely.

Capital Cost

A variety of responses have been developed to the capital constraint problem. These include: loan programs, rebates, tax credits, accelerated depreciation, or other financial assistance for efficient equipment; adjusting energy prices to reflect its full cost; direct installation of efficient equipment by the utility; alternative financing arrangements in order to eliminate the "disconnect" between the user and the utility or between the user and the owner; and other arrangements.

A number of rate incentive programs (adjusting electricity tariff rates in order to reduce peak loads, base loads, etc.) have been established in Brazil, Costa Rica, Indonesia, Pakistan, Uruguay, and other countries. Loans, grants, leasing arrangements, tax credits, and other financing schemes are being tried in Thailand, Brazil, China, India, and elsewhere.[161]

A broader range of barriers to investment in energy efficient technologies is given in table 4-6; a list of potential policy options is given in table 4-7. These are not directly matched up as many barriers have multiple policy responses and, correspondingly, many policy options address a variety of barriers. These barriers and policy responses are strongly influenced by the scale and type—small traditional to large modern—of the industry concerned.

AGRICULTURE

Relatively little energy is used directly in the agricultural sector, ranging from less than 1 percent of the national total in the United States to perhaps 5 to 8 percent of the national totals in developing countries such as Brazil, China, India, and Kenya (see tables 4-3, 4-29). Energy used to manufacture farm equipment and fertilizer, store and process food, or haul it to market are generally accounted for separately by these energy balances in the industrial, commercial, and transport sectors. Despite its relatively low energy use, agriculture is nevertheless a very important sector in the developing countries due to its social and economic significance: agriculture provides fully one-third of the Gross Domestic Product for the nearly 3 billion people in low income countries and it provides an even higher share of national employment (see figure 4-13).[162] In sub-Saharan Africa, for example, 75 percent of the work force is engaged in agriculture, compared to just 2 percent in the United States. Other comparisons of agriculture between developing and industrial countries are given in tables 4-30 and 4-31. Agriculture is also important in terms of its impact on the local, regional, and global environment.[163]

As for the residential, commercial, and industrial sectors, there are numerous opportunities for improving the efficiency of energy use in agriculture. These include improvements:

---

[161]Steven Nadel, Howard Geller, and Marc Ledbetter, op. cit., footnote 158.

[162]World Bank, *World Development Report 1990* (New York, NY: Oxford University Press, 1990).

[163]A more detailed discussion of these issues can be found in: U.S. Congress, Office of Technology Assessment, *Energy in Developing Countries*, OTA-E-486 (Washington, DC: U.S. Government Printing Office, January 1991).

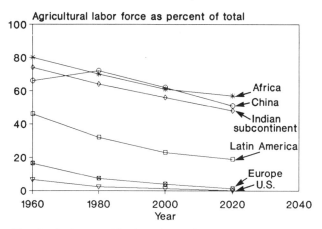

Figure 4-13—Agricultural Labor Force as a Percent of The Total Labor Force, 1960 to 2020

The developing countries have large shares of their total labor force in agriculture.
SOURCE: World Bank, *Agricultural Mechanization: Issues and Options* (Washington, DC: World Bank, 1987).

Photo credit: Appropriate Technologies, International

More efficient irrigation technologies could save energy and boost agricultural productivity.

- in the industrial energy used to produce farm implements, fertilizers and other chemicals (discussed above);
- in the application and utilization by the plants of fertilizers and other chemicals;
- in the pumping and application of irrigation water;
- in the efficiency of traction for cultivation, sowing, weeding, harvesting, and other operations as well as potentially minimizing the need for traction through low-tillage agriculture;
- in the efficiency of post-harvest crop drying (using solar energy or heat recovery systems) and storage (see commercial refrigeration, ch. 3);
- in the utilization of crop residues for energy production (see ch. 6); and
- in the transport of crops to market (see ch. 5), among others.

The discussion here focuses primarily on efficiency improvements in the use of commercial energy for irrigation and traction in developing countries. Improvements in the production of fertilizer were discussed above and transport is discussed in chapter 5. Discussion of agricultural practices, programs, or policies more generally are beyond the scope of this assessment (see box 4-A);[164] several recent Office of Technology Assessment publications examine these and related issues.[165]

Irrigation

Irrigation frees the farmer from dependence on irregular rains and raises yields, allowing double- and even triple-cropping. Some 160 million hectares of land in developing countries are irrigated. In Asia, 100 million hectares are irrigated, and this land produces roughly 60 percent of the region's food on just 45 percent of its cropped area.[166]

India, for example, nearly doubled its irrigated area between 1950 and 1984 in order to reduce its

---

[164]For a recent review, see: Ramesh Bhatia and Rishi Sharma, "Energy and Agriculture: A Review," Ashok V. Desai (ed.), *Patterns of Energy Use in Developing Countries* (New Delhi, India: Wiley Eastern Limited, and Ottawa, Canada: International Development Research Center, and Tokyo: United Nations University, 1990).

[165]U.S. Congress, Office of Technology Assessment, *Enhancing Agriculture in Africa: A Role for U.S. Development Assistance*, OTA-F-356 (September 1988); U.S. Congress, Office of Technology Assessment, *Grassroots Development: The African Development Foundation*, OTA-F-378 (June 1988); U.S. Congress, Office of Technology Assessment, *Technologies to Sustain Tropical Forest Resources*, OTA-F-214 (March 1984); U.S. Congress, Office of Technology Assessment, *Technologies to Maintain Biological Diversity*, OTA-F-330 (March 1987). See also the UNFAO and World Bank publications listed below.

[166]Montague Yudelman, "Sustainable and Equitable Development in Irrigated Environments," Jeffrey Leonard (ed.), *Environment and the Poor: Development Strategies for a Common Agenda* (New Brunswick, NJ: Transaction Books, 1989), pp. 61-85.

Table 4-30—Comparison of Agriculture in Developing and Industrial Countries, 1980

	Developing	Industrial
Share of world population	67%	33%
Share of world agricultural production	38%	62%
Production per agricultural worker	$550 (1975$)	$5,200
Arable land per agricultural worker	1.3 ha	8.9 ha
Fertilizer use (kg/ha) of agricultural land	9 kg/ha	40 kg/ha
Total daily food consumption	2,200	3,300
(low-Income)	2,000	NA
(middle-Income)	2,500	NA
Number of seriously malnourished	435 Million	NA

NA = Not available or not applicable.
SOURCE: "Agriculture: Toward 2000" (Rome, Italy: Food and Agriculture Organization of the United Nations, 1981).

Table 4-31—Agricultural Indicators for Various Developing and Industrial Countries, 1982/83

Country	Agricultural labor percent of total	Arable land Ha/capital	Fertilizer kg arable land	Rice yield kg paddy/ha
India	61	0.24	39	2,200
Bangladesh	83	0.10	60	2,050
China	57	0.10	181	5,070
Pakistan	52	0.22	59	NA
Sri Lanka	52	0.14	71	NA
Burma	50	0.27	17	3,090
Egypt	49	0.06	361	NA
Philippines	44	0.23	32	2,470
Brazil	36	0.58	37	NA
Iran	36	0.34	66	NA
Korea	35	0.05	345	NA
Mexico	34	0.32	61	NA
Italy	10	0.22	169	NA
Japan	9	0.04	437	5,700
France	8	0.34	312	NA
Canada	4	1.87	49	NA
Germany	3	0.12	421	NA
United Kingdom	2	0.12	375	NA
U.S	2	0.82	105	5,150

SOURCE: *TERI Energy Data Directory and Yearbook, 1988* (New Delhi, India: Tata Energy Research Institute, 1989).

vulnerability to poor monsoons.[167] More than 6 million electric and 3 million diesel pump sets have been deployed (see figure 4-14). Electric pumpsets consumed some 23,000 GWh of electricity in India in 1985-86, about 14 percent of total national electricity generation[168] and about two-thirds of rural electricity use.[169] Similarly, in China, irrigation consumes an estimated 70 percent of rural electricity, with the remainder used for food processing, various rural industries, and lighting.[170]

Failure to provide adequate power for pumping can have serious consequences for those regions dependent on it. The Sudanese lost some $100 million worth of agricultural output in 1984 due to a shortage of energy for pumping and traction. Similarly, the Somalis lost 40 to 60 percent of their irrigated crops in some regions during 1984 due to a lack of diesel fuel for irrigation.[171]

Irrigation is most commonly done with either electric motor or diesel driven pumps. Electric

---

[167] Tata Energy Research Institute, *TERI Energy Data Directory and Yearbook (TEDDY) 1988* (New Delhi, India: Tata Energy Research Institute, 1990), p. 128.

[168] Ashok Desai, "Energy Balances for India, 1985-86," contractor report prepared for the Office of Technology Assessment, 1990. This is equivalent to 125,000 GJ.

[169] S. Ramesh and T.V. Natarajan, "Policy Options in Rural Electrification in India," *Pacific and Asian Journal of Energy*, vol. 1, 1987, pp. 44-53.

[170] C. Howe, *China's Economy* (New York, NY: Basic Books, Inc., 1978), p. 88.

[171] U.S. Congress, House Select Committee on Hunger, *Hearings on Energy and Development: Choices in the Food Sector*, Serial No. 101-11, U.S. Government Printing Office, July 25, 1989.

> **Box 4-A—*Agricultural Energy Use in Context***
>
> A thorough discussion of agricultural energy use necessarily includes a host of important issues beyond the scope of this study. Other issues of importance in agriculture, but not discussed here, include:
>
> - proper pricing of agricultural products (allowing the market to work);
> - the international impact of agricultural subsidies by Europe, Japan, and the United States, particularly on domestic valuations of crops, land tenure, and subsistence agriculture in developing countries;
> - soil conservation and the proper accounting for the value of soil and other environmental assets;[1]
> - the agricultural potential of low-energy inputs such as microcatchments, improved pest management, intercropping and agroforestry, and improved post-harvest storage;
> - the role of higher value inputs (including information-intensive management as for improved pest management or intercropping, etc.) into agriculture to reduce agricultural expansion onto biologically rich or fragile lands and the corresponding environmental damage that can then result; and
> - the multiple roles and needs served by traditional technologies—draft animals can provide meat, leather, and dung in addition to traction, and cows also provide milk and reproduce while modern electric or engine-driven mechanical drive only provides traction.
>
> In the 1950s, most Western development economists did not view the agricultural sector as important for economic development.[2] Today, the view has been largely reversed, with agriculture widely seen as an important underpinning of an economy and some suggesting that, under certain conditions, agriculture could be an important driver of industrialization just as export markets have been for the newly industrialized countries.[3] Realizing this potential will require public sector support and substantial extension services due to the highly dispersed nature and small scale of agriculture.
>
> ---
>
> [1]Rattan Lal, "Managing the Soils of Sub-Saharan Africa," *Science*, vol. 236, May 1987, pp. 1069-1076.
>
> [2]John M. Staatz and Carl K. Eicher, "Agricultural Development Ideas in Historical Perspective," Carl K. Eicher and John M. Staatz (eds.), *Agricultural Development in the Third World* (Baltimore, MD: Johns Hopkins University Press, 1984) pp. 3-32.
>
> [3]Irma Adelman, "Beyond Export-Led Growth," *World Development*, vol. 12, No. 9, 1984, pp. 937-949; John W. Mellor, "Agriculture on the Road to Industrialization," John P. Lewis and Valeriana Kallab (eds.), *Development Strategies Reconsidered*, Overseas Development Council (New Brunswick, NJ: Transaction Books, 1986) pp. 67-90.

pumps are quite reliable (although subject to interruptions in the electric power grid) and convenient, and are often the lowest cost alternative. Diesel-electric pumping systems, in which diesel generators produce electricity that is then used to drive electric pumps, and direct diesel and gasoline-powered pumps are more often used where no electric grid is available. These are much less mechanically dependable than electric pumps. Other pumping systems include wind energy, photovoltaics, producer gas driven engines, and others. Only grid connected electric pumping will be examined here; other systems are explored in chapter 6.

As indicated above, numerous improvements are possible in pumping systems with potentially very large leverage on upstream utility investment. A field study in India found 90 percent of the agricultural pumps could be substantially upgraded. For example, reducing the friction in the piping system and foot valve in a field test of 25,000 pumps typically cost about $80, but provided energy savings of 30 percent for the same quantity of water pumped. Assuming 5 kW input motor/pump systems operating at an annual capacity of 10 percent,[172] this $80 investment in motor/pump system efficiency provides a $570 savings in utility capital investment (total, not annualized). Similarly, a field trial of 300 complete rectifications of pumping systems resulted in 50-percent energy savings or 10,600 kWh/year per pumping system. This is equivalent to 1.21 kW of firm capacity with an avoided cost of roughly $4,600 per pumping system (total, not annualized). In contrast, the investment per pumping system was about $600.[173] At a cost of $0.09/kWh, these

---

[172]These are nominal values as reported for more carefully documented field trials discussed by S. M. Patel, "Low-Cost and Quick-Yielding Measures for Energy Conservation in Agricultural Pumps," *Pacific and Asian Journal of Energy*, vol. 2, No. 1, 1988, pp. 3-11. and by S. Ramesh and T.V. Natarajan, "Policy Options in Rural Electrification in India," *Pacific and Asian Journal of Energy*, vol. 1, No. 1, 1987, pp. 44-53.

[173]S.M. Patel, "Low-Cost and Quick-Yielding Measures for Energy Conservation in Agricultural Pumps," *Pacific and Asian Journal of Energy*, vol. 2, No. 1, 1988, pp. 3-11.

Figure 4-14—Use of Agricultural Pumpsets in India, 1950 to 1990

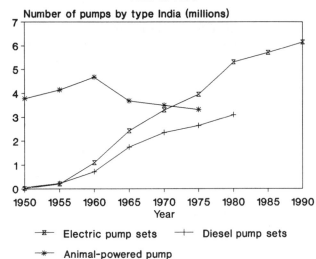

SOURCE: Tata Energy Research Institute, *TERI Energy Data Directory and Yearbook (TEDDY), 1988* (New Delhi, India: 1989).

investments would have a payback time of about 4 months. The electricity price paid by farmers, however, is substantially lower in India due to subsidies, with correspondingly longer payback times.

Similar efficiency improvements are possible elsewhere in the world. In the United States, for example, average agricultural pump efficiencies have been estimated at 55 percent, with the potential to increase them to an average of 82 percent.[174]

Other measures that could have a substantial impact on energy use for agricultural pumping include better techniques for delivering the water to the plant—such as drip irrigation, the use of sensors to monitor the actual need for water by plants, or the use of computer controls for scheduling irrigation.[175]

Traction

Many countries have made greater use of animal or mechanical traction for agricultural operations to ease seasonal labor demands. In Africa, the emphasis has been primarily on animal traction due to the frequent failure, for a variety of technical and institutional reasons, of projects attempting to introduce tractors.[176] Draft animals become increasingly difficult to support as pasture is converted to crops in the most densely populated regions, however, and feeding the animal over the course of a year can require the output of several times as much land as the animal can work during the short growing season.

Energy use (crop residues) for draft animals is not currently included in national energy balances, despite its important role.[177] If it were, the energy input into agriculture in some developing countries would increase dramatically. For example, it has been estimated that the energy use in animal traction in India during 1970/71 ranged from about 1 to 5 times greater (depending on assumptions and methodology) than commercial energy inputs into agriculture.[178] A variety of improvements in nutrition, harness design, and other factors can greatly improve the work output by draft animals.[179]

Mechanical traction comes in many forms, ranging from power tillers to large tractors. Large scale manufacturing and engineering innovations have also made many of these systems increasingly accessible to small farmers.[180] There is a wide variation between countries in the extent to which agriculture is mechanized (see figure 4-15). There is a similarly wide and inverse variation in the number of agricultural laborers (see figure 4-16).

In China the most popular tractor is probably the "Worker-Peasant," a 7-hp garden tractor. In Thai-

---

[174]Joseph T. Hamrick, "Efficiency Improvements in Irrigation Well Pumps," *Agricultural Energy: Volume 2, Biomass Energy Crop Production*, selected papers and abstracts from the 1980 American Society of Agricultural Engineers (St. Joseph, MI: American Society of Agricultural Engineers, 1981).

[175]Energetics, Inc., for the U.S. Department of Energy, Office of Industrial Technologies, "Industry Profiles: Agriculture," Contractor No. DE-AC01-87CE40762, December 1990.

[176]Hans P. Binswanger et. al., *Agricultural Mechanization: Issues and Options* (Washington, DC: World Bank, 1987).

[177]Arjun Makhijani, "Draft Power in South Asia Foodgrain Production: Analysis of the Problem and Suggestions for Policy," contractor report prepared for the Office of Technology Assessment, June 1990.

[178]Ramesh Bhatia, "Energy and Agriculture in Developing Countries," *Energy Policy*, vol. 13, No. 4, August, 1985, pp. 330-334.

[179]See, for example, Peter Munzinger, *Animal Traction in Africa* (Germany: Eschborn, 1982); Jane Bartlett and David Gibbon, *Animal Draught Technology: An Annotated Bibliography* (London: Intermediate Technology Publications, 1984). More recent and complete information can be obtained from: Tillers International, Kalamazoo, Michigan or from the Centre for Tropical Veterinary Medicine, University of Edinburgh, Scotland.

[180]Hans P. Binswanger et. al., op. cit., footnote 176.

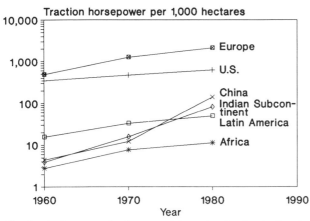

Figure 4-15—Mechanical Traction Per 1,000 Hectares of Agricultural Land

The developing countries have much lower levels of agricultural mechanization than the industrial countries.
Note the logarithmic scale.
SOURCE: World Bank, *Agricultural Mechanization: Issues and Options* (Washington, DC: World Bank, 1987).

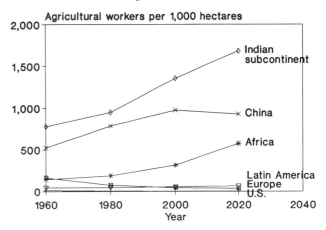

Figure 4-16—Agricultural Workers Per 1,000 Hectares of Agricultural Land

SOURCE: World Bank, *Agricultural Mechanization: Issues and Options* (Washington, DC: World Bank, 1987).

land, a versatile system known as the "iron buffalo" is widely used. This 8-hp system can alternately be used to plow (using 1 meter iron paddle wheels that are more stable in a paddy than rubber tires), pump water, power a cart, run a generator, or do other tasks. These iron buffaloes allow farmers to drain their rice fields, mechanically or hand-scatter the seeds and, when the rice sprouts, pump water back into the paddies. This avoids the traditional laborious hand planting of seedlings in the rice paddies. By 1984, some 400,000 of these systems were in use in Thailand.[181] In contrast to these smaller tractors, the most popular tractor in India, where the number of tractors almost doubled from 1972 to 1977,[182] is a 30-hp diesel.

Energy use by mechanical traction equipment varies widely by region and according to operating conditions and a host of other factors such as the degree to which the operation is power- or control-intensive. Some regions of Latin America are as mechanized as the industrial countries, while most African workers still rely primarily on hand tools.

Generally, diesel powered equipment is roughly 25-percent more efficient than gasoline powered equipment. Specific performance improvements in motors (for transport applications) are examined in detail in chapter 5. Mechanization of agriculture is one of several means of improving productivity; mechanization is not an end in itself.[183]

CONCLUSION

Industrial and agricultural energy use will continue to grow in developing countries in order to meet the aspirations of a growing population that is making the transition from a traditional rural to a modern urban lifestyle. Technologies either exist or can be readily adapted to substantially moderate these increased demands for industrial and agricultural energy use while still providing the energy services—manufactured goods and high quality foods—desired. These technologies can provide these energy savings at both a capital and an operating cost less than what is paid today. These savings can be an important source of capital to meet other pressing development needs and to spur more rapid economic development.

---

[181] Ben Barber, "Buffaloes Mooove With the Times," *Asian Wall Street Journal*, July 24-25, 1987.

[182] Tata Energy Research Institute, *TERI Energy Data Directory and Yearbook (TEDDY) 1988* (New Delhi, India: Tata Energy Research Institute, 1990), p. 137.

[183] Hans P. Binswanger et. al., op. cit., footnote 176.

Chapter 5

Energy Services: Transport

Photo credit: Thomas Burke

Contents

	Page
INTRODUCTION AND SUMMARY	145
TRANSPORTATION SYSTEMS IN DEVELOPING COUNTRIES	155
IMPROVING FREIGHT TRANSPORT ENERGY EFFICIENCY	156
The Truck Fleet	156
Improving Rail Efficiency	159
Modal Shifts	159
IMPROVING PASSENGER TRANSPORT ENERGY EFFICIENCY	161
Automobiles	161
Motorized Two- and Three-Wheelers	165
Buses	166
Modal Shifts in Urban Passenger Transport	166
THE DEMAND FOR TRANSPORT SERVICES	169
NONMOTORIZED MODES	169
ALTERNATIVE FUELS	171
CONCLUSION	176

Boxes

Box	Page
5-A. Singapore's Area Licensing Scheme: Carrot and Stick	167
5-B. Urban Design and Transport Energy Use in North America	170
5-C. Brazil's Ethanol Fuels Program: Technical Success, Economic Distress?	175

Chapter 5
Energy Services: Transport

INTRODUCTION AND SUMMARY

Transport services are a key component of economic development. As the economies of the developing world expand, transport services must also grow to supply the raw material, labor, food, and consumer goods needed by the growing economy, and to integrate rural areas into the larger economy. At the same time, higher standards of living associated with economic expansion lead to sharply rising demand for personal mobility, often by automobile.

Providing these services can be difficult and costly. The transport sector, in most developing countries, accounts for one-third of total commercial energy consumption (see table 5-1) and over one-half of total oil consumption. India and China are exceptions. In these countries transportation accounts for under 10 percent of commercial energy consumption.

As many developing countries have little or no domestic oil resources, purchases of imported oil to satisfy transport demand together with imports of transport equipment put considerable pressure on limited foreign exchange resources. In addition, capital requirements for road and railway construction and maintenance take up a significant share of development budgets.

Road vehicles are also major contributors to high levels of urban air pollution currently experienced in developing countries. In Indian cities, for example, gasoline fueled vehicles—mostly two and three wheelers—are responsible for roughly 85 percent of carbon monoxide and 35 to 65 percent of the hydrocarbons in the air from fossil fuels. Diesel vehicles—buses and trucks—are responsible for over 90 percent of nitrogen oxide emissions in urban India.[1] Emissions from the transport sector also account for a significant share of global greenhouse gas emissions.

Along with these high economic and environmental costs, the transport systems in developing countries are frequently unsatisfactory in terms of the quality and quantity of transport services they deliver, inadequately supporting the free flow of goods and people that is vital for economic development. Urban traffic is severely congested[2] and rural areas are poorly served.

This chapter examines ways in which the energy efficiency of the transport system could be improved and environmental pollution could be reduced while still providing the transport services needed for economic development. This can be done by improving the efficiencies of each transport mode, changing the modal mix, moderating the demand for transport through full costing of services, and improving land use planning (see table 5-2). This chapter focuses primarily on technological opportunities for improving transport efficiency.

Commercially available technologies, many of which are widely used in the industrialized countries, could significantly improve transport energy efficiencies in the developing countries. In freight transport, the existing developing-country truck fleet is generally older, smaller, and less technologically sophisticated than the truck fleet of industrial countries. A wide range of retrofits such as rebuilt motors incorporating improved diesel fuel injection systems, cab mounted front air deflectors to reduce wind resistance, turbochargers, and radial tires are available. The rapid diffusion of these and other technologies could yield substantial energy savings.

In passenger transport, automobiles and other modes could benefit from the use of commercially available technologies such as electronic control of spark timing, radial tires, improved aerodynamics, and fuel injection. Many more efficiency improvements are at various stages of development and commercialization.[3] Table 5-3 lists a few of these for the U.S. fleet. The energy efficiency of the rapidly

---

[1]Tata Energy Research Institute, *TERI Energy Data Directory and Yearbook 1988* (New Delhi, India: 1989), p. 250.

[2]Road vehicles in Lagos and Bangkok, for example, move at only half the speed of road vehicles in London or Frankfurt. World Bank, *Urban Transport*, World Bank Policy Study (Washington, DC: World Bank, 1986), p. 3.

[3]These technologies, their costs, and numerous controversies surrounding their implementation are explored in U.S. Congress, Office of Technology Assessment, *Improving Automobile Fuel Economy: New Standards, New Approaches*, OTA-E-504 (Washington, DC: U.S. Government Printing Office, October 1991).

Table 5-1—Total Delivered Energy by Sector, in Selected Regions of the World, 1985 (exajoules)[a]

Region	Residential/commercial		Industry		Transport		Total		Total energy
	Commercial fuels	Traditional fuels[b]	Commercial fuels	Traditional fuels[b]	Commercial fuels	Traditional fuels[b]	Commercial fuels	Traditional fuels[b]	
Africa	1.0	4.0	2.0	0.2	1.5	NA	4.4	4.1	8.5
Latin America	2.3	2.6	4.1	0.8	3.8	NA	10.1	3.4	13.5
India and China	7.3	4.7	13.0	0.2	2.0	NA	22.2	4.8	27.1
Other Asia	1.9	3.2	4.0	0.4	1.9	NA	7.8	3.6	11.3
Developing countries	12.5	14.5	23.1	1.6	9.2	NA	44.5	15.9	60.4
United States	16.8	NA	16.4	NA	18.6	NA	51.8	NA	51.8

NA = Not available or not applicable.
NOTES: This is delivered energy and does not include conversion losses from fuel to electricity, in refineries, etc. The residential and commercial sector also includes others (e.g., public services, etc.) that do not fit in industry or transport. Traditional fuels such as wood are included under commercial fuels for the United States.
[a]Exajoule (10^{18} Joules) equals 0.9478 Quads. To convert to Quads, multiply the above values by 0.9478.
[b]These estimates of traditional fuels are lower than those generally observed in field studies. See references below.
SOURCE: U.S. Congress, Office of Technology Assessment, *Energy in Developing Countries*, OTA-E-486 (Washington, DC: U.S. Government Printing Office, January 1991) p. 49.

Table 5-2—Improving the Efficiency of Transportation Services

Efficiency improvement
    Remarks

Vehicle technical efficiency:
  Automobiles
    The technical efficiency of automobiles can be improved by: reducing vehicle weight; reducing vehicle aerodynamic drag; improving engine performance—overhead cam engines, multipoint fuel injection, roller cam followers, low friction engine, etc.; improving vehicle drive trains and transmissions; reducing rolling resistance; etc. (See table 5-3 for additional detail).
  Two and three wheelers:
    The technical efficiency of two and three wheel vehicles can be improved by converting from two-stroke to four-stroke engines, using improved carburetors and electronic ignition, reducing rolling resistance; etc.
  Freight trucks:
    Freight truck efficiencies can be improved by using diesel rather than gasoline engines, using improved fuel injectors and injection pumps, using turbochargers, using improved lubricants, reducing aerodynamic drag, reducing rolling resistance, etc.
  Freight trains:
    Freight train efficiencies can be improved by using diesel-electric locomotives, lighter weight cars, low friction bearings, on-board flange lubricators, steerable trucks, computer directed operations, etc.

Vehicle load factor:
    Vehicle operational efficiencies can be improved by: increasing load factors through carpooling and other means for passenger vehicles and by maximizing loads and reducing empty or part-load back hauls for freight vehicles.

Vehicle operational efficiency:
    Vehicle operational efficiencies can be improved by: better driving habits; improved traffic control—high-occupancy-vehicle only lanes, timed traffic signals, limited access roads, and other traffic management systems, etc.—for passenger vehicles; efficient routing for freight trucks; sidewalks, bicycle paths, and other means to reduce road congestion; improved roads and related structures to allow higher average speeds, heavier loads, etc; and by other means.

Modal shifts:
    Efficiency gains by modal shifts require the movement of freight and passengers to more efficient transport systems. For freight, this typically includes from long-distance truck to rail; for passengers this includes from private automobile to commuter bus or rail mass transit, or to nonmotorized modes such as bicycling or walking.

Transport demand:
    The demand for transport services or their equivalent can be satisfied while the demand for transport energy can be dramatically reduced: by land-use planning to better match residences with jobs, schools, shopping, and transport corridors—aiding modal shifts to more efficient transit systems as well as potentially reducing infrastructure costs for water, sewage, etc. and possibly reducing loss of agricultural lands; by improved telecommunications technologies; by charging users the full cost of transport services—including roads, parking, death and injury, pollution, and other costs; etc.

NOTE: These efficiency improvement opportunities are often highly interdependent. For example, technical improvements to freight trucks such as reduced aerodynamic drag, etc. are of little use unless the road infrastructure allows medium to high speeds. The above list does not include air- or water-based transport systems or a variety of land-based modes. The same considerations, however, generally apply.
SOURCE: Office of Technology Assessment, 1992.

Table 5-3—Technologies for Improving the Fuel Economy of Automobiles

Weight reduction. Includes three strategies—substitution of lighter weight materials, e.g., aluminum or plastic for steel; improvement of packaging efficiency, i.e., redesign of drivetrain or interior space to eliminate wasted space; and technological change that eliminates the need for certain types of equipment or reduces the size of equipment.

Aerodynamic drag reduction. Primarily involves reducing the drag coefficient by smoothing out the basic shape of the vehicle, raking the windshield, eliminating unnecessary protrusions, controlling airflow under the vehicle (and smoothing out the underside), reducing frontal area, etc.

Front wheel drive. Shifting from rear to front wheel drive, which allows: mounting engines transversely, reducing the length of the engine compartment; eliminating the transmission tunnel, which provides important packaging efficiency gains in the passenger compartment; and eliminating the weight of the propeller shaft and rear differential and drive axle. Now in wide use.

Overhead cam engines. OHC engines are more efficient than their predecessor pushrod (overhead valve, OHV) engines through their lower weight, higher output per unit displacement, lower engine friction, and improved placement of intake and exhaust ports.

Four valve per cylinder engines. Adding two extra valves to each cylinder improves an engine's ability to feed air and fuel to the cylinder and discharge exhaust, increasing horsepower per unit displacement. Higher fuel economy is achieved by downsizing the engine; the greater valve area also reduces pumping losses, and the more compact combustion chamber geometry and central spark plug location allows an increase in compression ratio.

Intake valve control. Shift from fixed-interval intake valve opening and closing to variable timing based on engine operating conditions, to yield improved air and fuel feed into cylinders and reduced pumping loss at low engine loads.

Torque converter lockup. Lockup eliminates the losses due to slippage in the fluid coupling between engine and transmission.

Accessory improvements. Adding a two-speed accessory drive to more closely match engine output to accessory power requirements, plus design improvements for power steering pump, alternator, and water pump.

Four-and five-speed automatic transmissions, and continuously variable transmissions. Adding extra gears to an automatic transmission increases fuel economy because engine efficiency drops off when its operating speed moves away from its optimum point, and the added gears allow the transmission to keep the engine closer to optimal speed.

Electronic transmission control. Electronic controls to measure vehicle and engine speed and other operating conditions allow the transmission to optimize gear selection and timing, keeping the engine closer to optimal conditions for either fuel economy or power than is possible with hydraulic controls.

Throttle body and multipoint fuel injection. Fuel injection allows improved control of the air/fuel mixture and thus allows the engine to continually adjust this mixture for changing engine conditions. Multipoint also reduces fuel distribution problems. In wide use.

Roller cam followers. Most current valve lift mechanisms are designed to slide along the camshaft; shifting to a rolling mechanism reduces friction losses.

Low friction pistons/rings. Lower friction losses result from better manufacturing control of tolerances, reduced ring tension, improved piston skirt design, etc.

Improved tires and lubricants. Continuation of longstanding trends towards improved oil (in near-term, substitution of 5W-30 oil for 10W-40 oil), and tires with lower rolling resistance.

Advanced engine friction reduction. Includes use of light weight reciprocating components (titanium or ceramic valves, composite connecting rods, aluminum lifters, composite fiber reinforced magnesium pistons), improved manufacturing tolerances to allow better fit of moving parts, available post-1995.

Electric power steering. Used only for cars in the minicompact, subcompact, and compact classes.

Lean-burn engines. Operating lean (low fuel to air ratio) improves an engine's thermodynamic efficiency and decreases pumping losses. Requires a new generation of catalysts that can reduce NO_x in a "lean" environment.

Advanced two-stroke engines. Unlike a conventional engine, there is a power stroke for every ascent and descent of the piston, thus offering a significantly higher output per unit of engine displacement, reduced pumping loss, smooth operation, and high torque at low speeds, allowing engine downsizing and fewer cylinders (reduced friction losses). Also, operates very lean, with substantial efficiency benefits (if NO_x problems are solved). Compliance with stringent emissions standards is unproven.

Diesel engines. Compression-ignition engines, or diesels, are a proven technology and are significantly more efficient than gasoline two-valve engines even at constant performance; new direct injection turbocharged diesels offer a large fuel savings. Although the baseline gasoline engine will improve in the future, a portion of the improvements, especially engine friction reduction, may be used beneficially with diesels as well. Use may be strongly limited by emissions regulations and consumer reluctance.

Electric hybrids. Involves combining a small electric motor for city driving and a diesel for added power and battery charging. The small size of the diesel eases emissions limitations, and the substantial use of the electric motors reduces oil use.

SOURCE: U.S. Congress, Office of Technology Assessment, *Improving Automobile Fuel Economy: New Standards, New Approaches*, OTA-E-504 (Washington, DC: U.S. Government Printing Office, October 1991).

growing fleet of two and three wheelers could also be increased, notably through the use of improved carburetors, electronic ignition, and four-stroke rather than two-stroke engines.[4]

Developing countries currently have little influence over the development and commercialization of these technologies, due to the small size of their market. As their market is growing rapidly, however, their influence will most likely grow as well.

Greater use of these technologies could result in significant energy savings—estimated at about 20 percent over current levels[5]—and they are highly cost effective, with short payback periods. The environmental benefits would also be significant. Improved carburetors and electronic ignition in two wheelers, for example, could improve efficiency between 10 and 15 percent, and reduce hydrocarbon emissions by 50 percent. The use of four-stroke rather than two-stroke engines could increase energy efficiency by 25 percent and reduce hydrocarbon emissions by 90 percent.[6] Lower consumption of petroleum products would save foreign exchange (or in the case of oil exporting countries, increase the amount of petroleum products available for export).

Despite the multiple advantages of these technologies there are several factors impeding their diffusion or diluting their potential energy savings (see table 5-4):

- **Poor infrastructure.** The benefits of aerodynamic improvements and turbocharging accrue only at higher speeds, which are often not possible on the rougher, more congested roads in the developing world. Similarly, poor roadbeds deter the use of larger and therefore more energy efficient trucks. Many of the new technologies depend on high quality fuels and do not operate well with the variable quality fuels often encountered in developing countries.
- **Maintenance and training.** Several technologies require specialized skills for maintenance that may not now be generally available in developing countries. If truck owners need to seek out specialist firms for routine maintenance, "down time" is increased, thus reducing the benefits from improved energy efficiency. Poor driving habits can also reduce efficiency gains.
- **High first costs and high consumer discount rates.** As in other end-use sectors, potential users are deterred by the additional cost of the improved technology, which they are apparently willing to undertake only if they can recover their additional cost in a very short period of time. In many cases, the costs of energy are a relatively small part of total operating costs so that the expected benefit may not be large in relation to other considerations, including the effort involved in finding and maintaining a more efficient vehicle as well as the various attendant uncertainties. In transport modes where fuel is a large share of total costs (such as air and maritime transport) and where financial decisions are made on the basis of discount rates closer to commercial bank rates, energy efficient technologies have often been adopted more rapidly than in passenger road transport.
- **Fuel costs.** The length of the payback time is closely related to the costs of fuel. In oil exporting developing countries, prices of transport fuels are often kept well below world prices, offering little incentive to economize on their use. In the oil importing countries, gasoline prices are generally higher than international costs. Diesel prices, however, are often considerably lower than gasoline and/or international prices, which again discourages conservation.
- **Import duties.** High import duties on vehicles with higher initial costs due to more efficient equipment, and on retrofit equipment, deter their diffusion despite their potential to reduce oil (and/or refinery equipment) imports.
- **Low scrappage rates.** The high demand for road vehicles generally, relatively cheap labor for maintenance, expensive capital, and other factors have led to low scrappage rates in developing countries, and therefore an older (and less energy efficient) road vehicle fleet.

---

[4] "Big Bets On a Little Engine," *Business Week,* Jan. 15, 1990, pp. 81-83.

[5] World Bank, *Urban Transport,* World Bank Policy Study (Washington, DC: World Bank, 1986), p. 2; see also, Mudassar Imran and Philip Barnes, World Bank, *Energy Demand in the Developing Countries,* World Bank Staff Commodity Working Paper #23 (Washington, DC: World Bank, 1990).

[6] These improved two-stroke engines, currently under development for use in small automobiles or other applications, could also lower capital costs compared to the equivalent power four-stroke engine.

Table 5-4—Barriers to Transportation Efficiency Improvements

Technical

Availability

High efficiency vehicles and related technologies and their needed support infrastructure of skilled manpower and spare parts may not be locally available. Foreign exchange may not be available to purchase critical spare parts.

Infrastructure

The available infrastructure within a developing country may not be able to adequately support a particular high-efficiency technology. For example, the benefits of aerodynamic improvements and turbocharging accrue only at higher speeds, which are often not possible on the rougher, more congested roads in the developing world. Similarly, poor roadbeds deter the use of larger and therefore more energy efficient trucks. Many of the new technologies depend on high quality fuels, and do not operate well with the variable quality fuels often encountered in developing countries. The existing infrastructure might also impede the implementation of a more efficient system.

Information

Potential users of energy efficient vehicles may lack information on the opportunities and savings.

Reliability

Innovative high-efficiency vehicles and components may not have a well proven history of reliability, particularly under developing country conditions.

Research, development, demonstration

Developing countries may lack the financial means and the technical manpower to do needed RD&D in energy efficient vehicle technologies, or to make the needed adaptations in existing energy efficient technologies in use in the industrial countries to meet the conditions—such as relatively low highway speed and frequent speed changes—in developing countries.

Technical manpower

There is generally a shortage of skilled technical manpower in developing countries for installing, operating, and maintaining energy efficient equipment in vehicles. If truck owners need to seek out specialist firms for routine maintenance, "down time" is increased thus reducing the benefits from improved energy efficiency.

Training

Poor driving habits can also reduce efficiency gains. *See also* Behavior, below. Training in vehicle maintenance and driving habits could promote efficiency.

Financial/economic

Cost

As in other end use sectors, potential purchasers of high efficiency vehicles are deterred by the additional cost of the improved technology, which they are apparently willing to undertake only if they can recover their additional cost in a very short period of time. In many cases, the costs of energy are a relatively small part of total operating costs so that the expected benefit may not be large in relation to other considerations, including the effort involved in finding and maintaining a more efficient vehicle as well as the various attendant uncertainties. In transport modes (such as air and maritime transport) where fuel is a large share of total costs and where financial decisions are made on the basis of discount rates closer to commercial bank rates, energy efficient technologies have been adopted more rapidly than in passenger road transport.

Currency exchange rate

Fluctuations in the currency exchange rate raises the financial risk to firms who import high efficiency vehicles with foreign exchange denominated loans.

International energy prices

Falls in international oil prices below expected levels can reverse the profitability of investments in efficiency or alternative fuels. This has happened to some extent with the Brazilian ethanol program.

Scrappage rates

The high demand for road vehicles generally and for used vehicles in particular, combined with inflation (which protects the values of used cars), relatively cheap labor for maintenance, expensive capital, and other factors has led to low scrappage rates in developing countries, and therefore an older (and less energy efficient) road vehicle fleet. While old vehicles can be retrofit to improve fuel efficiency their owners are likely to be the least able to afford the additional costs. The replacement of old vehicles by new, though contributing to the improved fleet energy efficiency, would in the short run at least incur large foreign exchange costs.

Secondary interest

Energy efficiency is often of secondary interest to potential users. Vehicle acceleration, roominess, comfort, or accessories such as air conditioning may be of greater importance to the prospective purchaser.

Secondhand markets

Low-efficiency vehicles may be widely circulated in secondhand markets in developing countries, including "gifts" or "hand-me-downs" from industrial countries. Further, users who anticipate selling vehicles into the secondhand market after only a few years may not realize energy savings over a long enough period to cover the cost premium of the more efficient vehicle.

Subsidized energy prices

Energy prices in developing countries are often controlled at well below the long run marginal cost, reducing end-user incentive to invest in more efficient vehicles. Energy prices may be subsidized for reasons of social equity, support for strategic economic sectors, or others, and with frequent adverse results. In oil exporting developing countries, prices of transport fuels are often kept well below world prices, offering little incentive to economize on their use. In the oil importing countries, gasoline prices are generally higher than international costs. Diesel prices, however, are often considerably lower than gasoline and/or international prices, which again discourages conservation.

Taxes and tariffs

Unless appropriately designed, taxes and tariffs can bias purchasing decisions away from vehicles, components, or retrofit equipment with higher initial costs due to more energy efficient equipment despite their potential to reduce oil (and/or refinery equipment) imports.

(continued on next page)

Table 5-4—Barriers to Transportation Efficiency Improvements—Continued

Institutional

Behavior
   Users may waste energy, for example, by leaving vehicles on. In some cases, such seeming waste may be done for important reasons. Bus drivers in developing countries often leave their engines on for long periods, at a significant cost in fuel, in order to avoid jumpstarting their vehicle if the starter is broken, or to prevent customers from thinking (if the engine is off) that their vehicle is broken and going to a competitor whose engine is running.

Disconnect between purchaser/user
   In a rental/lease arrangement, the owner will avoid paying the higher capital cost of more efficient vehicles while the rentor/lessor is stuck with the resulting higher energy bills.

Integration
   Poor integration between modes—such as between rail and road systems—can result in shipping delays or simply not using the more efficient rail system due to its lower flexibility and difficulty in door-to-door delivery.

Land use planning
   Poor land use planning can prevent the establishment of efficient transportation corridors for mass transit options, and encourages low density urban sprawl dependent on private vehicles.

Political instability
   Political instability raises risks to those who would invest in more efficient vehicles that would only pay off in the mid to longterm.

Service
   Poor service—crowded, slow, infrequent, etc.—from, for example, buses in congested urban areas encourages people to purchase less energy efficient, but more convenient personal transport such as 2/3 wheelers or autos.

SOURCE: Office of Technology Assessment, 1992.

While old vehicles can be retrofit to improve fuel efficiency, their owners are likely to be the least able to afford the additional costs. The replacement of old vehicles by new, though contributing to the improved fleet energy efficiency, would—in the short run at least—incur large foreign exchange costs.

In addition to efficiency improvements within each transport mode, the energy efficiency of the transport sector as a whole (as measured by energy consumed per tonne-kilometer or per passenger-kilometer) could be improved by encouraging the movement of both freight and passenger traffic to the most energy efficient modes. In *freight* traffic, rail uses about one-fourth to one-third of the operating energy per tonne-kilometer that road vehicles use. If the energy embodied in transport equipment and associated railways and roads is included, the difference is reduced but railways continue to be much more energy efficient than roads for freight traffic. In addition, rail freight is usually cheaper per tonne-kilometer, at least for bulk commodities. This advantage is reduced, however, when door-to-door delivery is considered or as the size of the shipment decreases.

Despite its efficiency and cost advantages, rail's market share in those countries with rail networks has declined. This is due to a number of reasons, including: the greater convenience, flexibility, and reliability of truck transport; structural changes in the economy that favor road transport; and weakening of previously strong government support for rail systems, which has led to deterioration in the systems. Railroads serve well those markets with large volumes of commodities moved between fixed points, but are less effective where timeliness or flexibility in delivery are more important, as indicated in table 5-5. Technologies such as containerization and improved scheduling (aided by telecommunications and computing technologies) can help use all elements of the freight transport system in an integrated manner.

The energy efficiency of *passenger* transport can also be improved by greater use of more efficient modes. In the passenger transport sector there is a wide variation in energy use per passenger-kilometer with private auto by far the least efficient and

Table 5-5—Rail and Truck Shares for Different Commodities, India 1978-79

Commodity	Average distance hauled (kilometers)	Rail share (percent)	Road share (percent)
Iron ore	526	99.3%	0.7%
Coal	699	92.2	7.8
Cement	614	77.9	22.1
Fertilizers	794	71.9	28.1
Iron and steel	841	64.3	35.7
Stone and marble	290	49.4	50.6
Wood and timber	564	38.6	61.4
Building materials	242	13.1	86.9
Fruit and vegetables	532	8.9	91.1

SOURCE: Martin J. Bernard III, "Rail vs. Highway: A Difficult Intercity Transport Decision for a Developing Country," paper presented at the United Nations Center for Science and Technology for Development Workshop on New Energy Technologies and Transportation in Developing Countries, Ottawa, Canada, Sept. 20-22, 1989.

buses—the backbone of the urban passenger transport system in developing countries—the most energy efficient of the motorized forms of passenger transport. The inclusion of energy embodied in associated infrastructure further increases the attractiveness of buses (compared with cars, and two and three wheelers). In addition to the energy benefits of buses, capital costs of bus systems are lower than the alternatives. Despite these advantages, bus systems in developing countries often provide unsatisfactory service, not offering in their present form a sufficiently attractive alternative to private transport for those who can afford it.

Light rail systems, at present under construction or planned in 21 cities of the developing world, are generally more energy efficient than cars though less so than buses. According to a recent evaluation, light rail has achieved mixed results. Travel times were reduced, most systems provided high quality services, and ridership levels were high—although below forecasted levels. These systems require extensive subsidies, however, if fares are kept low enough to provide service to the poor.

Although important, technologies for transport efficiency are just one component of developing an efficient and cost effective transport system. Transport management schemes can also be a cost effective way of relieving congestion and thus improving the overall efficiency of the transport system. The success of such schemes in several cities of the developing world (e.g., Curitiba, Brazil and Singapore) testifies to their potential. Improved telecommunications can reduce the number of needed trips. Improved land use and transport planning—for example, by locating services (e.g., shopping) closer to intended customers, siting major freight terminals away from congested city centers, and promoting the integration of residential and employment centers—can reduce the demand for travel and thereby reduce transport energy use. These options are most promising for cities that are not yet fully developed but are growing rapidly—a common characteristic of cities in the developing world. Without a carefully implemented land use planning program, a private vehicle-based transport system is the likely end result by default. It is easy to add one automobile and one short stretch of road at a time; it is much more difficult to plan a comprehensive urban transport system.[7]

High and rising levels of urban air pollution are also focusing attention on fuels to replace gasoline and diesel in both industrial and developing countries. Several fuels, including CNG (compressed natural gas), ethanol, and methanol, not only reduce pollution but can often use domestic, rather than imported, energy resources. Several of these technologies are particularly suited to centrally fueled fleet vehicles, which form a larger share of the total fleet than in the industrial countries. CNG is an especially promising fuel for those countries with underutilized natural gas reserves. Alcohol fuels, although they are still more expensive than conventional fuels at current oil prices of $20/bbl, are becoming more cost competitive. Several of these fuel replacements are already widely used in some developing countries. Argentina had more than 37,000 CNG vehicles on the road as of July 1989—with the total increasing by 1,300 per month,[8] and Brazil had more than 4 million ethanol-fueled vehicles on the road in 1988.[9]

A substantial number of trips—performed largely by the urban and rural poor—for both passenger and freight traffic are by nonmotorized modes, including walking, push carts, animal drawn carts, and bicycles. Technologies exist—such as improved bullock carts and harnesses—to improve the efficiency of all of these modes. One obstacle to their more rapid diffusion is lack of capital for even the most modest improvements.

The incomplete diffusion of known, cost effective transport technologies with substantial energy and environmental benefits suggests the need for stronger

---

[7]Martin J. Bernard III, "Rail vs. Highway: A Difficult Intercity Transport Decision for a Developing Country," paper presented at the United Nations Center for Science and Technology for Development Workshop on New Energy Technologies and Transportation in Developing Countries, Ottawa, Canada, Sept. 20-22, 1989.

[8]Jorge Del Estado, "National Substitution Plan of Liquid Fuels: Compressed Natural Gas," paper presented at the United Nations Center for Science and Technology for Development Workshop on New Energy Technologies and Transportation in Developing Countries, Ottawa, Canada, Sept. 20-22, 1989.

[9]Jacy de Souza Mendonca, "The Brazilian Experience With Straight Alcohol Automobile Fuel," paper presented at the United Nations Center for Science and Technology for Development Workshop on New Energy Technologies and Transportation in Developing Countries, Ottawa, Canada, Sept. 20-22, 1989. Note, however, that the fluctuations in oil prices before, during, and after the gulf war have caused some conversions, when oil prices were low, back to gasoline and created some concerns about the direction of the Brazilian program and its cost effectiveness.

Photo credit: Ed Smith

Many people in developing countries rely on human muscle power for transportation energy.

incentives (or the removal of disincentives) for their adoption (see table 5-6). Such incentives could take the form of providing information on fuel efficient technology; higher fuel prices where appropriate; pricing policies and procedures that incorporate the environmental costs and benefits of transport fuels; acquisition and ownership taxes on motor vehicles inversely proportional to their energy efficiency; and setting standards on energy efficiency or environmental impacts (as seems to be the approach followed by cities such as Mexico City and Sao Paulo concerned about urban air pollution). Driver training and education could also improve energy operating efficiencies in many cases.

It is important to note, however, that people using energy ''inefficiently'' in a technical sense are nevertheless generally operating logically within their framework of incentives and disincentives. For example, bus drivers in developing countries often leave their vehicles idling for long periods, at a significant cost in fuel. This is not done through lack of training (as commonly suggested), but as a practical response to the problems they face. These problems include: 1) broken starters, requiring a difficult push start; and/or 2) the perception by potential customers that if the vehicle is not running then it must not be in service, resulting in them going to a competitors' vehicle.[10] Thus, as long as a driver's competitors keep their vehicles running, so must he.[11]

As the efficiency of transport energy use is closely correlated with the overall efficiency of the transport sector, policies to promote better and more efficient transport systems could simultaneously have economic and environmental benefits. Segregated bus lanes, truck climbing lanes,[12] traffic management systems, side walks, bicycle paths, and others can all reduce congestion, thus increasing the average speeds necessary for more fuel efficient operation and making bus services more attractive to potential patrons. These changes can also defer expensive construction of new roads. Making all users pay the full cost of using road space, parking space, accidental injury and death, environmental effects, and other impacts could also encourage more efficient use of existing road space. Integrated freight transport policy could combine the strengths of all the different modes—road, rail, and maritime transport where available.

U.S. policies and programs could assist in some of these initiatives:

- providing information and technical assistance on such topics as the technical and operational characteristics of road vehicles, the connection between transport energy efficiency and urban air quality, and rural transport technologies;
- as the multilateral development banks make major loans for transport infrastructure projects, U.S. influence on the development of these projects could ensure that energy efficiency and environmental considerations are taken into account; and
- although the main barriers to improved energy technology at present do not appear to be related to technology, improvements in fuel efficiency depend critically on developments in those countries—mainly Japan, Italy, France, Germany and the United States—that dominate global road transport technology. Developing countries share of the global market will, however, rise substantially in coming years, giving them increased influence on technology

---

[10]Personal communication, S. Padmanabhan, Energy and Environmental Policy, Innovation, and Commercialization (EPIC) Program, U.S. Agency for International Development, Office of Energy, June 9, 1991.

[11]This type of situation is explored in depth in studies of game theory.

[12]Special lanes for trucks on steep slopes enable other traffic to pass more quickly.

Table 5-6—Policy Options

Alternative financial arrangements
The high sensitivity of consumers to the initial capital cost of a vehicle often leads to lower levels of investment in efficiency than is justified on the basis of life cycle operating costs. Alternative financial arrangements to redress this might include low-interest loans for the marginal cost of efficiency improvements, rebates, or other schemes. These might be financed by various fuel or other taxes. Note that users that would have purchased efficient equipment anyway, however, also get the loan/rebate—the "free rider" problem. This reduces the effectiveness of the loan/rebate programs by raising the cost per additional user involved. This problem can be minimized by restricting the loans/rebates to only the very highest efficiency vehicles for which there is little market penetration.

Alternative fuels infrastructure investments
See Office of Technology Assessment, U.S. Congress, *Replacing Gasoline: Alternative Fuels for Light-Duty Vehicles*, OTA-E-364, September 1990.

Availability
Provide incentives to manufacturers to market high efficiency vehicles within the country when, otherwise, the potential market would be too small to justify the effort on the part of the manufacturer.

Data collection
The range of opportunities for energy efficient transport systems, end-user preferences, and operating conditions are not well known in many countries. Data collection, including detailed field studies, would help guide policy decisions.

Demonstrations
Many potential users of energy efficient vehicles remain unaware of the potential savings or unconvinced of the reliability and practicability of these changes under local conditions. Demonstration programs—such as government purchase of high efficiency vehicles—can show the effectiveness of the equipment, pinpoint potential problems, and in so doing convince potential users of the benefits of these changes.

Infrastructure investments
Improve roads, railways, etc. to allow higher average speeds, heavier loads, more timely delivery, etc.

Information programs
Lack of awareness about the potential of energy efficient vehicles and related equipment, traffic management improvements, modal shifts, land use planning, or other improvements can be countered through a variety of information programs; competitions and awards for efficiency improvements; etc.

Labelling programs
The efficiency of vehicles and related equipment can be labeled to provide purchasers a means of comparing alternatives. Measuring efficiencies, however, needs to be done in conjunction with standardized test procedures, perhaps established and monitored by regional test centers, rather than relying on disparate and perhaps misleading manufacturer claims.

Land use planning
Land use planning can be a particularly effective, but long term, means of improving system wide transport efficiencies. It does this by better matching residences to jobs, schools, shopping, and transport corridors, etc. to minimize transport distances and maximize the opportunities for using mass transport options. A phased approach—beginning with dedicated busways to hold and maintain right-of-ways and to encourage high density, mixed job and residential land use; followed by the replacement of buses with light-rail systems where necessary—might then be an effective approach (if combined with other policy measures such as appropriate land-use planning; full pricing of automobile use—fuel, roads, parking, congestion, injury and death, etc.; effective traffic control; and others) in developing a comprehensive and effective urban transport system.

Loans/rebates
See Alternative financial arrangements, above.

Marketing programs
A variety of marketing tools might be used to increase awareness of energy efficient vehicles and increase their attractiveness. These might include radio, TV, and newspaper ads, billboards, public demonstrations, product endorsements, and many others.

Non-motorized transport
Provide full consideration, including reasonable investment, in nonmotorized transport design and infrastructure—such as segregated bicycle paths, broad intersections, curb cuts, etc. Provide training to transport designers to assist consideration and incorporation of design features for nonmotorized transport.

(continued on next page)

Table 5-6—Policy Options—Continued

Pricing policies
Energy prices should reflect the full costs of supply. The structure of petroleum product prices should not be distorted; for example, gasoline prices should not be out of line with diesel or kerosene, or else inefficient and unplanned for substitutions will take place. Prices alone, however, are often insufficient to ensure full utilization of cost effective energy efficient vehicle technologies. There are too many other market failures as discussed above. As evidence of this, even the United States has adopted efficiency standards for vehicles.

R&D
Most vehicle R&D is done in the industrial countries with industrial country operating conditions and needs the primary considerations. As developing country markets for vehicles grow, however, they are likely to gain greater leverage over this research agenda.

Secondhand markets—standards, scrappage
Efficiency labels or standards might be set for secondhand vehicles and equipment. Alternatively, registration fees might be increased with the age of the vehicle, bounties might be offered for the oldest of vehicles, or other means employed—such as emissions standards—to retire the worst cars.

Social costs
Users of transportation should be assessed for the full costs of the transport services they receive. These include especially medical and economic costs (leave from job, disruption of office, etc.) associated with death or injury due to car accidents, as well as air pollution, and other social costs.

Standards for vehicles
Many industrial countries have chosen to use minimum efficiency standards for vehicles rather than market driven approaches, due to the difficulty of overcoming the numerous market failures noted above.

Tax credits, accelerated depreciation, tariffs
A variety of tax incentives—tax credits, accelerated depreciation, reduced import tariffs etc.—could be used to stimulate investment in energy efficient vehicles. Conversely, taxes or tariffs—such as gas guzzler taxes/tariffs on vehicle purchases or registration fees that increase as vehicle efficiency decreases—could be used to reduce investment in inefficient vehicles or equipment.

Traffic management
Improved traffic management, including, high occupancy vehicle lanes, segregated bus lanes, truck climbing lanes, bicycle lanes, timed lights, etc.

Training programs
Training programs may be needed in order to ensure adequate technical manpower for maintenance of advanced energy efficiency equipment in vehicles. Driver training programs can also be useful.

Transport planning
Transport planning can, in conjunction with land use planning, help ensure efficient choice of transport modes, effective integration of different modes, etc.

User costs
Users of transportation infrastructure—roads, parking, etc.—should pay the full cost of using these facilities.

SOURCE: Office of Technology Assessment, 1992.

Table 5-7—Road Vehicle Fleet Growth and Ownership in Selected Countries

Country	Annual growth (percent/year, 1982-86)				Ownership (1986 vehicles/1,000 people)
	Autos	Trucks buses	2 and 3 wheelers	Total	
Cameroon	11.8	29.5	9.1	13.1	16
Kenya	3.2	3.7	4.0	3.3	11
Bolivia	8.6	24.5	6.9	11.6	37
Brazil	8.9	7.3	35.6	9.8	106
Thailand	8.8	4.4	9.5	8.8	64
India	8.2	11.2	25.4	18.4	10
China	41.6	14.8	44.9	29.8	7
Japan	3.0	4.1	7.0	4.4	538
U.S.	2.4	3.5	-5.6	2.4	44
W. Germany	3.3	0.4	-2.2	2.6	511

Vehicle ownership includes autos, trucks, buses, and 2 & 3 wheelers. Nonmotorized transport not included.
SOURCES: Energy and Environmental Analysis (EEA), "Policy Options for Improving Transportation Energy Efficiency in Developing Countries," contractor report prepared for the Office of Technology Assessment, July 1990, pp. 2-15, 2-16.

development. Advances in alternative fuel technologies could help developing countries in their search for substitutes for petroleum products.

TRANSPORTATION SYSTEMS IN DEVELOPING COUNTRIES

With the important exceptions of rail in China and India, commercial transportation in the developing countries is road based—trucks, buses, motorcycles, and automobiles.[13] Road vehicle fleets are growing rapidly in the developing world (see table 5-7). The expansion of the road vehicle fleet is particularly rapid in India and China, where road transport has until recently played a relatively small role in overall transport services. In several countries, there has been a sharp increase in growth of two and three wheelers. These vehicles provide a high level of personal mobility at less cost to the purchaser than an auto. Ownership of road vehicles in developing countries is growing rapidly, but is still much lower than in industrial countries.

Over 90 percent of the transport energy used in the developing world is in the form of oil, with the remaining 10 percent largely coal for use in rail in India and China (see table 5-8).[14] Over half of the oil consumed for transport is in the form of diesel for freight transport (see table 5-9). This contrasts with the United States, which uses most of its transport fuel in the form of gasoline for autos.

The energy requirements of transport systems in developing Asian countries have grown at an average annual rate of 4.5 percent between 1980 and 1987.[15] According to a recent World Bank study,[16] this rapid growth could well continue, thus exacerbating financial and environmental problems.

Although the commercial transport sector attracts most attention, most of the population in developing countries live in rural and poor urban areas and are largely dependent on local informal transport systems—particularly walking, animal power, and bicycling. As a transport mode, walking is flexible and low cost, but it is slow, tiring, and very limited in hauling capacity. Animal technologies, such as bullock carts, have a much greater freight capacity than walking, but are more costly. They require considerable human effort to maintain the animals, capital investment in a bullock and cart, and operational costs for feed and veterinary services.

Bicycles are widely used. In China, for example, 50 to 90 percent of urban vehicle trips are made by bicycles. Although they are a highly energy efficient mode of transport, bicycles often require more roadway per passenger than buses due to their slower average speeds and less compact seating. Like other

---

[13] In China most (89 percent) rail energy use is in the form of coal, while in India rail energy use is split between coal (73 percent) and oil (22 percent).

[14] International Energy Agency, *World Energy Statistics and Balances 1971-1987* (Paris: Organization for Economic Cooperation and Development, 1989), data are for 1985, pp. 120, 124, 128.

[15] Ibid., pp. 118-128.

[16] M. Imran, op. cit., footnote 5.

Table 5-8—Transport Energy Use, 1985 (percent)

	China[a]	India	Brazil	Nigeria	U.S.
Fuel type (percent)					
Oil	52	79	100	100	100
Coal	47	20	0	0	0
Electricity	1	1	0	0	0
Mode (percent)					
Road	38	65	85	91	82
Rail	52	27	3	0	3
Air	1	7	7	9	14
Water	9	1	5	0	1
Total (exajoules)	1.03	0.92	1.13	0.27	18.65

[a]China mode percent data are for 1980, total is for 1985.

SOURCES: J. Yenny and L. Uy, *Transport in China*, World Bank Staff Working Papers Number 723, 1985, p. 35; International Energy Agency, *World Energy Statistics and Balances 1971-1987*, OECD, Paris, 1989, p. 366 (China), p. 784 (India), p. 256 (Brazil), p. 191 (Nigeria); International Energy Agency, *Energy Balances of OECD Countries 1970/1985*, OECD, Paris, 1987, p. 541.

Table 5-9—Annual Consumption of Diesel and Gasoline Fuel (1986)

Country	Diesel (1,000 metric tons)	Gasoline (1,000 metric tons)	Percent diesel	Diesel (kg per capita)	Gasoline (kg per capita)
Cameroon	296	276	52	28	26
Kenya	515	220	70	24	10
Bolivia	266	308	46	40	47
Brazil	18,004	10,182	64	130	74
Thailand	4,878	1,654	75	93	31
India	16,700	2,508	87	21	3
China	19,785	15,258	56	19	14
Taiwan	2,786	2,269	55	136	111
United States	68,056	290,113	19	282	1,201

Adapted from: Energy and Environmental Analysis (EEA), "Policy Options for Improving Transportation Energy Efficiency in Developing Countries," contractor report prepared for the Office of Technology Assessment, July 1990, pp. A-3, A-6.

forms of animate energy, the range and freight capacity of bicycles are limited, accounting for the decisive popularity of the internal combustion engine for both passenger and freight transport as soon as it can be afforded.

IMPROVING FREIGHT TRANSPORT ENERGY EFFICIENCY

In developing countries, freight transport accounts for a much higher share of total transport fuels—about one half—than in the industrial countries. Freight movement in the developing world is principally by truck, with the exception of India and China, which also use rail. Options for improving the energy efficiency of freight transport include:

1. improving the energy efficiency of trucks;
2. improving the energy efficiency of rail; and
3. greater use of the more efficient of these two modes.

The Truck Fleet

At first sight there appears to be considerable potential for improving the energy efficiency of developing countries' truck fleets. While there is wide variation in the types of trucks used in developing countries, their fleets are, in general, older, smaller, and less technologically sophisticated than trucks found in the industrialized countries—all factors that result in lower energy efficiencies (see table 5-10).

Developing-country truck fleets are older because vehicle scrappage rates (i.e., the fraction of the fleet that is scrapped per year) are much lower than in industrial countries. In developing countries, new trucks are expensive to buy, and if they need to be imported, foreign exchange may not be available. On the other hand, due to low labor costs, repair is relatively cheap. The result of these two reinforcing factors is that it is usually cheaper and easier to repair and patch up a vehicle than to replace it. In general, it is not easy to retrofit trucks with

Table 5-10—Energy Efficiency of Trucks in Selected Countries

Country/region	Truck name	Capacity (metric tons)	Energy consumption (megajoules per metric ton per kilometer)
OECD	Mercedes Benz 1217 (1979)	7.0	1.0
OECD	Man-VW 9136 (1980)	5.9	1.0
India	TATA 1201 SE/42	5.0	2.1
India	Ashok Leyland Beaver	7.5	1.6
China	Jiefang CA-10B	4.0	2.3
China	Dongfeng EQ140	5.0	1.8

NOTE: OECD and Indian trucks use diesel, Chinese trucks use gasoline.
SOURCE: J. Yenny and L. Uy, World Bank, *Transport in China*, Staff Working Paper No. 723, 1985, p. 70.

efficiency improvements once they are operating. An exception is periodic engine rebuilding. If this rebuild is done with more modern technology, such as improved fuel injectors and injection pumps, engine efficiency gains of up to 5 percent can be obtained.[17]

Many local shops in developing countries, however, do not have the technical expertise to rebuild engines with these improved technologies. This leads to increased downtime and transport costs as engines have to be sent to the factory for rebuilding. Increased attention to training in local shops could reduce these costs and make energy saving investments more attractive.

Low scrappage rates could be discouraged by policies such as the introduction of annual registration fees inversely proportional to vehicle age. Such policies could, however, involve heavy foreign exchange costs, at least in the short run. The long lifetimes of trucks in the developing world also emphasizes the importance of building efficiency into new trucks, as these trucks will continue to operate for many years.

The size of a truck has an important bearing on its overall energy efficiency. Small trucks, in particular, generally require more energy to move a ton of freight than large trucks. On average, developing world trucks are smaller than trucks found in the industrialized countries. Chinese trucks are mostly 4 to 5 ton,[18] and the largest Indian trucks are typically rated at 8 to 9 tons (although they routinely carry up to 14 tons).[19] These sizes are considerably less than the 20-ton trucks used in the United States. The poor highway infrastructure in many developing countries, however, constrains the use of larger trucks. Less than 20 percent of China's highways are paved, for example.[20] Increases in truck carrying capacity and the energy efficiency advantages resulting from this, cannot be implemented without corresponding improvements in road carrying capacity.

Developing world trucks are, on average, less technologically sophisticated than industrialized country trucks. Principal truck manufacturers in India build diesel engines comparable in technological sophistication and efficiency with those built in the industrialized countries in the 1960s. This is changing. For example, one of the dominant Indian truck manufacturers now offers an optional Japanese engine with improved fuel efficiency.[21] Much of China's truck fleet uses relatively inefficient gasoline engines.[22] The share of more energy efficient diesel trucks is growing,[23] however, and much of the vehicle-related foreign investment in China involves

---

[17] Energy and Environmental Analysis, Inc., "Policy Options for Improving Transportation Energy Efficiency in Developing Countries," contractor report prepared for the Office of Technology Assessment, July 1990, pp. 5-6.

[18] J. Yenny and L. Uy, "Transport in China," World Bank Staff Working Papers Number 723 (Washington, DC: World Bank, 1985), p. 39.

[19] Martin J. Bernard III, U.S. Department of Energy, Argonne National Laboratory, *Transportation-Related Energy Problems in India*, Report No. ANL/EES-TM-163 (Argonne, IL: Argonne National Laboratory, September 1981), p. 7.

[20] J. Yenny, op. cit., footnote 18.

[21] Energy and Environmental Analysis, Inc., op. cit., footnote 17.

[22] Ibid. In general, diesel engines are more efficient than gasoline engines, due in part to the thermodynamic advantages of the compression ignition (diesel) cycle over the spark ignition (gasoline) cycle.

[23] In 1978, 13 percent of the trucks in China used diesel; by 1988 this had increased to 18 percent. Motor Vehicle Manufacturers Association of the United States, Inc., *World Motor Vehicle Data, 1990 Edition* (Detroit, MI: Motor Vehicle Manufacturers Association of the United States, Inc., 1990), p. 46.

diesel truck engines.[24] Brazil assembles trucks designed in Western Europe and exports these relatively efficient trucks to all of Latin America. Many countries, particularly in Africa, import modern trucks from Europe and Japan, and therefore in principle have access to the most advanced truck technologies.

In operation, however, the energy benefits of these up-to-date technologies may not be fully realized under developing country conditions. The benefits of aerodynamic improvements and turbocharging accrue primarily at higher speeds, which are often not possible on the rougher, more congested roads in the developing world. High-efficiency engines are often dependent on high-quality fuels, and do not operate as well under the conditions of variable quality fuels frequently encountered in developing countries. Maintenance requirements are more complex and more critical to engine performance. As the share of developing countries in markets for trucks rise, they may be able to increase their leverage on manufacturers in the industrial world to provide technologies more suitable to the conditions of the developing world.

Further, the cost effectiveness of the various technological options may be compromised by energy pricing policies widely followed in developing countries. For social and development reasons, many governments keep diesel prices as low as possible. In oil exporting countries, prices are generally well below levels reflecting international costs plus distribution costs. In oil importing countries, diesel prices are usually near international prices or slightly higher but often do not bear the high taxes typically imposed on gasoline. The high cost of capital combined with relatively low diesel prices discourages energy efficiency improvements. For example, the installation of a turbocharged engine at a $800 to $1,000 premium for a relatively small (5 percent) benefit in fuel economy is cost effective, but only marginally so: a truck achieving 10 miles per gallon (23 liters/100 km), driven 40,000 miles (64,000 km) per year, and paying $1/gallon ($0.26/liter) for diesel will have a payback of about 4.7 years on such an investment.

Improved lubricants for the gearbox, engine, and axle can provide small benefits (1 to 2 percent) in fuel economy. At the low speeds typical in developing countries, however, aerodynamic improvements are not very effective in reducing fuel consumption. Improved drivetrain matching is also problematic because of the propensity of truck owners to overload vehicles. Tire improvements offer limited potential, as radial tires of modern design provide only a 2 to 3 percent efficiency benefit over bias-ply tires. Total improvements of 5 to 8 percent in diesel fuel economy may be all that is possible with current technologies under developing country conditions when using trucks.

Technology alone thus may have a limited role to play in improving truck efficiencies. Proven efficiency improvements—the use of larger trucks, travel at higher, constant speeds allowing the use of aerodynamic improvements and turbochargers—require smooth, uncongested, paved heavy duty roads. Furthermore the efficiency of diesel engines drops sharply under conditions of varying load and speed, such as are often found on the congested, poorly maintained roads of the developing world. Advanced truck technologies under development in the industrialized countries, such as further improvements in aerodynamics, low rolling resistance tires, and multiple turbochargers, also require smooth, high-speed roads.

Road construction and improvement is expensive, and the benefits are diffuse and often difficult to measure. For these reasons, the public sector, including the international development agencies, has long been the principal source of funds for road building and repair. The energy efficiency effects of improved roads are complex. In the short term, improved roads increase energy efficiency by allowing for higher, sustained speeds. By one estimate, heavy truck operating costs per kilometer (due largely to energy) drop 34 percent when a dirt road is paved.[25] In the longer term, improved roads allow for the use of larger, heavier trucks as well as the use of turbochargers and aerodynamic improvements. Improved roads may also encourage increased traffic (and therefore increased energy use), but this increased traffic may contribute to overall economic development in an efficient manner.

---

[24]C. Oman, *New Forms of Investment in Developing Country Industries* (Paris: Organization for Economic Cooperation and Development, 1989), p. 160.

[25]H. Adler, World Bank, *Economic Appraisal of Transport Projects* (Baltimore, MD: Johns Hopkins University Press, 1987), p. 120.

Operational improvements in trucking have the potential to improve freight energy efficiencies. Load factors are often low. Improved communications, route scheduling, and overall coordination of the freight transport system can improve energy efficiency by ensuring full loads and reducing waste. Part of this low load factor, however, is rooted in policy and institutional factors. Many enterprises in developing countries do their own shipping rather than rely on common carrier fleets that they find less reliable. These "own account" shippers are usually licensed to carry their own products only, resulting in empty backhauls and low average load factors.

Improving Rail Efficiency

Rail systems carry a significant fraction of freight only in India and China, but as these countries account for about half of the developing world's population, their rail systems represent a large part of the developing world's freight system.

The rail systems of India and China use a mix of coal, diesel, and electric locomotives, although coal locomotives are being phased out (see table 5-11). The substitution is more advanced in India than in China. Coal still accounts in China for about 80 percent of total rail freight tonne-kilometers.[26] The energy benefits of this technological change are considerable, as diesel locomotives are about five times more energy efficient than coal locomotives.[27] There appears to be no decisive energy benefit in switching from diesel to electric locomotives,[28] however, which in any event tend to be economically efficient only for high traffic densities.

In addition to the energy efficiency benefits associated with a conversion from coal locomotives, U.S. experiences suggest that there are other opportunities for energy efficiency improvements within the rail sector. Improved operations, such as reducing empty car miles and matching loads and speeds

Table 5-11—Characteristics of the Indian Rail System

	Number of locomotives	
	1971	1987
Steam (coal)	9,387	4,950
Diesel	1,169	3,182
Electric	602	1,366

SOURCE: Tata Energy Research Institute (TERI), *Teri Energy Data Directory and Yearbook,* New Delhi, 1988, pp. 165, 170.

to the locomotive's optimum operating characteristics, can reduce energy use. Lighter weight cars, low resistance bearings, and onboard flange lubricators all contributed to a 32-percent increase in energy efficiency (tonne-kilometers per unit energy) in U.S. freight railroads from 1970 to 1987.[29] Other promising technological innovations include steerable trucks for cars and locomotives. Computers to assist drivers in efficient operation can reduce consumption by up to 25 percent.[30]

Modal Shifts

In principle it is possible to improve the energy efficiency of the entire freight transport system by ensuring that the bulk of freight is carried in the most efficient mode. The relative operating energy efficiencies of rail and road for freight depend on several factors, including how heavily the vehicles are loaded and whether or not the vehicles are loaded for the return trip. Although estimates vary, it is generally agreed that, on average, rail is more energy efficient than road for freight.[31] The trend away from rail therefore represents a loss in energy efficiency. The magnitude of this loss depends on the specific system, but a rule-of-thumb number is that operating energy efficiencies in rail freight transport are 3 to 4 times as energy efficient as truck freight transport.[32] If the comparison of energy efficiencies between the two systems is expanded to include not only the amount of energy used for operating the train or

---

[26] J. Yenny op. cit., footnote 18.

[27] J. Dunkerley, *Trends in Energy Use in Industrial Societies,* Resources for the Future, Inc., research paper R-19, 1980, p. 149.

[28] The efficiency of electric locomotives depends on the fuel used to generate the electricity. For electricity produced from fossil fuels the energy efficiency of diesel and electric locomotives are comparable. See L. Alston, *Railways and Energy,* World Bank Staff Working Paper Number 634 (Washington, DC: World Bank, March 1984), p. 31.

[29] U.S. Department of Energy, Oak Ridge National Laboratory, *Transportation Energy Data Book: Edition 10,* Report No. ORNL-6565 (Springfield, VA: National Technical Information Service, September 1989), pp. 4-23. During this time period the average trip length increased 34 percent, therefore at least part of this increase in efficiency was probably due to the inherent efficiency of longer trips.

[30] L. Alston, op. cit., footnote 28, pp. 14-15.

[31] According to a World Bank report, "...railways have a substantial energy advantage for large volumes of bulk commodities, but for passenger transport they are generally no more energy-efficient than buses." Ibid., p. I.

[32] M. Bernard, op. cit., footnote 7. Also L. Alston, op. cit., footnote 30, p. 7.

Table 5-12—Systemwide Energy Use by Alternative Transportation Systems

	Operating energy	Linehaul energy	Modal energy
	United States		
Freight	(Btu/ton-mile[a])	(Btu/ton-mile)	(Btu/ton-mile)
Rail	660	1,130	1,720
Truck	2,100	2,800	3,420
Barge	420	540	990
Passenger	(Btu/pass-mile[b])	(Btu/pass-mile)	(Btu/pass-mile)
Auto	8,490	10,790	10,790
Bus	2,740	2,950	3,220
Metro	3,570	4,550	6,690
	Latin America		
Passenger	(Btu/pass-mile[b])	(Btu/pass-mile)	(Btu/pass-mile)
Auto	4,820	7,130	7,130
Bus	603	660	660
Metro	760	990	1,130

"Linehaul" is the total energy of the vehicle plus its related highway or rail infrastructure.
"Modal" includes the total energy to get door to door using primarily the mode listed. This also includes related infrastructure energy requirements as for linehaul.
[a]To convert to Joules/tonne-km, multiply by 595.
[b]To convert to Joules/passenger-km, multiply by 656, pass-mile=passenger mile.
SOURCE: Damian J. Kulash, "Energy Efficient Modes of Transportation: Comments on Urban Transportation in the United States and Latin America," (Washington, DC: Congressional Budget Office), February 1982, pp. 6, 220.

truck, but also the amount of energy contained in the manufacture of the transport equipment and associated railways and roads, and the amount of energy needed to gain access to each mode, then the gap between them is reduced. Railways then appear to be about twice as efficient as roads rather than the 3 or 4 times when only operating efficiencies are taken into account (see table 5-12).[33]

Despite the fact that rail transport is more energy efficient and cheaper than road freight transport, its share of total freight is declining in both India and China (see table 5-13). Road transport has the advantage of being more flexible and accessible. For rural areas it is essentially the only transport mode available. Door to door delivery possible with road transport avoids the risk of delay due to transshipment when using rail. As the demand for consumer goods increases, the demand for transport services that can deliver high value-added goods to many areas quickly increases.

Changes in government policy with regard to freight transport have also played a part in the relative decline of rail. In many countries throughout the world—in Europe as well as the developing countries—railways are State owned. Strenuous efforts have been made by governments through investment and pricing policies and regulations to reserve a dominant share for railways in the freight transport system. In recent years, however, this policy has been reversed. Investments in the railway system have failed to rise as they did previously. In India, for example, railroad's share of public investment in transport has decreased,[34] leading to poorer service and maintenance. At the same time some deregulation was introduced into the road transport system. Given the declining performance of rail freight and the growing opportunity to ship by road, customers have chosen road transport despite its frequently higher cost per tonne-kilometer (and its higher energy intensity).

The key issue is, however, to use all elements of the freight transport system in an integrated manner. The two systems have strong complementarities as railways (and waterways and coastal carriers where available) have a decisive advantage in the transport of large loads, especially of bulk commodities like coal, cereals, steel, fertilizer, etc., over long distances between fixed points, while road transport is preferred for the transport of small- and medium-sized loads over relatively small distances to dis-

---

[33]These concepts were developed and further explained in Damian J. Kulash, U.S. Congress, Congressional Budget Office, "Energy Efficient Modes of Transportation: Some Comments on Urban Transportation in the United States and Latin America" (Washington, DC: February, 1982), p. 6.
[34]Tata Energy Research Institute (TERI), op. cit., footnote 1, p. 161.

Table 5-13—Market Share of Road and Rail for Freight
(percent of freight ton-kilometers,
excluding water, air, and pipeline)

	1977		1983	
	Road	Rail	Road	Rail
China	5	95	14	86
India	37	63	51	49
U.S.	37	63	41	59

SOURCES: J. Yenny and L. Uy, *Transport in China*, World Bank Staff Working Papers No. 723, 1985, p. 4; Tata Energy Research Institute (TERI), *Teri Energy Data Directory and Yearbook*, New Delhi, 1988, p. 163. U.S. data are for intercity freight traffic in 1980 and 1987, from Association of American Railroads, *Railroad Facts*, 1989 edition, Washington, DC, November 1989.

persed recipients (see table 5-5). Technologies such as containerization, which facilitates transshipment between different modes, encourage such integrated systems. Historically, the separate elements of the freight transport system—road, rail, ports etc.—developed independently and often opposed integration. In many countries, major changes in institutions and administrative procedures will be required to fully implement efficient integrated systems.

IMPROVING PASSENGER TRANSPORT ENERGY EFFICIENCY

The demand for passenger transport, particularly private transport such as autos and two and three wheelers, rises rapidly as incomes grow (see figure 5-1). At the same time, there is wide variation in vehicle ownership rates even in countries with similar standards of living. Bolivia and Cameroon, for example, have similar standards of living (in terms of their purchasing power) but car ownership rates in Bolivia are twice as high as in Cameroon. Similarly car ownership in Thailand is more than nine times higher than in China, although Thailand's standard of living is only one-third higher (taking into account purchasing power). These disparities are due to a variety of factors—the degree of urbanization, the nature of the vehicle stock, and government polices towards motor vehicle ownership. Table 5-14 illustrates the mode share of motorized passenger trips in selected countries.

The energy efficiency of the passenger transport sector can be improved in three primary ways. First, the vehicles themselves can make use of technologies that increase energy efficiency. Second, their operational efficiencies can be improved by carrying larger loads or through improved traffic management. Third, there could be increased use of those modes, such as buses and rail-based systems or walking/bicycling that use less energy per passenger mile. These modes already provide a significant share of trips in some areas (see table 5-15). Land use planning and other government policies strongly influence the viability of these mass transit or nonmotorized modes.

Automobiles

The energy efficiency of automobiles—the most energy intensive of all forms of passenger transport—is of crucial importance due to the rapid rate of growth of automobile ownership. As for trucks, the average energy efficiency of the automobile fleet is held down in many developing countries by low scrappage rates. The reasons for this are much the same—low labor costs for repair, minimal quality requirements for annual registration, and the high cost and limited availability of new vehicles.[35] Measures to increase scrappage rates—through registration fees that are inversely proportional to age, offering bounties for old cars, or establishing safety emissions standards—would increase the average energy efficiency of the auto fleet, but at a financial cost to users. The long life of the average car in developing countries also puts a premium on high standards of energy efficiency in the new cars that are being added to the fleet.

The vehicles currently being sold in the developing countries vary widely in energy efficiency, depending largely on whether such vehicles are manufactured at home or imported. In India, for example, the most popular car until recently was the domestically produced ''Ambassador.'' This vehicle, based on a 1954 British Morris, is still being produced, although the fuel consumption is about twice that of a current and comparably sized Japanese or German car.[36] Energy efficiencies of cars produced in China are similarly low. In both countries, however, new automotive technologies

---

[35] Energy and Environmental Analysis, Inc., op. cit., footnote 17, p. 5-1.

[36] J. Sathaye and S. Meyers, ''Transport and Home Energy Use in Cities of the Developing Countries: A Review,'' *The Energy Journal*, vol. 8, special issue, 1987, pp. 85-103.

162 • *Fueling Development: Energy Technologies for Developing Countries*

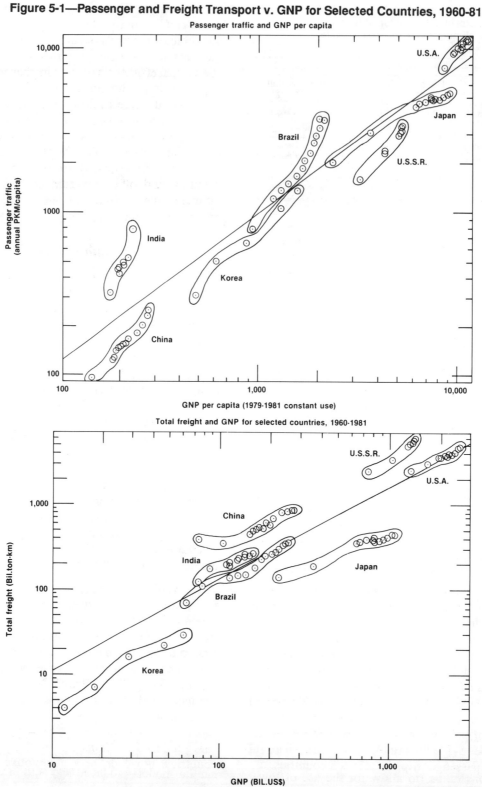

Figure 5-1—Passenger and Freight Transport v. GNP for Selected Countries, 1960-81

This figure shows how passenger and freight transport energy use have increased with GNP for seven countries. The individual data points are for specific years.

SOURCE: J. Venny and L. Uy, *Transport in China*, World Bank Staff Working Paper, No. 723 (Washington, DC: World Bank, 1985).

Table 5-14—Mode Share of Motorized Passenger Trips in Selected Cities, 1980

City	Modal Share (percent)					
	Auto	Taxi	Bus	Para-transit	Rail/subway	Other
Bangkok	25	10	55	10	NA	NA
Bombay	8	10	34	13	34	NA
Calcutta	NA	2	67	14	10	4
Hong Kong	8	13	60	NA	19	NA
Jakarta	27	NA	51	NA	1	21
Karachi	3	7	52	18	6	13
Manila	16	2	16	59	NA	8
Seoul	9	15	68	0	7	0
Bogota	14	1	80	0	0	5
Mexico City	19	NA	51	13	15	2
Rio de Janeiro	24	2	62	2	11	NA
Sao Paulo	32	3	54	NA	10	1
Abidjan	33	12	50	NA	NA	5
Nairobi	45	NA	31	15	0	9
Cairo	15	15	70	NA	NA	NA
Tunis	24	4	61	NA	10	NA
Amman	44	11	19	26	NA	NA
Ankara	23	10	53	9	2	2

NOTE: Excludes nonmotorized modes.
NA = not available.
SOURCE: A. Faiz, et al., *Automotive Air Pollution*, Infrastructure and Urban Development Department, The World Bank, WPS 492, August 1990, p. 39.

have been introduced in recent years, sharply increasing the fuel efficiencies of domestically manufactured cars. The fuel efficiencies of cars produced by other developing country producers—Mexico, Brazil, and South Korea—are at current international standards for their size, accessories, and other factors.

Most developing countries, however, import their autos from the industrialized world, either in finished form (as is the case with small African and Central American nations) or in the form of completely knocked down kits (CKD) from which cars and trucks are assembled. In the latter case, local industry provides many components such as tires, batteries, and light bulbs. Examples of CKD production include Thailand for cars, and Thailand and Taiwan for heavy-duty trucks.

Autos produced in the industrialized countries and then exported to the developing countries are similar but not identical to those sold in the industrialized countries. In general, models sent to the developing world have smaller engines, fewer luxury accessories (e.g., air conditioning, lower compressions ratios (to allow for the use of lower octane gasoline), and often do not use proven

Table 5-15—Motorized and Non-Motorized Mode Shares.

	Percent of passenger trips		
City	Walk	Bicycle	Motorized
Bangalore, India	44	12	44
Delhi, India	29	17	54
Jakarta, Indonesia	23	17	60
Shanghai, China	43	13	44

SOURCES: V. Setty Pendakur, "Urban Growth, Urban Poor and Urban Transport in Asia," Occasional Paper No. 39, The Centre for Human Settlements, The University of British Columbia, 1986, p. 33; J. Sathaye and S. Meyers, "Transport and Home Energy Use in Cities of the Developing Countries: A Review," *The Energy Journal*, Special LDC Issue, vol. 8, 1987.

efficiency technologies such as fuel injection and electronic engine controls. The lack of luxury accessories increases efficiency (air conditioning, for example, adds weight and requires engine power); however, the lack of electronic engine controls and other similar technologies decreases efficiency. The net effect is that autos produced in the industrialized countries and then exported to the developing countries are of comparable energy efficiency (in terms of kilometers per liter of fuel) to similar models produced and sold in the industrialized countries.[37]

---

[37] Energy and Environmental Analysis, Inc., op. cit., footnote 17, pp. 4-15.

These vehicles' energy efficiency could benefit from the use of readily available proven technologies, although at an increased first cost. Electronic control of spark timing and idle speed found in almost all industrialized country vehicles, for example, provides a 4 to 5 percent fuel efficiency gain at a cost of about $75 to $100.[38] Radial tires, improved aerodynamics, and fuel injection also offer similar energy savings. The cost effectiveness of such an investment is strongly dependent on retail fuel price—in Ethiopia, with high fuel prices, the payback is a relatively quick 2 or so years; in the United States, with low fuel prices, the payback is a little over 5 years. Table 5-16 gives payback times for the installation of electronic spark controls.

Such calculations incorporate only increased first costs and fuel savings, and do not account for the environmental and safety benefits. Nevertheless, it illustrates that investments in increased auto efficiency can provide a reasonable return. Evidence from the industrialized world, however, indicates that consumers often demand a payback of 2 years or less from efficiency investments,[39] and consumers in developing countries might be expected to have an even shorter payback period. Although gasoline prices in oil exporting countries are low (see table 5-17), in oil importing countries they are already well above international levels and there is a limit to how much further gasoline prices can be increased without unduly widening the gap between them and diesel prices. In most countries, purchase taxes on autos are already structured to discourage the purchase of large autos, suggesting a role for auto efficiency standards.

Some efficiency features, such as fuel injection, are considerably more complex than the present practice (carburetors), and therefore would require skilled (or differently skilled) labor for repair. Some efficiency features offer benefits in addition to fuel savings. Fuel injection is more reliable, does not require adjustment, and results in lower emissions than carburetors.[40] Radial tires offer improved handling and safety as well as increased tire life.

Developing countries are largely dependent on the industrialized countries for vehicle design. Thus, further improvements in fuel efficiencies will depend on advances in automotive technologies in the industrial countries. Vehicles currently being sold in both the developing and the industrialized world are not nearly as efficient as is technically possible. Several manufacturers, including GM, Volvo, and

Table 5-16—Simple Payback From the Addition of Electronic Spark Controls to Automobiles.

Country	Retail gasoline price (U.S. $/gallon,[a] Jan. 1990)	Simple payback for electronic spark controls (years)
Thailand	1.24	4.1
Brazil	1.41	3.6
India	1.92	2.6
Ethiopia	2.93	1.7
U.S.	1.04	4.9
U.K.	2.55	2.0
Japan	3.05	1.7

Assumptions: First cost is $87.50, initial mpg is 25; efficiency increase due to electronic control is 4.5%, annual mileage is 10,000 miles.
[a]To convert to $/liter, multiply by 0.2642.
SOURCE: Energy and Environmental Analysis (EEA), "Evaluation of Policies Influencing Road Transport Fuel Consumption in Developing Countries," Report to AID, March 1986, p. 4-17; Energy Information Agency, *International Energy Annual 1989*, DOE/EIA-0219(89), February 1991, p. 91.

Table 5-17—Diesel and Gasoline Prices in Selected Countries, as of Jan. 1, 1990, in U.S.$/gallon (including taxes)

	Gasoline	Diesel
Brazil	1.41	0.69
Ecuador[a]	0.52	0.37
Mexico[a]	0.88	0.67
India	1.92	0.78
Thailand	1.24	0.89
Ethiopia	2.93	1.44
Ghana	0.76	0.69
Venezuela[a]	0.24	0.06
Peru	1.28	0.39
Indonesia[a]	0.81	0.42
Pakistan	1.52	0.68
Japan	3.05	1.75
U.S.	1.04	0.99
W. Germany	2.72	1.91
Reference price[b]	0.87	0.70

[a]Oil exporters.
[b]USA refiner sales price of premium gasoline, excluding taxes, and #2 diesel, excluding taxes.
SOURCE: Department of Energy/Energy Information Administration, *International Energy Annual 1989*, February 1991, and Department of Energy/Energy Information Administration, *Annual Energy Review 1990*, May 1991, p. 157.

---

[38] Ibid., pp. 4-17.

[39] H. Ruderman, M. Levine, and J. McMahon, "The Behavior of the Market for Energy Efficiency in Residential Appliances Including Heating and Cooling Equipment," *The Energy Journal*, vol. 8, No. 1, 1987, p. 101.

[40] Fuel injection, however, requires gasoline with low levels of dirt and other contaminants; and the gasoline sold in some developing countries may not meet this requirement.

Volkswagen have built prototype automobiles that achieve from 66 miles per gallon (3.5 litres/100 km) to 70 miles per gallon (3.3 litres/100 km).[41] A prototype automobile introduced by Toyota in 1985 achieves 98 miles per gallon (2.4 litres/100 km, while providing room for 4 passengers. This vehicle uses a direct-injection diesel engine, a continuously variable transmission (CVT), and plastics and aluminum to reduce weight.[42] These vehicles are not in production; nevertheless, the long term technical potential for energy efficiency improvements in automobiles is large, and improvements made in the industries of the major producers will rapidly become the global standard. The rapid growth in the auto fleet in developing countries will give these countries increasing influence over the major car developers.

Motorized Two- and Three-Wheelers

The two-wheeler has appealed to many Asian and other developing nations as an inexpensive technology for providing personal transportation for a growing urban middle class. In many Asian cities, two- and three-wheelers vastly outnumber autos, and are responsible for a large fraction of total gasoline consumption (see table 5-18). The relative efficiency in terms of energy per passenger-mile of two-wheelers falls between that of autos and buses (see table 5-19).

Two-wheeler engines are either "two-stroke" or "four-stroke." In the early 1960s, virtually all but the largest motorcycles had two-stroke engines, since these engines are simple to manufacture and inexpensive. In addition, they produce more power for a given displacement and require little maintenance. Two-stroke engines, however, have emissions (largely unburned gasoline) 10 times greater and fuel efficiencies 20 to 25 percent lower than four-stroke engines of equal (or near equal) power (see table 5-20).

The problems associated with the two-stroke engine are still more acute for three-wheelers, of which India is a large producer. These vehicles are underpowered and the engine is usually operated at near wide-open throttle. Under these conditions, two-stroke engines produce high emissions and have poor fuel economy, with fuel consumption equivalent to that of some modern small cars. Yet there is a widespread perception in India and in some Asian countries that these "auto rickshaws" are very efficient modes of public transport. Two-stroke two-and three-wheelers are now a significant source of gasoline consumption and emissions in several developing countries.

Improved technologies are available, although at increased first cost to the user, which could drastically reduce emissions and fuel consumption. The use of improved carburetors and electronic ignition could improve efficiency 10 to 15 percent, and would reduce hydrocarbon (HC) emissions by 50 percent.[43] The use of four-stroke rather than two-stroke engines would reduce HC emissions by 90

Table 5-18—Estimated Fraction of Gasoline Consumption Attributable to Two- and Three-Wheelers (1987)

Country	Percent of gasoline
India	45
Thailand	30
Taiwan	50
China	12
Bolivia	8
Brazil	2
Kenya	<2
Cameroon	12
Japan	14
U.S.A.	<1

SOURCE: Energy and Environmental Analysis (EEA), "Policy Options for Improving Transportation Energy Efficiency in Developing Countries," contractor report prepared for the Office of Technology Assessment, July 1990.

Table 5-19—Relative Energy Consumption of Passenger Transport Technologies, United States

Technology	Energy use per passenger-mile (MJ/passenger-km)	Load factor assumed (passengers/vehicle)[a]
Private auto	2.5	1.7
Motorcycle	1.5	1.1
Rail	1.5	24.8
Bus	0.6	10.2

[a]Load factors in developing countries will generally be much higher than those listed here for the United States.

SOURCES: Oak Ridge National Laboratory, *Transportation Energy Data Book: Edition 10*, ORNL-6565, 1989, p. 2-23; M. Lowe, "Reinventing the Wheels," *Technology Review*, May/June 1990.

---

[41]M. Ross, "Energy and Transportation in the United States," *Annual Review of Energy*, vol. 14, 1989, p. 158.
[42]J. Goldemberg et al., *Energy for Development* (Washington, DC: World Resources Institute, 1987), p. 54.
[43]Energy and Environmental Analysis, Inc., op. cit., footnote 17, p. 3-11.

Table 5-20—A Comparison of the Performance Characteristics of a Two- and a Four-Stroke Motorcycle Engine

Engine size (cm³)	Engine type (stroke)	Emissions		Fuel economy (miles/gallon)
		HC (g/km)	NO$_x$ (g/km)	
400	2	11.1	0.1	50
500	4	1.2	0.1	67

SOURCE: Energy and Environmental Analysis (EEA), "Policy Options for Improving Transportation Energy Efficiency in Developing Countries," contractor report prepared for the Office of Technology Assessment, July 1990, p. 3-6.

percent and increase fuel efficiency by 25 percent (see table 5-20). The increased first cost for this technology is about $100.[44] The cost effectiveness of this investment depends on fuel price, but at a gasoline price of $1.50/gallon ($0.40/liter) the simple payback is about 1.6 years, even without taking into account environmental benefits.[45]

Buses

Buses are the backbone of urban passenger transport in the developing world, providing essential low cost transport, particularly for low income groups. They provide over half of all motorized trips in many cities (see table 5-14).

As for other road vehicles, there are a variety of technologies available to improve bus energy operating efficiencies, including turbochargers, smaller engines in some cases, automatic timing advances, lighter bodies through use of aluminum or plastic components rather than steel, and others. But the effectiveness of these technologies in conserving fuels is constrained by their operating environment. Most urban buses operate on congested streets, and the resulting low speeds and frequent speed changes are associated with low operating efficiencies. Turbochargers are often ineffective, for example, as they require higher, sustained speeds. In so far as unsatisfactory bus service has contributed to the rapid increase in ownership of autos and two and three wheelers, faster bus service could slow the increase in private vehicle ownership.

In an effort to improve bus services by increasing speed and reducing the number of stops, some cities have introduced exclusive or priority bus lanes. Abidjan has 5 kilometers of priority bus lanes,[46] Porto Alegre has exclusive busways, and Sao Paulo is paving and improving its bus routes in low-income areas.[47] In 1978, Porto Alegre, Brazil, with World Bank assistance, designated 30 kilometers of road for exclusive use by buses. The roads were paved, barriers to other vehicles were built, and bus signs and stops were added. The cost was about $500,000 per kilometer—or about 4 percent that of the Manila light rail system (see below). The system has a peak capacity of about 28,000 passengers per hour (slightly higher than the capacity of Manila's light rail system), and yielded an increased bus speed of about 20 percent.

Modal Shifts in Urban Passenger Transport

The high capital, energy, and environmental costs of a private auto-based urban transport system raises issues of how urban passenger traffic should be served, and what role different modes—buses, cars, two- and three-wheelers, light rail, walking, and bicycles—should play in an efficient urban transport system. These issues are closely interconnected with the broader issues of land-use and transport planning.

The energy intensities of the different passenger transport modes (measured by the amount of fuel required to move one passenger 1 kilometer) varies widely (see table 5-19). Buses use only one-fourth as much energy per passenger-mile as private cars according to U.S. data, and the difference is estimated to be much greater (one-eighth) in the developing countries because of the higher load factors on buses. Light rail fuel use is estimated to be some 30 percent higher per passenger-kilometer than buses, but still well below private passenger cars. If the indirect energy—the energy used in

---

[44] Ibid., pp. 3-5.

[45] Assuming 10,000 mi/year annual mileage, efficiency increase from 50 to 63 mpg.

[46] R. Barrett, *Urban Transport in West Africa*, World Bank Technical Paper Number 81 (Washington, DC: World Bank, 1988), p. 7.

[47] World Bank, *Urban Transport* (Washington, DC: World Bank, 1986), p. 30.

associated infrastructure such as roads and railbeds—is included, buses use less than one-tenth as much energy per passenger mile as private cars. Light rail, including energy in roadbeds and other indirect consumption, uses twice as much energy per passenger kilometer as a bus system but still one-fifth that used by private cars (see table 5-12).

Given the high energy and other costs of a private auto-based system, one of the most important and difficult issues facing those in charge of urban planning is the role of the private auto. The advantages of private cars—flexibility, comfort, and convenience—have led to unprecedented increases in car ownership. This has led to severe congestion in many areas. In many countries, car ownership and operation are already made expensive through high gasoline taxes, and high ownership and acquisition taxes. Demand has proven remarkably resilient in the face of such policies. A notable exception is Singapore, where an area licensing scheme has proven effective in controlling congestion in its central business district (see box 5-A).

A growing concern over the contribution of private cars to urban air pollution has led to new policy initiatives: both Mexico City and Sao Paulo, for example, have responded with mandatory installation of catalytic converters, bans on using cars on various days, and other measures. These actions could limit car use in the future.

In addition to saving energy and reducing pollution, a larger role for public transport could reduce transport costs. As table 5-21 shows, the total costs—including costs of road and metro construction, vehicles, etc.—are 2 to 3 times higher for private autos than for buses on segregated highways or paratransit[48] systems. Improvements in bus services could therefore release resources that could instead be used to finance improvements in efficiency, transport infrastructure, or other development needs. Buses have additional advantages. They are flexible in routing and scheduling. They can make use of existing roads or segregated "bus only" lanes. Buses usually use diesel engines, which are already familiar to most mechanics, and they are quite sturdy—able to go long distances with minimal repairs.

Despite the many advantages of buses, however, bus systems in developing (and industrialized)

> ### Box 5-A—Singapore's Area Licensing Scheme: Carrot and Stick
>
> Congestion in Singapore's central business district (CBD) led the government to institute an "area licensing scheme" in 1975. This program was intended to reduce congestion and increase bus ridership through the use of financial disincentives for private vehicle use. Private vehicles were charged an "entrance fee" (currently about $1.50 per day per vehicle) upon entering the central business district during peak hours.
>
> The initial effect of the program was a 51 percent drop in the number of cars entering the CBD in morning peak hours. From 1975 to 1989, the vehicle population in Singapore grew by 68 percent, yet traffic in the CBD is still below the levels found prior to the program's inception.
>
> The program has been modified a number of times. The peak hours have been changed, the restrictions have been extended to all vehicles, including motorcycles and taxis, the fees have changed, and the boundaries of the restricted zone have been enlarged. Although the program is not popular with motorists, it is seen as inexpensive and highly effective in reducing congestion. Energy savings are difficult to measure, but the modal shift from cars to buses, as well as the increased vehicle speeds, are thought to reduce energy use by about 30 percent. The fees charged to drivers more than cover the operating costs of the program.
>
> SOURCES: B. W. Ang, "Traffic Management Systems and Energy Savings: The Case of Singapore," paper presented at the New Energy Technologies Transportation and Development Workshop, Ottawa, Canada, September 1989, p. 13; Hagler, Bailly, and Co., "Road Transportation Energy Conservation Needs and Options in Developing Countries," contractor report to the U.S. Agency for International Development, Washington, DC, September 1986, p. A11.

countries have a mixed reputation. This is usually due to managerial, institutional, and infrastructural shortcomings rather than technical failures. Many systems are poorly run—overstaffed, inefficiently operated, and with very high costs. Some systems provide poor service—infrequent service at peak hours, overcrowded and unclean buses, and undependable schedules. Many publicly run bus systems require large government subsidies to operate.

---

[48]Paratransit is a blanket term for shared taxis, jitneys, minivans, and other similar systems that provide shared transport services.

The World Bank has examined bus services in cities of the developing world, and several World Bank reports point to the potential benefits of an increased private sector role in providing bus services. World Bank studies of public and private ownership found that the costs of private bus services were roughly half those of public systems, that little evidence could be found showing different levels of safety in the two systems, and that private systems offered more dependable and comfortable service.[49] Others argue, however, that safety and service to the poor are neglected by private operators; and that the need for central depots, special "bus only" lanes, and the difficulties involved in importing buses and spare parts all argue for a government role in urban bus systems.

In the longer term, bus services may find it difficult to compete effectively with automobiles due to their lower average speeds under typical urban conditions. A study of 32 cities in Asia, Australia, Europe, and North America found typical average bus speeds of about 20 kilometers per hour (km/h) compared to average speeds for automobiles of about 40 km/h in the United States and Australia.[50] However the use of segregated busways or bus only lanes can increase average bus speeds.

Rail-based light transit systems have become increasingly popular in North America, and are beginning to be seen in the developing world as well. Urban rail[51] systems—including metros, subways, and similar light rail systems—are operating, under construction, or being planned in 21 cities in the developing world. These systems offer the advantages of relatively high speed, smooth, and dependable service—but at a cost. According to a recent evaluation, rail-based transit systems in the developing world have achieved mixed results. Travel times were reduced, most systems provided high quality service, and ridership levels were high, although below forecasts. These systems are expensive, however. As fares are kept low to allow the poor access to the system, fare revenue cannot cover costs. As discussed above, however, a full accounting of the costs of a private auto-based system may increase the comparative economic attractiveness of a rail-based system.

Table 5-21—Total Costs of Passenger Transport Options, as Estimated by the World Bank

System type	Total costs in US$/passenger-km (1986)
Bus on segregated busway	0.05 to 0.08
Paratransit	0.02 to 0.10
Surface rail	0.10 to 0.15
Private car	0.12 to 0.24
Underground rail	0.15 to 0.25

NOTE: Total costs include initial and operating costs. Bus data include estimates for roadway construction.
SOURCE: A. Armstrong-Wright, *Urban Transit Systems*, World Bank Technical Paper No. 52, 1986, p. 49.

Manila's light rail system illustrates some of these points. Severe congestion and a growing urban population in Manila led the Philippine government to decide in 1980 to build a light rail system in Manila. This light rail system was completed in 1985, at an initial cost of $212 million, or about $14 million per kilometer. By 1988, 28 percent of urban trips were by light rail. Fares are set at a flat rate of 14 cents. The income from fares is sufficient to meet operating expenses, but does not cover interest and principal on the initial construction costs. It has been calculated that a fare of about 28 cents would be necessary to allow the system to be financially self-sufficient, however, it has also been shown that this fare would be unaffordable to most patrons. The Manila system, like many light rail systems, provides reliable, safe, and fast service at a reasonable price. However, it requires a large government subsidy; and the effects on urban congestion and energy use are unclear.[52]

A key issue is to use all elements of a transport system in an integrated manner. For example, the use of dedicated busways to hold and maintain right-of-ways, the encouragement of mixed employment and residential land use, and appropriate use of light rail systems could be considered. Related policy options might include full pricing of all transport options (including fuel, roads, parking, congestion, and injury), effective traffic control, and others; all with

---

[49] World Bank, *Urban Transport* (Washington, DC: World Bank, 1986), p. 23.

[50] Peter W.G. Newman and Jeffrey R. Kenworthy, *Cities and Automobile Dependence: A Sourcebook* (Brookfield, VT: Gower Technical, 1989).

[51] This discussion is based in part on, World Bank, "Metros in Developing Cities—Are They Viable?" *The Urban Edge*, vol. 14, No. 1, January/February 1990.

[52] A. Gimenez, "The Manila Light Rail System," paper presented at the New Energy Technologies Transportation and Development Workshop, Ottawa, Canada, September 1989.

the goal of developing a comprehensive and effective urban transport system.

THE DEMAND FOR TRANSPORT SERVICES

Although the emphasis in this report has been on improving the energy efficiency with which services are delivered, the high societal costs (as well as benefits) of the transport sector raise the question of whether or not economic growth could be sustained with lesser rates of growth in passenger-kilometers and freight tonne-kilometers. In the past, transport services have increased at about the same rate as economic growth or slightly faster (see figure 5-1). Although freight transport intensity (number of tonne-kilometers per unit of gross national product) declines at high levels of economic development, most of the developing countries are expected to continue experiencing rising or at least constant freight intensities as they continue the transformation from rural agricultural economies to urban industrial economies. The rapid rise in urban passenger transport services stems from the increase in urban population—between 1970 and 1980 the population of several major developing country cities doubled—and the spread of urban areas. Further, journeys tend to lengthen as cities grow; in Bogota, the average commuting distance increased by 13 percent between 1972 and 1978.

There are a number factors that could moderate these rapid increases. Improved communications (better telephone systems and fax machines, for example) could displace a certain amount of business travel.[53] Improved communications and overall logistics can ensure full loads and most efficient truck routing, and thereby improve overall efficiency. Traffic management schemes, such as segregated bus lanes, truck climbing lanes, and limited vehicle access to city centers, could reduce congestion and thus promote more fuel efficient vehicle operations. Several cities of the developing world have already instituted successful traffic management systems (see box 5-A).

In so far as an appreciable amount of demand for transport services arises because users do not pay the true costs of using the roads, efforts to reflect these costs (through road pricing, area licensing, user taxes on fuels, parking fees, import duties, sales taxes, and annual licensing taxes), could moderate the demand for transport services. Physical restrictions on access to certain areas have also been introduced.

In the longer term, careful attention to the physical layout of residences, employment centers, and services could allow for a less expensive and more efficient transport system. Locating employment closer to residences, locating public services (e.g., shopping and recreation) closer to intended users, siting major freight terminals away from congested city centers, and controlling the density of land occupation (as is done in Curitiba, Brazil and Bombay, India) could reduce travel needs. These options are especially promising for developing countries whose cities are not yet fully in place but are growing rapidly, in contrast to cities of the industrialized world, which are already largely built.

Without a carefully implemented land use planning program, however, a private vehicle-based transport system is the likely end result by default through "creeping incrementalism." It is easy to add one automobile and one short stretch of road at a time; it is much more difficult to plan a comprehensive urban transport system.[54] The end result can be an energy inefficient system that is difficult to change (see box 5-B).

NONMOTORIZED MODES

Much of the discussion on transport in the developing world focuses on motorized technologies—autos, trucks, trains, buses, and motorcycles. It is important to recognize, however, that many trips are by nonmotorized modes—walking, animal-drawn carts, or bicycles (see table 5-15)—used in rural areas and by a large part of the urban poor. While nonmotorized modes by definition do not use transport fuels, they have many shortcomings. They can be relatively uncomfortable under a variety of circumstances, they are slow, and they have limited freight capacity when compared to automobiles and trucks.[55]

---

[53] See, for example, P. Mokhtarian, "The State of Telecommuting," *Institute of Transportation Studies Review*, vol. 13, No. 4, August 1990). However, in most developing countries labor is largely physical rather than information-related, making telecommunications of less relevance.

[54] M. Bernard, op. cit., footnote 7.

[55] World Bank, "Gridlock Weary, Some Turn to Pedal Power," *The Urban Edge*, vol. 14, No. 2, March 1990.

> **Box 5-B—Urban Design and Transport Energy Use in North America**
>
> A private vehicle based transport system, such as that found in many U.S. cities, tends to have an urban core with few residents and many high-rise office buildings surrounded by farflung suburbs. Because there is little overlap between jobs and residences, most people travel relatively long distances to work; because suburban areas have low population densities, effective and low cost mass transit systems are difficult to support and, where available, may require driving to feeder stations. Finally, because the urban core must support a high density of roads and parking spaces for commuters, there is less space for parks, central plazas, or other amenities. This could encourage the flight of residents to the suburbs.
>
> In the low density suburbs of this system, the distances between houses and shops are great leading to dependence on private vehicles (rather than walking, bicycling, etc.) for transport to schools, shopping, and so on. Houses are often widely scattered, with the result that water, sewage, police, fire, and other public sector services must be extended long distances at considerable cost. Concerns of social equity could be raised, in that low suburban densities often require private vehicles, thereby excluding those not able to afford them. On the other hand, this form of land use has many well known attractions in terms of personal convenience, living space, and other factors.
>
> The private-vehicle based urban design that results by default from these processes has high capital costs and energy use. For example, per-capita fuel use for transport is four times greater for U.S. cities than for European cities (see figure 5-2). Numerous studies have examined various problems raised by the private vehicle-based urban design, and a number of cities in North America are now taking steps to reverse this trend. For example some 22 cities in North America now have light rail systems recently put into place, upgraded, or under construction.
>
> SOURCES: Peter W.G. Newman and Jeffrey R. Kenworthy, *Cities and Automobile Dependence: A Sourcebook* (Brookfield, VT: Gower Publishing Co., 1989); Jeffrey Mora, "A Streetcar Named Light Rail," *IEEE Spectrum*, February 1991, pp. 54-56; Transportation Research Board, National Research Council, "Light Rail Transit: New System Successes at Affordable Prices," Washington, DC, 1989.

The bicycle plays an important intermediate role between animate and motorized transport systems. In Beijing, for example, bicycles are the major form of passenger transport. Bicycles have many advantages. They do not contribute to air pollution, they are much less expensive per unit, and (in contrast to autos) they can often be produced domestically in relatively small quantities—providing local employment. Providing bicycles, and the infrastructure to support them (e.g., paved bike paths separated from motorized traffic) may provide improved transport to those who would otherwise walk. Such a policy could relieve pressure on overloaded buses, may delay or offset the transition towards motorized forms of private transport, or may supplement motorized transport in areas with appropriate land use planning.

Rural transport needs pose particular problems. Lack of transport services in rural areas means delays in delivering essential agricultural inputs such as seeds and fertilizers, and inability to bring harvested crops to market,[56] thus frustrating increases in agricultural productivity and rural indus-

Photo credit: Denise Mauzerall

Bicycles are a major form of transport in Beijing.

try. Technical improvements to cars, two and three wheelers, and buses and trucks also benefit rural areas, but these modes form only a small part of local rural traffic. In India, for example, trucks account for

---

[56] In some regions of Tanzania for example, the rudimentary state of the transport system has prohibited the development of cash crops. I.J. Barwell et al., International Labor Office, World Employment Program, *Rural Transport in Developing Countries* (Worcester, UK: Intermediate Technology Publications, 1985) pp. 93-108.

Figure 5-2—Urban Density v. Gasoline Use Per Capita, Adjusted for Vehicle Efficiency

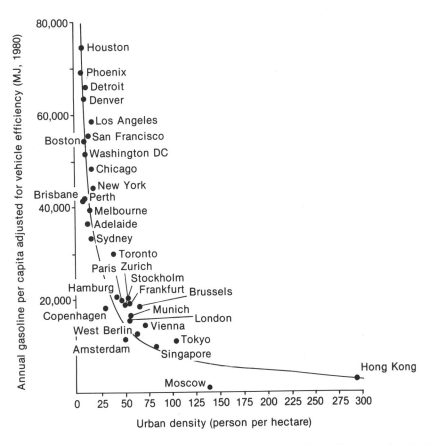

This figure shows a rapid decrease in transport energy requirements with increasing population density. Even for the same population density, however, transport energy requirements are shown to vary by a factor of two—Zurich v. Amsterdam—depending on local conditions.

SOURCE: Peter W.G. Newman and Jeffrey R. Kenworthy, *Cities and Automobile Dependence: A Sourcebook* (Brookfield, VT: Gower Publishing Co., 1989).

under 10 percent of total tonnage, and bullock carts for over two-thirds. Technologies for improving the efficiency of animal powered transport are available, but have diffused slowly due to the absence of rural credit.

ALTERNATIVE FUELS

Concern over high levels of urban air pollution and the high import costs of transport fuels have directed attention in developing countries to alternative transport fuels—fuels not derived from crude oil—especially those based on indigenous resources. Such a transition is desirable, in part, because of the emissions associated with gasoline and diesel combustion. Alternative fuels offer the potential to decrease the local and global pollution currently associated with vehicles.

According to the World Bank, "motor vehicles cause more air pollution than any other single human activity. The primary pollutants emitted by motor vehicles are hydrocarbons (HC) and nitrogen oxides (NO_x), the precursors to ground level ozone, and carbon monoxide."[57] In the most polluted American cities, highway vehicles alone are estimated to contribute about 40 to 45 percent of the total emissions of manmade volatile organic compounds

---

[57] A. Faiz et al., World Bank, Infrastructure and Urban Development Department, *Automotive Air Pollution: Issues and Options for Developing Countries*, working paper No. 492 (Washington, DC: World Bank, August 1990), p. 14.

(a key ingredient of smog). Although data are not widely available, the percentages may be even higher for a number of cities in the developing world. Gasoline combustion also results in other toxic emissions, including benzene, lead,[58] gasoline vapors, 1,3-butadiene, and polycyclic organic matter. Particulates emitted from vehicles cause visible air pollution and some adverse health effects, with diesel-fueled vehicles emitting 30 to 70 times more particulates than gasoline. Diesel vehicles also emit SO_x and NO_x. Due to their fewer numbers, however, diesel vehicles—in aggregate—emit less pollutants than do gasoline vehicles.[59]

Vehicle emissions have not only contributed dramatically to local air pollution problems in many developing countries, such as Mexico and India, but have also contributed to the global accumulation of greenhouse gases in the atmosphere. Of all the carbon dioxide resulting from the burning of fossil fuels, 15 to 20 percent is estimated to come from motor vehicles. Carbon monoxide, another vehicle emission, is likewise a greenhouse gas and also causes harmful local pollution.[60] Vehicles account for over 90 percent of the carbon monoxide pollution in many urban areas. Alternative fuels, with the exception of methanol from coal, would contribute less to the accumulation of greenhouse gases.

Chapters 6 and 7 discuss supplies and conversion of such fuels. Here their potential use in the transport fleet is examined. Although alternative fuels are often assessed as a replacement for gasoline, some types can also be used to replace diesel in trucks and buses. Indeed, for those developing countries with refinery capacity, the need to find a replacement for diesel is often more pressing than for gasoline. Changes in product demand in past years—a more rapid increase in demand for diesel compared with gasoline—has led to severe refinery imbalances. As a result, several countries have a surplus of gasoline that has to be exported—often at distress prices.

A recent OTA report[61] describes in detail the advantages and disadvantages of alternative fuels to replace gasoline used in private automobiles (see table 5-22). This report summarizes some of the issues raised in that report, and discusses issues specific to alternative fuels applications in developing countries, particularly for diesel vehicles. The focus is on transport fuels that can be produced from resources found in the developing world, with existing technology—methanol made from natural gas, ethanol from biomass, and natural gas used directly (either as CNG or LNG).[62]

Methanol is a liquid fuel that can be produced from natural gas, coal, or biomass; natural gas appears to be the most likely feedstock in the near future. One major advantage of methanol is that it would require fewer changes in vehicle design than would some other alternative fuels. Flexible-fueled vehicles, which can operate on methanol, ethanol, gasoline, or a mixture of these fuels, are already being produced in limited numbers in the United States.[63] The use of such vehicles would ease the transition away from gasoline. This may be less of a problem in developing countries, however, as their vehicle fleets are growing so rapidly and much of their infrastructure is only now being put into place—and could be more readily switched to alternative fuels or flexible-fuel capability.[64] Although methanol is frequently discussed as a replacement for gasoline, it can also be used to replace diesel. The cost for diesel engine modifications to accommodate methanol is estimated at $500 to $800 per vehicle.[65]

As discussed in chapter 6 of this report, the major disadvantage of methanol is its high cost of production. The cost of methanol made from natural gas is of course tied to the price of natural gas, but if a

---

[58] Although lead is no longer a major pollutant in the industrialized countries, many developing countries still use leaded gasoline.

[59] A. Faiz et. al., op. cit., footnote 57, pp. 41, 75; U.S. Congress, Office of Technology Assessment, *Replacing Gasoline: Alternative Fuels for Light-Duty Vehicles*, OTA-E-364 (Washington, DC: U.S. Government Printing Office, September 1990), p. 35.

[60] A. Faiz et al., op. cit., footnote 57, p. 25.

[61] U.S. Congress, Office of Technology Assessment, *Replacing Gasoline: Alternative Fuels for Light-Duty Vehicles*, OTA-E-364 (Washington, DC: U.S. Government Printing Office, September 1990).

[62] Other fuels often discussed as alternatives to gasoline and/or diesel include electricity, LPG, and hydrogen.

[63] U.S. Congress, Office of Technology Assessment, op. cit., footnote 61, p. 25.

[64] For example, as shown in table 5-7, India's vehicle fleet is growing at an annual rate of 18.4 percent. If all new vehicles were built to accommodate methanol, for example, then within about 4 years half the fleet would be methanol-fueled.

[65] Energy and Environmental Analysis, Inc., op. cit., footnote 17, pp. 4-13.

Table 5-22—Pros and Cons of Alternative Fuels

Fuel	Advantages	Disadvantages
Methanol	• Familiar liquid fuel. • Vehicle development relatively advanced. • Organic emissions (ozone precursors) will have lower reactivity than gasoline emissions. • Lower emissions of toxic pollutants, except formaldehyde. • Engine efficiency should be greater. • Abundant natural gas feedstock. • Less flammable than gasoline. • Can be made from coal or wood (as can gasoline), though at higher cost. • Flexfuel "transition" vehicle available.	• Range as much as one-half less, or larger fuel tanks. • Formaldehyde emissions a potential problem, especially at higher mileage, requires improved controls. • More toxic than gasoline. • M100 has nonvisible flame, explosive in enclosed tanks. • Costs likely somewhat higher than gasoline, especially during transition period. • Greenhouse problem if made from coal.
Ethanol	• Familiar liquid fuel. • Organic emissions will have lower reactivity than gasoline emissions (but higher than methanol). • Lower emissions of toxic pollutants. • Engine efficiency should be greater. • Produced from domestic sources. • Flexfuel "transition" vehicle available. • Lower carbon monoxide with gasohol (10 percent ethanol blend). • Enzyme-based production from wood being developed.	• Higher cost than gasoline. • Food/fuel competition possible at high production levels. • Supply is limited, especially if made from corn or sugar. • Range as much as one-third less, or larger fuel tanks.
Natural gas	• Excellent emission characteristics except for potential of somewhat higher nitrogen oxide emissions. • Gas is abundant worldwide. • Modest greenhouse advantage. • Can be made from coal.	• Dedicated vehicles have remaining development needs. • Retail fuel distribution system must be built. • Range quite limited, need large fuel tanks with added costs, reduced space (liquefied natural gas (LNG) range not as limited, comparable to methanol; LNG disadvantages include fuel handling problems and related safety issues). • Dual fuel "transition" vehicle has moderate performance, space penalties. • Slower recharging. • Greenhouse problem if made from coal.
Electric	• Fuel is domestically produced and widely available. • Minimal vehicular emissions. • Fuel capacity available (for nighttime recharging). • Big greenhouse advantage if powered by nuclear or solar. • Wide variety of feedstocks in regular commercial use.	• Range, power very limited. • Much battery development required. • Slow refueling. • Batteries are heavy, bulky, have high replacement costs. • Vehicle space conditioning difficult. • Potential battery disposal problem. • Emissions for power generation can be significant.
Hydrogen	• Excellent emission characteristics, minimal hydrocarbons. • Would be domestically produced. • Big greenhouse advantage if derived from photovoltaic energy. • Possible fuel cell use.	• Range very limited, need heavy, bulky fuel storage. • Vehicle and total costs high. • Extensive research and development effort required. • Needs new infrastructure.
Reformulated gasoline	• No infrastructure change except refineries. • Probable small to moderate emission reduction. • Engine modifications not required. • May be available for use by entire fleet, not just new vehicles.	• Emission benefits remain highly uncertain. • Costs uncertain, but will be significant. • No energy security or greenhouse advantage.

SOURCE: Office of Technology Assessment, 1992.

natural gas price of $1.00/mmBtu ($0.95/GJ) is assumed—not an unreasonable assumption for those countries with gas resources that are not presently fully used—then the wholesale methanol cost per gallon of gasoline equivalent could be at about $1.05 ($0.28/liter) if production volumes are high.[66] This compares with wholesale gasoline prices today in the range of 60 to 70 cents per gallon ($0.16 to 19/liter) for crude oil at $16 to $20 per barrel. Methanol from cheap natural gas is therefore substantially more expensive than gasoline from crude at present, but it does have environmental and supply security benefits. Methanol produced from coal or biomass is thought to be considerably more expensive.[67] By one estimate, manufacturing costs alone—excluding the costs of feedstock—for methanol from coal are about $1.00/gallon ($0.26/liter).[68]

Methanol could reduce air pollution, particularly urban smog. Tests have yielded mixed results, however, making the actual environmental benefits of methanol difficult to quantify.[69] Methanol may reduce carbon monoxide and nitrogen oxides, and may provide a small greenhouse gas benefit over gasoline. Any greenhouse gas benefits are highly dependent on the feedstock, however. Methanol from coal, for example, would result in higher greenhouse gas emissions.[70] Methanol does have some environmental disadvantages, particularly greater emissions of formaldehyde, which could require special emission controls. The liquid fuel itself is toxic, corrosive, and highly flammable, making methanol difficult to handle and distribute.[71]

Ethanol, like methanol, is a liquid fuel that can be used with minor modifications in gasoline engines. It can be produced from biomass—about one-third of Brazil's automobile fleet, for example, runs on straight ethanol produced from sugar (see box 5-C).[72] The vehicle-related technical issues for ethanol are essentially the same as with methanol—it requires only minor modifications for use in gasoline engines, but requires more complex changes for use in diesel engines. The major issue with ethanol is the cost of production. It is heavily dependent on the cost of the feedstock (corn in the United States, sugar in Brazil, molasses in Kenya) and on the market value of the byproducts. Research by the Solar Energy Research Institute, U.S. (now the National Renewable Energy Laboratory) and others into wood-to-ethanol processes is promising, but not yet commercial (see ch. 6).[73]

Ethanol is used either as an additive to gasoline or directly. As an additive, its primary environmental benefit is a reduction of carbon monoxide. Directly, ethanol may reduce concentrations of urban ozone, though probably not to the same extent as methanol. The net environmental effects of ethanol, however, are not yet clear.[74] If carefully done, ethanol would not add to net carbon dioxide emissions.

Compressed Natural Gas (CNG) is simply natural gas under pressure. It can be burned in gasoline engines with minor modifications, and in diesel engines with more complex modifications. Natural gas is a cleaner fuel than gasoline, with lower emissions of most pollutants. The major drawback of CNG as a transport fuel is the difficulty of transporting, storing, and delivering it.

The use of CNG in gasoline engines requires the installation of gas cylinders, high pressure piping, and appropriate fittings to the carburetor. In order to take full advantage of CNG, the compression ratio should also be raised to about 12:1.[75] An automobile designed for CNG would cost about $700 to $800 more than a comparable gasoline-fueled vehicle, due in large part to the pressurized tanks.

---

[66] U.S. Congress, Office of Technology Assessment, op. cit., footnote 61, p. 16. Note that these are wholesale prices, excluding taxes.

[67] Ibid., p. 13.

[68] Ibid., p. 78.

[69] Ibid., p. 61.

[70] Ibid., p. 71.

[71] U.S. Congress, Office of Technology Assessment, *Delivering the Goods: Public Works Technologies, Management and Financing*, OTA-SET-477 (Washington, DC: U.S. Government Printing Office, April 1991), p. 101.

[72] World Bank, "Alcohol Fuels from Sugar in Brazil," *The Urban Edge*, vol. 14, No. 8, October 1990, p. 5.

[73] U.S. Congress, Office of Technology Assessment, *Renewable Energy Technology: Research, Development, and Commercial Prospects*, forthcoming.

[74] U.S. Congress, Office of Technology Assessment, op. cit., footnote 61, p. 108.

[75] R. Moreno, Jr. and D. Bailey, World Bank, Industry and Energy Department, *Alternative Transport Fuels from Natural Gas*, World Bank Technical Paper No. 98, Industry and Energy Series (Washington, DC: World Bank, 1989), p. 11.

> ### Box 5-C—Brazil's Ethanol Fuels Program: Technical Success, Economic Distress?
>
> Brazil has long used ethanol as a replacement fuel for gasoline. As early as 1931, Brazil used ethanol-gasoline blends to fuel vehicles. In the early 1970s, the drop in world sugar prices combined with the 1973 jump in world oil prices led to an expansion of ethanol production from sugar cane. The program was intended to produce a mix of 20-percent ethanol and 80-percent gasoline. In response to the second oil price jump in 1979, the program was changed to promote vehicles that would operate on 100-percent ethanol.
>
> To encourage the purchase of ethanol-fueled vehicles, the government set up strong financial incentives. Ethanol prices were initially set at 65 percent of gasoline prices, yearly licensing fees were lower for ethanol vehicles, and generous vehicle financing was offered. Consumer acceptance was low at first, due to problems with cold-starting and fuel system corrosion, but by 1983, 89 percent of new light duty vehicles sold used ethanol. From 1979 to 1988, ethanol's share of total transportation fuel market jumped from 11 percent to 52 percent, and gasoline consumption in Brazil dropped 46 percent.
>
> The remarkable technical success of the program—fleet conversion to a domestically produced alternative fuel—has been overshadowed by a number of economic questions. World sugar prices have increased to the point where Brazilian sugar growers find it more profitable to export than to produce ethanol. This has led to shortages of ethanol, and has forced Brazil to import some ethanol. Ethanol shortages at the pump have angered consumers and led to a sharp drop in the sales of new ethanol-fueled vehicles. And domestic oil exploration has been successful, bringing into question one of the primary motivation of the program—to reduce dependence on imported oil.
>
> The program did reduce gasoline use, but at a cost—the economic losses of the program have been estimated as high as $1.8 billion, although these figures remain controversial. The program did demonstrate that it is technically feasible to convert a significant fraction of the fleet to an alternative fuel. Whether or not this was economically desirable is less clear.
>
> SOURCES: Jacy de Souza Mendonca, "The Brazilian Experience with Straight Alcohol Automobile Fuel," paper presented at the Ottawa Workshop on New Energy Technologies Transportation and Development Workshop, September 1989, p. 9; S. Trindade, "Non-fossil Transportation Fuels: The Brazilian Sugarcane Ethanol Experience," paper presented at the Energy and Environment in the 21st Century Conference, Cambridge, MA, March 1990, p. A-142; World Bank, "Alcohol Fuels from Sugar in Brazil," *The Urban Edge*, October 1990, p. 5.

Using CNG in diesel engines is more complex. Retrofitting existing diesels to run on CNG is possible but difficult. A more practical option is modifying the design of diesels to use CNG. Buses in Hamilton, Ontario use diesel engines redesigned to use CNG; this engine is estimated to cost about 10 percent more than the standard diesel engine.

Transporting, storing, and delivering CNG will be the major barrier to widespread CNG use. Natural gas, unlike liquid fuels, cannot be easily moved by truck or ship.[76] Most developing countries do not have a natural gas supply infrastructure, and the construction of such a system is expensive. Using the gas for other end uses, such as process heat or electricity generation, would improve the economic attractiveness of pipeline construction. Transferring natural gas from a pipeline to a vehicle can also be expensive, as a compressor is required. For fleet vehicles that make use of central refueling facility, the cost per vehicle for the compressor system will be reduced somewhat.

Unlike methanol and ethanol, natural gas reduces emissions contributing to urban smog, although it may increase nitrogen oxide emissions. Natural gas vehicles also should contribute less to greenhouse gases than petroleum or coal-based transport fuels. The lower emissions of carbon monoxide and carbon dioxide by natural gas compared to petroleum or coal may be offset, however, by methane leaks—methane is a potent greenhouse gas—during the production and distribution of gas. Although natural gas presents some special handling problems, it is neither toxic nor corrosive, unlike methanol and gasoline.

Liquefied Natural Gas (LNG) is natural gas that has been liquefied by cooling it to −161 °C. The advantage of LNG over CNG is its energy density— a given volume of LNG will provide about 3 times the vehicle range between refuelings than the same volume of CNG.[77] The liquefaction process is expensive, however, and the fuel must be kept at −160 °C to prevent boiling off. The practical

---

[76] Unless it is converted to LNG.

[77] U.S. Congress, Office of Technology Assessment, op. cit. footnote 61, p. 99.

difficulties of maintaining these low temperatures, along with the high cost of containers capable of storing LNG, make LNG less promising as a fuel for road vehicles.[78] Furthermore, LNG requires an expensive conversion process to liquefy it from natural gas. This costs about $1 to $3 per thousand cubic feet of natural gas, plus the process consumes about 10 percent of the incoming gas as fuel.[79]

Electric vehicles are under development in the United States, and several prototypes exist. Major limitations are short range due to limited battery storage, and the cost of electricity. Advanced battery technologies, although not yet commercially available, may allow for greater energy storage and longer battery life. Many developing countries are now electricity-short and electric vehicles would aggravate this problem. If in the future electricity supplies are ample then electric vehicles may be a more realistic option. Electric vehicles hold special promise in congested urban areas, where most trips are short and therefore the limited range of electric vehicles is less of a problem.

Electric vehicles have essentially no direct emissions and therefore may alleviate urban air quality problems. The overall contribution to pollution depends, however, on the nature of the electricity generation process. Electricity generated from a coal-fired power plant will contribute significantly to local and global pollution. Vehicles powered with electricity from nuclear, hydroelectric, or solar technologies, however, will generate less pollution than conventional gasoline-powered vehicles. In any case, electric vehicles may contribute less to urban pollution as power plants are frequently located outside of urban areas.

Electric vehicles may pose an additional environmental hazard, unique among the alternative fuels. The batteries required by electric vehicles typically have short lifetimes and present a disposal problem. The battery technologies under development also require special disposal procedures for production wastes as well as for spent batteries.[80]

Hydrogen has been discussed as an alternative fuel for transport. Hydrogen is an extremely clean fuel,[81] and is compatible with internal combustion engines. The two largest barriers to widespread use are storage and costs of production. Hydrogen has a very low energy density. Therefore, a hydrogen-fueled vehicle would require very large on-vehicle storage tanks. Hydrogen could be produced from natural gas or coal; however, a more environmentally appealing idea is to produce hydrogen from electrolysis of water using renewable energy generated electricity. This would, like electric vehicles, require an inexpensive source of electricity. Due to its various limitations, hydrogen is a speculative and very long term option.

CONCLUSION

The demand for transport services will continue to grow rapidly in developing countries, driven by population growth, economic growth, structural change, such as the transition to urban industrial economies, and other factors. The energy needed to meet these transport service demands, however, can be moderated through improvements in vehicle technical and operational efficiencies, and shifts to more energy efficient transport modes. Land-use planning can also substantially reduce the underlying need for transport services and assist the movement to more efficient modes by better matching residences with jobs, schools, shopping, and transport infrastructures. In contrast to the industrial countries whose infrastructure is largely in place, developing countries are only now building their infrastructures for the next century. Thus, they have a particular opportunity to redirect urban growth with appropriate incentives and disincentives towards more efficient and environmentally sound forms.

---

[78] R. Moreno, Jr. and D. Bailey, op. cit., footnote 75, p. 11.

[79] U.S. Congress, Office of Technology Assessment, op. cit., footnote 61, p. 103. Natural gas at the wellhead costs about $1/million Btu, or $1/thousand cubic feet, so the LNG process will, at a minimum, double the fuel cost.

[80] U.S. Congress, Office of Technology Assessment, op. cit., foonote 61, p. 119.

[81] Hydrogen-powered vehicles emit only water vapor and small amounts of nitrogen oxides. U.S. Congress, Office of Technology Assessment, op. cit., p. 128; also J. MacKenzie and M. Walsh, *Driving Forces* (Washington, DC: World Resources Institute, December 1990), p. 42.

Chapter 6
Energy Conversion Technologies

Photo credit: Appropriate Technologies, International

Contents

	Page
INTRODUCTION AND SUMMARY	179
Introduction to the Energy Supply Sector	179
Energy Conversion Technologies	180
THE POWER SECTOR IN THE DEVELOPING WORLD: IMPROVEMENTS TO EXISTING SYSTEMS	183
Generating System Rehabilitation	183
Transmission and Distribution Systems	185
System Interconnections	187
Improved System Planning Procedures	188
Improved Management	188
Environmental Considerations	189
TECHNICAL OPTIONS FOR NEW ON-GRID GENERATION	190
Clean Coal	190
Fluidized Bed Combustion (FBC)	191
Integrated Gasification Combined Cycle (IGCC)	192
Applications to the Developing World	193
Large Hydro	193
Nuclear Power	194
Gas Turbines	196
Cogeneration	197
Geothermal	198
Solar Thermal	200
Fuel Cells	200
Comparing the Technologies	200
OPTIONS FOR RURAL ELECTRICITY SERVICE	201
Wind Turbines	202
Photovoltaics (PVs)	204
MicroHydropower	206
Engine Generators	209
Comparing the Technologies	209
Grid Extension	211
OIL REFINING	212
BIOMASS	214
Gases and Electricity From Biomass	215
Gas Cost Comparisons	222
Liquid Fuels From Biomass	222

Boxes

Box	Page
6-A. Thermal Plant Rehabilitation in Kenya	186
6-B. Opportunities for Interconnection, Regional Integration, and Joint Development of Electricity Systems in the Developing World	189
6-C. Training Programs for Utility Planning	190
6-D. Nuclear Power in China	196
6-E. Wind Turbines for Water Pumping	202
6-F. Photovoltaic Powered Vaccine Refrigerators	206
6-G. Residential Photovoltaic Systems in the Dominican Republic	207
6-H. Micro-Hydropower in Pakistan	208
6-I. Gasifier-Engine Implementation in the Phillippines and India	219
6-J. Current Gas Turbine Research	220

Chapter 6
Energy Conversion Technologies

INTRODUCTION AND SUMMARY

Introduction to the Energy Supply Sector

The previous chapters analyzed the services that energy provides to consumers in the major end-use sectors. The analysis identified many cost-effective and often capital-saving technical opportunities for improving the efficiency of energy use in all sectors. Despite these benefits, institutional problems, a variety of market failures, and many other factors frequently discourage the adoption of these energy efficient technologies even when they are mature and well known.

Even under the most optimistic assumptions about improvements in energy efficiency, however, energy supplies will need to increase. Rapid population growth, economic growth, and structural change are creating a demand for energy services (see ch. 2) that cannot be met by efficiency gains alone. According to the U.S. Agency for International Development, for example, approximately 1,500 gigawatts (GW) of new electricity generating capacity could be needed in the developing world by 2008. Under an aggressive program of efficiency improvements in both the supply and demand sectors, this growth in demand could be reduced to about 700 GW, which would still require more than a doubling of present day capacity.[1]

Chapters 6 and 7 therefore turn to the ways in which energy is supplied in developing countries—the processes and technologies by which energy is produced, converted from one form into another, and delivered to users. Parallel to the analysis of end-use sectors in chapters 3 to 5, a range of energy supply technologies are examined, particularly in relation to their suitability for the special circumstances and opportunities of developing countries, the problems that could impede their adoption, and the policy issues involved. Many of these technologies have also been discussed in other recent OTA reports,[2] so the discussion here will be primarily limited to issues of particular relevance to developing countries.

Chapter 6—the conversion sector—covers that part of the energy sector devoted to turning primary or raw energy, such as coal, crude oil, gas, biomass, and other resources, into high quality forms for end users. Currently, the major conversion processes include electricity generation, oil refining, and, for biomass fuels, the production of charcoal from wood and ethanol from sugar cane (largely confined in the developing world to Brazil). In the future, the biomass conversion sector could expand to include a wider range of liquid fuels, gases, and electricity.[3]

Chapter 7—primary energy supplies—completes the exploration of the energy system back upstream to its sources: the exploration and mining of fossil fuels and the growth and collection of biomass. The energy resources available to a country invariably determine the composition of its fuel supply.

Table 6-1 and figure 6-1 illustrate the current structure of primary energy supplies in developing countries. According to these data, biomass is the most important fuel, closely followed by oil and coal. There is considerable variation between countries. For example, coal accounts for 70 percent of energy use in China and almost 40 percent in India but is little used elsewhere. For the rest of the developing countries, oil is the major source of commercial primary energy. Natural gas accounts for a relatively small share in all countries. Primary electricity (e.g., hydroelectricity) is an important component of the energy supply mix. A large quantity of electricity is also generated from fossil fuels. Developing countries meet a much higher share of their needs with biomass and much less with natural gas than the industrial nations. The share of biomass in the total energy supply mix of developing

---

[1] U.S. Agency for International Development, Office of Energy, *Power Shortages in Developing Countries: Magnitude, Impacts, Solutions, and the Role of the Private Sector* (Washington, DC: U.S. Agency for International Development, March 1988). Forecasts quoted here are for the medium economic growth rate scenario.

[2] See, for example: U.S. Congress, Office of Technology Assessment, *Energy Technology Choices: Shaping Our Future*, OTA-E-493 (Washington, DC: U.S. Government Printing Office, July 1991).

[3] See also, U.S. Congress, Office of Technology Assessment, *Renewable Energy Technology: Research, Development, and Commercial Prospects* (forthcoming).

Table 6-1—1985 Primary Energy Supplies (exajoules)

	Coal	Oil	Gas	Primary electricity	Total commercial	Biomass	Total energy
World	88.7	104.6	58.2	33.0	284.5	36.9	321.3
Industrial countries	63.5	77.0	51.7	26.6	218.7	5.5	224.2
Developing countries	25.2	27.7	6.5	6.4	65.7	31.3	97.1
Share of industrial countries	72%	74%	89%	81%	77%	15%	70%
Share of developing countries	28%	26%	11%	19%	23%	85%	30%

NOTE: As in table 2-1, the values reported for developing country biomass are too low. Field surveys indicate that biomass accounts for roughly one-third of the energy used in developing countries.
SOURCE: World Energy Conference, *Global Energy Perspectives 2000-2020*, 14th Congress, Montreal 1989 (Paris: 1989).

countries generally declines as the standards of living and the extent of urbanization rises.

The developing world as a whole produces more energy than it consumes, and significant amounts of both oil and gas are exported. Again, there are large disparities among countries. Only a few developing countries, primarily the OPEC nations, export energy; most are heavily import dependent.

Reliable and affordable supplies of energy are critical for economic and social development. Conversely, inadequate or unreliable energy supplies frustrate the development process. Electricity supplies in many developing countries are characterized by disruptions, including blackouts, brownouts, and sharp power surges. Lost industrial output caused by shortages of electricity have had noticeable detrimental effects on Gross Domestic Product (GDP) in India and Pakistan. Supplies of household fuels are notoriously intermittent, leading households to install a wide range of cooking systems in order to ensure against the shortage of any one fuel. Transportation services are similarly subject to disruption because of unreliable fuel supplies. Unreliable supplies of high quality fuels and electricity also impede the diffusion of improved technologies that are sensitive to fuel or power quality.

On the other hand, energy supply systems are expensive to build and maintain. Capital intensive electricity generating stations and petroleum refineries already account for a large part of all public investment budgets in developing countries, with electric utilities taking as much as 40 percent of public investment in some.[4] Overall, annual power sector investments would have to double to meet rapidly growing demands. This would take up virtually the entire projected annual increase in the combined Gross National Product (GNP) of the developing countries, leaving little for other pressing development needs. Further, a large part of the investment in capital equipment for energy facilities and in fuel to operate them must be paid for in scarce foreign exchange, which is already under pressure in many countries to service foreign debt. Similarly, there is often a shortage of local currency to pay for energy development due to inadequate revenues from existing operations. The energy supply sector also relies heavily on other scarce resources, such as skilled labor and management, and can cause environmental damage.

Energy Conversion Technologies

The conversion sector covers the processing of primary—or raw—energy such as coal and crude oil into forms (e.g., electricity, gasoline, and diesel) required by end users. Historically, most fuels passed from the production phase to final consumers with minimum processing or conversion; this remains the case today in many developing countries. As development takes place, however, an increasing amount of processing takes place in order to make these fuels cleaner and more effective, notably in the share of fossil fuels converted into electricity. If biomass based fuels are to provide an increasing share of developing country energy supplies, they too will have to undergo further processing into convenient forms, such as liquids, gases, and electricity with properties that can enable them to compete with energy products based on fossil fuels.

---

[4] U.S. Congress, Office of Technology Assessment, *Energy in Developing Countries*, OTA-E-486 (Washington, DC: U.S. Government Printing Office, March 1991), p. 93.

Figure 6-1—1985 Energy Consumption in Developing Countries

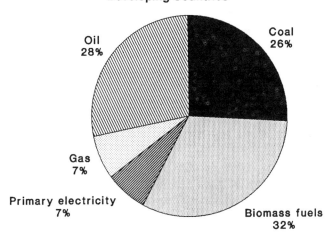

Developing nations
Total = 97 exajoules

SOURCE: World Energy Conference, *Global Energy Perspectives 2000-2020*, 14th Congress, Montreal 1989 (Paris: 1989).

The following survey of the three major conversion sectors—electricity; oil refining; and liquids, gases, and electricity from biomass—indicates that there are many technologies presently available or under development to meet the rapidly growing needs of this sector. There are problems, however, in financing the expansion of these sectors on the scale projected, and in improving the poor technical performance that dogs the electricity and refining sectors in many developing countries. These problems also could impede the timely adoption of energy efficient equipment.

Electricity

Electric utilities in developing countries face a rapidly growing demand for electricity, stimulated in many cases by low, subsidized prices. With large segments of the population typically still without electricity, the political and social pressures for system expansion are strong. Many utilities, however, have difficulty meeting even present day demand. Electricity systems in many developing countries are poorly maintained, resulting in unreliable service and frequent system breakdowns. The operating efficiency of electricity generating equipment in developing countries—with some notable exceptions—is often substantially below that of industrialized countries with similar technology, and the financial performance of many developing country utilities is deteriorating. Management attention must often focus on short term remedial measures, to the detriment of sound long term planning. Industry is frequently forced into self generation on a large scale, often with diesel generators dependent on high-cost imported oil.

Although the electricity systems of the developing world are as diverse as the countries themselves, several common issues underlie their frequently poor technical and financial performance:

- overstaffing, but shortages of trained manpower;
- lack of standardization of equipment;
- limited system integration and planning;
- political and social obligations to provide parts of the population with electricity at less than cost;
- shortages of foreign exchange to buy spare parts; and
- a regulatory framework that discourages competition.

Poor current performance raises doubts about the ability of the system to meet the projected rapid rise in demand even if the financial resources were available for capacity expansion. The effective deployment of technology will therefore depend on addressing these related financial, policy, and institutional issues.

Given low operational efficiencies, technologies relating to plant rehabilitation, life extension, system interconnections, and improvements in transmission and distribution (T&D) systems often offer higher returns to capital investment than new generating technologies. Plans for system expansion, reflecting indigenous energy resources, center on coal and hydro, followed by gas, nuclear, oil, and geothermal. Developing countries could benefit from several technologies in this expansion.

Fluidized bed combustion (FBC) with its greater tolerance for low quality coal could improve plant availability—offsetting its higher initial capital cost—as well as reduce SO_x and NO_x emissions. Combined cycle coal plants, specifically integrated gasification combined cycle (IGCC) plants, have much higher efficiencies than conventional coal plants, but uncertainties remain over IGCC's performance with lower quality coals. Gas turbines operated in a steam injection mode also promise to be attractive for electricity generation in the developing world. Gas turbines are small and modular with short construction lead times and high operating efficiencies.[5] These characteristics make gas turbines particularly attractive to private power producers with limited capital. Natural gas or gasified biomass also have environmental benefits relative to coal and oil.

Although hydropower is a viable option for many developing countries, concerns about environmental and social impacts are coming to the fore. Growing knowledge of the environmental impacts of large scale hydro projects can, however, contribute to better project design with lesser environmental impacts and a longer productive life for the plant.

Much of the planned expansion in nuclear power (in countries with large systems such as South Korea, China, and Taiwan) is based on existing nuclear technology. Smaller, modular, and safer nuclear power technologies under development in the industrialized countries might extend the market for nuclear power in developing countries depending on their ultimate cost and performance characteristics. The technical skill requirements of nuclear power may continue to limit its use in the developing world, as may the larger problems of nuclear proliferation and waste disposal.

Rural electrification is an important component of economic and social development in the developing world. While costs are variable, site specific, and poorly documented, the costs of stand alone renewables—notably microhydro, wind turbines, and photovoltaics—are competitive in many cases with diesel generators and grid extension, and in addition have significant environmental benefits compared to conventional systems.

Thus, there are many opportunities for technical improvement—in improving the existing system, in system expansion, and in rural electrification—but there are serious financial and institutional impediments. There are several ways in which the United States might help. The United States could work with the multilateral development agencies, which are major providers of finance to the electricity sector, to:

- improve the operational and financial efficiency of the sector;
- encourage the use of integrated resource management approaches, giving equal weight to conservation and renewables as alternatives to conventional supply expansion; and
- stimulate, through institutional and other reform, greater participation by the private sector.

Lack of training and familiarity with modern concepts of utility management could be addressed in part through training programs and ''twinning'' arrangements with U.S. and other utilities.

Oil Refining

Although there is a wide variation among countries, refinery operations in many developing nations are substantially below international norms in terms of efficiency and cost. Average refinery operating costs in Africa, for example, are $2 per barrel compared with $0.75 in the rest of the world.

A variety of technologies are available today that can improve refinery performance. Examples of cost-effective retrofits include heat recovery from stack gas and recuperation of flared relief valve and other gas. By the end of the decade, new catalysts are expected to improve product yields. These technology improvements would increase operational efficiencies and somewhat reduce the adverse environmental impacts of oil refining.

Problems hindering investment in the rationalization of refinery operation in the developing world include: lack of foreign exchange for parts; distorted pricing structures and earmarked government subsidies that hide the inefficiency of the plant; global overcapacity; and, in some countries, local markets that are too small to support efficient equipment.

An alternative to refineries for those countries with natural gas is methanol conversion. New technologies permit the direct conversion of methane to methanol near the well head cheaply enough to be used for such energy services as transport.

Biomass Fuels

Increases in energy consumption in developing countries are spurring efforts to use domestic resources, including biomass. If biomass fuels are to gain consumer acceptability, however, they must be converted into clean and convenient fuels. Biogas is, after a rocky start, becoming an established technol-

---

[5]Note that the operating efficiencies of conventional combustion turbines are lower than many other options; in contrast, the steam-injection mode, particularly with intercooling, provides higher efficiencies.

ogy, although there are significant doubts about its financial viability in small scale (household) applications. Small scale producer gas made from biomass is relatively low in capital cost and can produce a moderate quality gaseous fuel at competitive prices under a range of conditions. These gases can be used directly for process heat or to generate electricity. Electricity from internal combustion engine systems fueled with biogas or larger sized producer gas systems may now be marginally competitive with central station power at remote sites—if the avoided costs of transmission and distribution are taken into account. Larger scale biomass based gas turbine electricity generation is nearing commercial readiness and is expected to be competitive with many central station power plants if adequate biomass feedstocks are available at reasonable prices.

While liquid fuels from biomass are not competitive at the oil prices of today, current or near commercial technologies could be competitive at oil prices of perhaps $30 a barrel—a price that is widely projected to be attained by the end of the century. Technologies now under development, particularly ethanol by enzymatic hydrolysis, may provide liquid fuels at competitive prices in the future.

Biomass technologies offer the prospect of using domestic resources at competitive costs. There are other benefits as well. Together with other renewable energy technologies such as wind energy, photovoltaics, and microhydro, these small scale systems can bring high quality energy to rural areas—creating jobs and providing important environmental benefits.

On the other hand, there are a variety of obstacles to the introduction of renewables. In some cases, further RD&D are needed; subsidies to other forms of energy may discourage their development; credit may not be as readily available for renewables as for other more traditional energy supply systems; and there may be a lack of information on these alternatives, their costs, and their benefits at the local level. For biomass, in particular, land costs may become too high—due to competition between producing fuels and electricity for the wealthy versus producing food for the poor; environmental impacts on land and water could be excessive if biomass is grown too intensively or poorly managed; and infrastructure needs (notably roads) for large scale bioenergy plants could be high. The United States could help in overcoming these obstacles in a number of ways—through supporting RD&D, encouraging financial institutions to lend to such projects, and encouraging the transfer of these and related technologies through the private sector or by other means.

THE POWER SECTOR IN THE DEVELOPING WORLD: IMPROVEMENTS TO EXISTING SYSTEMS

Electric utilities in most developing countries face a rapidly growing demand for electricity; many, however, have difficulty even meeting present demands due to poor technical, operational, and institutional performance. Improvements in existing electricity systems should therefore be considered before or in conjunction with system expansion. Options for system improvement discussed here include:

- Rehabilitation of existing generating plants,
- Transmission and distribution systems,
- System interconnections,
- System planning, and
- Management.

Generating System Rehabilitation

The operating efficiency of the electricity generating equipment in developing countries—with some notable exceptions—is often substantially below that achieved in the industrialized countries, despite the fact that the basic technology is the same. An excellent example is the use of low speed diesels for power generation. This technology is widely reputed to be highly reliable, yet in developing countries many such units are out of service for much of the time.

Thermal efficiency is an accepted indicator of the ability of a utility to maintain its thermal power stations in good working order. As can be seen from table 6-2, thermal efficiencies in some Asian countries are comparable to those achieved in the United States. Performance in India and Pakistan, however, is quite poor.

Table 6-2—Thermal Efficiency of Selected Electric Power Plants

Country	Fuel type	Thermal efficiency
Korea	oil	35.4
	coal	30.6
Indonesia	oil	35.3
Hong Kong	oil	34.5
Malaysia	oil	30.0
Pakistan	gas	26.0
	oil	28.8
India	coal	25.9
Thailand	brown coal	23.8
	oil	33.8
Typical American Plants (with initial year of operation)		
Canal, Massachusetts (1,072MW, 1968)	oil	36.4
Ghent, Kentucky (2,226MW, 1973)	coal	33.5
Pleasants, W. Virginia (1,368MW, 1979)	coal	33.0
Chesterfield, Virginia (1,353MW, 1952)	coal	34.8
Tradinghouse, Texas (1,380MW, 1970)	gas	33.0
Northport, New York (1,548MW, 1963)	oil	33.8

SOURCE: International Development and Energy Associates (IDEA), Inc. "Improving Power Sector Efficiency in Developing Countries," November 1990, OTA contractor report, p. 22; based on *Technology Policy in the Energy Sector: Issues, Scope and Options for Developing Countries,* United Nations Conference on Trade and Development, UNCTAD/TT/90, 1990. American Data from *Electric Plant Cost and Power Production Expenses, 1988,* Energy Information Administration, Department of Energy, DOE/EIA 0455/88, Washington, DC, 1989.

A second important measure of maintenance efficiency[6] is the forced outage rate (FOR).[7] Over the past 10 years, there has been considerable progress in reducing forced outage rates in those developing countries that have had unusually high rates in the past. In India, for example, the average FOR for coal fired units in the 1983 to 1984 time period was 24 percent; in 1988, the average had improved to 16 percent.[8]

Over the past decade, the cost effectiveness of generation system rehabilitation has become recognized in the United States and Europe, where a great deal of attention has been placed on what has become known as "life extension" or "life optimization." The average cost of utility refurbishment projects in the United States has been reported to be considerably below the cost of new generating units.[9] Improved procedures for predictive and preventative maintenance have also contributed to the extended life of many industrialized-country power plants.

A number of major generation rehabilitation projects are now underway in developing countries. For example, a rehabilitation project in Pakistan, funded in part by the World Bank,[10] will refurbish 15 steam and combustion turbine units at Sukkur, Guddu, Faisalabad, Quetta, and Multan.[11] This project will provide additional capacity of 120 megawatts (MW) by restoring original ratings, in addition to improved heat rates and forced outage rates (see table 6-3). The cost for this rehabilitation is about $110 million. The potential for rehabilitating generation plants in the smaller systems of Africa is illustrated in box 6-A for Kenya.

---

[6]It might be noted that the use of the plant load factor (PLF), which is often used as a yardstick for comparison of the relatively poor performance of developing country generating units, is not a very useful measure for this purpose, since there may be sound economic dispatch reasons for a plant not to be used for all of the hours in which it is available.

[7]FOR is defined as (forced outage hours)/(service hours + forced outage hours).

[8]O.P. Dhamija, *Operation, Maintenance and Rehabilitation of Thermal Power Plants in India* (India: Uttar Pradesh State Electricity Board, 1988).

[9]A. Armor et al., "Utility Implementation of Life Optimization Programs," paper presented at the GEN-UPGRADE Conference, Washington, DC, March 1990, table 2.

[10]In fact the financing plan, as is increasingly the case, involves cofinancing by other entities, including the Overseas Development Administration of the U.K. (ODA), USAID (from its Energy Commodities Equipment Program), and the Government of Pakistan.

[11]IDEA, "Clean Coal Technologies for Developing Countries," contractor report prepared for the Office of Technology Assessment, May 1990, p. 72.

Table 6-3—Impact of Generating Plant Rehabilitation in Pakistan

Plant	Heat rate before rehabilitation Btu/kWh	Heat rate after rehabilitation Btu/kWh	FOR[a] before rehabilitation percent	FOR after rehabilitation percent
Multan, steam/fuel oil	12,808	11,280	13	6
Faisalabad, steam/fuel oil	12,850	11,290	13	6
Guddu Steam/fuel oil	13,250	11,113	10	6
Quetta Steam/coal	20,850	20,850	13	8
Faisalabad gas turbines	16,700	15,113	10	5

[a]FOR—forced outage rate.
SOURCE: International Development and Energy Associates (IDEA), Inc., "Improving Power Sector Efficiency in Developing Countries," November 1990, OTA contractor report, p. 72, based on *Integrated Operations Study for WAPDA and KESC*, report by Coopers & Lybrand to the Asian Development Bank, April 1990.

Transmission and Distribution Systems

System losses are perhaps the most common indicator used to describe the overall efficiency of the transmission and distribution system. T&D losses in the 6 to 9 percent range are regarded as good. In a recent compilation of system losses by the World Bank,[12] however, few developing countries experienced total system losses of below 10 percent. Several other developing countries had considerably higher losses (see table 6-4).[13] The most common way of measuring system losses is to compare generation at the busbar with sales (usually obtainable from billing records). This method measures both technical losses and nontechnical losses— which reflect the failure of many developing country utilities to meter and/or bill consumers and their failure to control illegal connections.

Over the next 20 years there will likely be a substantial effort in developing countries to reduce T&D system technical losses. A number of power system efficiency studies have been conducted by the World Bank and others, and T&D system rehabilitation has been recommended in almost all cases. Often, the recommendations are not just for hardware, but for technical assistance to strengthen T&D departments.

Losses are particularly high at the end of the distribution system, especially in low tension feeders and distribution transformers (see table 6-5). Technical improvements to reduce T&D losses include: providing adequate capacity in overloaded T&D system; adding power factor correction capacitors or overexcited synchronous motors to correct

Photo credit: *U.S. Agency for International Development*

In some cases, it may be more cost effective to improve transmission and distribution lines than to add new generating capacity.

---

[12] J. Escay, *Summary Data Sheets of 1987 Power and Commercial Energy Statistics for 100 Developing Countries*, Industry and Energy Department Working Paper, Energy Series Paper 23 (Washington, DC: World Bank, 1990).

[13] One might expect slightly higher losses in developing countries than in Western Europe, due to the lower voltages and greater number of rural networks in the developing world; but the loss levels shown in table 6-5 are much larger than would be considered desirable.

> **Box 6-A—Thermal Plant Rehabilitation in Kenya**
>
> The Kenya Power and Lighting Co. operates a generally well run and efficient power system, serving about 170,000 customers in Nairobi and other urban centers. The Kenya Power System Efficiency Assessment found that its main hydro plant (335 MW on the Tana River) operates with a high degree of reliability, and is relatively trouble-free. The 2 × 15 MW Geothermal plants at Olkaria, commissioned in 1981-1982, were found to be in ''exceptionally good condition,'' and operate as base-load units at their fully rated capacity.
>
> The Kipevu thermal power station in Mombasa, however, was in need of immediate rehabilitation. The 12 MW combustion turbine has been derated to 6 MW due to basic design shortcomings. Installation of a forced-air-finned cooler system and forced ventilation systems would permit uprating. The condensers and sea-water cooling systems are in severe disrepair due to corrosion and lack of proper maintenance: the cathodic protection system has not functioned for many years. The lack of a chlorination system has resulted in extensive fouling of lines and condensers with clams, worms, and other marine life. The 7 steam turbine units are in various stages of disrepair; 1 is no longer functional, 2 of the 12 MW units are derated to 8 and 10 MW, respectively, and even the 2 most recent units, aged 11 and 7 years, require rehabilitation because critical monitoring and metering equipment is nonfunctional. The costs of rehabilitating the power station are estimated at about $2.8 million, with a resulting benefit valued at about $1.8 million per year. This is a payback of about 1.5 years.
>
> SOURCE: IDEA, ''Improving Power Sector Efficiency in Developing Countries,'' contractor report prepared for the Office of Technology Assessment, October 1990, p. 73.

the power factor; using low loss distribution transformers;[14] and changing distribution system design.

A number of distribution configurations are used in developing countries. The traditional European system was designed to serve high density urban areas of Europe: the entire system is three phase, and is not the most appropriate for low density rural areas. The modified European system, however, provides good flexibility for single phase extensions to serve lightly loaded areas. The North American system carries a neutral wire along with the three phase wires. Single phase lines consist of a phase wire and an uninsulated neutral conductor, and can be cheaper than the European systems. The single wire earth return system uses a single conductor with the earth as return path; it is especially suitable for very lightly loaded systems involving long distances. The first reported use of such a system in a developing country is in a project in the Ivory Coast, partly funded by the World Bank.

High Voltage Direct Current (HVDC)

Capital investment in developing country transmission systems will need to increase over the next decade. An important innovation likely to come into more widespread use in developing countries is HVDC. HVDC has proven economic for moving large blocks of power over long distances (500 kilometers (km) or more), and increasingly has been used in the United States. Major HVDC projects in developing countries are +500 kilovolts (kV) links in China and India. The Indian project is a 900 km link between the Singrauli minemouth generating complex and Delhi that will provide a transfer capability of about 1,500 MW. The Chinese HVDC project links the central and East China Regional Grids.

A more immediate use for HVDC in many situations is as asynchronous links between neighboring systems that use different frequencies and voltages. Again in India, such an HVDC link has been made to connect the Northern and Western

Table 6-4—Technical and Nontechnical Losses
(in percent of net generation)

	Total	Technical	Nontechnical
Sri Lanka	18	14	4
Panama	22	17	5
Sudan	31	17	14
Bangladesh	31	14	17
Liberia	35	13	22
Malaysia	28	11	17
Ivory Coast	12	8	4

SOURCE: M. Munasinghe et al., *A Review of World Bank Lending for Electric Power*, World Bank, Washington, DC, Energy Series Paper No. 2, p. 60.

---

[14]Samuel F. Baldwin, ''Energy-Efficient Electric Motor Drive Systems,'' Thomas B. Johansson, Birgit Bodlund, and Robert H. Williams (eds.), *Electricity: Efficient End-Use and New Generation Technologies, and Their Planning Implications* (Lund, Sweden: Lund University Press, 1989).

Table 6-5—Distribution of Technical Energy Losses (percent)

	Madagascar	Kenya	Bangladesh	Target[a]
Power plant transformers	2.2	0.5	0.5	0.3
Transmission lines	2.7	5.2	2.2	3.8
Substations	0.8	0.9	1.1	0.3
Primary lines	2.1	2.1	4.2	2.5
Distribution transformers and low tension network	4.2	6.3	6.0	1.5
Total	12.0	15.0	14.0	8.3

[a]Target levels are those found in a relatively efficient, well-run system.
SOURCE: IDEA, "Improving Power Sector Efficiency in Developing Countries," draft contractor report to the Office of Technology Assessment, October 1990, p. 26.

regions as the initial step in the establishment of a national grid. This link provides for the exchange of up to 500 MW of power between the two regional grids.

Compact Design

Transmission line research over the past decade has indicated substantial potential for reducing transmission line cost, lessening visual impact, and maximizing use of existing rights of way by compact construction involving much lower line spacings than has been traditional practice. The most dramatic reductions have been at the 115 to 138 kV level where traditional clearance specifications were established long before technical requirements were clearly understood.

In the United States, a number of 138 kV lines have been uprated to 230 kV without any change in either conductor or insulation system, a practice that has been made possible because of better understanding of insulation and clearance requirements, and the need for large design margins correspondingly diminished. Particularly in the 50 to 230 kV range, attractive opportunities for uprating will often exist. Voltage uprating of a line greatly increases its load carrying capacity.

System Interconnections

The systems of India, Brazil, and China, in particular, are not single integrated systems, but consist of a set of regional systems, each no more than about 10 to 15 GW in size. The Chinese system, for example, consists of six regional grids—of which four exceed 10 GW in size—plus seven provincial grids. In this respect, these countries are similar to the United States, where individual power pools are the basic dispatch entities.[15] While interconnections do exist for power transfers to exploit interregional diversity, or for emergency assistance, a national power grid for large scale transfers of power does not yet exist.

The effects of capacity shortages and skewed financial incentives on economic dispatch and regional operation of interconnected systems are well illustrated in India.[16] Regional electricity boards were established in India as the primary mechanism for integrated regional operation. Each regional board consists of several State Electricity Boards, which are, in theory, subordinate to the regional dispatch centers but, in practice, often resist dispatch instructions.[17] The practice of allocating shares in large central sector plants to individual States creates particular difficulties, because during peak periods States tend to overdraw power from the central stations, which results in the frequency of the system falling from the nominal 50 Hz to as little as 48 Hz. During off-peak hours power is dumped into the system, resulting in frequency increases to 51 Hz or higher. Grid management problems arise when frequency falls outside the normal range of 50 ± 0.2 Hz, including the potential of grid collapse and damage to large thermal and nuclear generating sets.

The difficulties of backing down local plants during off-peak periods, even though these are less

---

[15]A detailed study of system interconnection issues can be found in U.S. Congress, Office of Technology Assessment, *Electric Power Wheeling and Dealing: Technological Considerations for Increasing Competition*, OTA-E-409 (Washington, DC: U.S. Government Printing Office, May 1989).

[16]Swarup, K., "Inter-state Power Exchanges and Integrated Grid Operations Proceedings," *Planning for the Power Sector* (Agra, India: September 1989).

[17]Another reason for the reluctance of State Electric Boards to back down units is the increasing use of the PLF as a yardstick of plant performance in India; the higher the Plant Load Factor (PLF), the "better" the performance of the plant. As noted in footnote 6, however, use of the PLF for such purposes is questionable.

efficient than the large central sector plants, is largely a matter of tariffs that provide few incentives for efficient operation. Moreover, because of financial difficulties, States are reluctant to permit net transfers of energy for fear of not getting paid by the recipient.

Trade in electricity among developing countries is very small compared to that in Europe and North America and far less than the technical and economic potential. The largest import dependency among African countries appears to be that of Zimbabwe on Zambia, reaching 25 percent of supply in the early 1980's. One major reason for the low level of intercountry trade in electricity is the question of payment. Exporting countries expect to get paid in a timely way, and in foreign exchange—not in inconvertible national currency. National security concerns also play a role. Box 6-B describes opportunities for interconnection, regional integration, and joint development of electricity systems in the developing world.

Improved System Planning Procedures

In the past, power system planning in the developing world has meant finding the least expensive generating mix to meet forecasted demand.[18] Although such least-cost analysis is essential, these methods conventionally have not incorporated demand side options and environmental impacts. There are three basic improvements that could enhance system planning in developing countries. First, the analysis itself could account for uncertainty, particularly in availability of foreign exchange and in demand growth. This could help avoid the sort of overbuilding characteristic of some Central American hydropower facilities in the last decade. Second, demand reduction and load management could be considered as investment opportunities and even as alternatives to new generation. Finally, the analysis could incorporate the social and environmental costs of power supply, which often are not reflected in cost estimates and regulations.

In the United States, many utility- and State-level regulators are paying considerable attention to the need for improved system planning. Although a single integrated approach to system planning has not been agreed on, many of the utilities and State regulators have adopted the concept of "integrated resource planning" (IRP) (see box 3-G on IRP in ch. 3). IRP allows for consideration of both demand and supply side investments and externalities. These planning methods could help promote energy efficient technologies in the developing world as well. Integrated resource planning may be particularly appropriate for developing countries, where there are often severe capital constraints and an untapped potential for demand reduction. The United States (particularly utilities and State regulatory agencies) could help the developing countries by providing policy advice, training, and technical assistance.

Improved Management

The issue of effective management of electric utilities is not just one of public versus private sector ownership, since there are many public sector utilities throughout the world that are well run, and private sector utilities whose technical and financial management has been consistently poor. In the case of government-owned entities, which applies to most developing countries, top management of utilities is sometimes appointed primarily for political reasons. In other cases, appointments are made on the basis of technical competence, and managers are held accountable in exchange for a certain degree of autonomy.

The Volta River Authority (VRA) of Ghana is a good example of a well run facility. The VRA was established primarily to own and operate the 1,000 MW hydro facility at the Akosombo Dam. The 1961 Volta River Development Act requires VRA to operate its plants according to sound public utility practices, a requirement that has in fact been met. In addition, the international institutions that financed the project insisted from the beginning on high standards of technical management. The tradition of technical competence in management has endured, coupled with a degree of autonomy from government interference.

Although competent management at the head of a utility is a key condition for good performance, it alone is not sufficient. A utility must also build up and maintain a competent technical staff (see box 6-C). Training is required at all levels—college level

---

[18] D. Jhirad, "Power Sector Innovation in Developing Countries: Implementing Multifaceted Solutions," *Annual Review of Energy*, vol. 15, 1990, p. 377.

> **Box 6-B—Opportunities for Interconnection, Regional Integration, and Joint Development of Electricity Systems in the Developing World**
>
> A 330 kV interconnection between the Ivory Coast, Ghana, Togo, Benin, and Nigeria has long been discussed.[1] The Nigerian system, with a strong potential for low cost gas fired generation, would complement the hydro-based system in Ghana. Even with the large storage capacity at Akosombo in Ghana, 2 or 3 wet years in a row necessitates spilling, which could instead be used to produce power for export to Nigeria. On the other hand, dry years cause serious problems for Ghana, since the 1,072 MW Akosombo project accounts for all but 50 MW of the installed capacity of Ghana. Shortfalls could be filled in with thermal power from Nigeria.
>
> The possibility of linking Burkino Faso with Ghana has also been examined as part of extending the Ghanaian grid to its northern regions. For Ghana, electricity sales would generate sufficient export revenues to justify the investment, and enable its northernmost loads to be served more economically by increasing the capacity utilization of the grid extension. Economic analyses of this proposal indicate rates of return of about 15 percent on the cost of the grid extension.
>
> Although cooperation among the East African countries has been impeded for many years by political factors, the potential for joint development of hydroelectric resources remains. Tanzania plans to develop significant hydro capacity on the Rufiji river, and could export expected surpluses in the late 1990s. This would require construction of a 220 kV intertie between Mombasa in Kenya and Tanga in Tanzania, and would enhance reliability in both systems as well. In Uganda, there is a potential for 500 MW of hydro at Ayago, which could be developed jointly with Kenya.
>
> Two separate systems serve Pakistan: the smaller Karachi Electric Supply Corp. (KESC) and the larger Water and Power Development Authority (WAPDA). The former is a 100 MW all thermal system serving the Karachi area (93 percent of whose shares are owned by government) and the latter is an autonomous government body with an installed capacity of 2,900 MW of hydro and 2,000 MW of thermal, serving the rest of Pakistan.
>
> At present two single circuit 132 kV lines and one double circuit 220 kV line interconnect these two systems. While exchanges between the two systems do occur, the interconnected system is still operated on the basis of avoiding load shedding, rather than on economic dispatch of an integrated system. A recent study found that integrated system operation could provide savings in:[2]
>
> - capital costs on new generating capacity—because a larger system needs less total capacity to meet an identical reliability criterion than the sum of two separate systems;
> - fuel costs—achieved through least cost dispatch of available plants; and
> - energy—achieved through more effective use of plants to reduce load shedding.
>
> The economic analysis indicated overall benefits of about $1 billion (expressed as a present worth at 10-percent discount rate) over the 20-year planning period, equal to about 5.7 percent of total system costs of separate KESC and WAPDA systems. About 90 percent of the benefits are in fuel cost savings.
>
> ---
> [1] See, for example, World Bank, *Ghana: Issues and options in the Energy Sector*, Report 6234-GH (Washington, DC: November 1986).
>
> [2] Coopers and Lybrand, *Integrated Operations Study for WAPDA and KESC*, Report to the Asian Development Bank (April 1990).

training[19] for electrical engineers and in-country, on-the-job training programs, perhaps in conjunction with utilities from the industrial countries.

Environmental Considerations

The degree to which electricity generating equipment in the developing world currently incorporates pollution control equipment varies. Modern electrostatic precipitators (ESP) for particulate control are now routinely fitted to new coal-fired power plants in developing countries: a recent survey of World Bank financed projects indicates that over the past decade, all such plants were so equipped. The extent to which such equipment is properly maintained, however, is not clear. If the plant as a whole is in poor condition, ESPs may be among the first items

---
[19] College level training is often done at institutions in the United States or Europe. In an attempt to ensure that foreign trained nationals return home, some developing country institutions now require the posting of bonds prior to departure on overseas University courses to ensure return, sometimes in amounts of as much as 3 months salary. Yet with a degree in hand, graduates discover that industrial country salaries are so attractive in comparison that such bonds can easily be forfeited.

of equipment not to be properly maintained as their failure does not require the plant to shut down. Moreover, where ash contents are far in excess of design norms (as is the case at many Indian plants, for example), failure rates of ESPs will be high.

Environmental considerations are now playing a larger role in system expansion decisions. For example, concerns over the environmental impacts of large hydropower development have influenced system planning in India, China, and Brazil. Environmental issues are likely to be even more important in the future as concerns over the regional and global environment grow.

TECHNICAL OPTIONS FOR NEW ON-GRID GENERATION

The improvements to the existing system described above could lead to substantial increases in supplies from existing capacity. Sooner or later, however, additional capacity will be required.

The selection of a technology for electricity generation is based on many factors, but a principal factor is and will continue to be availability of the fuel needed to power the technology. Several Asian countries have access to coal and therefore can be expected to continue to build new coal-fired generation. Hydropower resources in many areas, notably Latin America, are abundant and will probably continue to be exploited for electricity generation. Oil is a less popular fuel for new on-grid electricity generation due to its volatile price. Even those developing countries with domestic oil reserves may prefer to export their oil rather than consume it domestically. Natural gas is still unavailable in many areas, although the gas resource base is considerable (see ch. 7).

Much of the planned expansion in electricity generating capacity in the developing world reflects these resource considerations (see table 6-6). Coal and hydro will supply much of the expected expansion, followed by gas, nuclear, oil, and geothermal.

In this section, the technologies for expanding the supply of on-grid electricity generation in the developing world are reviewed. Many technologies could potentially play a role, but the analysis here will be limited by two principal criteria: those likely to play a large role in near term expansion (coal, hydro, etc.); and those playing a smaller role in current plans but with significant environmental or

> ### Box 6-C—Training Programs for Utility Planning
>
> There are several examples of successful training programs in utility planning procedures. The training program at Argonne National Laboratory, financed for more than a decade by the International Atomic Energy Agency (IAEA) in Vienna, covers many aspects of power sector planning, and has played a significant role in raising the level of power sector planning in developing countries. A second example is the Brookhaven/Stony Brook Energy Management Training Program, sponsored for many years by Agency for International Development in Washington, whose objective was to train energy sector planners. Some 40 percent of the more than 350 senior individuals who attended the course were from electric utilities. In some countries almost the entire cadre of senior energy sector planners attended the course. The energy planning program at the University of Pennsylvania, as well as activities of the National Rural Electric Cooperative Association (NRECA), are two more successful and well-regarded training programs.
>
> Developing effective programs requires an institutional commitment over substantial time periods, sufficient for adequate curriculum development and for the planning philosophy to become established. For demand-side management and integrated resource planning to gain widespread acceptance, a long-term training effort must be launched. A major training effort will similarly be necessary if new agencies with environmental planning and regulatory functions are to be adequately staffed.

other benefits (gas turbines, renewables) and substantial long term potential. The focus is on those technologies that are currently commercially available or expected to become so in the near future. Due to the diversity of the developing world, it is not appropriate to identify a specific technology as the best for all situations; nevertheless, the analysis identifies issues influencing the choice of technologies for electricity supply expansion in the developing world.

Clean Coal

The conventional technology for utilizing coal to produce electricity—the pulverized coal boiler—is reliable and technically straightforward, but results in relatively high emissions of NO_x, SO_x, particulates, and solid wastes. These pollutants can have adverse impacts on human health, ecosystems, and

Table 6-6—Planned Expansion in Electric Power Facilities of the Developing World, 1989-99

Plant type	Planned new capacity (GW)	(percent)
Coal thermal	172	45
Hydro	137	36
Gas thermal	34	9
Nuclear	24	6
Oil thermal	14	4
Geothermal	3	1
Total	384	100

NOTE: Data reflect official country plans, and do not reflect capital or other constraints.

SOURCE: World Bank, *Capital Expenditures for Electric Power in the Developing Countries in the 1990s*, Industry and Energy Department Working Paper, Energy Series Paper No. 21, February 1990, p. i.

structures such as bridges and buildings. Coal burning also releases significant amounts of the greenhouse gas CO_2: about 11 percent more than oil combustion; and 67 percent more than natural gas combustion per unit of energy output.[20]

As a result of the Clean Air Act Amendments of 1977, coal burning power plants in the United States were fit with flue gas desulfurization (FGD) or "scrubbers," a technology for removing SO_x from the post-combustion gas stream. SO_x scrubbing techniques currently in use include lime/limestone and double alkali scrubbing in which SO_x is captured from the gas stream and consolidated as sludge. Disposal of the sludge remains an environmental problem, though in some cases the sludge can be regenerated. There are also wet scrubbers and fabric filters for removing particulates from the flue gas.[21]

FGD is effective at removing SO_x, but is expensive, reduces overall plant efficiency, and does little to reduce NO_x, CO_2, and other emissions. A recent Indian study concluded that the addition of a FGD system would increase costs per kWh output by about 15 percent.[22] Although common in the United States,[23] FGD is almost unknown in India and China[24] as their coals are, on average, relatively low sulfur.

The high cost of FGD in the United States led to a search for other methods to burn coal with reduced emissions. The resulting technologies are often called "clean coal" technologies. They include improved coal processing before combustion (see ch. 7), improved ways to burn coal, and improved ways to treat waste gases. Some "clean coal" technologies offer important benefits in addition to reduced emissions, such as greater tolerance for low quality coal, but their major benefit is reduced emissions and this usually comes at a cost in comparison to conventional pulverized coal technologies.

Fluidized Bed Combustion (FBC) [25]

Fluidized bed combustion combines pulverized coal with limestone particles in a hot bed fluidized by upflowing air. Calcium in the limestone combines with sulfur in the coal to reduce SO_x emissions, and the relatively low combustion temperatures also reduce NO_x formation. FBC systems can be either pressurized (PFBC—operating at about 10 atmospheres air pressure) or atmospheric (AFBC—operating at ambient air pressure). PFBC is still in the development stage, although there are demonstration projects running in Stockholm and in Ohio.[26] AFBC technology can be retrofit onto an existing power plant, and appears to be quite tolerant of low quality coals.

Many AFBC units are in operation worldwide, including about 280 in the industrialized countries (some for industrial use).[27] India and China both

---

[20] United Nations Department of Technical Cooperation for Development, *Energy and Environment: Impacts and Controls* (New York, NY: October 1990), p. 10.

[21] United Nations Department of Technical Cooperation for Development, Ibid., pp. 24-26.

[22] IDEA, "Clean Coal Technologies for Developing Countries," contractor report prepared for the Office of Technology Assessment, May 1990, p. 25.

[23] As of 1988, about 22% of coal-fired power plants in the United States used FGD systems. U.S. Department of Energy, Assistant Secretary for Fossil Energy, *Clean Coal Technology*, DOE/FE-0149 (Washington, DC: U.S. Department of Energy, November 1989), p. 10.

[24] IDEA, "Clean Coal Technologies for Developing Countries," contractor report prepared for the Office of Technology Assessment, May 1990, p. 25.

[25] For more information on FBC technologies, see U.S. Congress, Office of Technology Assessment, *New Electric Power Technologies*, OTA-E-246 (Washington, DC: U.S. Government Printing Office, July 1985).

[26] PFBC Industry Newsletter, "Clean Coal Technology," No. 1, 1991.

[27] World Bank, *The Current State of Atmospheric Fluidized-Bed Combustion Technology*, World Bank Technical Paper Number 107, September 1989, p. 13.

Table 6-7—Costs of AFBC and PC Coal-Burning Power Plants

Cost component	Conventional PC	AFBC	PC with FGD	PC with 15 percent lower availability
Total costs (cents/kWh)	5.5	6.1	6.2	6.4

NOTES: Estimates are for a single 60 mW unit, and include levelized capital costs. PC=pulverized coal; AFBC=atmospheric fluidized bed combustion; FGD=flue gas desulfurization.

SOURCE: IDEA, "Clean Coal Technologies for Developing Countries," contractor report prepared for the Office of Technology Assessment, May 1990, p. 32.

have hundreds of small AFBC bubbling bed plants.[28] Most of the small Indian and Chinese plants are used for steam heat rather than for electrical generation. A few, such as the 30 MW plant in Trichy, India, do generate electricity.

The advantages of AFBC over traditional pulverized coal (PC) technology are reduced emissions and increased tolerance to a wide range of low quality fuels. PC plants tend to break down when the ash in low quality fuels melts and clogs the boiler machinery. AFBC plants have reported successful burning of fuels with ash contents of from 4 to 40 percent,[29] and AFBC can also burn rice husks and other agricultural wastes. The costs of AFBC are thought to be "competitive" with PC plants, depending on the local situation.[30] A cost analysis for AFBC plants in U.S. applications found that AFBC had lower capital requirements and a similar cost of electricity when burning high quality fuels, when compared to pulverized coal with flue gas desulfurization.[31] Few plants in India or China presently use FGD, however, so this comparison may be misleading.[32]

The benefits associated with the greater flexibility of AFBC can be evaluated by comparing the costs of AFBC to those of a conventional plant burning low quality coals. AFBC is more expensive than a conventional (PC) plant, but less expensive than a PC plant with FGD (see table 6-7). If the reduced availability of a PC plant due to low quality fuel is taken into account, AFBC becomes less expensive. This analysis assigns no value to the reduced air emissions of AFBC in comparison to PC. Despite increasing use of FBC plants, they are a relatively new, still evolving technology. There is, therefore, a certain amount of uncertainty about the future costs of FBC technologies.

Integrated Gasification Combined Cycle (IGCC)[33]

Integrated Gasification Combined Cycle (IGCC) combines several advanced technologies: a gasifier that reacts coal with oxygen to produce a mixture of combustible gases, a gas cleaning process to remove sulfur and other pollutants from the gas prior to combustion, a turbine that burns the gas to produce electricity, a waste heat recovery boiler that produces steam, and a steam turbine to generate additional electricity. Several demonstration projects in the 100 to 200 megawatts electric (MWe) size are being designed or operated.

IGCC systems have several advantages. IGCC plants have somewhat higher efficiencies than conventional pulverized coal plants. Unlike most coal-fired electricity technologies, relatively small (100 MW) IGCC plants can be built that are similar in cost per kW to large (500 MW) IGCC plants; and

---

[28] Stratos Tavoulareas, Senior Energy Consultant, U.S. Agency for International Development, unpublished memorandum, Apr. 26, 1991.

[29] World Bank, *The Current State of Atmospheric Fluidized-Bed Combustion Technology*, World Bank Technical Paper Number 107, September 1989, p. xviii.

[30] Ibid., p. xviii.

[31] Ibid., p. 52.

[32] Differences in plant size also complicate the comparison. AFBC plants have been built up to 200 MWe in size, while traditional pulverized coal plants are often 300 MWe or larger.

[33] For more information on IGCC technologies, see U.S. Congress, Office of Technology Assessment, *New Electric Power Technologies*, OTA-E-246 (Washington, DC: U.S. Government Printing Office, July 1985).

[34] IDEA, "Clean Coal Technologies for Developing Countries," contractor report prepared for the Office of Technology Assessment, May 1990, p. 54. The efficiency gain of the IGCC process comes from the use of a combined (2-stage) cycle—a concept that can also be applied to oil and natural gas plants.

Table 6-8—Incremental Effects of Improved Coal Technologies Relative to a Conventional (PC) Plant Without FGD

Technology	SO_2 reduction (percent)	NO_x reduction	Plant efficiency	Total electricity costs (cents/kWh)
PC w/out FGD	—	—	—	5.5
FGD	90	No change	Decrease	6.2
AFBC	85-90	Moderate	No change	6.1
IGCC	95-99	Moderate	Small increase	6.4

NOTES: AFBC = atomospheric fluidized bed combustion; FGD = flue gas desulfurization; IGCC = integrated gasification combined cycle; PC = pulverized coal.
SOURCE: IDEA, "Clean Coal Technologies for Developing Countries," contractor report prepared for the Office of Technology Assessment, May 1990, pp. 32, 47.

IGCC plants have low SO_2 emissions.[34] The tolerance of IGCC plants to lower quality coals is not yet clear. IGCC plants have been operated with coals with up to 28 percent ash content,[35] but not at the ash levels of 40 percent typical of Indian coals.

Applications to the Developing World

AFBC and IGCC technologies have higher first costs than conventional coal plants—but they offer several advantages, including reduced emissions, increased efficiency for IGCC, and increased fuel quality tolerance for AFBC (see table 6-8). Both AFBC and IGCC are promising technologies for those developing countries with domestic coal reserves, though improved coal washing alone would do much to improve plant availability and efficiency in conventional PC plants in India and China (see ch. 7). AFBC would allow use of lower grade, relatively dirty coals and would result in reduced SO_2 and NO_x emissions.

Coal burning technologies allow use of a vast, inexpensive resource but have environmental impacts and costs. The use of AFBC and/or IGCC technologies reduces both SO_2 and NO_x emissions, but CO_2 and the remaining NO_x emissions will still be considerable. The costs of these technologies (see app. B, at the end of this report) are relatively high as well, depending on plant availability and other factors.

Large Hydro

Hydroelectric power plants currently supply a significant fraction of the developing world's electricity (see table 6-9). By one estimate, an additional 137 GW of hydroelectric power may be added in the developing world by 2000.[36] Although hydropower has many advantages, notably use of an indigenous renewable resource, it often has relatively high capital costs (see app. B). A number of specific sites, such as the Narmada River project in India and the Three Gorges project in China, are also controversial due to adverse environmental and social impacts.

Although regions differ in their hydropower resources, according to one estimate, less than 10 percent of the technically usable hydropower potential in developing countries has been developed.[37] There are many valid reasons why this huge technical potential is not being fully utilized. In some cases, existing markets are too small. Zaire, for example, has 120 GW of technical hydro potential—about 50 times their current system capacity.[38] The estimates for technical hydro potential do not account for costs,[39] and many of the potential sites are located in remote areas, making construction costs unaffordably high and markets for the power distant. Construction times can be long. Large projects in China, for example, can require up to 7 to

---

[35] Government of India, Ministry of Energy, Department of Coal, *IGCC Power Generation 100-120 MWe Demonstration Plant Based on High Ash Indian Coals*, January 1988, p. VIII.

[36] E. Moore and G. Smith, World Bank, Industry and Energy Department, *Capital Expenditures for Electric Power in the Developing Countries in the 1990s*, Industry and Energy Department Working Paper, Energy Series Paper No. 21, February 1990, p. i.

[37] J. Besant-Jones, World Bank, *The Future Role of Hydropower in Developing Countries*, Energy Series Paper No. 15, April 1989, p. 15.

[38] World Bank, *A Survey of the Future Role of Hydroelectric Power in 100 Developing Countries*, Energy Department Paper No. 17, August 1984, Annex II. World Bank, *Summary Data Sheets of 1987 Power and Commercial Energy Statistics for 100 Developing Countries*, Industry and Energy Department Working Paper No. 23, March 1990.

[39] As much of the costs of hydropower construction are site-specific and for locally available labor and materials, construction costs vary widely. The average is typically $1,500-$2,000/kW. See World Bank, ibid., p. 19

Table 6-9—Hydroelectric Power Generation in Developing World Countries

Country	Hydroelectric power generation (terawatthours)	(percent of total generation)
Africa	50	18
Asia	241	22
Latin America	330	63
U.S.	250	10

SOURCE: International Energy Agency, *World Energy Structure and Balances 1971-1987* (Paris: OCED, 1989); Energy Information Administration, *Annual Energy Review 1989* DOE/EIA-0384(89), p. 203.

10 years for construction.[40] In many cases, these factors constitute a major drain on developing country economies. Other problems include seasonal fluctuations in precipitation, which may limit plant output, and accommodating competing uses for water.

Large hydropower can also have significant environmental costs. Large land areas are cleared for access, construction materials, or for the reservoirs themselves and are then flooded. This can affect the watershed, displace people and flood fertile agricultural land, and destroy forests and wildlife habitats.[41] If the area to be flooded is not adequately cleared, the resulting rotting vegetation can release large quantities of methane, a greenhouse gas. Environmental problems also arise when a dam is operational. Dams may cause detrimental downstream effects due to changes in the level of sedimentation and flood patterns. New reservoirs can also mean an increase in diseases, such as schistosomiasis, which flourish in still water.

Nuclear Power

Nuclear power currently supplies a relatively small amount of electricity in the developing world—less than 5 percent of total generation.[42] Both South Korea and Taiwan make heavy use of nuclear power, and are expected to continue to do so. Several developing countries are currently building nuclear power plants (see table 6-10). Other countries with nuclear power programs, including Argentina, India, and the Philippines, have had problems similar to those of the United States[43]—high capital costs, construction delays, cost overruns, low reliability, and concerns over waste disposal. Several countries, including Argentina, Brazil, Mexico, Pakistan, and the Philippines have scaled back or canceled their nuclear power plant construction plans.[44]

Most existing nuclear power plants are of two main types: boiling water reactors (BWRs) and pressurized water reactors (PWRs). BWRs use the heat of fission to cause water to boil, and the resulting steam is then used to drive the turbine. PWRs keep the primary cooling water at high pressure to prevent boiling, and have a secondary water system that absorbs heat from the cooling water and generates steam, which then drives a turbine. Both systems are in widespread use. Heavy water reactors, which use water in which hydrogen contains an additional neutron to moderate the fission rate, are used in Canada, Argentina, and India. Gas-cooled reactors, which use a gas as a coolant, are used in the United Kingdom. Advanced technologies, including breeder reactors and fusion, are not currently in commercial use.

Recent R&D efforts have focused on making reactors safer, simpler, and of a more standardized design. These designs vary, but common features include smaller unit size, use of so called "passive" safety features (which do not require external power, signals, or forces to operate), and overall simplified design, potentially reducing costs and opportunities for operator error.[45] Data on costs and performance however, are not yet available. The proponents of these new technologies argue that they will offer high performance and safety at a reasonable cost, while others suggest that costs will be high and

---

[40] H. Yicheng, "Relying on Reform to Speed up Hydropower Development," *JPRS Report, Science and Technology-China: Energy*, JPRS-CEN-90-014, Nov. 13, 1990, p. 11.

[41] "Impacts of Hydroelectric Development on the Environment," *Energy Policy*, vol. 10, No. 6, 1982, p. 351.

[42] U.S. Congress, Office of Technology Assessment, *Energy in Developing Countries*, OTA-E-486 (Washington, DC: U.S. Government Printing Office, January 1991).

[43] U.S. Congress, Office of Technology Assessment, *Nuclear Power in an Age of Uncertainty*, OTA-E-216 (Washington, DC: U.S. Government Printing Office, February 1984).

[44] J. Perera, "Power Generation," *South*, No. 102, April 1989, p. 65; John Surrey, "Nuclear Power: An Option for the Third World?," *Energy Policy*, vol. 16, No. 5, October 1988, pp. 461-479.

[45] C.W. Forsberg and A. Weinberg, "Advanced Reactors, Passive Safety, and Acceptance of Nuclear Energy," *Annual Review of Energy*, vol. 15, 1990, p. 140.

Table 6-10—Status of Nuclear Power Plants in the Developing World
(as of December 1988)

Country	Electricity supplied by nuclear power plants		Nuclear power plants under construction
	(GWh)	(percent of total)	(MWe)
Argentina	5,100	11.2	692
Brazil	600	0.3	0
China	0	0.0	2,148
Cuba	0	0.0	816
India	5,400	3.0	1,760
Mexico	0	0.0	1,308
Pakistan	200	0.6	0
S. Africa	10,500	7.3	0
S. Korea	38,000	46.9	900
Taiwan	29,300	41.0	0
U.S.	526,900	19.5	7,689

SOURCE: B. Semenov et al., "Growth Projections and Development Trends for Nuclear Power," IAEA Bulletin, vol. 31, No. 3, 1989, p. 7.

reliability low. Historically, nuclear power systems have cost more and operated at lower capacity factors than anticipated, suggesting that it may be appropriate for risk-averse potential users of nuclear power with limited capital to wait for these systems to operate commercially, thereby demonstrating their cost and performance characteristics.

There are several issues related to nuclear power in the developing countries that differ from those in the industrialized countries. These include scale, technical skill requirements, environmental constraints, and proliferation.

- **Scale:** it is sometimes argued that nuclear power is too large for developing countries. If the rule-of-thumb that no single facility should supply more than 10 percent of total system electricity is used, then a 600 MW nuclear power plant would be appropriate only for systems 6,000 MW or larger. According to official capacity expansion plans,[46] over 20 developing countries will have systems this size or larger by 1999. Although the World Bank projections are essentially a "business-as-usual" scenario, excluding aggressive efficiency improvements, it appears that 600 MW power plants, nuclear or otherwise, are not oversized for at least the larger developing countries. Moreover, the new nuclear plants currently under development are expected to be about 150 MWe in size, making them a feasible size for some smaller countries as well.

- **Technical Skill Requirements:** Nuclear technology is complex, requiring a high level of skilled personnel. Few developing countries currently have sufficient domestic training facilities to produce these personnel, and would therefore be dependent on other countries for both the equipment and the operation of nuclear power plants. This could further aggravate foreign debt problems. This is not true for those countries, like China, that have considerable experience with and infrastructure for nuclear technologies (see box 6-D).

- **Environmental Aspects:** Nuclear power releases little of the air pollution associated with fossil fuel combustion. Every stage of the nuclear fuel cycle, however, poses the risk of releasing radioactive pollutants. Wastes vary from mining dust with a low level of radiation to the highly radioactive spent fuel rods, requiring careful handling and long term disposal strategies.[47] The probability of a large scale nuclear accident—such as Chernobyl—may be small, but poses serious implications for the environment and human health.

- **Proliferation:** Concern over nuclear weapons proliferation limits the attractiveness of export-

---

[46]E. Moore and G. Smith, World Bank, Industry and Energy Department, *Capital Expenditures for Electric Power in the Developing Countries in the 1990s*, Industry and Energy Department Working Paper No. 21, February 1990, p. 49.

[47]United Nations Department of Technical Cooperation for Development, *Energy and Environment: Impacts and Controls* (New York, NY: October, 1990), p. 33.

ing nuclear power technologies. The links between nuclear power and nuclear weapons are subject to some dispute, but the basic issue is straightforward. Nuclear power requires nuclear fuels, facilities to process them, and trained personnel, and these facilities hypothetically could be used to produce nuclear weapons.[48] India's first demonstration of a nuclear weapon, detonated in 1974, was believed to be derived from research facilities.[49] More recently, the discovery of Iraq's nuclear weapons program has cast serious doubts on the ability of the existing nonproliferation regime or even much more stringent criteria[50] to control the spread of nuclear weapons.[51]

Gas Turbines

A basic gas turbine is a conceptually straightforward device. Air is compressed and then mixed with natural gas. The mixture is ignited, and the expanding hot gas turns a turbine. The turbine drives both the air compressor and an electric generator.

Gas turbines have long served as peaking power plants for electric utilities in the United States. A number of innovative technologies have increased the efficiency of gas turbines from about 25 percent for a standard gas turbine[52] to up to 47 percent for the most advanced turbine systems using steam injection and other improvements.[53] These efficiency improvements are lowering costs to the point that gas turbines are no longer necessarily limited to peaking. Gas turbine costs are also relatively insensitive to scale,[54] and therefore can be sized to meet relatively small loads with only a modest cost penalty.

> ### Box 6-D—Nuclear Power in China
>
> China suffers from severe power shortages, and coal and hydropower reserves, although plentiful, are expensive to use, aggravate environmental problems, and are located far from population centers. China, unlike other developing countries, also has a relatively large and experienced nuclear industry. These factors have led to ambitious plans for nuclear power generation. As with other technologies, China has shown a clear preference for domestic production over imports. In the case of nuclear power, China is both importing selected technologies where necessary while also working to develop indigenous production capabilities.
>
> China's first nuclear power plant was a 300 MW PWR reactor using mostly indigenous technology located near Shanghai. Two 900 MW units, using imported technology, are under construction in Daya Bay. It is expected that much of the electricity from these units will be sold to Hong Kong. Also planned are two 600 MW units using domestic technology and two 1,000 MW units using Soviet technology.
>
> SOURCES: Foreign Broadcast Information Service, JPRS Report, *Science and Technology—China: Energy*, JPRS-CEN-89-005, May 17, 1989. Y. Lu, "The Prospects and Economic Costs of the Reduction of the CO_2 Emissions in the PRC," presented at the Global Climate Change: The Economic Costs of Mitigation and Adaptation Conference, Dec. 4-5, 1990, Washington, DC. U.S. Congress, Office of Technology Assessment, *Energy Technology Transfer to China*, A Technical Memorandum, OTA-TM-ISC-30 (Washington, DC: U.S. Government Printing Office, September 1985).

One effective modification to the basic gas turbine is to use its hot waste gas to heat water and power a separate steam turbine. This combined cycle

---

[48] Proliferation concerns are discussed in more detail in U.S. Congress, Office of Technology Assessment, *Energy Technology Transfer to China*, OTA-TM-ISC-30 (Washington, DC: U.S. Government Printing Office, September 1985); U.S. Congress, Office of Technology Assessment, *Nuclear Power in an Age of Uncertainty*, OTA-E-216 (Washington, DC: U.S. Government Printing Office, February 1984), pp. 200-203; U.S. Congress, Office of Technology Assessment, *Nuclear Proliferation and Safeguards* (New York: Praeger Publishing Co., June 1977).

[49] U.S. Congress, Office of Technology Assessment, *Nuclear Power in an Age of Uncertainty*, Ibid., p. 202.

[50] R.H. Williams and H.A. Feiveson, "Diversion-Resistance Criteria for Future Nuclear Power," Center for Energy and Environmental Studies, Princeton University, May 26, 1989.

[51] R. Jeffrey Smith, "Iraq's Secret A-Arms Effort: Grim Lessons for the World," *The Washington Post*, Aug. 11, 1991, p. C1; see also, George W. Rathjens and Marvin M. Miller, "Nuclear Proliferation after the Cold War," *Technology Review*, August/September 1991, pp. 25-32.

[52] Electric Power Research Institute (EPRI), *TAG-Technical Assessment Guide, Electricity Supply 1989*, EPRI P-6587-L (Palo Alto, CA: Electric Power Research Institute, September 1989), p. 7-56, average annual heat rates. All heat rates given here are measured at higher heating value (HHV), that is they include the latent heat of condensation of the water produced.

[53] Natural gas fired ISTIG. See R. Williams and E. Larson, "Expanding Roles for Gas Turbines in Power Generation," Thomas B. Johansson, Birgit Bodlund, and Robert H. Williams (eds.), *Electricity: Efficient End-Use and New Generation Technologies, and Their Planning Implications* (Lund, Sweden: Lund University Press, 1989), p. 524.

[54] Ibid., p. 515.

process increases overall efficiency but at an increased capital cost (see app. B).[55] Construction can be phased, whereby the gas turbine can be built and operated, and the steam turbine added when needed.

Alternatively, the hot waste gas from the gas turbine can be used to heat water and the resulting steam injected directly back into the turbine. Again, this increases efficiency, but at some increase in capital cost (see app. B).[56] Several steam injected gas turbines, or STIGs, are already in use for industrial cogeneration.[57] Further modifications, such as intercooling, increase the efficiency even more,[58] but costs are uncertain.

Gas turbines are a promising technology for the developing world. As discussed in chapter 7, many developing countries have untapped natural gas reserves that could be a supply for future electricity needs. Gas turbines have a relatively low initial cost, and when combined with low cost domestically produced natural gas could produce electricity at a low total cost. Gas turbines can be sized to fit a wide range of needs, and there is little cost penalty for installing smaller units. Construction times can be quite short—less than 1 year.[59]

The technological sophistication of advanced gas turbines poses potential difficulties in maintenance for developing countries, but these could be overcome if modular aircraft derivative turbine designs are used. In this case, spares could be kept at a central regional facility and quickly transported to the site and swapped for a turbine needing maintenance.

Concerns over air quality and global warming also favor gas turbines, as they give off little SO_x and up to 60 percent less CO_2 per kWh produced than coal-fired plants.[60] NO_x emissions, however, are a continuing concern: technologies for reducing NO_x emissions, such as water or steam injection, are being explored.

Cogeneration

Cogeneration is the combined production of electric or mechanical power plus thermal energy in a single process and from the same primary energy resource. A typical cogeneration system will produce electricity for running factory motors or for sale to the electric utility grid, and at the same time use the waste heat from electricity production to provide process heat for the plant or, in some cases, to be sold to neighboring industry, commerce, government buildings, or residential housing for heating. The passage of the Public Utility Regulatory Policies Act (PURPA) in 1978 has led to a substantial increase in cogeneration in the United States with large sales of privately cogenerated or otherwise produced electricity to the utility grid.[61]

Cogeneration is also heavily used in some developing countries. In China, for example, 4.6 GW of steam extraction turbines for cogeneration were on line in 1980, representing 11 percent of all steam turbine units. The thermal output from central station power plants used for cogeneration totaled about 11 percent of total national heating demand; 85 percent of this was used for process heat, 15 percent for space heat. For example, roughly 7 percent of building space heating in Beijing is supplied by cogeneration facilities.[62]

Yet cogeneration faces substantial obstacles in many countries.[63] In India, for example, many state

---

[55] Electric Power Research Institute (EPRI), *TAG-Technical Assessment Guide, Electricity Supply 1989*, EPRI P-6587-L (Palo Alto, CA: Electric Power Research Institute, September 1989), p. 7-61.

[56] Ibid.,, pp. 7-57.

[57] R. Williams and E. Larson, "Expanding Roles for Gas Turbines in Power Generation," Thomas B. Johansson, Birgit Bodlund, and Robert H. Williams (eds.), *Electricity: Efficient End-Use and New Generation Technologies, and Their Planning Implications* (Lund, Sweden: Lund University Press, 1989), p. 529.

[58] Ibid., p. 531.

[59] Electric Power Research Institute (EPRI), *TAG-Technical Assessment Guide, Electricity Supply 1989*, EPRI P-6587-L (Palo Alto, CA: Electric Power Research Institute, September 1989), p. 7-58.

[60] For ISTIG. R. Williams and E. Larson, "Expanding Roles for Gas Turbines in Power Generation," Thomas B. Johansson, Birgit Bodlund, and Robert H. Williams (eds.), *Electricity: Efficient End-Use and New Generation Technologies, and Their Planning Implications* (Lund, Sweden: Lund University Press, 1989), p. 539.

[61] U.S. Congress, Office of Technology Assessment, *Industrial and Commercial Cogeneration*, OTA-E-192 (Washington, DC: U.S. Government Printing Office, February 1983).

[62] Qu Yu, "Cogeneration in the People's Republic of China," *The Energy Journal*, vol. 5, No. 2, 1984, pp. 133-137.

[63] An OTA report has examined many of the issues surrounding cogeneration; U.S. Congress, Office of Technology Assessment, *Industrial and Commercial Cogeneration*, OTA-E-192 (Washington, DC: U.S. Government Printing Office, February 1983).

Table 6-11—Geothermal Electricity Generation in Developing Countries (1990)

Country	Geothermal capacity (MW)	Technology
Philippines	894	Single flash
Mexico	725	Dry steam, single flash, double flash
Indonesia	142	Dry steam, single flash
El Salvador	95	Single flash, double flash
Nicaragua	70	Single flash
Kenya	45	Single flash
Argentina	<1	Binary
Zambia	<1	Binary
U.S.	2,827	All

SOURCE: R. DiPippo, "Geothermal Energy," on *Proceedings of the Energy and Environment in the 21st Century Conference,* Mar. 26-28, 1990, Cambridge, MA.

electricity boards may refuse to take privately generated power at all; others may impose a sales tax on self-generated electricity; some may decrease the maximum power available to industries with on-site generation capabilities, and then may be reluctant to provide backup power for occasions when the cogeneration systems are down.[64] Responses to these barriers include requirements for state electricity boards to purchase self-generated power at reasonable—i.e., avoided cost with some small adjustments—rates as in the United States under PURPA regulations; the development of generic contract forms for cogeneration arrangements with state or national grids; and the provision of backup power.

Geothermal [65]

Heat inside the Earth can be used to produce electricity. Worldwide geothermal generating capacity, which includes several developing countries, is almost 5 GW (see table 6-11).[66] Recent technological advances may allow for the expanded use of geothermal energy. Worldwide geothermal resource estimates are very uncertain and must generally be quantified locally by drilling wells (see table 6-12).

Technologies for utilizing geothermal resources include direct steam, single-flash, dual-flash, and binary systems. The simplest technology is direct steam, in which steam is piped directly from underground reservoirs and used to drive turbines. The initial capital cost of a direct steam geothermal unit is estimated at about $1,100/kW.[67] Single-flash units are quite similar, except they make use of underground hot water that is "flashed" into steam. Most existing geothermal plants are direct steam or single-flash. A dual-flash system uses a second flash-tank to capture energy that would otherwise not be utilized. This can increase the overall efficiency by up to one-fifth, but at a somewhat higher cost—capital costs for dual-flash units are estimated at $1,600 to 1,900/kWe.[68] Several dual-flash systems are currently operating.[69]

A new technology of special relevance to developing countries is the binary cycle unit. This technology uses an intermediate working fluid to transfer energy from the geothermal resource to the turbine. This allows for the use of relatively low temperature resources, typically 170 to 180 °C.[70] Binary plants can be quite small (typically 5 to 10 MWe) and can be erected and operated in as little as

---

[64] Ahmad Faruqui et al., "Application of Demand-Side Management to Relieve Electricity Shortages in India," contractor report prepared for the Office of Technology Assessment, April 1990.

[65] This discussion does not include geopressured or hot dry rock resources, as these resources cannot be utilized for electricity generation with existing technology.

[66] R. DiPippo, "Geothermal Energy: Electricity Production and Environmental Impact A Worldwide Perspective," paper presented at the Energy and Environment in the 21st Century Conference, Cambridge, MA, Mar. 26-28, 1990, p. DIII-21.

[67] Electric Power Research Institute (EPRI), *TAG-Technical Assessment Guide, Electricity Supply 1989*, EPRI P-6587-L (Palo Alto, CA: Electric Power Research Institute, September 1989), pp. 7-73. 1990 dollars.

[68] U.S. Congress, Office of Technology Assessment, *New Electric Power Technologies*, OTA-E-246 (Washington, DC: U.S. Government Printing Office, July 1985), p. 99, in 1990$. These costs do not include costs to extract, pump, transport, or reinject the water/steam.

[69] For a more detailed discussion of the dual-flash technology, see ibid., p. 97.

[70] The Heber binary-cycle demonstration plant in Northern California makes use of brine at 170-180 °C. See Electric Power Research Institute (EPRI), *TAG-Technical Assessment Guide, Electricity Supply 1989*, EPRI P-6587-L (Palo Alto, CA: Electric Power Research Institute, September 1989), p. 7-75.

Table 6-12—Geothermal Resources Suitable for Electricity Generation in Selected Developing Countries

Country	Geothermal electric power potential (GW)	Country	Geothermal electric power potential (GW)
China	160	Bolivia	22
Indonesia	130	Philippines	20
Peru	86	Venezuela	12
Mexico	77	Vietnam	11
Chile	54	Nicaragua	10
Ethiopia	50	Chad	9
Ecuador	30	Zambia	8
Brazil	29	Guatemala	7
Kenya	24	Korea	7
Colombia	23	U.S.	150

SOURCE: Electric Power Research Institute, "Geothermal Energy Prospects for the Next 50 Years," EPRI ER-611-SR, February 1978. Estimates are uncertain, and do not consider costs.

100 days.[71] Costs for a large (54 MWe) plant are estimated at about $1,800/kWe[72], smaller plants are estimated to cost $1,800 to $2,400/kWe.[73] Other costs associated with geothermal energy include drilling wells, pumping, and reinjecting water. These costs are heavily site dependent, but are estimated at 2.4 to 8.5 cents/kWh.[74]

The major advantages of geothermal technologies are its use of an indigenous resource (see table 6-12), its low land requirements, and, if modular technologies such as binary systems are used, short construction lead times.

The major constraints on greater use of geothermal resources for electricity production are resource limits and costs. Although geological studies can help to locate geothermal potential, this resource can only be quantified by expensive drilling.[75] The subsequent resource extraction also requires technical expertise and can be costly. Finally, geothermal energy can be depleted if oversubscribed, as has occurred at The Geysers in northern California.[76]

The environmental effects of geothermal power include CO_2 and H_2S emissions and high water consumption. The CO_2 emissions are dependent on the CO_2 content of the resource, but are on average about 5 percent of the CO_2 emissions of a coal plant, per kWh output. H_2S emissions are regulated in the United States to 30 parts per billion. There are several H_2S control technologies in use in the United States, while many other countries do not currently control H_2S emissions. Binary technology has no emissions of CO_2 or H_2S. Water requirements vary, depending on the plant design. Large binary plants can require as much as five times as much water, per kW, as a coal plant. Small binary plants, however, can use air-cooled condensers.[77]

Site specific environmental problems can include subsidence of land overlying wells; contamination of water supplies by saline (and sometimes toxic) geothermal fluids and reinjected water; and the generation of surplus high temperature liquid effluents containing metals and dissolved solids.[78] These problems can be controlled.

---

[71]U.S. Congress, Office of Technology Assessment, *New Electric Power Technologies*, OTA-E-246 (Washington, DC: U.S. Government Printing Office, July 1985), p. 101. This does not include the time required to locate or tap the resource.

[72]Electric Power Research Institute (EPRI), *TAG-Technical Assessment Guide, Electricity Supply 1989*, EPRI P-6587-L (Palo Alto, CA: Electric Power Research Institute, September 1989), pp. 7-73.

[73]U.S. Congress, Office of Technology Assessment, *New Electric Power Technologies*, OTA-E-246 (Washington, DC: U.S. Government Printing Office, July 1985), p. 102, in 1990$.

[74]Ibid., p. 309.

[75]By one estimate, geothermal drilling costs in Kenya are roughly $250/foot. From E. Eliasson, "Geothermal Conversion and Its Role," paper presented at The Advanced Technology Alert System Meeting on New Energy Technologies, Moscow, U.S.S.R., October 1988, p. 8.

[76]Richard A. Kerr, "Geothermal Tragedy of the Commons," *Science*, vol. 253, No. 5016, pp. 134-135, July 12, 1991.

[77]R. DiPippo, "Geothermal Energy: Electricity Production and Environmental Impact A Worldwide Perspective," paper presented at the Energy and Environment in the 21st Century Conference, Cambridge, MA, Mar. 26-28, 1990, pp. DIII-13, DIII-16.

[78]U.S. Congress, Office of Technology Assessment, *Energy in Developing Countries*, OTA-E-486 (Washington, DC: U.S. Government Printing Office, January 1991), p. 130.

Solar Thermal

Solar thermal electric technologies use sunlight to heat a fluid, and then use the hot fluid and steam to turn a turbine that generates electricity. The parabolic trough is the most mature of the solar thermal electric technologies. Parabolic troughs are long channels that focus sunlight onto a pipe at the center, heating a liquid (oil, water, or brine) that is then used to heat steam, powering a turbine generator. Several hundred MW of solar thermal electric capacity using the parabolic trough design is now in place in California. These systems are gas and solar hybrids, with the gas supplementing solar radiation at night or during cloudy periods.[79] Other technologies include parabolic dishes and central receivers. Their costs and performance, however, remain uncertain.[80]

Although the costs for solar thermal electric technologies remain somewhat higher than conventional fossil steam plants, they have been declining in recent years due to improvements in the technology gained largely through demonstration projects in southern California.[81]

Off-grid generation of electricity through solar thermal-electric technology has yet to be demonstrated, but such an application is thought possible.[82] The modular nature of the technology could lessen problems of economies of scale that often inhibit electrification in developing countries. For many developing countries, the abundance of sunshine and the potential for off-grid generation make these technologies promising. Since the areas with the highest incidence of solar radiation tend to be dry and hot (see table 6-13), water used as a transfer medium or to operate the turbine generator may be a scarce resource. Without further research, development, and demonstration, however, these technologies may not be cost effective for grid applications in many developing countries in the near term.

Table 6-13—Solar Radiation in Selected Countries

Region	Solar radiation (kWh/m^2/year)
Mali	2,490
Niger	2,450
Mexico	2,080
Sierra Leone	2,000
Venezuela	2,000
India	1,950
Brazil	1,880
Chile	1,630

SOURCE: California Energy Commission, *Renewable Energy Resources Market Analysis of the World*, P500-87-015, pp. 14-16. Average annual horizontal insolation.

Fuel Cells

Fuel cells are another promising future electric power source for developing countries. Fuel cells are conceptually similar to a battery in that they convert chemical energy in a fuel to direct current electricity, achieving high efficiencies (40 to 60 percent, and 85 percent with cogeneration[83]) with low emissions. The cells are modular and can be sized to fit different and changing applications. This highly sophisticated technology is still being developed and is just beginning to see commercialization in industrialized countries. Further commercial experience is needed to resolve some of the uncertainties about cost and performance before fuel cells can be used with confidence in developing countries.

Comparing the Technologies

If choice of technology were based solely on the levelized costs of generation, abstracting from site specific factors that can strongly influence the costs, then many technologies are available within the same broad cost range (see figure 6-2 and app. B at the back of this report). Even some of the highest cost options, such as decentralized renewables, may still be competitive in many situations as they avoid transmission and distribution losses and costs.

---

[79]U.S. Department of Energy, Solar Energy Research Institute, *The Potential of Renewable Energy: An Interlaboratory White Paper*, Report No. SERI/TP-260-3674 (Golden, CO: Solar Energy Research Institute, March 1990), p. E-1; Electric Power Research Institute (EPRI), *TAG-Technical Assessment Guide, Electricity Supply 1989*, EPRI P-6587-L (Palo Alto, CA: Electric Power Research Institute, September 1989), p. 7-82.

[80]Charles Y. Wereko-Brobby and Frederick O. Akuffo, *Solar Power Generation: Status and Prospects for Developing Countries*, unpublished paper (October 1988), p. 24, table 3.

[81]U.S. Department of Energy, Solar Energy Research Institute, *The Potential of Renewable Energy: An Interlaboratory White Paper*, SERI/TP-260-3674 (Golden, CO: Solar Energy Research Institute, March 1990), p. E-6.

[82]Ibid., p. E-5.

[83]Fred Kemp, Program Manager, International Fuel Cells, presentation to the Office of Technology Assessment, May 22, 1991.

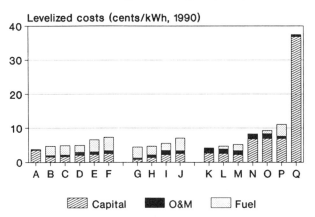

Figure 6-2—Levelized Costs of Electricity Generating Technologies (1990)

Legend: Capital / O&M / Fuel

Technologies A-Q are described in app. B. Technologies A-F are conventional technologies: A = hydroelectric power; B = natural gas-fired steam plant; C = oil-fired steam plant; D = coal-fired steam plant; E = diesel engine generator; F = combustion turbine. Technologies G-J are improvements to existing technologies: G = distillate-fired combined-cycle plant; H = natural gas-fired steam-injected gas turbine; I = coal-fired fluidized bed; J = natural gas-fired advanced combustion turbine run on a simple cycle. Technologies K-Q are innovative technologies: K = binary geothermal; L = advanced nuclear; M = integrated gasification combined cycle; N = wind turbine; O = solar thermal/natural gas hybrid; P = fuel cell; and Q = photovoltaic. Note that these costs do not include other costs associated with operations on a systemwide basis. See app. A and app. B for details.

SOURCE: U.S. Congress, Office of Technology Assessment, 1992. See app. B for more detail.

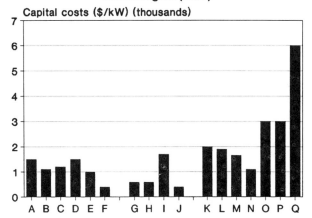

Figure 6-3—Capital Costs of Electricity Generating Technologies (1990)

Technologies A-Q are described in app. B. Technologies A-F are conventional technologies: A = hydroelectric power; B = natural gas-fired steam plant; C = oil-fired steam plant; D = coal-fired steam plant; E = diesel engine generator; F = combustion turbine. Technologies G-J are improvements to existing technologies: G = distillate-fired combined-cycle plant; H = natural gas-fired steam-injected gas turbine; I = coal-fired fluidized bed; J = natural gas-fired advanced combustion turbine run on a simple cycle. Technologies K-Q are innovative technologies: K = binary geothermal; L = advanced nuclear; M = integrated gasification combined cycle; N = wind turbine; O = solar thermal/natural gas hybrid; P = fuel cell; and Q = photovoltaic. Note that these costs do not include other costs associated with operations on a systemwide basis. See app. A and app. B for details.

SOURCE: U.S. Congress, Office of Technology Assessment, 1992. See app. B for more detail.

The initial capital cost of the different technologies is an important factor for many developing countries. With capital costs, the range is much wider; gas turbine and oil combined cycle are on the low end and hydro and nuclear are on the high end (see figure 6-3). Though developing countries generally are capital constrained, the electricity sector may be less affected than others due to the availability of relatively low cost public sector funds for supply expansion. If the role of private generators, —dependent on raising funds from local markets— grows, initial capital costs may become of mounting importance in technology choice.

Technology choice is also affected by the resource and other characteristics of the individual country. Most countries will wish to adopt technologies that use domestic resources, especially when these resources (e.g., coal and hydro) are of limited commercial value in other uses. In some cases, however, development of domestic resources may be limited by their inaccessibility. The size of the country and its grid influence technology choice. Countries with small grids cannot accommodate technologies (e.g., present day nuclear facilities) that are inherently large scale. Uncertainty over load growth can lead to favoring small scale technologies that are well known, such as diesel generators.

OPTIONS FOR RURAL ELECTRICITY SERVICE

The provision of electric service to rural areas is an important aspect of social and economic development.[84] The traditional means of rural electrification is extension of the existing electricity grid. An alternative method is to produce electricity off-grid, using diesel or gasoline powered engine generators.

---

[84] A recent World Bank report states, "most of the rural population in developing countries are not served by electricity and even in those countries where there have been rural electrification programs over the last 10-20 years, only a few serve more than 20% of their rural population." From M. Mason, *Rural Electrification-A Review of World Bank and USAID Financed Projects,*" background paper for the World Bank, April 1990, p. 1.

This avoids the need for expensive transmission lines to remote areas, but engenders problems with expensive, scarce, and unreliable fuel supplies.

Renewable electricity generating technologies—small wind turbines, photovoltaics, and microhydro—may be able to play an important role here. They are not dependent on fossil fuels and recent advances have reduced their costs to competitive levels for many remote power generation applications.

Wind Turbines

Wind power has long been used to pump water, grind grain, and meet other mechanical needs (see box 6-E). The amount of electricity generated by wind remained small, however, until the early 1980s when a combination of new legislation and tax credits in the United States led to the installation of a large number of wind turbines. Many of these were large (100+ kW) grid-connected units in California,[85] but smaller, off-grid turbines were installed as well. Although many of the tax credits have been eliminated, wind turbine technology has continued to progress and now appears to be competitive with traditional generation in some applications. The major constraints on widespread use of wind turbines are wind resource limitations and backup requirements.

Wind turbines come in all sizes, from units as small as 100 watts used in China to 100+ kW units used for utility scale generation. Furthermore, units can be combined into large wind farms like the Altamont Pass area in California, with a generating capacity of over 600 MW.[86]

Wind resources are distributed very unevenly over the Earth's surface and are strongly influenced by climate and terrain. Several studies have used spot measurements and climatological data to identify regions with high windpower potential. A recent study for the World Bank identified developing countries[87] that would be appropriate for grid-connected wind turbines. The criteria for inclusion included sufficient wind resource within 50 km of the existing electricity grid, so this list is not appropriate for identifying all off-grid potential sites. A separate study identified the western and eastern coasts of Africa, eastern Asia, and western and southern South America as most promising for wind-generated electricity.[88]

There are two principal components of costs for wind turbines—the initial capital costs, and the operation and maintenance (O&M) costs. The capital costs of small turbines varies from about $5,500/

Box 6-E—Wind Turbines for Water Pumping

The Naima region of Northeast Morocco is very dry and the soil is relatively unfertile, making water the key to sustaining life. Donor aid had provided a diesel pumping system that tapped a local spring and pumped the water to holding tanks—but the local residents found the operations, maintenance, and fuel costs of the system made it all but unaffordable.

The need for a sustainable system, which would not be dependent on the residents' scarce cash income, was needed. In 1986 the Moroccan Government, along with the U.S. Agency for International Development (AID), implemented the Naima Wind Project. This project used two 10 kW wind turbines, designed and manufactured in the United States, to replace the diesel pumps.

After several delays and problems (including poor timing-Ramadan, the Islamic month of fasting, began 1 day after the installation team arrived), the wind pumping system began operations in 1990. Reliability has been quite high (88 to 100 percent), and replication costs are estimated at about $2,500/kW. A detailed economic analysis found that the lifecycle costs for the wind system were lower than for a comparable diesel system—and the benefits to the local residents of a dependable, low-cost water supply, although difficult to measure, are large.

SOURCE: M. Bergey, "Sustainable Community Water Supply: A Case Study from Morocco," paper presented at the American Wind Energy Association National Conference, September 1990.

---

[85] By 1985, California wind turbines were generating 670 GWh/year. D.R. Smith, "The Wind Farms of the Altamont Pass Area," *Annual Review of Energy*, vol. 12, 1987, p. 146.

[86] D.R. Smith, "The Wind Farms of the Altamont Pass Area," *Annual Review of Energy*, vol. 12, 1987, p. 145, as of December 1986.

[87] Strategies Unlimited, "Study of the Potential for Wind Turbines in Developing Countries," contractor paper prepared for the U.S. Department of Energy and the World Bank, March 1987. The identified countries include Argentina, Brazil, Chile, China, Colombia, Costa Rica, India, Kenya, Pakistan, Peru, Sri Lanka, Tanzania, Uruguay, Venezuela, Zambia, and Zimbabwe.

[88] Based on data collected by Battelle Pacific Northwest Laboratory, and reported in California Energy Commission, *Renewal (sic) Energy Resources Market Analysis of the World*, CEC P500-87-015, p. 34.

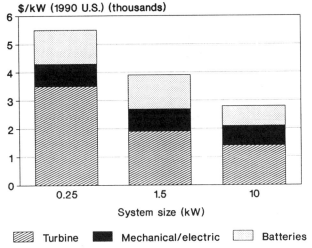

Figure 6-4—Capital Costs of Wind Turbines

Turbine / Mechanical/electric / Batteries

SOURCE: U.S. Congress, Office of Technology Assessment, 1992.

kW for a small 250 W system to about $2,800/kW for a larger 10 kW system, including storage (see figure 6-4).[89] The cost of large wind turbines for on-grid applications (not including storage) are much lower than this. By some estimates, the capital cost of large turbines could be as low as $600 per installed kilowatt by the mid-1990s.[90]

The operations and maintenance costs of intermediate size turbines are estimated at 0.7 to 1.5 cents per kWh.[91] Data on O&M costs for small turbines are not well documented. Data from a 5-year field test of a 10 kW turbine conducted by Wisconsin Power and Light Co. suggest O&M costs as low as 0.3 cents per kWh.[92] The levelized costs of wind-generated electricity are compared with those of other generating technologies in figure 6-5.

Wind turbines are dependent on a resource that, unlike fossil fuels, cannot be stored. The best wind turbines in California achieve a 90 percent reliability,[93] but this translates to a capacity factor of 35 percent (see glossary for definitions) because the wind does not blow steadily year round. Although the reliability of the wind turbines has improved considerably, resource limits will continue to be the major constraint to widespread use of this technology.

Stand alone applications often require a backup generating system or a storage system depending on the specific application and the characteristics of the local wind resource. Small wind turbines are used for battery charging in Inner Mongolia, for example,[94] and for this application short term fluctuations in wind are less important. For services such as lighting or refrigeration, however, a battery or other backup supply is needed, considerably adding to the system cost. For grid-connected wind turbines, fluctuations in output due to changes in wind speed can be moderated by other generating units in the grid and no storage is necessary.

The land requirements of wind turbines can be large. A typical 100 kW turbine in California, for example, requires about 1.2 hectares (ha) of land, or 12 ha/MW.[95] Crop production and livestock grazing can still be done on this land, however. Otherwise, wind turbines have few adverse environmental impacts. There are no direct air emissions or water requirements, and noise levels are generally low (except in the immediate vicinity of the turbine). Concerns have been raised, however, about the impact of wind turbines on local bird populations.

The design and fabrication of the wind turbine itself is somewhat complex and may not be readily done in many developing countries. The manufac-

---

[89] These costs are for the entire wind-electric system, including batteries.

[90] Judith M. Siegel and Associates, "Wind Electric, Photovoltaic, and Micro-Hydro Technologies for Developing World Applications," contractor report prepared for the Office of Technology Assessment, October 1990, p. 22.; Michael L.S. Bergey, Bergey Windpower Co., *Comments on the Maturation and Future Prospects of Small Wind Turbine Technology*, paper presented at the A.S.E.S. Solar '90 Conference, Norman, OK, Mar. 22, 1990; Michael L.S. Bergey, *Small Wind Turbines for Rural Energy Supply in Developing Countries* (Norman, OK: Bergey Windpower Co., June 1988, revised June 1989).

[91] Judith M. Siegel and Associates, "Wind Electric, Photovoltaic, and Micro-Hydro Technologies for Developing World Applications," contractor report prepared for the Office of Technology Assessment, October 1990, p. 23.

[92] Michael L.S. Bergey, Bergey Windpower Company, *Comments on the Maturation and Future Prospects of Small Wind Turbine Technology*, paper presented at the A.S.E.S. Solar '90 Conference, Norman, OK, Mar. 22, 1990, page 6.

[93] T. Moore, "Excellent Forecast for Wind," *EPRI Journal*, vol.15, No. 4, June 1990, p. 19.

[94] Judith M. Siegel and Associates, "Wind Electric, Photovoltaic, and Micro-Hydro Technologies for Developing World Applications," contractor report prepared for the Office of Technology Assessment, October 1990, p. 28

[95] D.R. Smith, "The Wind Farms of the Altamont Pass Area," *Annual Review Energy*, vol. 12, 1987, p. 145.

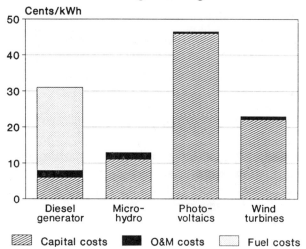

Figure 6-5—Levelized Cost of Remote Electricity Generating Technologies

This figure compares nominal costs for different technologies where good wind, hydro, and solar resources are present. Note that these figures do not include system losses of power conditioning or battery storage and will vary widely depending on local conditions.
SOURCE: U.S. Congress, Office of Technology Assessment, 1992.

turing and assembly of most other components—such as the tower, cabling, etc., however, account for half the initial cost of the system and might be done within the country.[96]

Photovoltaics (PVs)

Photovoltaics convert sunlight directly into electricity with no moving parts and no direct fuel consumption. Their modular design lends itself to both small and large scale applications, they are reliable, and they require little maintenance. The major disadvantages of photovoltaics are high first cost and, as in wind turbines, the intermittent nature of the resource. There are several types of photovoltaics, with varying costs, performance, and potential for future cost reductions.

Crystalline silicon photovoltaic cells have been the dominant technology since the first solar cell was produced in 1954, and still account for a large fraction of commercial photovoltaic production. The efficiencies of these cells are as high as 23 to 24 percent in the laboratory (with no concentration) and typically 12 to 13 percent in commercially available form. The manufacturing costs and material requirements of crystalline silicon photovoltaics cells are relatively high, which has led to research into using other manufacturing approaches and materials. Polycrystalline thin films of silicon are lower cost than single crystal silicon, but have somewhat lower laboratory and commercial unit conversion efficiencies.[97]

Amorphous silicon photovoltaic cells are used in consumer products, such as watches and calculators. Efficiencies of these cells in commercial modules are typically 4 to 5 percent after several months of use, although laboratory efficiencies as high as 9.5 percent have been achieved.[98] Present designs suffer from light induced degradation to about five-sixths of their initial efficiency in the first year,[99] although alternative designs to reduce this are being explored. Amorphous silicon cells have the potential for low cost manufacturing.

Compound semiconductors, made up of combinations of semiconductor materials other than silicon, are also produced, although in considerably lower volumes than silicon based solar cells. Recent laboratory breakthroughs have pushed the efficiency of thin film cadmium telluride cells to roughly 14.5 percent. The required materials for compound photovoltaics could be environmentally harmful if mishandled.

Concentrators use mirrors or lenses to concentrate solar radiation onto photovoltaic cells. Laboratory efficiencies of more than 28 percent have been achieved with concentrators using single crystal

---

[96] The turbine blades, however, would probably need to be imported, but they account for only about 5-15% of total material costs. Michael L.S. Bergey, Bergey Windpower Co., letter to Office of Technology Assessment, Feb. 6, 1991.

[97] Department of Energy, "Photovoltaic Energy Program Summary: Volume I: Overview, Fiscal Year 1989"; D. Carlson, "Photovoltaic Technologies for Commercial Power Generation," *Annual Review of Energy*, vol. 15, 1990, pp. 85-119. Also T. Moore, "Thin Films: Expanding the Solar Marketplace," *EPRI Journal*, March 1989, p. 4. Efficiency, as used here, is the amount of electric output divided by the amount of energy, as sunlight, received as input. It is a useful criterion because the cells are the major expense of PV systems, and greater efficiency means less cell area required to produce a given electrical output. PV efficiencies given here are for commercial products, which can be considerably lower than that achieved in the laboratory under ideal conditions.

[98] Department of Energy, "Photovoltaic Energy Program Summary: Volume I: Overview, Fiscal Year 1989"; D. Carlson, "Photovoltaic Technologies for Commercial Power Generation," *Annual Review of Energy*, vol. 15, 1990, p. 85-119. Also T. Moore, "Thin Films: Expanding the Solar Marketplace," *EPRI Journal*, March 1989, p. 4.

[99] T. Peterson, "Amorphous Silicon Thin-Film Solar Cells," *EPRI Journal*, vol. 13, No. 4, June 1988, p. 42.

silicon cells.[100] The additional cost of the mirrors or lenses can be offset by the smaller cells; concentrators may, however, require more frequent cleaning, adding to the O&M costs of the system.

There are two components of photovoltaic system capital costs—the photovoltaic cells and the balance-of-system (BOS) components, which includes everything else—the structures holding the photovoltaic cell, power conditioning equipment to convert direct current electricity to commonly used alternating current, controls, possibly battery storage, and others. Costs of photovoltaic cells have dropped dramatically due to technical advances, but still remain higher than many alternatives. Current prices are about $4,000 to $6,000/kW$_p$, excluding balance-of-system costs.[101] Prices are expected to continue to fall, due to advances in photovoltaics design, advances in manufacturing, and economies of scale from increased production volumes.[102]

Balance-of-system costs depend on the specific application. AC systems require an inverter to convert the photovoltaics DC current to AC current. There are, however, lights, refrigerators, and other appliances that can use DC current directly. Small, remote systems commonly use batteries to store power. The size (and therefore cost) of the battery system will depend on the load and on how long one requires the backup system to produce electricity. Shown in table 6-14 are some representative costs for specific systems.

Operating and maintenance (O&M) costs for photovoltaic systems are in general quite low. O&M costs for existing systems can usually be traced to problems with the BOS components. By one estimate, O&M costs for small photovoltaic systems are about 0.5 cents/kWh.[103] Systems with battery storage require battery replacement every 3 to 5 years. The Electric Power Research Institute examined O&M costs for utility scale systems, and found that flat plate photovoltaic systems had actual O&M costs of 0.39 to 1.44 cents/kWh.[104]

Photovoltaic systems have been found to be highly reliable in developing countries,[105] as well as the United States. For example, a survey of residential photovoltaic systems in the United States found most systems were operating properly 90 percent or more of the time.[106]

Table 6-14—Representative Balance of System (BOS) Costs for Specific Photovoltaic Applications

	Dominican Republic
Application:	Small residential system, 38 W$_p$
BOS costs:	Battery $1,050/kW (lasts 3-5 years) Electronic control equipment $1,000/kW
	United States
Application:	Residential mixed use system, 200 W$_p$
BOS costs:	Battery $1,400/kW (lasts 3-5 years) Mounting hardware $800/kW Electronic control equipment $1,800/kW

SOURCE: Data for Dominican Republic from D. Waddle, R. Perlack, and R. Hansen, "Rural Based Photovoltaic Systems in the Dominican Republic: An Agenda for Future Activities," Report to the U.S. DOE by ORNL. Data for United States from Real Goods Trading Company, Ukiah, CA.

---

[100]Department of Energy, "Photovoltaic Energy Program Summary, Volume I: Overview, Fiscal Year 1989,".

[101]D. Carlson, "Photovoltaic Technologies for Commercial Power Generation," *Annual Review of Energy*, vol. 15, 1990, p. 92, estimates $4,000/kWp in 1988. United Nations, Committee on the Development and Utilization of New and Renewable Sources of Energy, "Direct Solar-to-Electrical Energy Conversion," Report of the Secretary-General, Mar. 8, 1990, estimates price for crystalline silicon of $4,000/kW$_p$ for large orders, excluding delivery and taxes. A retailer in the U.S. sells PV panels for about $6,000/kW$_p$ in small quantities. Current price in the Dominican Republic for a 38 watt (peak) panel is $226, or about $6,000/kWp. See D. Waddle, R. Perlack, and R. Hansen, U.S. Department of Energy, Oak Ridge National Laboratory, *Rural Based Photovoltaic Systems in the Dominican Republic: An Agenda for Future Activities*, Report No. DE-840R21400 (Oak Ridge, TN: Oak Ridge National Laboratory).

[102]U.S. Department of Energy, Solar Energy Research Institute, *The Potential of Renewable Energy*, SERI/TP-260-3674 (Golden, CO: Solar Energy Research Institute, March 1990), reports PV industry representatives forecasting $1,000/kWp by 1995. Electric Power Research Institute (EPRI), *TAG-Technical Assessment Guide, Electricity Supply 1989*, EPRI P-6587-L (Palo Alto, CA: Electric Power Research Institute, September 1989), p. 7-81 forecasts $1,500/kWp in 1994 for a large flat-plate system.

[103]Meridian Corp., "Evaluation of International Photovoltaic Projects, Volume II: Technical Report," prepared by Sandia for DOE and AID, SAND85-7018/2, September 1986, p. 13-4.

[104]Lynette and Associates, "Photovoltaic Operation and Maintenance Evaluation," GS-6625 (Palo Alto, CA: Electric Power Research Institute, December 1989).

[105]Meridian Corp., "Evaluation of International Photovoltaic Projects, Volume II: Technical Report," prepared by Sandia for DOE and AID, SAND85-7018/2, September 1986. Reliability refers to the fraction of time the system is operating correctly, and does not include resource limits. A PV system with 100% reliability will have zero output at night, and diminished output when solar radiation levels are low.

[106]Electric Power Research Institute, "Survey of U.S. Line-Connected Photovoltaic Systems," EPRI GS-6306 (Palo Alto, CA: Electric Power Research Institute, March 1989), pp. 3-7.

> **Box 6-F—Photovoltaic Powered Vaccine Refrigerators**
>
> Vaccines must be kept cold to maintain effectiveness. Vaccine refrigerators powered by fossil fuels are dependent on expensive and erratic fuel supplies, leading to loss of vaccines. The World Health Organization (WHO) has examined the potential of photovoltaic-powered refrigerators for vaccine storage and is installing them in rural health care centers and hospitals in several developing countries. Photovoltaic powered refrigerators offer independence from fossil fuels, longer system life, improved temperature control, and greater mechanical reliability.
>
> In Zaire, domestically designed and produced photovoltaic powered refrigerators using imported compressors and batteries have been placed in rural areas and have been reported as more reliable than kerosene refrigerators. In India, there are more than 200 rural health care centers without grid access, and photovoltaic powered refrigerators are now being tested under Indian conditions. Mali, Sudan, Indonesia, and several other countries are now using photovoltaic powered vaccine refrigerators.
>
> Although the initial capital costs of the photovoltaic system are high, the longer life, fuel cost savings, and increased reliability of the photovoltaic system lead to a cost of about 32 cents per dose for the photovoltaic system, as compared to 47 cents for the kerosene system.
>
> SOURCES: United Nations, Committee on the Development and Utilization of New and Renewable Sources of Energy, "Direct Solar-to-Electrical Energy Conversion," Report of the Secretary-General, Mar. 8, 1990. Government of India, Ministry of Energy, Department of Non-Conventional Energy Sources, *Annual Report 1988-89*, p. 53. See also C. Rovero and D. Waddle, *Vaccine Refrigeration Technologies and Power Sources*, Report No. ORNL-6558 (Oak Ridge, TN: Oak Ridge National Laboratory, April 1989).

The lifetimes of photovoltaic cells vary. Cells used for vaccine refrigeration systems are expected to last 15 years (see box 6-F).[107] Utility scale systems are forecasted to last 20 years.[108] Balance-of-system components, particularly batteries, rather than the cells themselves, may often be the limiting factor in system lifetimes.

Photovoltaic systems can be sized to fit any application. Installed photovoltaic systems range from 38 W household systems found in the Dominican Republic (see box 6-G) to a 5.2 MW grid-connected system in California.

The operation of photovoltaics requires no fuel, and therefore avoids the detrimental environmental effects of fuel combustion. Manufacturing some types of photovoltaics does require the use of potentially harmful chemicals, if released.[109]

Electricity generation by photovoltaics is not land constrained. For example, India's entire electricity needs could be met with photovoltaic cells covering 1,120 km^2, or about 0.034 percent of the land.[110] Some argue that traditional centralized generation, such as coal and nuclear, requires more land than do photovoltaics if mining, transportation, and waste disposal are taken into account.[111]

MicroHydropower

Microhydroelectric plants[112] are common in several Asian countries—notably China and India—and are also found in some African and Latin American countries. Like all renewable resources, hydropower requires no direct use of fossil fuels and therefore has low operating costs. Although microhydropower systems can have relatively high initial capital costs, much of this is in labor and locally available materials for site preparation.

---

[107] United Nations, Committee on the Development and Utilization of New and Renewable Sources of Energy, "Direct Solar-to-Electrical Energy Conversion," Report of the Secretary-General, Mar. 8, 1990, p. 9.

[108] Electric Power Research Institute (EPRI), *TAG-Technical Assessment Guide, Electricity Supply 1989*, EPRI P-6587-L (Palo Alto, CA: Electric Power Research Institute, September 1989), p. 7-81.

[109] D. Carlson, "Photovoltaic Technologies for Commercial Power Generation," *Annual Review of Energy*, vol. 15, 1990, p. 90.

[110] Electricity consumption in 1987 from International Energy Agency, *World Energy Statistics and Balances 1971 to 1987* (Paris: OECD, 1989). Land area from World Bank, *World Development Report 1989* (New York, NY: Oxford University Press, 1989). Ten percent conversion efficiency assumed.

[111] H.M. Hubbard, "Photovoltaics Today and Tomorrow," *Science*, vol. 244, No. 902, Apr. 21, 1989, p. 300.

[112] This discussion focuses on small systems of up to 100 kW, called "micro-hydro," which are suitable for off-grid applications. This is a subset of "small hydro", which refers to systems of up to 15 MW.

> **Box 6-G—Residential Photovoltaic Systems in the Dominican Republic**
>
> A simple, low-cost residential photovoltaic system, costing about $500, now supplies electricity for lights and radios in households in the Dominican Republic. Enersol, a U.S.-based nonprofit organization, began photovoltaics demonstrations in the Dominican Republic in 1984, and has since installed the systems in over 1,000 households. This has been done with minimal outside funding, and has made use of local resources and the private sector wherever possible.
>
> The photovoltaic system consists of a 38 watt (peak) photovoltaics array, a 12-volt car battery, associated electrical components, and lights and other appliances. The photovoltaics array is imported from the United States and faces a 70-percent import tax. The battery is manufactured domestically and is relatively inexpensive, although it must be replaced every few years. Service centers and trained technicians provide system repair when needed (most commonly blown fuses and dead batteries). A revolving loan program is used to provide the capital for initial purchase.
>
> SOURCE: R. Hansen and J. Martin, "Photovoltaics for Rural Electrification in the Dominican Republic," *Natural Resources Forum*, vol. 12, No. 2, 1988, pp. 115-128.

Unlike large hydropower projects, microhydropower can often be installed quickly and usually does not require the flooding of large areas.[113] Sites for microhydro are limited, however, and hydro resources are often highly seasonal. With sufficient long term, site specific water flow data, electrical output can be predicted with reasonable accuracy. These data can be expensive to collect, however, adding to the total cost of the system.

Hydropower potential depends on annual runoff, seasonal distribution of the runoff, topography, and other factors. Some estimates of the potential for small hydro (up to 15 MW) have been made. For example, a recent report identified a small hydropower potential of over 40 GW in the developing world.[114] The fraction of this potential suitable for microhydro (less than 100 kW) is unclear. Furthermore, these data are based only on large scale hydrologic information and do not consider local factors, such as the distance between the potential system installation site and the area needing the power.

Hydropower converts the kinetic energy of falling or flowing water to drive a turbine connected to an electrical generator. The hydropower potential of a specific site is proportional to the water flow rate and the "head," which is the vertical distance the water falls.[115] The system design varies, depending on whether or not storage is incorporated and on the type of turbine used.

The simplest hydropower system is a waterwheel. These have long been used to produce mechanical power for crushing, grinding, and other similar mechanical tasks. Waterwheels are simple to construct, and require little in the way of special tools or materials, but their slow speed makes them unsuitable for generating electricity.

A complete hydropower system consists of a river and dam (if used), a pipe carrying water to the turbine, the turbine itself, the generator, various controls, and other components.

For hydroelectricity generation, either impulse turbines—which use a nozzle to direct the water at high speed against the turbine blades—or reaction turbines—that are powered by the pressurized flow of water through the unit—can be used. Reaction turbines are more common in low-head[116] microhydro applications.

Hydropower systems to generate electricity can be divided into "run-of-river" systems that do not incorporate storage; and storage systems, which use a dam or basin to store water. A run-of-river system is less expensive to build and has reduced downstream impacts as the river flow will be affected less significantly. The electricity output will fluctuate with changes in river flow, however, requiring storage or other systems to offset variations in output. Alternatively, a dam or basin can be used to

---

[113] In aggregate, however, microhydro systems may flood substantial areas.

[114] California Energy Commission, "International Market Evaluations: Small-Scale Hydropower Prospects," Report No. P500-87-005, p. 25.

[115] The governing equation is $P = 9.81 QHe$, where P is the power output of the system in kW, Q is the water flow rate in m^3/sec, H is net head in meters, and e is the conversion efficiency.

[116] Low-head means that the water falls less than about 15 meters (50 feet).

store water so that electric output can be tailored to meet demand.

The initial capital costs for microhydropower systems vary widely, depending on the site characteristics and how one accounts for local labor and materials. Reported system costs range from as low as $290/kW in Pakistan[117] (see box 6-H) to $2,350/kW in Thailand.[118] Typical costs are in the range of $1,000 to $2,000/kW.[119] Site specific construction costs are 15 to 45 percent of total costs (see table 6-15). If these costs are met with low cost local labor and with locally available materials (concrete, wood, etc.), then total system costs can be quite low while benefiting the local economy and saving foreign exchange.

Costs for the mechanical components—the turbine, generator, and associated electronic control equipment—benefit from economies of scale, making larger systems less expensive per kW than small systems. Mechanical equipment costs in the United States range from about $2,800/kW for a small, 500 W, 12 V DC system to about $700/kW for a larger 50 kW 120/240 V AC system.[120] These costs are not expected to change significantly in the future, as the technology for microhydropower is in general already mature and little R&D work is being done.[121]

Costs for operation and maintenance of microhydro systems are not well known. O&M costs of about 2 cents/kWh are reported,[122] but are highly variable. Routine operation and maintenance may only involve periodic cleaning of water intakes. If major repair is needed, however, the costs of skilled labor may be high, especially in remote areas. Microhydro systems can be quite reliable—many plants installed in the 1940s and 1950s are still operating.[123]

The environmental impacts of microhydro systems can include flow disruption due to diversion of water from the stream and sedimentation due to changes in river flow. Depending on the design of the system, water quality and aquatic organisms can be affected—for example, a system using a dam provides a habitat for mosquitoes, which can spread malaria or other diseases.

> **Box 6-H—Micro-Hydropower in Pakistan**
>
> Less than one-quarter of rural villages in Pakistan have electric service. Extending the electric grid to these remote villages is prohibitively expensive, as is diesel fuel for diesel generators. The Appropriate Technology Development Organization (ATDO), an agency of the Government of Pakistan, has pursued the use of micro-hydropower projects to supply electricity to remote villages.
>
> The program emphasis is on the use of local resources. All decisions regarding project size, location, operation, etc. are made by the villagers, The ATDO staff work with the villagers, and encourage the use of locally available materials and labor wherever possible. Costs are shared and the community decides on the electricity rates to users. The end result is rural electrification at a very low cost—typically $250 to $400 per kW. Efforts are continuing to further reduce costs with local fabrication of generators and penstock piping.
>
> SOURCE: M. Abdullah, "Micro-Hydroelectric Schemes in Pakistan," *Small-Scale Hydropower in Africa-Workshop Proceedings*, Abidjan, Ivory Coast, March 1982, pp. 27-30.

---

[117] M. Abdullah, "Micro-Hydroelectric Schemes in Pakistan," *Small-Scale Hydropower in Africa-Workshop Proceedings* (Abidjan, Ivory Coast: March 1982), p. 29.

[118] P. Clark, "Cost Implications of Small Hydropower Systems," *Small-Scale Hydropower in Africa-Workshop Proceedings* (Abidjan, Ivory Coast: March 1982), p. 100.

[119] United Nations Conference on Trade and Development, *Technology Policy in the Energy Sector: Issues, Scope and Options for Developing Countries*, UNCTAD/TT/90 (Geneva, Switzerland: United Nations, November 1989), p. 80, reports an average cost of $2,350/kW for small (6-21 kW) systems in Nepal. P. Clark, "Cost Implications of Small Hydropower Systems," *Small-Scale Hydropower in Africa-Workshop Proceedings* (Abidjan, Ivory Coast: March 1982), p. 106 reports an average cost of $1,000/kW for 7 projects under 50 kW in size in Asia.

[120] Dan New, Canyon Industries, Deming, WA, personal communication, November 1990. Five kW systems are about $1,400/kW, 20 kW are about $1,000/kW. All prices include turbine, generator, and electrical equipment.

[121] Improved materials and manufacturing processes, and large volume production, however, might impact costs.

[122] M. wa Kabasele, "Small Hydroelectric Power Development in Zaire," *Small-Scale Hydropower in Africa-Workshop Proceedings* (Abidjan, Ivory Coast: March 1982), p. 36; J. Fritz, "Cost Aspects of Mini-Hydropower Systems," *Small-Scale Hydropower in Africa-Workshop Proceedings* (Abidjan, Ivory Coast: March 1982), p. 116; California Energy Commission, "International Market Evaluations: Small-Scale Hydropower Prospects," Report No. P500-87-005, p. 19. Fifty percent capacity factor assumed.

[123] California Energy Commission, "International Market Evaluations: Small-Scale Hydropower Prospects," Report No. P500-87-005, p. 18.

Table 6-15—Breakdown of Micro-Hydro Capital Costs

Item	Percent of cost
Turbine-generator	18-39
Other electrical and plant equipment	7-16
Site-specific construction	15-45
Engineering	20
Other	10

SOURCE: P. Clark, "Cost Implications of Small Hydropower Systems," *Small-Scale Hydropower in Africa—Workshop Proceedings*, March 1982, Abidjan, Ivory Coast, p. 106.

Table 6-16—Some Typical Fuel Consumption Rates for Engine Generators

Size (kW)	Fuel	Consumption rate liters/kWh	Fuel operating costs $/kWh
4.0	gasoline	0.71	0.71
7.5	gasoline	0.63	0.63
20	diesel	0.40	0.20
50	diesel	0.28	0.14
100	diesel	0.24	0.12
1,000	diesel	0.28	0.14

Fuel priced at 50 cents/liter.
SOURCE for fuel consumption rates: Real Goods Trading Co., Ukiah, CA; Onan Corp., Minneapolis, MN.

Engine Generators

The most common technology today for remote generation of electricity is usually an engine generator—a small internal combustion engine using gasoline or diesel fuel to generate electricity. This technology is popular for good reason—it has a relatively low initial cost, it is widely available, it can be installed anywhere, and it uses a technology already familiar in the form of cars, trucks, and buses. It is dependent on scarce and expensive fossil fuels, however, which are often imported. The costs and other characteristics of engine generators[124] are summarized here, as they are the technology against which off-grid renewables must compete.

Small (5 to 10 kW) engine generators typically use gasoline and have initial costs of $600 to $700/kW. Medium sized systems (10 to 100 kW) usually use diesel fuel and have initial costs of $200 to $700/kW. Very large systems (250 kW and more) exclusively use diesel and have initial costs of about $125/kW.[125]

Operation and maintenance costs for engine generators can be divided into fuel and nonfuel costs. Nonfuel costs include routine maintenance, such as oil and filter changes, as well as repair. A considerable amount of scheduled maintenance is required, which costs about 1.3 to 2 cents/kWh.[126] In addition, a complete overhaul is required about every 10,000 hours of operation, at a cost of about 15 percent of the capital cost. The lifetimes for engine generators are quite short—about 20,000 hours for diesel units and 3,000 to 5,000 hours for gasoline units.[127] Engine generators have a reputation for high reliability as long as maintenance requirements are met.

Fuel costs depend on fuel prices (table 6-16). In many rural areas, fuel supplies for engine generators are unreliable and expensive due to difficult and time consuming transport. For oil importing countries, fuel expenditures require a large fraction of foreign exchange. As long as fuel supplies are available, engine generators can operate at very high capacity factors. As the fuel supply can be stored, no battery backup is required for engine generators. In addition, no site surveys are needed to measure the resource, and the engine generator can be moved to a new location if desired.

The environmental effects of engine generators include noise and air pollution. Noise effects can be minimized by locating the engine away from residences, however, and in rural areas at least the local air quality effects may be readily dispersed.

Comparing the Technologies

There is no one "best" technology: each specific application must be considered on its own. Costs are also uncertain, changing—rapidly in some cases—over time, and are site dependent. The values given here are approximate.[128] Capital, O&M, fuel costs,

---

[124]The term "engine generator" is used to mean an integrated system, including an engine, an electrical generator, and associated electrical equipment.

[125]Prices are approximate, and are from Real Goods Trading Company, Ukiah, CA, and from Curtis Engine Equipment, Inc., Baltimore, MD. Prices quoted here are retail prices in the U.S.—prices in developing countries will differ due to transport costs, import duties, and other factors.

[126]Typical requirements include checking engine oil every 8 hours, changing engine oil and oil filter every 150 hours, and changing fuel and air filter every 500 hours. The cost for these parts in the U.S. is about 2 cents/kWh. Others estimate maintenance and repair at 1.3 cents/kWh (Sandia National Laboratories, *Evaluation of International Photovoltaic Projects Volume II: Technical Report*, SAND85-7018/2, September 1986, p. 13-5.)

[127]Ibid., 8-7.

[128]See also, U.S. Congress, Office of Technology Assessment, *Renewable Energy Technologies: Research, Development, and Commercial Prospects*, forthcoming.

Table 6-17—Initial Capital Costs of Electricity Generating Systems

Technology	Size (kW_p)	Initial capital cost ($/kW_p$)
Engine generator:		
Gasoline	4.0	760
Diesel	20.0	500
Microhydro	10-20	1,000-2,400
Photovoltaic	0.07	11,200
Photovoltaic	0.19	8,400
Wind turbine	0.25	5,500
Wind turbine	4	3,900
Wind turbine	10	2,800

NOTE: Costs are for entire system, including conversion device, electric generator, and associated electrical equipment. Prices for wind turbine and photovoltaic systems include batteries. kW_p ratings are for peak output, average output will be somewhat lower depending on the resource. Prices for wind turbine, photovoltaic, and engine generators are actual retail prices in the U.S. in 1990. Prices for microhydro are averages across several installations.

SOURCES: Bergey Windpower Co., Norman, OK; Onan Corp., Minneapolis, MN; Real Goods Trading Co., Ukiah, CA.

Table 6-18—Operating, Maintenance, and Fuel Costs (diesel fuel price of $0.50/liter assumed)

Technology	O&M costs (cents/kWh)	Fuel costs (cents/kWh)
Engine generator	2	20 (diesel)
Micro Hydro	2	0
Photovoltaics	0.5	0
Wind turbines	1	0

SOURCE: Office of Technology Assessment, 1992.

Table 6-19—Approximate Lifetimes of Off-Grid Renewable Generating Technologies

Technology	Lifetime (years)
Engine Generator	8-10 (diesel)
Micro Hydro	20-30
Photovoltaics	20-30
Wind Turbine	15-25
Batteries	3-5

SOURCE: Office of Technology Assessment, 1992.

equipment lifetimes, and other data for these technologies are shown in tables 6-17 to 6-19. Capital costs—of particular concern for developing countries—are lowest for engine generators. Life cycle operating costs for a 10 kW system—enough to supply a small village—however, indicate that for a variety of applications, various renewable energy technologies may be more attractive than engine generators (see app. B).

Obtaining the results shown in figure 6-5 required making a number of assumptions. These results are particularly sensitive to the assumed discount rate and fuel price (see app. B). Access to good quality resources that are reliable throughout the year, or at least when power is needed, is also important.[129] At high discount rates, those technologies with a high initial cost—notably photovoltaics—look less attractive (see app B). Individual consumers in developing countries often can borrow money only at very high rates—40 percent per year in the Dominican Republic,[130] for example, while governments and large institutions can often borrow capital at much lower rates—5 to 10 percent. If the selection of remote generating technologies is made by individuals facing capital constraints and high interest rates, renewable technologies with high capital costs will be less attractive. If these decisions are made by larger institutions and governments, these technologies are more financially attractive.

The cost of fuel is an important determinant of the cost of diesel generation. At 50 cents/liter for diesel, the cost of generating electricity with diesel sets is two-thirds the cost achieved of photovoltaic power, while at $1/liter, engine generated electricity costs about 10 percent more than that from photovoltaics (see figure 6-6 and app. B).

Given the large uncertainty in the cost data, it is not appropriate to interpret these cost data as applying to any specific situation Other factors, such as those listed in table 6-20, may be as or more important than cost. Nevertheless, the following tentative conclusions can be drawn:

- With reasonable assumptions concerning discount rates, capacity factors, and fuel cost, microhydro and wind turbines can have the lowest life cycle costs in locations where the resource is sufficient.
- Diesel generators have by far the lowest initial capital cost, but when fuel and O&M costs are considered, diesel generators are of comparable expense to renewable technologies—more expensive than wind turbines and microhydro, and less expensive than photovoltaics. The cost

---

[129] For example, the best hydropower resources are during the rainy season when irrigation pumping is not needed. Conversely, when irrigation pumping is needed—during the dry season—hydropower resources are usually at their lowest.

[130] R. Hansen and J. Martin, "Photovoltaics for Rural Electrification in the Dominican Republic," *Natural Resources Forum*, vol. 12, No. 2 1988, p. 120.

Figure 6-6—Costs of Remote Generating Technologies

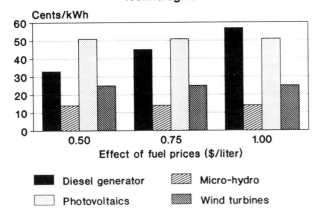

10 kW systems with storage

This figure shows the relative cost of remote generating technologies for three different diesel fuel prices.

SOURCE: U.S. Congress, Office of Technology Assessment, 1992. See app. B for more detail.

Table 6-20—Non-Cost Attributes of Off-Grid Electricity Generating Technologies

Engine generators
- Independent of renewable resource fluctuations
- Dependent on fuel supply system
- Susceptible to fuel price increases

Micro-hydro
- Large part of costs for construction, can be paid with local currency
- Proven, mature technology, highly reliable and long lifetimes
- Output dependent on water flow

Photovoltaics
- Low maintenance requirements
- Low availability due to solar resource dependence
- Requires batteries for energy storage

Wind turbines
- Low maintenance requirements
- Suitable only for areas with sufficient wind resource
- Requires batteries for energy storage

SOURCE: Office of Technology Assessment, 1992.

of electricity production from diesel engine generators is heavily dependent on fuel prices and quality of maintenance.

- Photovoltaics are expensive, due largely to the cost of the panels themselves. Panel costs are expected to continue to drop in the near future, however. Further, the analysis presented here is for a 10 kW system. Photovoltaic systems are more economically attractive for smaller applications, such as household or agricultural pump (less than 1 kW) systems.

Grid Extension

The alternative to remote generation of electricity, either by renewable technologies or by diesel generators, is to extend the existing electricity grid. Grid extension costs have several components: the capital cost of extending the electricity lines themselves; the operations and maintenance costs of the lines; and the cost of generating the electricity. The cost of electricity distribution lines varies with line voltage and capacity, topography, and other factors, and is estimated at $4,600 to $12,700 per kilometer, with an average of about $9,000/km in the United States.[131] The operations and maintenance costs of these lines are estimated at 2 to 4 percent of the capital costs per year,[132] or about $270/km year for the average line. The cost of generating the electricity varies widely, depending on fuel used, power plant efficiency, and other factors. Long run marginal costs of electricity generation, which include fuel costs, operation and maintenance costs, and levelized capital costs, were estimated at from 4.4 cents/kWh to 13 cents/kWh, with an average of 9 cents/kWh, in a recent study of developing countries.[133]

The cost of grid extension increases with distance. In contrast, the costs of remote generation are not affected by distance from the grid. At some distance, called the "break even" distance, the costs of grid extension exceed those of remote generation. The results of such a calculation, first considering only capital costs and then considering life cycle costs, are shown in table 6-21. Calculations and assump-

---

[131]For single phase systems. From National Rural Electric Cooperative Association, *Understanding Electric Utility Operations* (Washington, DC: National Rural Electric Cooperative, 1989), p. 106. There are alternatives to this technology—for example, a single-wire earth return system—common in rural Australia, could be used in developing countries, and costs as little as half as much as a traditional single-phase system (Allen Inverson, NRECA, personal communication, 12/4/90).

[132]M. Mason, *Rural Electrification-A Review of World Bank and USAID Financed Projects*, Background Paper for the World Bank, April 1990, p. 27.

[133]Ibid., p. 27. The median long-run marginal cost would be somewhat lower.

Wind turbines may be less expensive than grid extension in some areas.

tions are given in appendix B.[134] For the assumed system size, the break even distance varies from less than 2 km for microhydro to 14 km for photovoltaics when life cycle costing is used. Smaller system loads than the assumed 10 kW reduce this break even distance sharply.[135] Different assumptions of grid extension costs also change this break even distance. For example, assuming the typical U.S. cost of $9,000/km (rather than $4,500/km) the break even distance for photovoltaics drops to 7km (from 14). In contrast to remote generation, however, costs for grid extension can be spread over several villages if multiple service drops to several villages are possible.

OIL REFINING

Petroleum products are obtained in the developing world through a combination of direct import of refined products and local processing of crude oil—usually imported—in local refineries. About one-quarter of world refining capacity is located in the developing world.

The scale and other characteristics of refineries in the developing countries varies widely—from the large state-of-the-art refineries of the major oil producers, such as Venezuela and Mexico, to the small uneconomic refineries operated in several of the smaller African countries. One common characteristic, however, is widespread state ownership. The government, generally acting through a parastatal enterprise, such as the national oil company, takes a majority interest or even owns the facility outright. A foreign company is generally involved in designing, building, and sometimes managing the refinery under a service contract. Such companies often provide access to crude oil feedstock for the refinery from their integrated international system, though some state owners of refineries rely mostly on government-to-government supply agreements.

The major oil producers apart, refineries in developing countries face a number of problems. In many cases, developing country refineries do not produce a range of light and heavy products appropriate to the consumption patterns in the country. This comes about for two reasons. First, much of refinery technology is based on the product demand of the industrial countries. In developing countries, between 60 and 70 percent of refinery demand is diesel and residual, compared with 30 percent for the United States. On the other hand, gasoline accounts for about half of U.S. production, compared with about 20 percent in developing countries. Further, developing-country consumers do not use the petroleum products designed for use in cold climates.

Second, most of the plants use the simplest refining process—primary distillation (or hydroskimming)—which is limited in its flexibility. When crude oil prices rose in the 1970s, demand for petroleum products shifted. Developing-country refineries, which typically do not have secondary conversion technologies, could not adjust to these changes. The only flexibility available in such cases (without huge investments) lies in varying the quality of the imported crude feedstock. Many developing countries now import crude oil, refine it locally, and re-export the excess (typically residual fuel oil and gasoline). As the quantities exported are small, they are often sold for prices well below the world average.

Diseconomies of scale cause refining costs to be much higher than in the developed world. Average refinery operating costs in Africa are $2 per barrel, compared to $0.75 in the rest of the world. Very few developing country refineries can process crude for less than $1 per barrel, except the large export refineries in the Far East and the Caribbean. Refinery losses, which should not exceed 1 percent

---

[134]See also, Chandra Shekhar Sinha and Tara Chandra Kandpal, "Decentralized v. Grid Electricity for Rural India," *Energy Policy* June 1991, pp. 441-448.

[135]"On-site Utility Applications for Photovoltaics," *EPRI Journal*, March 1991, p. 26.

Table 6-21—Break-Even Distances for Comparing Grid Extension to Remote Generation, (10 kW system operating at 20% capacity factor)

Considering only initial costs	
Technology	Break-even distance
Diesel generator	1.4 km
Micro-hydro	3.5 km
Photovoltaics	18.0 km
Wind turbine	7.9 km
Considering initial, operating and maintenance, and fuel costs	
Technology	Break-even distance
Diesel generator	8.1 km
Micro-hydro	2.1 km
Photovoltaics	13.9 km
Wind turbine	5.6 km

SOURCE: Office of Technology Assessment, 1992. See app. B for details.

in a properly maintained and operated refinery, often exceed 2 or even 4 percent in developing-country facilities.

Technology could improve the performance of the refinery sector in developing countries. This is of particular importance as old and inefficient refineries in many countries are among the largest single consumers of petroleum fuels.

Besides repair and replacement of worn out or obsolete equipment, the main targets of increased efficiency are connected with the heat used to process crude. For example, reducing excess air mixing into the flue gas can result in energy savings of about 2 percent in the heaters and 30 percent in the catalytic reformer. This technology has been common in industrial-country refineries since the 1970s, is relatively low cost, but is still the exception in developing countries. Heater efficiency can also be improved through the installation of heat exchangers to recover heat from the stack gas, where temperatures can reach 1,000 °F. Retrofitting to take advantage of previously flared relief valve and other gas, reducing evaporation from storage tanks, and other measures, can result in substantial savings at moderate cost.

These technology improvements could also curtail some of the adverse environmental impacts of oil refining. Obsolete and poorly maintained equipment can lead to increased leaks of oil and a number of toxic emissions, with negative effects on soil, water, and air quality. Refineries also have solid and liquid wastes that require special handling and disposal. Many of the pollutants associated with refining could be mitigated through properly maintained or improved equipment.

More generally, prospects are for a continual incremental improvement in oil refinery technology and energy efficiencies, using improved separation techniques and more intensive use of computer technology to improve product quality. A major change could be in new catalysts for the light end of petroleum fraction and natural gas liquids. Present day catalytic crackers produce excessive quantities of the light ends such as off gas, methane, and ethane. By the end of the decade, new catalysts are expected to improve the yield of gasoline from these light ends. Membrane separation technologies are under development but appear to be more suitable for use in large scale refineries rather than the smaller operations typical of many developing countries.

The low level of refinery efficiencies in developing countries is due largely to lack of investment. Problems hindering investment in the rationalization of refinery operation in the developing world include lack of foreign exchange for parts; artificial refinery profits stemming from subsidized feedstock, distorted pricing structures and earmarked government subsidies; and noneconomic reasons such as "security of supply."

It is estimated that between 1991 and 1995, refinery rehabilitation and conservation in the developing world will require an investment of almost $2 billion.[136] The extent of the problems existing in some countries and the large amounts of money needed for rehabilitation raise the question of whether some refineries should not be closed down, meeting local needs through a combination of regional restructuring and direct import of refined products. Refining is capital rather than labor intensive and so often requires highly paid expatriate personnel and creates relatively few spinoffs for the local economy. Where a small local market constrains capacity utilization and/or the facility is old and/or decrepit and inefficient, the decision to close the facility, politically painful as that may be, would reduce economic losses.

---

[136] Theodore J. Gorton, "Oil and Gas Development in the Third World," contractor report prepared for the Office of Technology Assessment, June 1990.

Generalization is, however, difficult. Some developing country refineries benefit from strategic regional location and good management and are clearly success cases. For example, the Societe Ivoirienne de Rafinnage (SIR) in Abidjan, Cote d'Ivoire, exports products to neighboring countries and consistently makes a foreign-exchange profit for its public and private shareholders.

An alternative for refineries in those countries with natural gas is methanol conversion. There are technologies to convert gaseous methane to liquid methanol, allowing smaller scale liquefaction to a versatile and easily transportable fuel. The disadvantages of methanol include the energy cost of natural gas conversion to methanol and the low volumetric energy content of methanol—about half that of gasoline. Methanol can be exported, however, getting around the problem of how to generate foreign exchange in a project and allowing foreign investors to remit profits. A recent example of a methanol export project is that of the government of Chile, the U.S. company Allied Signal, and the World Bank affiliate—the International Finance Corp. (IFC).[137] In 1987, the project set up a methanol plant worth $298 million at Cabo Negro in the remote south of Chile on a "non recourse" basis.[138] This project is, however, probably more suitable for countries with large gas reserves, such as Nigeria.

BIOMASS[139]

Biomass is the principal energy source for the half of the world's population living in rural areas and accounts for about one-third of all primary energy used in developing countries today. As presently used, biomass is an inconvenient and inefficient fuel, which contributes to high levels of indoor air pollution and resource depletion.

To play a larger role in supplying energy services, biomass must be converted into cleaner and more convenient fuels (gases, liquids, electricity). A modern biomass industry would have many advantages. In some cases, there would be a decrease in the amount of biomass needed for a given energy service. For example, cooking with a gas produced efficiently from biomass could use less total biomass than cooking directly with wood in a traditional stove (see ch. 3).[140] If operated on a renewable basis, biomass makes no net contribution to atmospheric CO_2. Biomass use could have economic benefits; as an indigenous energy source, biomass could save valuable foreign exchange. Establishing bioenergy industries also would bring increased activity and jobs into rural areas.

On the other hand, establishing a modern biomass industry presents challenges. The physical and chemical composition of biomass feedstocks varies widely, potentially requiring the tailoring of conversion technologies to specific biofuels. The relatively low bulk densities of biomass and possibly large required collection areas limit the amount of biomass at any given site. This constrains the size of individual conversion systems and limits the extent to which economies of scale in capital and other costs can be captured.[141] Although there are a large number of biomass technologies, the selected few examined here appear to hold particular promise for the future. This review does not include mature technologies such as charcoal production or the use of wood or agricultural wastes in boilers to produce process heat and electricity. Some of these applications were already discussed in chapters 3 and 4.

---

[137] Some of these are analyzed in G. Greenwald, "Encouraging Natural Gas Exploration in Developing Countries," in *Natural Resources Forum*, vol. 12, No. 3, 1988.

[138] This means that the operator puts in a modest level of equity capital and gives a completion guarantee for the facilities, but does not assume any further risk; the project financing is arranged solely on the basis of the assets of the project, in this case substantial gas reserves.

[139] This section is drawn primarily from Eric D. Larson, "A Developing Country-Oriented Overview of Technologies and Costs for Converting Biomass Feedstocks into Gases, Liquids, and Electricity," contractor report to the Office of Technology Assessment, September 1991.

[140] G.S. Dutt and N.H. Ravindranath, "Alternative Bioenergy Strategies (Direct Use of Biomass, Charcoal, Biogas, Producer Gas, Alcohol) for Cooking," draft manuscript for *Fuels and Electricity from Renewable Sources of Energy*, Johansson, Kelly, Reddy, Williams (eds), 1991 (forthcoming).

[141] Typical rates of biomass fuel production or use at individual sites range from 1-4 kW_{FUEL} (0.2 to 0.4 kg/hr of dry biomass, assuming a dry-biomass energy content of 20 MJ/kg) for residential cooking up to a maximum of some 300-400 MW_{FUEL} (54-72 dry tonnes/hr) at large factories that produce biomass as a byproduct and use it for energy (e.g. cane sugar and kraft pulp factories) (This can be compared to the 800 to 4,000 MW of coal consumed at central station electric power plants.) Larger concentrations of biomass could be made available, e.g. from plantations dedicated to producing biomass for energy. Under such schemes, transportation costs and land availability will be limiting factors on the quantity of biomass that can be concentrated at a single site.

Gases and Electricity From Biomass

Combustible gas can be produced from biomass through thermochemical processes (producer gas) or through biological processes (biogas). Both can be burned directly to produce heat, e.g., for residential cooking or industrial process heating, and can be used to produce electricity.[142]

Producer Gas[143]

Producer gas is a long established technology, though largely abandoned with the development of inexpensive petroleum supplies after the second world war.[144] Since the early 1970s, there have been many efforts to resurrect producer gas technology, largely for small scale use in the rural areas of developing countries.[145]

Gasification systems can be classified as either small scale—those with a fuel input of less than about 2 GJ/hr (100 kg/hr dry biomass)—or large scale. Economies of scale permit large scale systems to be generally more technologically sophisticated. Two basic gasifiers, updraft and downdraft, are used in small scale applications using raw biomass. In an updraft unit, biomass is fed in the top of the reactor and air is injected into the bottom of the fuel bed. Updraft gasifiers have high energy efficiencies, typically 80 to 90 percent (chemical energy in gas output divided by feedstock energy input), due to the efficient counter-current heat exchange between the rising gases and descending solids. However, the tars produced by updraft gasifiers mean that the gas must be cooled before it can be used in internal combustion engines. Thus, in practical operation, updraft units are used almost exclusively for direct heat applications. For use in internal combustion engines, downdraft technologies are needed. Even so, fairly elaborate additional gas cleaning systems are required in the downdraft technology, resulting in lower overall energy efficiencies of 60 to 70 percent.

Large scale applications include more elaborate versions of the small scale updraft and downdraft technologies, and fluidized bed technologies. The superior heat and mass transfer of fluidized beds leads to relatively uniform temperatures throughout the bed, better fuel moisture utilization, and faster reactions than in fixed beds, resulting in higher throughput capabilities.[146] Fluidized beds are, however, generally more expensive than fixed beds below a fuel input rate of 35 to 40 GJ/hr (1.8 to 2.0 tonnes/hr of dry biomass), due to the high unit cost of blowers, continuous feed systems, and control systems and other instrumentation.[147]

Direct Heat Applications—Producer gas has been used most widely (the country with the most experience is Brazil) for direct heating in industrial applications. In part, this is because such applications are often characterized by relatively high capacity utilization rates and, more importantly, because biomass is often available relatively inexpensively, e.g., as in the forest products industries. For relatively small capacity units, the estimated cost of gas ranges from $3.20 to $4.80 per GJ ($3.40 to $5.10/million Btu) for wood-fueled systems and $4.70 to $7.40 per GJ ($5.00 to $7.80/million Btu) for charcoal-fueled systems, competitive with fuel oil when crude is priced at about $38 per barrel (see table 6-22).[148] Producer gas is generally not used for cooking,[149] but it would offer advantages over raw biomass—higher systems efficiencies, reductions in indoor smoke and particulates, and reduced collec-

---

[142]Thermochemical gasification is also the first step in producing methanol, a liquid fuel, from biomass.

[143]Producer gas derives its name from the ''gas producer'' in which it is made and is a combustible mixture consisting primarily of carbon monoxide, hydrogen, carbon dioxide and nitrogen, and having a heating value of 4 to 6 MJ/Nm³, or 10% to 15% of the heating value of natural gas. It can be made from essentially any carbon-containing feedstock, including woody or herbaceous biomass (lignocellulose), charcoal, or coal.

[144]The use of producer gas dates back well into the 1800s when coal-derived producer gas was used in a number of cities worldwide for cooking and heating. This ''town gas'' is still used in Calcutta, Beijing, and Shanghai, where about half of all households use it for cooking. Producer gas from wood-charcoal was a prominent civilian fuel in Europe during the Second World War running several hundred thousand vehicles and powering industrial machinery.

[145]G. Foley and G. Barnard, ''Biomass Gasification in Developing Countries,'' Earthscan Technical Report No. 1 (London: 1982).

[146]E.D. Larson, P. Svenningsson, and I. Bjerle, ''Biomass Gasification for Gas Turbine Power Generation,'' Thomas B. Johansson, Birgit Bodlund, and Robert H. Williams (eds.), *Electricity: Efficient End-Use and New Generation Technologies, and their Planning Implications* (Lund, Sweden: Lund University Press, 1989), pp. 697-739.

[147]Ibid.

[148]Assuming that the per barrel wholesale price of residual fuel oil is 0.87 times the refiner acquisition cost of crude oil (characteristic for the United States) and that a barrel of residual oil contains 6.6 GJ.

[149]See, Samuel F. Baldwin, ''Cooking Technologies,'' OTA staff working paper, 1991.

Table 6-22—Comparative Summary of Calculated Costs for Modern Energy Carriers Produced From Biomass

	Production capacities		Installed capital costs		Total production costs[b]	
	10^3 GJ/yr	kW	$ per GJ/yr	$/kW	$/GJ	$/kWh
Biogas						
Domestic (E)	0.016-1.2	0.50-38	30-12	950-375	11-5	NA
Industrial (E)	3.6-167	114-5290	13-2.5	410-80	1.4-2.5	NA
Producer gas[c]						
Small (E)	1-12	32-380	2-0.7	65-20	5-3	NA
Medium (E)	20-200	634-6,340	6-3	190-95	NA	NA
Electricity						
Steam-turbine (E)		5-50,000	60	1,900	NA	0.05-0.07[d]
Biogas-IC engine (E)	0.16	5	38	1,200	NA	0.10
Producer gas-						
IC engine (E)	0.16-3.2	5-100	22-13	680-420	NA	0.24-0.15
Gas turbines						
(NC)	1,580	50,000	36	1,150	NA	0.04-0.05[d]
(Y2)	3,260	100,000	28	890	NA	0.03-0.04[d]
Methanol						
(CR)	10,000	316×10^6	30-50	950-1,580	10-13	NA
(Y2)	10,000-40,000	316×10^6 1,268×10^6	30-12	630-315	5-9	NA
Ethanol from cane						
(E)	1,000	32×10^6	10	315	10	NA
(Y2)	2,000	63×10^6	9	280	8.5[e]	NA
Ethanol acid hydrolysis						
(CR)	1,000-2,000	32-63×10^6	80-60	25-1,890	19-21	NA
(Y2)	2,000	63×10^6	50	1,580	15	NA
Enzymatic						
(NC)	5,500-27,000	142×10^6	18-27	1,260	9-11	NA
(Y2)	6,500-12,700	205×10^6	11-15	500	6-6.5	NA

NA = not applicable or not available.
[a] All figures are approximate for purposes of cross-fuel comparisons. Total production costs assume a 7% discount, $2/GJ for biomass, and capacity utilization rates as discussed in appropriate sections of the report. Blanks indicate no estimate was made in this study.
[b] Units are $/kWh where the product is electricity and $/GJ where the product is a gas or a liquid.
[c] Wood fuel (not charcoal).
[d] Low cost is for electricity production via cogeneration at industrial sites. High cost is for stand-alone electric power generation.
[e] Assumes use of biomass-gasifier/gas turbine cogeneration, with export of excess electricity, the revenues from which are credited against the cost of ethanol production.
NOTE: (E) indicates the results are based on operating experiences; (CR) on commercially-ready, but not commercially implemented technologies; (NC) on technologies that are near commercialization; and (Y2) technologies which could become available by the year 2000 with a concerted RD&D effort. All costs are in 1990 U.S. dollars.[a]
SOURCE: Eric D. Larson, "A Developing-Country-Oriented Overview of Technologies and Costs for Converting Biomass Feedstocks into Gases, Liquids, and Electricity," September 1991. Office of Technology Assessment, contractor report, p. 74.

tion times. Cooking with producer gas also involves disadvantages, such as increased risk of exposure to carbon monoxide in poorly maintained or operated systems. Because of its low capital intensity, the cost of producer gas is strongly affected by the cost of the feedstock.

Internal Combustion Engine Applications— Producer gas from downdraft gasifiers can be used in either compression ignition (diesel) engines or spark ignition (gasoline) engines. These stationary engines, particularly diesel engines, have a proven record in developing countries as a versatile, relatively durable technology for producing mechanical drive and small increments of stationary power, e.g., for irrigation pumping, lighting, cottage industries, and rural processing facilities. In India alone, some 4 million small diesel engines are used solely to drive irrigation pumpsets,[150] with each engine consuming energy at about the same rate as the average automobile in the United States—1,500 liters of diesel fuel annually. Producer gas can typically

[150] B.C. Jain, "Assessment of Current Status and the Potential for Commercial Exploitation of Biomass Gasification in India," *Biomass*, vol. 18, 1989, pp. 205-19.

replace 70 to 80 percent of the diesel fuel that would be used in normal operation of a diesel engine. Some diesel fuel is still needed because the low energy density of producer gas prevents it from self-igniting under compression.

Reported unit capital costs for producer gas engine generator electricity plants are generally below $1,000/kW, even for very small (5 kW) units. Calculated costs of electricity production with gasifier engine systems, assuming a 50 percent capacity factor, range from $0.13 to 0.24/kWh for small units (4 to 5 kW capacity) and from $0.10 to 0.15/kWh for units larger than about 100 kW (see table 6-20). This compares with an average, highly subsidized, price of $0.05/kWh currently being charged in the developing countries. The most important cost component for small units is labor; for larger units it is fuel.

The costs of electricity from producer gas are thus higher than the busbar cost of electricity from new central stations, but for remote use a more appropriate comparison would include transmission and distribution costs. Such an analysis for the state of Karnataka, India indicated that gasifier engine systems could produce electricity on a competitive basis with a large coal and nuclear central station power plants.[151] Electricity from producer gas could have additional advantages for developing countries, including lower capital intensity, shorter lead times (6 months versus 3 to 6 years for coal-fired central station plants), and the use of indigenous rather than imported fuel. Counting only investment costs and foreign exchange expenditures for fuel (zero for biomass),[152] the Karnataka analysis indicated that irrigation pumping could be about 10 percent less costly with gasifier-based systems than through extension of the national grid.

Despite these advantages, obstacles have been encountered to more rapid diffusion. It has proven difficult to reduce tar to tolerable levels even in downdraft technologies (tar entering an engine is deposited on components, causing loss of performance and, if unchecked, complete engine failure[153]). Both fundamental and applied research efforts are continuing.[154] For example, a 5 kW gasifier engine generator system developed by the Centre for the Application of Science and Technology to Rural Areas (ASTRA) relies on strict specification of the feedstock and careful design, operation and maintenance of the gasifier and gas cleaning system. The design has proven to be a technical success in the field.

The importance of institutional and management issues in introducing new technologies is illustrated by comparing major gasifier engine implementation efforts undertaken beginning in 1981 in the Philippines with a more recently initiated effort in India (see box 6-I).

Gas Turbine Applications—Producer gas can also serve as a feedstock for gas turbines that operate at much larger scales than internal combustion engines. In a biomass gasifier/gas turbine system (BIG/GT), biomass is gasified in a pressurized air-blown reactor and the products cleaned of particulates and other contaminants before being burned in an efficient power cycle based on aeroderivative gas turbines, such as the steam injected gas turbine (STIG), intercooled STIG (ISTIG), or a

---

[151] A.K.N. Reddy et al., "Comparative Costs of Electricity Conservation: Centralized and Decentralized Electricity Generation," *Economic and Political Weekly*, June 2, 1990, pp. 1201-1216.

[152] B.C. Jain, "Assessment of Current Status and the Potential for Commercial Exploitation of Biomass Gasification in India," *Biomass*, vol. 18, 1989, pp. 205-19.

[153] Tar problems with gasifier-engine systems have been widely reported. See: N. Coovattanachai, W. Chongchareon, and C. Kooptarnond, "The Feasibility of Producer Gas in Electricity Generation," *Reg. J. Energy Heat Mass Transfer*, vol. 4, No. 4, 1982; S. Kumar, et al., "Design and Development of a Biomass Based Small Gasifier-Engine System Suitable for Irrigational Needs in Remote Areas of Developing Countries," *Energy from Biomass and Wastes VIII* (Chicago, IL: Institute of Gas Technology, 1984), pp. 723-745; T. Zijp and H.E.M. Stassen, "Operating Experience with Producer Gas Plants in Tanzanian Villages," Department of Chemical Engineering, Twente University of Technology (Enschede, The Netherlands: 1984).

[154] See A. Kaupp, "Gasification of Rice Hulls: Theory and Praxis," German Appropriate Technology Exchange, Eschborn, Germany, 1984; M.J. Groeneveld, *The Co-Current Moving Bed Gasifier*, Ph.D. thesis for the Department of Chemical Technology, Twente University (Eindhoven, The Netherlands: 1980); H. Susanto, *Moving Bed Gasifier with Internal Recycle and Separate Combustion of Pyrolysis Gas*, Ph.D. thesis for the Institut Teknologi (Bandung, Indonesia: 1984); T.B. Reed and A. Das, *Handbook of Biomass Downdraft Gasifier Engine Systems* (Golden, CO: Solar Energy Research Institute, 1988).

combined cycle.[155] Hot gas cleanup avoids cost and efficiency penalties, and pressurized gasification avoids energy losses associated with compressing the fuel gas after gasification. While particulate cleanup requirements are much stricter than for heating applications and comparable to those for internal combustion engines, tar removal is not required. As with hot gas cleanup, the tar would remain in its vapor state until it is burned in the combustor. Thus, the fundamental technical problem that has plagued gasifier internal combustion engines would not be present in gas turbine applications. A complication, however, is the need to remove trace amounts of alkali vapor from the gas before it enters the gas turbine. There appears to be a basic understanding of the means for adequately cleaning gases for gas turbine applications with either fluidized bed gasifiers[156] or updraft gasifiers, although there has been no commercial demonstration of alkali removal. Box 6-J gives a description of current efforts in gas turbine research.

Given the lack of commercial operation, cost estimates must necessarily be tentative. BIG/GTs are characterized by high conversion efficiencies and low expected unit capital costs ($/kW) in the 5 to 100 MW_e size range.[157] The upper end of this range is probably near the practical upper limit on the size of a biomass installation. The expected performance and costs compare favorably with direct combustion steam turbine systems and with much larger central station fossil fuel and nuclear plants. The higher fuel cost figure corresponds to a target fuel cost for biomass (after processing for use in the gasifier) from energy plantations in the United States. Plantations in many developing countries could probably produce biomass at a lower cost than this.

While the economics of plantation based BIG/GT power generation appear attractive, initial BIG/GT applications are likely to be at industrial sites where biomass processing residues are readily available today, such as at cane sugar processing mills[158] and mills in the forest products industry[159]—replacing the currently used biomass-fired steam turbine cogeneration systems. BIG/GT systems have much higher electrical efficiencies that would permit them to meet onsite electricity needs and produce excess electricity that could be sold to utilities. Because BIG/GTs would produce less steam than steam turbines, however, steam use efficiency would generally need to be improved in a factory to enable BIG/GT systems to meet onsite steam needs.

BIG/GT systems have a number of characteristics that make them particularly attractive for developing country applications, including their low anticipated capital costs and high share of local content. With the primary exception of the high technology core of the gas turbine, most components could probably be manufactured locally. U.S. companies appear to have a strong competitive advantage in this technology.

The maintenance characteristics of the high technology core of the aeroderivative gas turbines at the heart of a BIG/GT system are also attractive. Their compact, modular nature makes it possible to replace failed parts and even whole engines quickly, with replacements flown or trucked in from centralized maintenance facilities. The required maintenance network is already largely in place in most developing countries that have their own commercial airlines. The scale characteristics of these systems are also well suited to developing countries.

---

[155] See E.D. Larson and R.H. Williams, "Steam-Injected Gas Turbines," *ASME Journal of Engineering for Gas Turbines and Power*, vol. 109, No. 7, 1987, pp. 55-63; R.H. Williams and E.D. Larson, "Expanding Roles for Gas Turbines in Power Generation," T.B. Johansson, et al., (eds.), *Electricity: Efficient End Use and New Generation Technologies, and Their Planning Implications* (Lund, Sweden: Lund University Press, 1989), pp. 503-553; R.H. Williams and E.D. Larson, "Advanced Gasification-Based Biomass Power Generation and Cogeneration," *Fuels and Electricity from Renewable Sources of Energy*, T.B. Johansson et al., (eds), (forthcoming).

[156] E. Kurkela, et al., "Removal of Particulates, Alkali, and Trace Metals from Pressurized Fluid-Bed Biomass Gasification Products—Gas Cleanup for Gas Turbine Applications," paper presented at Conference on Energy from Biomass and Wastes XV, Washington, DC, Mar. 25-29, 1991.

[157] See E.D. Larson and R.H. Williams, "Biomass-Gasifier/Steam-Injected Gas Turbine Cogeneration," *Journal of Engineering for Gas Turbines and Power*, vol. 112, April 1990, pp. 157-63; P. Elliott and R. Booth, "Sustainable Biomass Energy," Selected Paper (London: Shell International Petroleum Co., Ltd., December 1990).

[158] See E.D. Larson et al., "Biomass-Gasifier Steam-Injected Gas Turbine Cogeneration for the Cane Sugar Industry," *Energy from Biomass and Wastes XIV*, Elsevier Applied Science, 1991; J.M. Ogden, R.H. Williams, and M.E. Fulmer, "Cogeneration Applications of Biomass Gasifier/Gas Turbine Technologies in the Cane Sugar and Alcohol Industries," *Energy and Environment in the 21st Century* (Cambridge, MA: MIT Press, 1990).

[159] E.D. Larson, "Biomass-Gasifier/Gas Turbine Cogeneration in the Pulp and Paper Industry," paper presented at 36th ASME International Gas Turbine and Aeroengine Congress, Orlando, FL, June 1991.

> **Box 6-I—Gasifier-Engine Implementation in the Philippines and India**
>
> A 1981 presidential decree in the Philippines called for a strong national commitment and effort at reducing dependence on imported oil through use of gasifier-engine systems. Irrigation pumpsets were identified as an important target market, and a goal was set of replacing 1,150 diesel fueled units with biomass/diesel dual fuel gasifier-engine systems. By 1985, when implementation efforts were halted, 319 units were installed, and by 1987 an estimated 99 percent of these were nonfunctioning, primarily due to lack of maintenance. According to a recent analysis,[1] the fundamental reasons for the failure of the program were institutional and management-related rather than technical problems. Political pressures pushed an un-debugged technology (charcoal gasifiers) into the field prematurely and without a full understanding of the users needs.[2] A single quasigovernment agency (the Farm Systems Development Corp.—FSDC) was given responsibility for technology development, dissemination, financing, and maintenance, as well as for implementation of a new fuel (charcoal) supply infrastructure. Together with its other responsibilities, the under-funded, over-burdened FSDC was unable to provide adequate service to the user: training of users was inadequate; monitoring equipment needed for proper operation and maintenance was not installed in order to reduce costs; and the charcoal production system was insufficiently developed to meet demand, which resulted in high charcoal prices and thus marginal or negative fuel cost savings to farmers. Furthermore, gasifiers were produced by a single quasi-governmental company (GEMCOR) controlled by the FSDC, so that competitive market pressures were absent in the program.
>
> There is a strong national commitment to gasifier systems in India, as there was in the Philippines, but the Indian program appears to be proceeding at a more deliberate pace, with a keen appreciation of the need for evolutionary development of the technology. The prospects for developing sound technologies is auspicious in India, given its generally strong technological infrastructure. Also, market pressures are present, as there are several competing manufacturers of gasifier-engine systems. Some 250 units are now installed (mostly for small—3.5 to 7.5 kW—irrigation engine-pumpsets), with hardware monitoring/feedback efforts in-built in many cases. Carefully measured efforts to understand and meet user needs are ongoing.[3] Furthermore, many of the constraints to commercial implementation that were recognized only in retrospect in the Philippines appear to be well understood in India: the need for sound technology and a maintenance infrastructure, the need for committed and trained users, the need to understand and meet user demands, and the need for financing to help small farmers.
>
> ---
>
> [1] F.P. Bernardo and G.U. Kilayko, "Promoting Rural Energy Technology: The Case of Gasifiers in the Philippines," *World Development*, vol. 18, No. 4, 1990, pp. 565-574.
>
> [2] For example, farmers generally saw less benefit in irrigation than presumed. Thus, the actual number of hours per year a typical farmer irrigated his land was relatively low, contributing to marginal cost effectiveness of switching from diesel fuel to biomass.
>
> [3] R. Bhatia, "Diffusion of Renewable Energy Technologies in Developing Countries: A Case Study of Biogas Engines in India," *World Development*, vol. 18, No. 4, 1990, pp. 575-90.

Biogas

Biogas is produced by the biological process of anaerobic (without air) digestion of organic feedstocks. It consists primarily of methane and carbon dioxide and has a heating value of about 22 megaJoules/normal cubic meter or (590 Btu/ft^3). In the absence of oxygen, organic matter introduced into the digester is degraded by the action of three classes of bacteria. Proper operation of the digester relies on a dynamic equilibrium among the three bacterial groups. This balance, and hence the quality and quantity of gas produced, are affected by changes in digester temperature and acidity, and by the composition and rate of loading of the feedstock.

Two basic digester designs have been used most widely in developing countries. The floating cover digester (India) and the fixed dome digester (China). A third design, the bag digester, is gaining in popularity. In the floating cover digester, a gas holder floats on a central guide and provides constant pressurization of the gas produced. The reactor walls are typically brick or concrete. Traditionally the cover is made of mild steel, though more corrosion resistant materials are also being used. The digester is fed semi-continuously, with input slurry displacing an equivalent amount of effluent sludge. The primary drawback of the floating cover design as developed by the Indian Khadi Village Industries Commission (KVIC) is the high cost of the steel cover. The floating cover digester is suitable for both household size and larger scale community or commercial operation.

In the Chinese fixed dome digester, biogas collects under a fixed brick or concrete dome,

> ### Box 6-J—Current Gas Turbine Research
>
> Biomass integrated gasification/gas turbines (BIG/GT) systems are likely to be available by the mid-1990s, based on development efforts ongoing in Scandinavia, Brazil, and the United States.[1] Ahlstrom, a Finnish producer of biomass gasifiers, plans to build a 6-10 MW_e BIG/GT plant in southern Sweden in collaboration with Sydkraft, a major Swedish electric utility.[2] The plant will operate in a cogeneration mode, using an Ahlstrom pressurized circulating fluidized-bed gasifier and a sophisticated hot gas cleanup system, including ceramic filters for particulates.
>
> In Brazil, a major electric utility has an ongoing R&D program to develop biomass from planted forests as a major fuel source for power generation, with conversion to electricity using BIG/GT units.[3] The utility is currently planning to build an 18 MW_e demonstration BIG/GT plant. The overall program goal is commercial implementation of plantation-based BIG/GT systems starting in 1998.
>
> The U.S. Department of Energy (DOE) announced in late 1990 a major new program initiative to commercialize BIG/GT technology[4] by the late 1990s. The DOE recently selected the pressurized bubbling-fluidized bed RENUGAS gasifier developed by the Institute of Gas Technology (IGT) for a large scale gasification demonstration.[5] A scaled-up unit will be built in Hawaii and be run initially on sugarcane bagasse (50 tonne per day capacity). Start-up is anticipated in late 1992. Also in the United States, the Vermont Department of Public Service, in cooperation with in-state electric utilities, is exploring possibilities for a commercial demonstration of BIG/GT technology fueled by wood chips derived from forest management operations.[6] The DOE, U.S. Environmental Protection Agency, U.S. Agency for International Development, and GE are also participating.
>
> ---
>
> [1] E.D. Larson, "Biomass-Gasifier/Gas Turbine Cogeneration in the Pulp and Paper Industry," *Journal of Engineering for Gas Turbines and Power*, forthcoming.
>
> [2] Ahlstrom Corp. and A.B. Sydkraft, "Finland Goes Ahead with Unique Gasification Process," press release, Helsinki, Finland and Malmo, Sweden, Nov. 7, 1990.
>
> [3] E.A. Carpentieri, Chief of Alternative Energy Research, Compania Hidroeletrica do Sao Francisco, Recife, Brazil, personal communication, August 1991.
>
> [4] R. San Martin, U.S. Department of Energy, Deputy Assistant Secretary, Office of Utility Technologies, Division of Conservation and Renewables, "DOE Research on Biomass Power Production," presentation at Conference on Biomass for Utility Applications, Tampa, Florida, Oct. 23-25, 1990.
>
> [5] S. Babu, Institute of Gas Technology, Chicago, IL, personal communication, November 1990.
>
> [6] R. Sedano, Department of Public Service, State of Vermont, Montpelier, VT, personal communication, January 1991.

displacing effluent sludge as the gas pressure builds. The dome geometry is used to withstand the higher pressures generated. This technology has been widely used for small household scale units. Relatively few large scale units have been built, due to the difficulty of constructing large domes. A major shortcoming of the fixed cover units, even in small sizes, has been the difficulty of constructing leakproof domes. However, a number of improved versions of the fixed dome design have been introduced, including those designed to operate with plug flow conditions and/or with storage of gas in variable volume "bags."[160]

The typical digester feedstock in India is wet cattle dung mixed with water in a 1:1 ratio. A typical yield of gas with the KVIC digester is 0.02 to 0.04 m^3 per kg of fresh manure input at a design ambient temperature of 27 °C.[161] A family of 5 would therefore need the dung output from a minimum of 2 to 3 animals to meet their cooking fuel needs.[162] This is beyond the means of the majority of rural Indian households. In China, the feedstock is typically a mix of nitrogen-rich pig manure, cow manure and night soil, and carbon-rich straw and grass[163] with water added to achieve a total input solids concentration of about 10 percent. This technology

---

[160] Wu Wen, "Biomass Utilization in China," draft paper for Energy Research Group, Ottawa, Canada, July 1984.

[161] C. Kashkari, "Biogas Plants in India," *Energy in the Developing World: The Real Energy Crisis* (Oxford, United Kingdom: Oxford University Press, 1980), pp. 208-14.

[162] Assuming a daily gas yield of 0.2 m^3 per m^3 of digester volume.

[163] See Wu Wen, "Biomass Utilization in China," draft paper for Energy Research Group, Ottawa, Canada, July 1984; Q. Daxiong, et al., "Diffusion and Innovation in the Chinese Biogas Program," *World Development*, vol. 18, No. 4, 1990, pp. 555-563.

is therefore accessible to a larger share of households than for the KVIC digester.

There is general agreement that the capital costs of fixed dome units are significantly lower than floating domes, at least for household scale digesters, primarily because they do not require a steel cover. Capital costs for both technologies are declining,[164] indicating significant learning from the experiences of the 1970s and 1980s. Labor is required in the production of biogas to collect water and dung, mix and load inputs, distribute sludge, clean out the plant, and for maintenance. Total costs (both capital and operating) of community sized digesters (100 GJ/yr to 1,000 GJ/yr capacity) are estimated at $9/gigajoules to $5/GJ ($9.50 to $5.25/MMBtu). At the household scale, the cost appears to be higher, about $11/GJ[165] These high costs greatly limit the applicability of biogas units where justified by energy output alone. At the present time, much of the labor involved in small scale units may be performed by household members, lowering the financial cost; in the future, such low cost or "free" labor will not be so readily available at either the household or community scale and may limit biogas operations to locations where large quantities of waste materials are already collected for sanitation or other purposes.

A major advantage of biogas over the direct use of raw biomass is that valuable nutrients are retained in the slurry, instead of being partially lost in the combustion process. For example, the wet slurry output from a digester fed with fresh cattle dung has essentially the same nitrogen fertilizer value as the input dung. Because water is also added to the digester, however, about twice as much fresh digester effluent is needed to supply the same amount of nitrogen.

Biogas digesters also have an important sanitation advantage, reducing or eliminating pathogens present in animal and human wastes. Significant declines in parasite infections, enteritis, and bacillary dysentery have been noted in areas following installation of digesters.[166] Air drying or composting of digester effluent can further reduce or completely eliminate pathogens. An additional health benefit where biogas replaces wood used in traditional cook stoves is the elimination of noxious gases and particulates from wood fires.

Biogas, like producer gas, can be used to fuel either compression or spark ignited internal combustion engines, which can provide shaft power or drive an electrical generator. Gas from an innovative floating cover gasifier developed by researchers from[167] ASTRA of the Indian Institute of Science (Bangalore) is used to replace about 70 percent of the diesel fuel needed to run a 5 kW diesel engine generating electricity used for pumping water and providing lighting. Villagers are paid a fee of $0.0016/kg for delivered dung and are also returned digester sludge in proportion to their dung contribution. The sludge is passed through a simple sand bed filter system to concentrate the solids content before it is returned. The village biogas engine operation employs two village youths full time. With the current operating hours of the system (4.3 hours/day), corresponding to a capacity factor of about 18 percent, the total levelized cost of electricity is about 15 cents/kWh.

There are an estimated 5 million digesters operating in China today, mostly at the household scale

---

[164] The five sets of data referred to here are all based on experiences with digesters in India, and thus comparisons among the data sets are meaningful. The Rijal and Orcullo estimates are from experiences in Nepal and the Philippines and are thus probably not strictly comparable to the Indian data. See M.T. Santerre and K.R. Smith, "Measures of Appropriateness: The Resource Requirements of Anaerobic Digestion (Biogas) Systems," *World Development*, vol. 10, No. 3, 1982, pp. 239-61; N.A. Orcullo, "Biogas Technology Development and Diffusion: the Phillipine Experience," *Biogas Technology, Transfer and Diffusion, El-Halwagi (ed)* (London: Elsevier Applied Science Publishers, 1986), pp. 669-86; K. Rijal, "Resource Potential and Economic Evaluation of Bio-Gas Utilization in Nepal," *Energy*, vol. 11, No. 6, 1986, pp. 545-50.

[165] For household-sized units, an alternative perspective on production cost might be more appropriate, however. Householders would probably use family labor to operate and maintain a digester and might not consider this a cost. In addition, capital costs converted using a purchasing-power-parity (PPP) exchange rate might better represent the capital cost for a rural dweller with little or no access to hard currency. Also, capital is generally likely to be scarce for the household, which would be reflected by a much higher discount rate than the 7% assumed above. Neglecting labor and maintenance costs, converting capital costs to U.S.$ using a PPP exchange rate, and applying a 30% discount rate would result in biogas costs up to five times those shown for household-sized units (25 GJ/yr), corresponding to a gas cost of perhaps $25/GJ (R. Summers and A. Heston, "A New Set of International Comparisons of Real Product and Price Levels Estimates for 130 Countries, 1950 to 1985," *Review of Income and Wealth*, March 1988, pp. 1-25 (with accompanying data diskettes).

[166] C.G. Gunnerson and DC Stuckey, *Integrated Resource Recovery: Anaerobic Digestion*, Technical Paper 49, World Bank, Washington, DC, 1986.

[167] A.K.N. Reddy et al, "Studies in Biogas Technology. Part IV. A Novel Biogas Plant Incorporating a Solar Water-Heater and Solar Still," Proceedings of the Indian Academy of Sciences, C2(3), Bangalore, India, Sept. 1979, pp. 387-93.

and some 300,000 in India.[168] The efforts in China and India to popularize biogas have been very different in nature historically, which permits some lessons to be drawn on implementation. These include:

- the need for national commitment (this has helped address key problems such as distorted user economics due to subsidized prices for electricity and alternative fossil fuels, valuing of nonpecuniary benefits such as improved sanitation, and supporting R&D efforts aimed at cost reduction and technology improvement);
- the need for several stages of development, including a strong technology base, and an experimental and limited field test stage before large scale dissemination;
- an interdisciplinary approach;
- the importance of training of disseminators and users;
- a mix of centralized and decentralized institutions; and
- competitive market-type forces.

Gas Cost Comparisons

From our analysis of costs we conclude that among gas generating technologies, biogas systems are an order of magnitude more capital intensive than producer gas systems up to the quite high production level of 1,000 GJ/yr, when producer gas and biogas are about equal in cost due to the high feedstock cost for producer gas. At a very small scale (less than 20 GJ/yr) biogas may likely be more costly than producer gas, but the health and fertilizer benefits of biogas technology are not included in the biogas cost, and further reduction in costs of floating cover digesters can be expected.

Liquid Fuels From Biomass

The production of liquid fuels from biomass, with the exception of ethanol from sugarcane and corn, has not been widely implemented commercially because of their high cost. In the case of ethanol from cane and corn, government subsidies have supported commercial production. Research and development work to reduce costs and improve yields have been modest, except in the case of ethanol from sugarcane in Brazil. Advances may make liquid biofuels more competitive with fossil fuels over the next decade.

This section discusses three alternative liquid fuels from biomass: methanol from lignocellulose (any woody or herbaceous biomass), ethanol from sugarcane (ethanol from corn is generally not considered a practical option for most developing countries), and ethanol from lignocellulose.

Methanol From Lignocellulose

Methanol is produced today primarily from natural gas. But it can also be produced from coal and, through a similar process, from lignocellulosic biomass feedstocks.[169] Biomass-to-methanol plants would typically convert 50 to 60 percent of the energy content of the input biomass into methanol, though some designs have been proposed with conversion efficiencies of over 70 percent.

Three basic thermochemical processes are involved in methanol production from biomass:

1. A "synthesis gas" (a close relative of producer gas) is produced via thermochemical gasification, but by using oxygen rather than air in order to eliminate dilution of the product gas with nitrogen (in air). Oxygen plants have strong capital cost scale economies, which contributes to most proposed biomass-to-methanol facilities being relatively large (typically 2,000 tonnes/day or more input of dry biomass). Biomass gasifiers designed for methanol production are not commercially available. A number of pilot and demonstration scale units were built and operated in the late 1970s/early 1980s,[170] but most of these efforts were halted when oil prices fell. Work on one (a fluidized bed unit developed by the Institute

---

[168]Q. Daxiong et al., "Diffusion and Innovation in the Chinese Biogas Program," *World Development*, vol. 18, No. 4, 1990, pp. 555-563; D.L. Klass, "Energy from Biomass and Wastes: 1985 Update and Review," *Resources and Conservation*, vol. 15, Nos. 1 and 2, 1987, pp. 7-84.

[169]C.E. Wyman, et al., "Ethanol and Methanol From Cellulosic Materials," *Fuels and Electricity from Renewable Sources of Energy*, T. Johansson et al., (eds), (forthcoming).

[170]See A.A.C.M. Beenackers and W.P.M. van Swaaij, "The Biomass to Synthesis Gas Pilot Plant Programme of the CEC: A First Evaluation of Results," *Energy from Biomass, 3rd EC Conference* (Essex, United Kingdom: Elsevier Applied Science, 1985), pp. 120-45; E.D. Larson, P. Svenningsson, and I. Bjerle, "Biomass Gasification for Gas Turbine Power Generation," T.B. Johansson et al., (eds.), *Electricity: Efficient End-Use and New Generation Technologies, and their Planning Implications* (Lund, Sweden: Lund University Press, 1989), pp. 697-739.

of Gas Technology[171]) has recently been revived, with the construction of a bagasse-fueled demonstration unit now being planned. In addition, there are some commercial gasifiers originally designed for coal that could be used for biomass use.[172]

2. The synthesis gas is cleaned and its chemical composition is adjusted to produce a gas consisting purely of hydrogen (H_2) and carbon monoxide (CO) in a molar ratio of 2:1. The specific equipment configuration in the second step in methanol production will vary depending on the gasifier used. A reactor common to all systems is a "shift" reactor used to achieve the desired 2:1 ratio of H_2 to CO by reacting steam with the synthesis gas. The shift reactor is a commercially established technology. Other processing may be required before the shift stage, however, depending on the composition of the synthesis gas leaving the gasifier. For example, tars contained in the synthesis gas must be removed or cracked into permanent gases.

3. The gas is compressed and passed through a pressurized catalytic reactor that converts the CO and H_2 into liquid methanol. A variety of commercial processes can be used.

As biomass to methanol plants are not yet commercially available, costs are uncertain. From scattered cost data, mainly based on U.S. experience, it is estimated, however, that methanol could be produced for about $11 to $14/GJ ($11.50 to $14.75/MMBtu) with commercially ready technology in a plant with a capacity of about 10 million GJ/yr (about 500 million liters/yr). At larger scale—40 million GJ/year—costs would be $7/GJ to $8/GJ ($7.50-$8.50/MMBtu). Capital represents the largest fraction of the total cost of methanol produced in small plants, while feedstock is the dominant cost in large plants. Thus, capital cost reductions will be most important in reducing methanol costs from small plants while increases in biomass conversion efficiency will be most important at large scale.

Ethanol From Biomass by Fermentation

Two varieties of ethanol are being produced today from sugarcane for use as fuel in developing countries: anhydrous ethanol—essentially pure ethanol—and hydrous ethanol containing about 5 percent water. Anhydrous ethanol (apart from Brazil the most common in developing countries) can be blended with gasoline up to a maximum ethanol content of about 20 percent without need for modifying conventional spark ignition vehicle engines.[173] Hydrous ethanol cannot be blended with gasoline, but can be used alone as a fuel in engines specifically designed for ethanol. Most national anhydrous ethanol programs are small, due to uncertainty over oil prices and the limited size of market (20 percent of gasoline consumption). In Brazil, however, 90 percent of new cars are designed to use hydrous ethanol.

At an autonomous alcohol distillery (i.e., not associated with sugar production), raw sugarcane is washed, chopped, and crushed in rolling mills to separate the sugar laden juice from the bagasse, the fiber portion of the cane. The raw cane juice, containing over 90 percent of the sucrose in the cane, is filtered and heated, and is either sent directly to a fermentation tank after cooling, or limed, clarified, and concentrated before fermentation. The fermented mixture contains water and ethanol in about a 10:1 ratio. The mixture is then distilled, typically through two distillation columns to concentrate the ethanol.[174] A typical yield of hydrous alcohol is 70 liters per tonne of cane processed. Stillage, a potassium-rich liquid, is drained from the bottom of the distillation columns.

The bagasse, accounting for about 30 percent of the weight of fresh cane, or about 60 percent of the cane's energy content,[175] can be used to produce both electricity and steam for process heating. The

---

[171] R.J. Evans et al., *Development of Biomass Gasification to Produce Substitute Fuels*, PNL-6518, Battelle Pacific Northwest Laboratory, Richland, WA, 1988.

[172] See Chem Systems, "Assessment of Cost of Production of Methanol from Biomass" (draft) (Golden, CO: Solar Energy Research Institute, December 1989); E.D. Larson, P. Svenningsson, and I. Bjerle, "Biomass Gasification for Gas Turbine Power Generation," Thomas B. Johansson, Birgit Bodlund, and Robert H. Williams (eds.), *Electricity: Efficient End-Use and New Generation Technologies, and their Planning Implications* (Lund, Sweden: Lund University Press, 1989), pp. 697-739; V. Brecheret Filho, A.J. Ayres Zagatto, "Methanol From Wood in Brazil," presented at the 179th American Chemical Society National Meeting, Houston, TX, Mar. 23-28, 1980;

[173] World Bank, *Alcohol Production from Biomass in the Developing Countries* (Washington, DC: The World Bank, 1980).

[174] Additional distillation steps are needed to produce anhydrous ethanol.

[175] About 35% of the original energy in the sugar cane stalk brought to a mill is converted to alcohol, 59% to bagasse, and 6% to stillage.

potential for selling excess electricity from alcohol distilleries could be high as a result of new process technologies that reduce onsite energy needs, and new cogeneration technologies that increase the ratio of electricity to heat produced.[176] Credits for these sales could significantly improve the economics of ethanol production. Stillage, which contains about 6 percent of the energy contained in the cane, also has economic potential as a fertilizer for sugarcane[177] and as a feedstock for biogas production. If sugarcane tops and leaves (typically burned in the fields and contributing to local air pollution) could also be economically used for electricity production, ethanol production costs would be further reduced.

In a distillery annexed to a sugar factory, the fermentation feedstock is typically molasses produced as a minor byproduct of sugar processing. Some Brazilian factories are designed to use either molasses or a mixture of molasses and raw cane juice as the fermentation feedstock, thus permitting them to vary their production of sugar and ethanol to better match market demands.[178]

Estimating costs of ethanol from annexed distilleries is complicated by the multitude of options for feedstocks and relative product mixes.[179] The application for which hydrous ethanol is most often considered is as an automotive fuel to replace gasoline. The cost of hydrous alcohol in the most efficient mills in Brazil is about $9.00/GJ ($0.20/liter). Correcting both for the lower energy content of ethanol and its higher thermodynamic efficiency in an engine, ethanol at this price would be competitive with gasoline when crude oil prices are about $30 per barrel. An average production cost, including some less efficient plants, would be $0.22 to $0.26 per liter ($10.00 to $11.80 per GJ), corresponding to an equivalent crude oil price of about $30 to $35 per barrel (see table 6-23).

Costs have been falling since the inception of the program, due in part to more efficient distillery operation (increased liters of ethanol per tonne of cane) and, more importantly, increased land productivity. The cost of delivered cane is the largest single determinant of ethanol costs. Cane growing, harvesting, and transporting costs vary significantly from one region of the world to another with Brazilian costs among the lowest because of the large scale of production and emphasis on cane varieties and cultivation practices to maximize yield.[180] Production costs in other countries would be higher, leading to higher ethanol prices, which in turn would only become competitive with gasoline prices at crude oil prices over $35.

The economics of ethanol could be improved by more intensive use of byproducts such as bagasse, stillage, and sugarcane leaves. Crediting revenues from the sale of electricity or stillage for fertilizer would reduce the cost of the ethanol operation. A recent study illustrates some of the possibilities, based on Brazilian conditions and assuming the use of three different cogeneration technologies.[181] This study found that with state-of-the-art steam turbine technology (CEST), ethanol costs could be reduced to levels competitive with gasoline, but the cost of exported electricity would not be competitive with most central station alternatives. More importantly, the study found, the use of advanced gas turbine cogeneration technologies (BIG/STIG and BIG/ISTIG) could also lead to ethanol production costs substantially lower than today's level, and simultaneously to electricity production costs that would be competitive in many cases with central station alternatives.

---

[176] M. Fulmer, "Electricity-Ethanol Co-Production from Sugar Cane: A Technical and Economic Assessment," MSE thesis, Mechanical and Aerospace Engineering Department, Princeton University, Princeton, NJ, January 1991.

[177] J. Goldemberg and L.C. Monaco, "Ethanol as Alternative Fuel: Successes and Difficulties in Brazil," draft manuscript for *Fuels and Electricity from Renewable Energy Sources*, T. Johansson et al., (eds), forthcoming.

[178] J.M. Ogden and M. Fulmer, "Assessment of New Technologies for Co-Production of Alcohol, Sugar and Electricity From Sugar Cane," PU/CEES Report 250, Center for Energy and Environmental Studies, Princeton University, Princeton, NJ, 1990.

[179] M. Fulmer, "Electricity-Ethanol Co-Production From Sugar Cane: A Technical and Economic Assessment," MSE thesis, Mechanical and Aerospace Engineering Department, Princeton University, Princeton, NJ, January 1991.

[180] J.M. Ogden and M. Fulmer, "Assessment of New Technologies for Co-Production of Alcohol, Sugar and Electricity from Sugar Cane," PU/CEES Rpt. 250, Center for Energy and Environmental Studies, Princeton University, Princeton, NJ, 1990.

[181] See J.M. Ogden, R.H. Williams, and M.E. Fulmer, "Cogeneration Applications of Biomass Gasifier/Gas Turbine Technologies in the Cane Sugar and Alcohol Industries," *Energy and Environment in the 21st Century* (Cambridge, MA: MIT Press, 1990); M. Fulmer, "Electricity-Ethanol Co-Production from Sugar Cane: A Technical and Economic Assessment," MSE thesis, Mechanical and Aerospace Engineering Department, Princeton University, Princeton, NJ, January 1991.

Table 6-23—Required Production Cost of Ethanol for Competition With Wholesale Gasoline[a]

Crude oil price ($/bbl)	Gasoline price[a]		Equivalent hydrous ethanol price[b]		Equivalent anhydrous ethanol price[c]	
	($/bbl)	($/lit)	($/lit)	($/GJ)[d]	($/lit)	($/GJ)[d]
20	28	0.176	0.148	6.73	0.204	8.71
25	35	0.220	0.185	8.42	0.255	10.91
30	42	0.264	0.222	10.10	0.306	13.09
35	49	0.308	0.259	11.78	0.357	15.27
40	56	0.352	0.296	13.47	0.408	17.45

NOTE: All dollars are U.S. 1990$.
[a]Gasoline price relative to crude price is based on Brazilian conditions [96].
[b]Assumes 1 liter of hydrous ethanol as a neat fuel is worth 0.84 liters of gasoline [123].
[c]Assumes 1 liter of anydrous ethanol is worth 1.16 liters of gasoline when the ethanol is used as an octane-boosting additive [124].
[d]Higher heating value basis.
SOURCE: Eric D. Larson, "A Developing-Country-Oriented Overview of Technologies and Costs for Converting Biomass Feedstocks into Gases, Liquids, and Electricity," September 1991. Office of Technology Assessment contractor report, p. 71.

Much of the controversy over alcohol programs centers on its indirect social and economic impacts. The Brazilian PROALCOOL program—the only large scale program yet in developing countries—has been considered successful in achieving three of its major initial goals: reduced dependence on foreign oil, increased employment, and expanded capital goods (distillery equipment) production capabilities in Brazil.[182]

Its success in creating jobs is, however, ambiguous. On the one hand, the current program supports about 800,000 direct jobs, or 570,000 full time equivalent jobs. The labor intensity of the industry is much greater in the Northeast, the area of greatest unemployment, than in the South-Central region. The capital invested to create these jobs—some $32,000 per job in the South-Central region and $8,200 per job in the Northeast—has been relatively small compared to other industries. The average for all industry in Brazil is some $53,000; for the paper and pulp industry, $88,000; and for petrochemicals, $250,000. On the other hand, the quality of jobs is more debatable. The large component of seasonal labor has led in some cases to low wages, poor working and living conditions, and a lack of social benefits for workers,[183] especially in the Northeast. One strategy has been to extend the harvest period from 6 to 8 or 9 months by planting cane varieties that mature at different times.[184] In addition to extending the period of employment, a longer production season increases output and improves capital utilization.

The impact of the ethanol program on land use is similarly debatable. Some 4.3 million hectares are currently planted with sugar cane in Brazil,[185] compared with a total crop area of 52 million hectares[186] and a total potential agricultural area of 520 million hectares.[187] About half the cane area is devoted exclusively to ethanol production.[188] Many analysts appear to agree that cane production has not displaced domestic crop production.[189] On the other hand, production of export crops such as soybeans has been growing faster than domestic crops, due to agricultural pricing policies that make export crop production more attractive to farmers.[190]

---

[182]J. Goldemberg, L.C. Monaco and I. Macedo, "The Brazilian Fuel Alcohol Program,"*Fuels and Electricity from Renewable Energy Sources*, T. Johansson et al., (eds), (forthcoming).

[183]H.S. Geller, "Ethanol from Sugarcane in Brazil," *Annual Review of Energy*, vol. 10, 1985, pp. 135-64.

[184]Ibid.

[185]*Anuario Estatistico do Brasil*, IBGE, Rio de Janeiro, 1989.

[186]*Agricultural Land Use Census*, IBGE, Rio de Janeiro, 1985.

[187]F. Rosillo-Calle, "Brazil: A Biomass Society," *Biomass: Regenerable Energy* (Chichester, United Kingdom: John Wiley & Sons, 1987), pp. 329-348.

[188]J. Goldemberg, L.C. Monaco and I. Macedo, op. cit., footnote 188.

[189]See H.S. Geller, "Ethanol from Sugarcane in Brazil," *Annual Review of Energy*, vol. 10, 1985, pp. 135-164; J. Goldemberg, L.C. Monaco and I. Macedo, Op. cit., footnote 188, F. Rosillo-Calle, "Brazil: A Biomass Society," *Biomass: Regenerable Energy* (Chichester, United Kingdom: John Wiley & Sons, 1987), pp. 329-48.

[190]H.S. Geller, "Ethanol from Sugarcane in Brazil," *Annual Review of Energy*, vol. 10, 1985, pp. 135-64.

Ethanol From Lignocellulose

The high costs of sugarcane (and corn in the United States) has motivated efforts to convert lower cost biomass, primarily woody and herbaceous materials, into ethanol. These feedstocks are less costly largely because they do not compete as food crops. However, they are more difficult (and to date more costly) to convert into ethanol. Woody and herbaceous biomass, referred to generally as lignocellulosic materials, consist of three chemically distinct components: cellulose (about 50 percent), hemicellulose (25 percent), and lignin (25 percent).[191] Most proposed processes involve separate processing (either hydrolysis or enzymatic) of these components. In the first step, pretreatment, the hemicellulose is broken down into its component sugars and separated out. The lignin is also removed. The cellulose is then converted into fermentable glucose through hydrolysis. Following fermentation, the products are distilled to remove the ethanol. Byproducts of the separation process, such as furfural and lignin, can be used as fuel.

Acid Hydrolysis—A number of variants on the basic process have been proposed, each typically involving use of a different acid and/or reactor configuration.[192] One system incorporates two stages of hydrolysis using dilute sulfuric acid. In the first step, the acid breaks the feedstock down into simple sugars. However, the acids also degrade some of the product sugars so that they cannot be fermented, thus reducing overall yield. R&D effort has been aimed at improving the relatively low yields (55 to 75 percent of the cellulose) through the use of other acids.[193] Low cost recovery and reuse of the acids is necessary to keep production costs down,[194] but has yet to be commercially proven.

The conversion process becomes more competitive when furfural production from the hemicellulose fraction is maximized[195] and sold, but the furfural market is too small to support a large scale fuel ethanol industry. Byproduct electricity could also offset ethanol costs, but the amounts of exportable electricity coproduced in process configurations to date are relatively small. This situation might change if more advanced cogeneration technologies are considered.

Enzymatic Hydrolysis—In enzymatic hydrolysis, biological enzymes take the place of acid in the hydrolysis step. Enzymes typically break down only the cellulose and do not attack the product sugars. Thus, in principle, yields near 100 percent from cellulose can be achieved. Typically, a feedstock pretreatment step is required since biomass is naturally resistant to enzyme attack. The most promising of several options for pretreatment appears to be treatment by a dilute acid,[196] in which the hemicellulose is converted to xylose sugars that are separated out, leaving a porous material of cellulose and lignin that can more readily be attacked by enzymes.

A number of bacteria and yeasts have been identified and tested as catalyzers of cellulose hydrolysis, of which three process configurations have received the most attention from researchers.

- In the Separate Hydrolysis and Fermentation (SHF) of cellulose, three separate operations are used to produce enzymes, hydrolyze cellulose, and ferment the glucose. The presence of glucose produced during hydrolysis slows the catalytic effect of the enzymes, thus increasing costs of ethanol production. Some enzymes have been identified that are less susceptible to end product inhibition, but the improvement in overall economics of the SHF process are relatively modest.
- A more promising modification of the SHF process involves simultaneous saccharification and fermentation (SSF), permitting higher prod-

---

[191] J.D. Wright, "Ethanol from Lignocellulose: An Overview," *Energy Progress*, vol. 8, No. 2, 1988, pp. 71-78.

[192] C.E. Wyman, Op. cit., footnote 175.

[193] J.D. Wright, "Ethanol from Lignocellulose: An Overview," *Energy Progress*, vol. 8, No. 2, 1988, pp. 71-78; See J.D. Wright, A.J. Power, and P.W. Bergeron, "Evaluation of Concentrated Halogen Acid Hydrolysis Processes for Alcohol Fuel Production," SERI/TR-232-2386 (Golden, CO: Solar Energy Research Institute, 1985); C.E. Wyman, Op. cit., footnote 175.

[194] J.D. Wright, A.J. Power, and P.W. Bergeron, "Evaluation of Concentrated Halogen Acid Hydrolysis Processes for Alcohol Fuel Production," SERI/TR232-2386 (Golden, CO: Solar Energy Research Institute, 1985).

[195] See P.W. Bergeron, J.D. Wright, and C.E. Wyman, "Dilute Acid Hydrolysis of Biomass for Ethanol Production," *Energy from Biomass and Wastes XII* (Chicago, IL: Institute for Gas Technology, 1989), pp. 1277-1296; M.M. Bulls et al., "Conversion of Cellulosic Feedstocks to Ethanol and Other Chemicals Using TVA's Dilute Sulfuric Acid Hydrolysis Process," *Energy from Biomass and Wastes XIV* (London: Elsevier Applied Science, 1991) pp. 1167-1179.

[196] J.D. Wright, "Ethanol from Biomass by Enzymatic Hydrolysis," *Chemical Engineering Progress*, August 1988, pp. 62-74.

uct yield and use of a single reaction vessel. This improves the economics substantially.[197] However, the yeasts and bacteria used in SSF processes cannot ferment xylose sugars to ethanol. Incorporating xylose fermentation with SSF cellulose fermentation promises significant reductions in cost for ethanol from biomass.[198]

- Single reactor Direct Microbial Conversion (DMC) combines enzyme production, cellulose hydrolysis and glucose fermentation in a single process. In limited efforts to date, however, DMC ethanol yields have been lower than for the SHF or SSF processes, and a number of undesired products in addition to ethanol have been produced.

The next anticipated advancement is increased xylose conversion to ethanol, which would double efficiencies to 50 percent and reduce capital costs. If this target can be achieved at a commercial scale, hydrous ethanol from enzymatic hydrolysis might become competitive with gasoline when the crude oil price is as low as $20 per barrel, with most of the cost in the feedstock. Advances in this technology and in biotechnology more generally, suggest economically competitive commercial systems might be developed by the year 2000.

---

[197] J.D. Wright, C.E. Wyman, and K. Grohmann, *Simultaneous Saccharification and Fermentation of Lignocellulose: Process Evaluation* (Golden, CO: Solar Energy Research Institute, 1988).

[198] L.R. Lynd, et al., ''Fuel Ethanol From Cellulosic Biomass,'' *Science*, vol. 251, No. 4999, March 1991, pp. 1318-1323.

Chapter 7
Energy Resources and Supplies

Photo credit: Jennifer Cohen

Contents

	Page
INTRODUCTION AND SUMMARY	231
OIL AND GAS SUPPLY	233
Oil and Gas Reserves	234
Oil and Gas Exploration and Production Technologies	236
Institutional Issues	241
The Special Characteristics of Gas	243
Policy Options	243
COAL	244
Coal Reserves and Consumption in the Developing World	244
Coal Production	245
Coal Cleaning Techniques	246
BIOMASS RESOURCES	247
Agricultural/Industrial Residues	248
Forest Management	250
Energy Crops	251
Conclusions	258

Chapter 7
Energy Resources and Supplies

INTRODUCTION AND SUMMARY

Developing countries are projected to require substantial increases in primary[1] energy supplies. The World Energy Conference predicts that with an annual economic expansion of 4.4 percent, the developing countries will require more than a three-fold increase in commercial energy supply by 2020.[2] An increase on this scale raises issues of the availability of financial resources, the extent of the domestic energy resource base, and the environmental impacts of rapidly expanding energy production.

Attention has already been drawn to the large investments in the electricity sector. Roughly similar amounts are invested in fossil fuel production, mainly oil and gas. As in electricity, much of the investment is in the form of foreign exchange, though unlike electricity, private foreign investment plays a much greater role. In addition to investments in domestic energy production, developing countries also incur large foreign exchange costs for imported energy, in some cases accounting for 40 percent and more of total export earnings. As consumption continues to rise, import dependence could rise, or in the case of exporting countries, the export surplus could disappear.[3]

These considerations focus particular attention on technologies that could develop domestic energy resources, while minimizing the environmental impacts of rising domestic energy production. Coal, already the largest single source of fossil fuels in the developing countries, presents both opportunities and problems. Coal resources are much more abundant than are oil and gas, though coal is widely used in relatively few developing countries—notably India, China, and South Korea. As these countries are among the biggest energy consumers of the developing world, coal accounts for almost one-third of total developing-country energy consumption. Many other countries in Asia, Africa, and Latin America also have significant proved or probable coal reserves.

In addition to its abundance, coal has other advantages. Per unit of heat value, coal is cheaper than oil and, in most cases, gas. Coal is a familiar fuel with a long established technology. Finally, coal mine capital costs are low and in many countries have a high content of locally manufactured goods and services.

At the same time, coal suffers from serious disadvantages. As a solid fuel it is difficult to handle and transport, and is less versatile than oil. Its variable and frequently poor quality discourages the use of advanced combustion technologies and may contribute to poor power plant performance. There are also significant environmental drawbacks to all phases of coal production, from mining to combustion. Mining operations can cause local air and water pollution and are associated with severe occupational health hazards. Coal combustion results in large amounts of solid wastes and airborne pollutants, including acid rain precursors and the greenhouse gas, carbon dioxide. Carbon dioxide emissions from coal are higher per unit energy than those from other fuels so that the projected rapid expansion in coal use will contribute to high levels of greenhouse gas emissions.

Technologies exist to mitigate at least some of these problems. The introduction of mechanized open cast and longwall mining into those parts of the still unmechanized Indian and Chinese industries could improve productivity, reduce occupational hazards, and reduce incombustible waste. The simplest coal cleaning techniques (physically removing impurities such as dirt and stones) could result in

---

[1] The term "primary energy" includes fossil fuels (such as coal, crude oil, gas) and biomass in their crude or raw state before processing into a form suitable for use by consumers. The term also includes electricity generated by geothermal, wind and solar resources, and nuclear power. Electricity generated from fossil fuels or biomass is not included in primary energy to avoid double counting.

[2] World Energy Conference, *Global Energy: Perspectives 2000-2020*, Conservation and Studies Committee 14th Congress Montreal 1988 (London, England: World Energy Conference, 1988), tables 3 and 5.

[3] According to one study of 15 major energy consuming developing countries, if the trends of the 1980s and 1990s continued into the future, oil import dependence in the five oil importing countries would reach around 88 percent of the total oil needs by 2010 (compared with 37 percent in 1988) and three oil exporting developing countries could become oil importers (Mudassar Imran and P. Barnes, *Energy Demand in the Developing Countries: Prospects for the Future*, World Bank Staff Commodity Working Paper No. 23 (Washington, DC: World Bank, 1990).

important benefits such as reduced airborne particulates, lower transport cost, increased plant availability, and the use of higher performance coal burning equipment.

Despite the promising resource situation, there exist several obstacles to the further development of coal in many of the developing countries. In many cases, governments (especially in countries without domestic resources) give coal a low priority, preferring other forms of energy. Both China and India are exceptions; with well established industries, they are committed to a rapid expansion in output (doubling by 2010). Shortages of capital, the ready availability of low-cost labor, and other factors, however, may deter the introduction of improved production technologies. Finally, global climate change negotiations may ultimately limit the use of coal due to its high level of carbon dioxide emissions per unit energy.

The United States is a leader in coal production and combustion technologies. Recognizing these opportunities, the U.S. Department of Energy (DOE) has established programs to explore export opportunities for American coal-based technologies in developing countries.

For most of the developing countries (India and China are the major exceptions), oil is the mainstay of their commercial energy supplies, accounting for about two-thirds of the total. Oil is easy to transport and versatile in use in all sectors, at all scales of operation. These qualities led to an average annual growth of 4.5 percent in oil consumption in the developing countries from 1971 through 1987. This growth is expected to continue at almost the same rate (4 percent) in the future. In the absence of sizable increases in domestic production, import dependence will rise sharply.

The developing world[4] possesses only limited crude oil reserves with a reserves/production ratio of 26 years, compared with a worldwide ratio of 43 years. These reserves are concentrated in a few countries; one-half of the developing countries do not have any discovered recoverable reserves. The industry consensus is that oil reserves likely to be proved in developing countries will be relatively small. Development of such fields, while traditionally not attractive to the major oil companies, is important for the developing countries themselves, especially the poorer ones.

A number of recent technical developments in oil exploration and development may reduce risks and costs, thus making small field development more attractive than before, particularly for smaller oil companies. These technologies include 3-D seismic modeling, advanced drill bits, slim hole drilling, directional drilling, and innovations in offshore production. These technologies could also benefit exploration for gas, a relatively undeveloped resource.[5] Many developing countries, including several poor sub-Saharan African countries,[6] have significant natural gas reserves, and the number of nations with proven gas resources is following an upward trend.

Despite promising geological prospects, the degree of exploration activity (density of wells drilled) in the developing countries is much lower than the world average, and is concentrated in countries where resources have already been developed. In the developing world, investment in petroleum exploration and development is carried out (with the exception of a few, large-population, countries) almost exclusively by international oil companies. Thus the fiscal and contractual arrangements between country and company are important in providing the appropriate incentives. These incentives have traditionally been biased in favor of large, low cost rather than small, higher cost fields. Gas development faces additional obstacles. Unlike oil, markets have to be developed simultaneously with the resource, thus adding to the start-up costs and complexity of gas projects. Moreover, gas sold in local markets does not directly generate the foreign exchange needed to repatriate profits to foreign investors.

Efforts are already being made to overcome some of these obstacles. The multilateral development banks, notably the World Bank, finance preinvestment studies, help countries towards agreements with foreign investors, and encourage the development of fiscal systems and financial agreements

---

[4]The definition of ''developing world'' (see ch. 2) includes several OPEC countries but not the high income Organization of Petroleum Exporting Countries (OPEC) members, such as Saudi Arabia.

[5]The gas reserves/production ratio for the developing world is about 88 years, which is significantly higher than the 32.4 years for crude oil.

[6]Mozambique, Ethiopia, Somalia, Madagascar, Côte d'Ivoire, Equatorial Guinea, Sudan, Senegal, Tanzania, and Namibia.

recognizing the special characteristics of gas. Given the importance of oil and the great potential of gas, there is room for further extension of such programs.

For the longer term, renewable forms of energy offer the potential for sustainable, domestically produced energy supplies. As discussed in chapter 6, extensive hydro, wind, solar, and geothermal resources are present in many countries, and indeed are particularly abundant in some of the poorest. Chapter 6 also showed that there are opportunities to convert biomass fuels into modern efficient energy carriers. The question here is to what extent the feedstock for a modern biomass conversion industry can be supplied at a competitive price on a truly sustainable basis, without major adverse environmental and social impacts.

Developing countries have the potential to greatly improve utilization of their biomass energy resources. Large scale use of modern biomass energy technologies could have major advantages for developing countries, and for the world. If produced on a sustainable basis, biomass would not add to net greenhouse gas emissions. If substituted for fossil fuels, sustainably grown biomass could actually decrease greenhouse gas emissions as well as improve local environments through reduced emissions of SO_2 and NO_x. As an indigenous resource, biomass could reduce energy import dependence and stimulate rural development.

In practical terms, there are several obstacles to the development of dependable and large scale biomass supplies for a modern biomass industry. Data on the extent of forest area and the annual increment of forest growth are sparse and unreliable, creating uncertainty over how much biomass can be produced and collected. There is also uncertainty over costs; current estimates suggest that woody biomass could be available at the competitive price of $2 per gigajoule (GJ) ($1.90 per million British thermal units (BTUs)), but this estimate will be subject to wide local variations. The basic incentives to growing biomass, however, are often absent. Improvements in forest management are notoriously difficult to achieve, and the introduction of high yield field crops will require long term sustained efforts. The long term environmental impacts of sylvan monoculture and high yield crops are unknown.

There are also difficulties setting up a large scale commercial biomass feedstock industry in rural areas with inadequate infrastructure, especially when national pricing policies often favor established energy supply industries. Meshing a commercial biomass industry with the current, largely noncommercial usage of biomass could raise problems of access to traditional fuel supplies by the poor and issues of equity between landholders and the landless. In many developing countries, energy plantations might compete with food crops for limited land resources, particularly as populations continue to grow. Existing biomass resources (dung, agricultural residues) may have important uses other than fuel, including livestock feed, fiber, and fertilizer. Any diversion of land or resources could adversely affect the poor, through decreased availability of food and previously "free" fuel, fodder, fiber, and fertilizer.

In the technical arena, several advances have been made in improving plant productivity in recent years. Physiological knowledge of plant growth processes have improved, particularly through biotechnology. These and other efforts have identified and developed fast growing species. Planting methods, such as intercropping, have also improved. Successful experiments with crop residue densification have increased energy content, reducing transport and handling costs. The United States has contributed to these and other technology improvements through experimental trials funded by the U.S. Department of Energy and the U.S. Department of Agriculture, and has unrivaled experience with agricultural extension. Many institutional issues and uncertainties will have to be addressed, however, before large scale biomass feedstock supplies can be developed beyond existing agricultural and forestry residues.

The most promising innovations for biomass resources have been in waste-to-energy technologies. Agricultural and industrial wastes, such as sugar cane residues (bagasse, etc.) and sawdust, are readily available energy resources in developing countries. Many developing countries do not now use these resources, or use them inefficiently.

OIL AND GAS SUPPLY

For most of the developing countries (India and China are the major exceptions), oil is the mainstay of their commercial energy supplies, accounting for about two-thirds of the total. Oil is easy to transport and versatile in use in all sectors and at all scales of operation. These qualities led to an average annual growth of 4.5 percent in oil consumption in the developing countries from 1971 through 1987. Projections of energy consumption in the developing countries[7] foresee a continuation of this rapid rise, though at a slightly lower rate—about 4 percent. Domestic production of oil in the developing countries is projected to stabilize or even decline, so that the level of imports rises sharply. For five major oil importing countries (Brazil, India, Pakistan, Philippines, and Thailand) oil import dependence is projected to rise from 37 percent now to over 80 percent by 2010. Over the same time period, a group of oil exporting developing countries (China, Indonesia, Malaysia) could become net importers.

Such developments would impose severe strains on the foreign exchange budgets of these countries. Half of the oil-importing developing countries already must import over three-quarters of their commercial energy requirements in the form of petroleum products.[8] For many of the poor African countries, foreign exchange budgets are already strained by oil imports, which account in several cases for 50 percent and more of total foreign exchange earnings.

This imbalance between projected consumption and production underlines the need for improving energy efficiencies, and the substitution of other cheaper fuels for oil where possible. The imbalance also suggests that opportunities to increase domestic production of oil, gas, and other fuels should be pursued. Many of the developing countries have under-utilized known reserves of oil and natural gas. Recovery of these reserves and resources could bring in foreign exchange through exports, or could supply domestic energy markets. Many factors are involved in the successful exploration, production, and marketing of oil (and even more with gas). These include the size and nature of the resource base, the availability of suitable technology to develop these resources, and the complex institutional factors that determine the incentives for investors to engage in oil and gas development.

Oil and Gas Reserves

The starting point is the resource base.[9] The world's proved recoverable crude oil reserves[10] total approximately 932 billion barrels (see table 7-1). The Middle East accounts for two-thirds of the total. Outside the Middle East, the former Soviet Union has the largest share—about 80 billion barrels. The developing world possesses only 17.5 percent of world crude oil reserves. About half of these are in Latin America of which three-quarters is in just two countries, Mexico and Venezuela; the other half is divided almost equally between Asia and Africa. Only one-half of the developing countries have discovered recoverable reserves.

Assuming a static rate of production and discovery, and of population growth, the developing world has a reserves/production ratio of 26 years, com-

---

[7]Mudassar Imran and Philip Barnes, op. cit., footnote 3.

[8]T. Gorton, "Petroleum in the Developing World," contractor report prepared for the Office of Technology Assessment, July 1990, p. 1.

[9]The estimation of hydrocarbon reserves involves more than a geologically based analysis of the original or remaining hydrocarbon endowment. While the amount of oil originally generated or thought to have been trapped and potentially available is a scientific question (though still subject to differences of opinion), there are additional factors involved in assessing actual *reserves*: how much oil and gas has actually been proven to exist at the present time? How much is economically producible (that is, the value of the production brought to market exceeds the costs of producing and transporting it)? At what point will discovered but currently uneconomic reserves become viable, if ever? These nongeological factors include the price that reserves produced in the middle to distant future will fetch, and even the uses to which it will be put and the political relations along the spectrum of producers and consumers. There are a multitude of uncertainties in estimates of the petroleum resource base.

[10]*Proved recoverable reserves* are defined by the Society of Petroleum Engineers as follows. Proved reserves can be estimated with reasonable certainty to be recoverable under current economic conditions. Current economic conditions include prices and costs prevailing at the time of the estimate. Proved reserves may be developed or undeveloped. In general, reserves are considered proved if commercial producibility of the reservoir is supported by actual production or formation tests. The term proved refers to the estimated volume of reserves and not just to the productivity of the well or reservoir.

The salient points in this definition are the word "reasonable" applied to the certainty required; and the criterion of "current economic conditions." This recognizes the fact that certainty is unattainable in this endeavor; and that reserves will not be counted if they are not economic to produce *at the present time*. Discovered reserves that are expected to become economic under the market conditions they expect to prevail at a given time in the *future* are not included; besides most of the shale oil, tar sands, and deep offshore oil (which may or may not be exploited even under future high-price scenarios) this excludes reserves closer to viability such as the Orinoco Heavy Oil Belt in Venezuela. While this is a restrictive definition in some respects, it has the advantage of economic realism at least over the next 30 years or so.

Table 7-1—World Oil and Natural Gas Resources

	Proved oil reserves		Proved gas reserves	
	Billion barrels	Percent share of world	Trillion cubic feet	Percent share of world
World	932	100	3,997	100
Developing countries	163	17	727	18
Latin America	(81)	(9)	(240)	(6)
Asia	(40)	(4)	(247)	(6)
Africa	(42)	(4)	(240)	(6)

SOURCE: T. Gorton, "Oil and Gas Development in Developing Countries," OTA contractor report, 1991.

pared with a worldwide ratio of 43 years.[11] There are wide variations among developing countries: Mexico is expected to produce oil at current rates for another 35 years (though there is much controversy over South American reserves in general and Mexican ones in particular), while Pakistan will exhaust current reserves in 9 years, Cameroon in 8, Gabon in 12, and Peru in 7 years.

Estimates of world oil reserves change over time as new reserves are discovered or reevaluated. According to the World Energy Conference, as well as other authoritative current estimates, global proved reserves have been revised upward by 32 percent over the past 3 years,[12] due to successful exploration and exploitation activities.[13] Decreases in drilling and other oilfield costs over the past 5 or 6 years also have led to reserve increases. In so far as future developments in technology (see below) lead to lower exploration and drilling costs, reserves will further increase. The industry consensus is that oil reserves likely to be proved in developing countries will be relatively small in size. Although such fields may not add much to the overall global supply, they are very important to the developing countries themselves.

While gas is a fuel and feedstock that in many ways is as useful as crude oil, it is relatively undeveloped. Worldwide proved reserves of natural gas amount to some 4,000 trillion cubic feet (TCF), providing 1988 production of about 67 TCF per year.[14] At those rates, the world's natural gas reserves will last for 60 years, about half again as long as global proved crude oil reserves. The developing nations account for 18 percent of global raw natural gas reserves, but only 13 percent of global natural gas consumption. This suggests that these countries are not taking full advantage of their endowment of natural gas resources. The reserves/production ratio for the developing world is about 88 years, which is significantly higher than the 32.4 years for crude oil in these countries.

At least 52 developing countries have significant natural gas reserves (about 10 more than the number of nations that possess oil reserves). Several poor sub-Saharan African countries that are greatly in need of additional energy supplies contain undeveloped gasfields: Mozambique, Ethiopia, Somalia, Madagascar, Cote d'Ivoire, Equatorial Guinea, Sudan, Senegal, Tanzania, and Namibia. Furthermore, the number of nations with discovered reserves has

---

[11] It should be kept in mind here, as for the discussion on Natural Gas reserves below, that the definition of reserves does not include undiscovered resources or growth of existing fields through investment or reevaluation. This figure is also distorted by the enormous reserve "overhang" from Middle East OPEC countries (with an average oil reserve/production ratio of over 100 years even by conservative estimation). This figure also obscures that the position of the industrialized oil producers is much further along the decline-and-depletion curve.

[12] World Energy Conference, op. cit., footnote 2, p. 35.

[13] In the mid-1970s, the non-OPEC developing nation Oman, for example, was expecting its production to level off at about 330,000 barrels per day (b/d) in 1977 and decline to depletion within 5 years. Production is now currently over 600,000 b/d and holding steady.

[14] These figures do not include "natural gas liquids" (NGLs) or liquid hydrocarbons which, though generally gaseous under original reservoir conditions of temperature and pressure, may be separated or extracted from the gas at the wellhead and produced as condensate (also called "natural gasoline") and liquefied petroleum gas (LPG), the latter referring to butane and propane fractions that are separated from condensate and sold in bottled form as fuel. Though NGLs share many of the uses and economic properties of crude oil, NGLs are generally considered together with natural gas reserves (of which they are a constituent part in the reservoir) for the purposes of global or national energy supply analysis. This is defensible because all the economic properties of NGLs produced from a given gas field are determined by the general reservoir and other parameters of that field, so that the "wetness" or richness in NGLs of a given gas field is best treated as part of the quality description of the gas field rather than as a separate resource. For project-specific economic analysis, however, the NGL production stream is always given separate treatment because it is generally marketed very differently from the "dry" gas left over after the liquids have been removed. Worldwide NGL reserves total about 65 billion barrels.

followed an upward trend. While only 40 nations of the world claimed to possess natural gas resources in 1960, 85 countries claimed natural gas reserves by 1987.[15] Some observers have noted that natural gas reserves in developing nations are underestimated because the majority of discoveries have been a by-product of the search for crude oil. There has been little exploration for gas itself. The opportunities for discovering non-associated gas in the developing world have yet to be fully explored,[16] especially as gas costs are competitive and expected to remain so.[17] In the developing nations, a significant amount of natural gas is lost through flaring processes; in 1975, 80 percent of all natural gas produced in oil producing developing nations was flared.[18] In Nigeria, 97 percent of natural gas is currently flared.

Aside from its relative abundance and competitive costs, gas also is attractive for its environmental considerations. While coal produces 25 kilograms (kg) of carbon per GJ (this number is a weighted average of industry and utility), oil produces 19 kg of carbon, and natural gas only 13.6 kg.

However, geologic promise is not a sufficient condition for investment in exploration activity.[19] Whether such promises lead to exploration and production depends on a number of factors, including developments in technology and economic incentives.

Oil and Gas Exploration and Production Technologies

In the developing world, investment in petroleum exploration and development is carried out (with a few exceptions) by international oil companies, many of which are American. American companies are, therefore, key players in determining the type of oil and gas supply technologies used. At the same time, the developing countries offer a major market for U.S. hydrocarbon development services.

The two major and complementary components of exploration for oil and gas are characterization and exploration drilling. Characterizing a reservoir helps to determine the best area to drill, and material uncovered by exploratory drilling helps characterize in detail the subsurface.

Drilling can be the most expensive component of exploration and development, costing anywhere from 15 to 40 percent of offshore development costs, and as much as 80 percent of the less expensive land development costs.[20] Disposal of wastes accounts for a portion of these costs. Exploration and production activities produce brine, drilling muds—a combination of water, clay, and various chemicals—and rock cuttings that must be discarded. There may be as much as 50 tons of mud to dispose of by the time a well is completed.[21] Offshore, these muds are generally discharged into the surrounding waters.[22] Onshore, the mud wastes are disposed of on land or through reinjection processes. Land disposal of wastes has to be managed carefully to avoid contamination of ground and surface waters.

Technological advances have decreased the high costs and mitigated the environmental problems associated with exploration and drilling. These exploration technologies are applicable to developing countries and are already used in some places. United States companies enjoy the technological lead in this area, particularly in identifying smaller reservoirs.[23]

Characterization technologies consist of compiling and analyzing geological and geophysical infor-

---

[15] World Energy Conference, *Survey of World Resources 1989*, p. 61

[16] Afsaneh Mashayekhi "Natural Gas Supply and Demand in Less Developed Countries," *Annual Review of Energy*, vol. 13, 1988, p. 122

[17] Ibid.

[18] Jae Edmonds and John M. Reilly, *Global Energy: Assessing the Future* (New York: Oxford University Press, 1985).

[19] See discussion in Harry G. Broadman "Determinants of Oil Exploration and Development in Non-OPEC Developing Countries," Discussion Paper D-114, Resources for the Future, 1983, and "An Econometric Analysis of the Determinants of Exploration for Petroleum Outside North America," unpublished manuscript.

[20] Shell Briefing Service, *Producing Oil and Gas* (London: Group Public Affairs, Shell International Petroleum Company, Ltd. 1989), p. 2.

[21] United Nations Department of Technical Cooperation for Development, *Energy and Environment: Impacts and Controls* (New York, NY: October 1990), p. 15.

[22] U.S. Department of Energy, Assistant Secretary for the Environment, Office of Technology Impacts, *Environmental Data Energy Technolog Characterizations: Petroleum*, April 1980, p. 1-9, 1-10.

[23] RCG/Hagler, Bailly, Inc., Washington, DC, "U.S. Exports of Oil and Gas Exploration and Production Equipment and Services," a draft pape prepared for the Agency for International Development, Office of Energy, 1990, p. 6.

mation on a given reservoir. Characterization enhances three important stages of exploration. In the first stage, specialists choose the location for "appraisal wells" (wells that define a reservoir and provide data for more detailed characterization). In the second, they perform the detailed geological characterization and analysis, including the likelihood of finding a commercially viable well. And finally, if hydrocarbons are found, they assess the properties of the reservoir that will affect production.

Advances in these three stages of geophysical research greatly enhance the likelihood of successful drilling. The costs of this sophisticated research have declined so that these technologies increasingly are used worldwide.[24] Lowered costs may allow marginal or small fields, often found in developing countries, to be economically viable. In addition, detailed knowledge of a reservoir before production can lessen negative environmental impacts, including contamination of subsurface water supplies.

The principle characterization tool in the three stages of exploration is Computer-Aided Exploration and Development (CAEX), also called 3-dimensional seismic modeling. 3-D seismic modeling is based on information on regional and reservoir geology, as well as material samples from a particular site. Petrophysical studies provide data on types, porosity, hydrocarbon saturation, permeability and capillary pressure of the reservoir rock, as well as on the nature of trapped hydrocarbons.[25] A detailed analysis of extremely small fossils, found on the surface or in wells, helps identify precise rock intervals and ages. Geochemical tests are used to predict the existence and thermal maturation qualities of rocks that may hold oil.[26]

The 3-D seismic studies take this geophysical and geological data, process them on a minicomputer, and turn the data into "reservoir maps." The maps describe the "trends in sand quality, structural properties, and tectonic history" of an area, providing a 3-D picture of the reservoir.[27] A second "reservoir simulation" model, derived from the geological model, tests and forecasts the production performance of a reservoir. Although the simulation model usually requires computers that may not be widely available to developing countries, other stages of characterization are within reach. In addition, material left-over from older, "unsuccessful" drilling can be used in the analysis, expanding the knowledge base of a reservoir without the expense of acquiring new raw data.[28]

The oil business is almost by definition a risky one. No matter how sophisticated the techniques for evaluating the possibility of a find, the final answer can only come through the expensive process of drilling a well. Worldwide, about 9 out of 10 exploratory wells are dry or recover only subcommercial quantities or qualities of hydrocarbons. Improved drilling technologies, by reducing the element of risk, can enhance the discovery of commercially viable oil and gas wells. Exploration drilling has especially benefited from advanced drill bits, slim hole drilling, directional drilling, and measurements-while-drilling.

- *Advanced drill bits*. Drills must be sharp and durable to cut through rock, and they must also carry the rock cuttings and drilling muds out of the well to the surface. A sharp drill bit allows a high penetration rate, and a faster drilling process. Drill bit durability also speeds the process by cutting the number of times drilling must stop for replacement of worn bits. In the past, there was a trade-off between durability and sharpness, but advanced drill bits allow for both, decreasing overall drilling time and costs. American companies have the technological lead in advanced drill bits.[29]
- *Slim hole drilling*. A simple and highly cost-effective drilling technology is slim hole drilling. Reducing the size of a bore hole reduces the costs of drilling a well for exploration or production by up to 50 percent.[30] Bore holes are lined with heavy steel pipes, which are cemented into place to prevent collapse or con-

---

[24]Gorton, op. cit., footnote 8, p. 11.
[25]Shell Briefing Service, op. cit., footnote 20, p. 1.
[26]Gorton, op. cit., footnote 8, p. 11.
[27]Shell Briefing Service, op. cit., footnote 20, p. 1.
[28]Gorton, op. cit. footnote 8, p. 11.
[29]RCG/Hagler, Bailly, Inc., op. cit., footnote 23, p. 8.
[30]Shell Briefing Service, op. cit., footnote 20, p. 2.

tamination from surrounding fluids. Both the process and materials can be costly, and a smaller hole cuts costs. Although some of the detritus brought up out of a bore hole is analyzed, much is waste. Slim hole drilling is environmentally desirable as it reduces the amount of waste muds and rock cuttings—as much as 70 percent. Characterization information can be used to determine the smallest hole that can be drilled without compromising effectiveness in gathering core material or conveying gas and oil.

- *Directional drilling.* In the past, drilling was only vertical, chiefly leading to the discovery of deposits oriented along the vertical bore hole. Some deposits, however, deviate from the vertical pattern. New drilling technologies allow for flexible directional or horizontal drilling, which allows the drill to probe horizontally into previously inaccessible wells. Directional drilling is already being used in developing countries. For example, the American company Amoco extensively used horizontal drilling offshore in China in 1989.[31]

In exploratory drilling, whether exploring for appraisal or for an actual deposit, constant measurements are necessary. The angle, depth, and diameter of the bore hole must be monitored, mostly by tracking the position of the bit.[32] Information on the condition at the far end of the hole also needs to be measured. Conventionally, drilling is periodically halted and directional surveys and wireline logging (lowering special measuring instruments into the drill hole) are used to determine when drilling is complete. A new technology for this process is "measurement-while-drilling." With this technology, the measurement instruments are incorporated into the drill above the bit, transmitting information to the surface while the drill is in operation. Measurement-while-drilling (MWD) allows for continuous drilling, which saves money, and provides more information than conventional wireline logging. MWD can provide detailed directional surveys and data on the environment surrounding the drill, including electrical resistivity, density, natural gamma ray radiation, and porosity.[33]

Drilling technologies benefit exploration, but in many cases, they also benefit production. Directional drilling is a particularly useful new technology for recovering oil and gas. When a bore hole passes through a deposit, oil or gas flows through a perforated casing at the bottom of the pipe. Conventional vertical drilling may bypass horizontally deposited wells, causing only a small amount of the deposit to flow through the perforations. Such fields are therefore abandoned. If further recovery is attempted with the conventional technology, numerous vertical bore holes have to be drilled, driving up production expenses and increasing environmental risks. With directional drilling, however, these horizontal deposits can be recovered by drilling one well, making previously unproductive fields more accessible and attractive.

For offshore production, extended-reach wells are drilled diagonally from a platform, allowing access to a larger area than vertical drilling allows. Ocean platforms are expensive to construct and maintain, especially in deep water. Extended-reach can cut costs by reducing the number of platform installments required for one field.[34]

A key stage in the production process is recovery, or drawing the deposit from underground to the surface. Oil and gas will flow from a reservoir at varying rates, depending on the environment around the deposit and the bore hole. An ideal reservoir environment would include: a fairly simple layout that enables easy drilling access to the deposit; highly permeable rock; crude oil or gas with low viscosity; and natural pressure in the reservoir exceeding that in the bore hole.[35]

In such an ideal case, oil or gas can flow unaided to the surface. This "primary recovery" involves natural drive mechanisms, naturally occurring factors that help push oil and gas upward. Such mechanisms "account for more oil production than all other recovery methods combined."[36] In order for gas or oil to flow into the bore, the pressure in the

---

[31]Gorton, op. cit., footnote 8, p. 11.
[32]RCG/Hagler, Bailly, Inc., op. cit., footnote 23, p. 7.
[33]Shell Briefing Service, op. cit., footnote 20, p. 2.
[34]RCG/Hagler, Bailly, Inc., op. cit., footnote 23, p. 7.
[35]Shell Briefing Service, op. cit., footnote 20, p. 4.
[36]Ibid.

surrounding rock must be more than that in the bore. This natural drive can be enhanced by artificial lift, which consists of reinjecting gas into the oil flowing in the bore. Injecting gas enhances the pressure in the reservoir, reducing the amount of natural drive needed for an upward flow.

Well-stimulation technologies, such as fracture stimulation and matrix stimulation, by improving the permeability of the reservoir rock, also enhance drive. Fracture stimulation involves actually cracking the rock by forcing fluid into the well at high pressures. The cracks are propped open with a fine sand, keeping the channels to the well open when the applied pressure is released.[37] Matrix stimulation involves injecting a chemical solution into the rock to dissolve any material blocking the pores. Fracture stimulation is very expensive, and both techniques must be applied carefully to avoid environmental damage.

Many oil and gas fields in developing countries do not have ideal conditions. A "tight" or less than ideal reservoir may require extra energy to move the hydrocarbons to the surface. A growing proportion of the world's oil supply, particularly in the developing countries, comes from such "tight" reservoirs.[38] Conditions also become tight in older reservoirs, as natural drive flags. A deposit extracted in these cases is considered a "secondary recovery."[39] Although the oil industry has the know-how to make a tight reservoir more productive, these technologies are not often applied in developing countries, for economic rather than technical reasons.[40]

Secondary or enhanced oil recovery technologies include thermal, gas, chemical, and microbial methods. Thermal oil recovery is the most common, and has been used in some developing countries. In some fields, the crude oil can be too viscous to flow out of the rock formation. Steam or hot water can be injected down the well, and the heat will decrease oil viscosity and increase the flow rate. The gas method is used when medium or light viscosity oil gets trapped in pores during oil flows. The injection of a gas into the bore can displace the residual oil, and move it upward. Enriched hydrocarbon gas, carbon dioxide, and nitrogen have all been used.

Chemical enhanced oil recovery is less common. The natural drive mechanisms begin moving oil when a well is drilled; underlying water displaces the oil, providing the drive upward. A gas cap lying above the oil may similarly force the oil upward in the well. When entrained water is not providing enough natural pressure (in an old well, for example), chemicals can be added. The chemicals modify the water, changing the way the oil is displaced and moved through the reservoir rock. These chemical methods are technically difficult to execute, and can be especially threatening to the subsurface environment.[41]

Although the United States leads in enhanced oil recovery technologies, low oil prices have limited the market for such techniques. In some cases, however, foreign projects have been led by foreign companies. For example, steamflood enhanced recovery in Venezuela is operated by the Venezuelan companies Lagoven, Maraven, and Corpoven.[42]

Offshore Production

Since the construction of the first offshore platform in 1947, offshore oil production has come to account for a growing share of world production. A number of innovations are lowering the costs and difficulties of offshore production, and the United States has the lead in most of these technologies.[43] There are offshore fields in many developing countries, and in particular, Brazil and Mexico have used new technologies.

---

[37]Ibid.

[38]Ibid.

[39]The distinction between primary and secondary recovery is not always clearcut. For example, artificial lift and well stimulation are sometimes considered technologies for secondary recovery.

[40]Gorton, op. cit., footnote 8, p. 13.

[41]There are other less common techniques. Microbial technologies are still in the laboratory stage, though there may be some limited field testing. This method consists of injecting microbes into the reservoir. These microbes generate carbon dioxide vapor pressure that forces the oil out. This technology could be difficult in practice as conditions in reservoirs—no air and the presence of metals—are hostile to microbes. There is no evidence to date to suggest that this could be done cost effectively. Another enhanced recovery technology that is not widely used is in situ combustion. Similar to the thermal technologies, this involves burning some of the oil in the reservoir to generate heat and decrease the viscosity of heavy crude. One method for doing this is to inject air underground, causing the combustion of a small amount of the oil.

[42]RCG/Hagler, Bailly, Inc., op. cit., footnote 23, p. 9.

[43]The United States does not have a technological lead in multiphase pumping. RCG/Hagler, Bailly, Inc., op. cit., footnote 23, p. 11.

Conventional offshore technologies can be very expensive, especially in deep or rough water. Traditionally, offshore production has required the installation of permanent platforms with heavy equipment and crew accommodations. Vertical drilling brings up deposits of oil, mixed with the gas and other liquids often found in association with oil. Oil has to be separated on the platform and then pumped through underwater pipes back to shore for further refining. Undersea maintenance of the platforms also is costly. Wastes from offshore drilling, including residual oil, are often dumped back into the ocean. There is some evidence that drilling and production waste may be harmful to aquatic life. Oil leaking and spilling from these operations makes up about 10 percent of all oceanic oil pollution.[44]

Almost all aspects of offshore production have been affected by recent innovations. In some developing countries, production costs may be reduced by using tankers as floating storage for offshore production. This system is called "single-point mooring." Single-point mooring may be especially valuable for the small or marginal fields that become unfeasible with the added costs of subsea pipelines. This system may, however, increase the likelihood of minor oil spills as oil is transferred from the platform to the ship.

Platforms no longer have to be rigid structures attached to the sea floor. Lighter platforms can float on the surface, held in place by cables fastened to the sea floor. This tension leg platform design with a simplified deck or "topsides" reduces the cost and complexity of offshore production. Temporary drilling, also called jack-up rigs, packaged rigs, or semisubmersible tenders, are particularly cost-effective. Such rigs reduce capital costs by up to 25 percent, and operating costs by up to 40 percent.[45]

Staffing an ocean platform is expensive, and the work is difficult and involves risks. The introduction of lighter and less complex rigs allows for easier operation and maintenance. In addition, two technologies allow for automated platforms: remotely operated vehicles and multiphase pumping. Remotely operated vehicles are underwater maintenance robots that repair subsea pipes and valves. Multiphase pumping eliminates the need for crews to separate the oil from its accompanying liquids. Multiphase pumps can pump oil, gas, and other condensates back to land-based production facilities. With such pumps, small platforms can be located further from shore.[46]

Weather satellites can also lower the costs of offshore production. Strong ocean currents and inclement weather can halt drilling and maintenance programs. Advanced weather and ocean current forecasting can help prevent "downtime in drilling and production operations."[47]

Natural Gas

Hydrocarbons that exist in a gaseous state at atmospheric conditions of temperature and pressure are called natural gas. While gas is in many ways as useful as crude oil, it is definitely the poor relation of the hydrocarbon family, with relatively low exploitation in the developing countries. Frequently, natural gas is discovered in connection with oil. In many developing countries, however, there is little production infrastructure for gas. Gas may then be burned off, or a well yielding gas may be treated as a "dry hole." The most significant technological change for developing countries may be developing the infrastructure for recovering and refining natural gas.

The high start-up costs of conventional gas production technologies have discouraged many developing countries from utilizing gas reserves. Imminent technological innovations may lower these costs, however. Characterization and exploration technologies as well as multiphase pumping already have contributed to lower costs. One technology that may encourage the use of gas is conversion of gas to electricity through combined-cycle or advanced steam injected gas turbines (see ch. 6). Conversion to electricity eliminates the need for expensive pipelines and can help turn resources into commercially viable reserves.

Natural gas liquids (NGLs) or liquid hydrocarbons, though generally gaseous under original reser-

---

[44]United Nations Department of Technical Cooperation for Development, *Environment and Energy: Impacts and Controls* (New York, NY: October 1990), p. 15.

[45]Shell Briefing Service, op. cit., footnote 20, p. 3.

[46]Ibid.

[47]RCG/Hagler, Bailly, Inc., op. cit., footnote 23, p. 11.

voir conditions of temperature and pressure, may be separated or extracted from the gas at the wellhead and produced as condensate (also called "natural gasoline"). Lower cost and smaller scale technology is becoming available for removing natural gas liquids from gas that was previously being flared or simply shut in. The presence of such liquids is of considerable economic importance to a domestic gas project as they can be sold as oil and thus generate foreign exchange earnings. Natural gas liquids can also be produced as liquefied petroleum gas (LPG) and sold in bottled form as fuel. Both natural gas liquids and liquefied petroleum gas can be used for conventional services such as heating and transportation. The Nigerian government is studying plans for a dramatic increase in liquefied petroleum gas supplies, and countries such as Equatorial Guinea, although with much more limited supply and market possibilities, also are investigating such possibilities.

There are a few projects in the United States for extracting coal-bed methane gas that could be applied in the developing world. This technology consists of drilling just above a coal bed and injecting a mix of water and gas into the well. The injection creates a cavity at the bottom of the well. Once the water is pumped out, methane flows in through a perforated casing that allows gas in while keeping coal out.[48] Currently, however, this technology is costly and is supported in the United States through special tax credits.

Whether burned as NGL or in a gaseous state, gas retains an important distinction from other hydrocarbons. Natural gas combustion releases less greenhouse gas, such as carbon dioxide, and less localized pollution than do crude oil and other fossil fuels. Natural gas will gain attention as a relatively clean fuel, since "environmental concerns are likely to be among the strongest influences acting on the oil and gas industry in the next decade."[49]

Institutional Issues

Despite the geological promise for oil and gas discoveries in the developing countries, the density of wells drilled averages just 7 per 1,000 square miles in non-OPEC developing countries compared with a world average of 109.[50] Furthermore, within the developing countries, much of the exploration takes place in countries where resources already have been developed. One recent study of exploration by the international industry in the 66 Lome Convention countries (which constitute a representative sample of the African, Caribbean, and Pacific (ACP) countries) reported that exploration in ACP countries is almost wholly confined to countries already producing oil, particularly Nigeria, Cameroon, Congo, Gabon, Angola, and Trinidad.[51] Only 12 of the 66 countries studied had produced oil to date, and those 12 received all but a negligible amount of the total investment in seismic surveys and exploration drilling. Although they overlie extensive sedimentary basins, the ACP countries remain, with some notable exceptions, among the least explored countries.[52]

In the developing world, investment in petroleum exploration and development is carried out, with the exception of a few large-population countries like Mexico and Venezuela, by international oil companies. The cash and foreign exchange position of low-reserve developing countries does not lend itself to such capital-intensive high-technology and high-risk investments.

During the high-energy-price era of the mid-1970s and early 1980s, a number of developing countries with proven or potential hydrocarbon reserves formed or expanded national oil companies such as CEPE in Ecuador, YPF in Argentina, Braspetro in Brazil, Petronas in Malaysia, Pertamina in Indonesia, and NNPC in Nigeria. In the mid-1980s, however, increasing debt and general budgetary pressures led many of these governments to moderate their policies towards foreign investment. Algeria and Burma have invited the international industry to apply for exploration and production

---

[48]Hayes, Thomas, "Drillers Find Coalbeds Yield Gas and Profits," *The New York Times*, Dec. 26, 1990, p. D3.

[49]Shell Briefing Service, op. cit., p. 9.

[50]R. Vedavelli, *Petroleum and Gas in Non-Opec Developing Countries: 1976-1985*, Staff Working Paper No. 289 (Washington, DC: World Bank, 1978), p. 9.

[51]A. Fee, "Oil and Gas Exploration and Production in the ACP Countries," *OPEC Review*, vol. XIII, No. 2, summer 1989, pp. 137-151.

[52]Ibid, p. 141.

rights, and Venezuela has begun to discuss the possibility of allowing foreign companies to form joint ventures for the implementation of secondary recovery schemes on some of the country's more mature oilfields. Algeria has also expressed interest in similar schemes.

There are several factors affecting companies' decisions to invest. Oil is a risky business: no matter how sophisticated the techniques used, oil and gas can only be found by expensive drilling. In the long run, the value of a successful case must be large enough to accommodate the exploration risk and cost of roughly nine dry wells.

In addition to the technical risks there is a host of risks referred to as "country," or political risk. These risks, which are not intrinsic to the oil business, include "war, riot, and civil commotion." Violent local turmoil can damage property, halt or disrupt operations, or threaten the safety of a company's employees. Nationalization can mean the sudden involuntary divestiture of the foreign assets of a corporation in favor of the local government. Private interests also are wary of "creeping nationalization," the gradual process of the host state exerting greater control over operations and extending its share of ownership and/or of profits.

There are dramatic examples of how "war, riot, and civil commotion" have disrupted oil and gas development. Chevron has been waiting since the early 1980s for the situation in southern Sudan to stabilize to begin the development of substantial oil reserves discovered there. Meanwhile, the nearly $1 billion invested to date by Chevron and former consortium members is immobilized. Exxon, Royal Dutch/Shell and others are similarly engaged in Chad. Instability in Mozambique has hindered exploration efforts in that country, and in neighboring landlocked Malawi. Operations in northern Somalia have been suspended pending stabilization of the situation there.

On the other hand there are several examples where development has continued throughout periods of instability. Chevron has been producing oil offshore at Angola's Cabinda enclave throughout the past decade of instability and guerrilla warfare. Nigeria's frequent coups d'etat have taken place without disturbing the activities of the numerous oil multinationals producing Nigeria's high quality crude oil. Though an oil company is likely to prefer a stable political situation over an unstable situation, one analysis concluded[53] that political risk has only a slight influence on decisions to engage in geophysical testing and exploratory drilling.

The other major factor in deciding to undertake investment in exploratory activities in developing countries is the economic incentive. Most modern legislative systems reserve the State's sovereignty over petroleum operations. Most governments, however, reserve the right to contract with foreign companies for exploration and production of petroleum. The contract also includes terms under which country and company will share the net income, or rent, of the project.

The **concession** agreement was the most common arrangement before and immediately following the Second World War. More recently, these agreements have given place to the **production sharing agreement**, which provides that hydrocarbons extracted by the private companies be shared with the government or the national oil company according to predetermined rates. Other countries, notably in Latin America, have adopted a third category of arrangement called the **service contract**, which specifies that the contractor will spend certain sums to explore and, if successful, develop and produce hydrocarbons. In return, he will recover his expenses plus some profit by means of a service fee, which is sometimes calculated as a percentage of costs incurred or as a percentage of production achieved.[54] Companies now generally prefer production-sharing when given the choice.

In the industrial oil-producing countries (e.g., the United Kingdom or Norway), taxation of petroleum activities is now almost wholly profit-related. Profit-related taxation has an important advantage for developing countries as it favors exploration for and development of small, high-cost, or otherwise marginal deposits—precisely the type of resource typical of oil-importing developing countries. As the investor does not begin paying a significant amount of tax or other consideration to the state until he has a positive cumulative cashflow (with allowance for

---

[53] Broadman, op. cit., footnote 19.

[54] The latter formula occurs mainly in the so-called "service contracts with risk," and the bottom-line economic difference between them and production-sharing agreements is negligible.

depreciation), he will have an incentive to develop almost any field that has economic value. Such a system could also be important to the increasing numbers of smaller independent oil companies, which seek low-cost licenses and smaller developments at lower costs than would the majors.

The measures currently used by most developing countries (royalties or flat proportions of gross production), however, bear no relation to profitability.[55] The effect of profitability-insensitive systems is to achieve higher effective tax rates from small or marginal fields than from large, profitable ones. As a result, small fields go undeveloped and declining fields are prematurely abandoned. This factor is all the more important as oil companies in general have little interest in pursuing the smaller end of the probability spectrum and tend to be willing to forgo the modest benefits to be derived from developing marginal accumulations.

The Special Characteristics of Gas

Despite the obvious advantages of gas, industry and lesser developed countries' (LDC) governments have found it difficult to take the steps necessary to bring about a significant increase in the development of gas reserves in countries not now consuming natural gas. There are a number of reasons for this low level of resource development:

- due to its physical characteristics, gas is expensive to transport, requiring high front-loaded capital investment (pipelines from producing to consuming regions, or costly facilities and tankers to liquefy and transport the gas internationally).
- whereas markets for oil are well developed, markets for gas have typically to be developed at the same time as the resource is developed, adding to the total costs and complexity of the project.
- the sale of the gas gives rise to revenues in local currency rather than foreign exchange, leaving investors uncertain about their ability to repatriate profits. This last difficulty is especially acute in the case of highly indebted developing countries, where the bulk of scarce hard currency is earmarked in advance for debt service.
- in many such countries, the fiscal/contractual terms under which the gas was discovered by foreign operators are inappropriate for the special characteristics of gas. As a result, companies tend to treat a gas discovery as a "dry hole" from their economic point of view.

Together, these factors lead many developing countries to import large quantities of crude or fuel oil to generate electricity, even though they possess reserves of natural gas that could do this more economically and with less harm to the global and local environment.

Policy Options

Exploration and production of oil and gas in developing countries involves several main factors, including resources, technological advances, contractual arrangements, and the special problems of gas development. There are a number of possibilities for stimulating or improving oil and gas development.

Oil companies are reluctant to engage in exploration in countries that do not now have production. Part of this reluctance may be based on lack of geological and other preliminary knowledge about the resource base. The World Bank has addressed this barrier by helping countries prepare prospectuses based on existing geological data and other relevant material, thereby saving prospective investors the time and expense of collecting such information. This could be particularly useful for small independents.

Some investors may shy away from developing countries due to nontechnical risks such as nationalization and currency conversion and transferability. Insurance against such risks can be obtained by U.S. investors from the Overseas Private Investment Corporation and from the Export-Import Bank through its agent, the Foreign Credit Insurance Association. The Multilateral Investment Guarantee Agency, a new World Bank unit, also offers investor insurance with fewer limitations on the nationality of the investor and the nature of the investment.

Contract/fiscal questions often have been contentious. There may be room for the development banks to act as impartial arbiters of disputes and to help countries not experienced in oil and gas negotiations to understand the issues so that they can work towards satisfactory, long term agreements.

---

[55] A. Kemp, "Petroleum Exploitation and Contract Terms in Developing Countries After the Oil Price Collapse," *Natural Resources Forum*, vol. 13, No. 2, May 1989, pp. 116-126.

If gas can displace more expensive fuels or if there is potential for an export project, then a favorable environment for investors is especially important. Egypt, Tunisia, Pakistan and other countries have adopted improved fiscal systems that have brought about a dramatic increase in exploration specifically for gas. The lowest income countries and highly indebted countries will need to seek the assistance of the international aid donors in order to secure the "lumpy," front-end investments needed to develop gas fields. In 1989, the World Bank created a Natural Gas Unit to bring increased gas projects to developing countries through preinvestment support.

Ultimately, the problem lies in altering a well-established system to take account of new needs. In the oil and gas sector, the need for timely development of gas resources for local markets is foremost. Closer collaboration among energy firms, governments, national operating entities, and providers of finance and other services could facilitate the process.

COAL

Coal is the most abundant fossil fuel in the world. According to the United Nations, coal reserves are currently about 10 times as large in energy content as world oil and gas reserves combined.[56] Due to its abundance and relatively low price, coal is widely used in those countries with domestic coal reserves. Coal does have some drawbacks, however; it is expensive to transport, it cannot easily be converted to a liquid transport fuel,[57] and it is a major contributor to several environmental pollutants, including NO_x, SO_x, CO_2, and particulates. These pollutants have a deleterious effect on local and global environmental quality and on human health.

Due to its availability and low price, coal will continue to be widely used as a fuel in the near future. Currently, China and India—by far the largest coal producers in the developing world—produce and consume large amounts of relatively low quality coal. From 1983 to 1986, coal production in China and India increased on average about 6.6 and 7 percent a year respectively, higher than the

Table 7-2—Developing Countries With Significant Coal Reserves or Coal Consumption

Country	Proved reserves in place (million metric tons)	Consumption (PJ/year)
China	737,100	17,858
India	27,912	4,178
South Africa	115,530	3,111
Korea, DPR	2,300	1,457
South Korea	200	912
Brazil	3,098	395
Mexico	2,401	189
Viet Nam	312	147
Colombia	2,073	135
Zimbabwe	2,500	120
Indonesia	23,232	66
Mongolia	12,000	61
Chile	4,579	55
Swaziland	2,020	NA

Criteria for inclusion: Proved reserves in place of greater than 2,000 million metric tons, or consumption greater than 100 PJ/year. Proved reserves in place include anthracite, bituminous, sub-bituminous, and lignite.
NA = not available.
SOURCE: United Nations, *1986 Energy Statistics Yearbook*, United Nations, New York, NY, 1988, pp. 58-85, 424-427.

world average annual production increase of about 4.1 percent over that same period. Rapid increases in production have been due to growing populations and lack of energy alternatives. Many other developing countries with coal reserves are facing similar circumstances, and projections show coal use in the developing world doubling over the next 30 years.

Coal Reserves and Consumption in the Developing World

Coal reserves worldwide are quite concentrated, with just three countries—China, the former Soviet Union, and the United States—possessing about two-thirds of the world's coal.[58] In the developing countries, large coal reserves are found in China, India, Indonesia, and S. Africa, with the reserves of these four countries accounting for about 97 percent of developing-country proved recoverable reserves. Other developing countries with significant coal reserves include Mongolia, Chile, Brazil, Zimbabwe, Mexico, North Korea, Colombia, and Swaziland (see table 7-2). Coal consumption does not correlate directly with coal reserves, as some coun-

---

[56]United Nations, *1986 Energy Statistics Yearbook* (New York, NY: United Nations, 1988), pp. 424-426. Proved reserves of anthracite, bituminous, sub-bituminous, and lignite.

[57]Coal liquefaction, widely used in S. Africa, converts coal into a liquid fuel which can be used for transport. This process is relatively expensive and complex.

[58]United Nations, *1986 Energy Statistics Yearbook*, op. cit., footnote 56, pp. 424-426. Proved reserves of anthracite, bituminous, sub-bituminous, and lignite.

Photo credit: U.S. Agency for International Development

India and China account for most developing-country coal use, though several other countries have deposits.

tries with coal (e.g., Indonesia) have not so far developed it extensively. China, India, South Africa, North Korea, and South Korea are the largest coal consumers in the developing world.

Coal varies considerably in quality. Hard coal (anthracite) and soft coal (bituminous) have higher calorific values than lignite. While the Central African Republic, Ecuador, Ethiopia, Haiti, and Mali all possess coal resources, all has been identified as lignite. The seventh largest coal producer in the developing world, Thailand, only produces lignite coal. For many developing countries, therefore, other resources may be more attractive than coal. Brazil, for example, has substantial recoverable coal reserves, but produces a relatively small amount. Brazil's coal is low quality lignite, with little export potential. Moreover, for domestic use, Brazil has access to other high quality resources, such as oil and hydropower.

The relative importance of coal as a commercial energy source is shown in table 7-3. In China, India, and South Africa, coal is the single largest commercial energy source. Electricity generation and industry are typically the large coal users. In some countries, coal is also used in transportation. In South Africa, for example, coal liquefaction (a chemical process for turning coal into a liquid fuel) accounts for 21 percent of coal use. Liquefaction is too expensive and complex, however, to be a viable technology for most developing countries. In India and China, consumption by steam locomotives accounts for a substantial part of coal use. In China, the residential sector uses about 35 percent of China's coal (mostly for space heating and cooking).

Coal Production

As the bulk of coal use in the developing world occurs in India and China, this discussion focuses on coal production in these two countries. Coal mining and processing in China varies in scale and technological sophistication. Large, state-run mines account for about 44 percent of coal production. These mines are becoming mechanized rapidly—in 1981, 18 percent of these mines were fully mechanized, by 1988 this number had climbed to 31 percent.[59] The remaining 56 percent of China's coal comes from village and individual mines, which are typically small, labor-intensive, dangerous to work in, and environmentally damaging.

Coal mining affects land, air, and water quality. The extent of the environmental impacts depends largely on the mining techniques used, though in any case, fugitive dust emissions and leaching from tailings contribute to local air and water pollution. Mining methods are selected according to the depth of the coal, the thickness of the seams, and/or the availability of capital and equipment. The underground mining that accounts for about 95 percent of China's coal production[60] can cause land subsidence and acid drainage, which can contaminate local water supplies and damage aquifers. In addition, underground mines involve hazardous conditions for miners.

China has plans to increase the relative share of surface mining. Surface mining can require the removal of large amounts of top soil and overburden, leading to soil erosion, siltation, and water contamination. In many cases, soil productivity permanently diminishes. The increase in surface mining planned in China could lead to an annual destruction of nearly 150,000 ha by the year 2000.[61] Fortunately, much of China's planned increase in mining is in

---

[59] Ministry of Energy, People's Republic of China, ''Energy in China,'' 1989.
[60] Vaclav Smil, ''China's Energy,'' contractor report prepared for the Office of Technology Assessment, 1990, p. 62.
[61] Ibid.

Table 7-3—Coal Use (1985)

Country	Coal as a percent of total commercial energy use	Coal use breakdown (percent)			
		Electric	Industry	Residential	Other
China....................	76.8	18	46	25	11
India.....................	56.5	40	39	2	19
South Africa.............	85.3	49	17	1	33
United States............	24.7	83	13	<1	4

SOURCE: International Eneregy Agency, *World Energy Statistics and Balances 1971-1987* (Paris: Organization for Economic Cooperation and Development, 1989), pp. 366, 384, 204; International Energy Agency, *Energy Balances of OECD Countries 1970/1985* (Paris: Organization for Economic Cooperation and Development, 1987), p. 541.

arid or semi-arid areas that have little or no arable farmland.[62] Even though the surface mines will lead to significant land loss and degradation, such surface mines could be an improvement over underground mines currently in production. Although these mines produce badly needed coal with relatively little capital, they produce coal of poor quality at great risk to laborers and make inefficient use of the coal resource.

Coal in China is typically low sulfur (less than 2 percent) and high ash (more than 20 percent).[63] Low sulfur coal emits less SO_x, a component of acid rain. Ash is incombustible material (rocks, dirt, and other contaminants), which requires transportation and disposal but gives no energy. Due to the low mechanization rate and existence of many small, manual mines, Chinese coal has as much as 30 to 40 percent incombustible waste.[64] Simple coal cleaning to remove gross impurities could decrease transportation costs and allow for the use of more advanced coal burning technologies.

In India, over 80 percent of total coal output is produced by Coal India Limited—a government agency. The remainder is produced by both private and government-owned mines.[65] Mining in India has traditionally been similar to that in China; highly labor-intensive and underground. India has, however, made a shift to mechanized, open-cast mining and is beginning to move to longwall mining.[66] These methods of mining are feasible as much of India's coal is in thick seams near the surface. The sulfur and ash content of India's coal varies, but on average Indian coal is medium sulfur and high ash—rocks, dirt, and other incombustible material.

The expansion of longwall mining in India has been slow. In addition to unexpected geological difficulties, longwall mining in India may have lacked sufficient scale and concentration to provide adequate learning and adapting of the needed equipment to Indian conditions. It also appears that the transfer of this technology (under the British bilateral aid program) was ineffective as it did not include adequate incentives for the primarily British equipment manufacturers to participate and ensure success of the technology. Even more important in the poor performance of the Indian coal industry in general have been other management and engineering failures, including extensive political interference in managerial decisions, inadequate capital investment, and lack of wage incentives.[67]

Coal Cleaning Techniques

Coal cleaning refers to processing of the coal before it is burnt. Coal can be cleaned at the mine, at the power plant, or in-between, using a variety of methods. If a significant fraction of the overall weight is removed in cleaning, transport costs can be reduced. Environmental costs likewise can be lessened through cleaning; fly ash is emitted during combustion as particles, some of which contain harmful trace elements, sulfates, and nitrates.[68] Particulates contribute to haze and poor local air

---

[62]Ibid.

[63]IDEA, "Clean Coal Technologies for Developing Countries," contractor report prepared for the Office of Technology Assessment, May 1990, p. 8.

[64]Vaclav Smil, op. cit., footnote 60, p. 23.

[65]A. Desai, "Energy, Technology and Environment in India," contractor report prepared for the Office of Technology Assessment, 1990, p. 16.

[66]Ibid.

[67]Ibid.

[68]United Nations Department of Technical Cooperation in Development, op. cit., p. 49.

quality and they can be a health risk, causing pulmonary irritation and respiratory disease. Cleaning prior to combustion can reduce the amount of airborne particulates. Coal cleaning also can improve power plant availability,[69] as the use of dirty coal contributes to power plant breakdowns.[70] In one test, a 1 percent drop in ash due to coal cleaning led to a 2 percent increase in plant availability.[71] In another example in India, plant availability increased from 73 to 96 percent when coal with 40 percent ash was cleaned to 32 percent ash.[72]

Coal cleaning processes are, in order of increasing complexity and cost, physical, chemical, and biological. Physical cleaning is simply removing the noncoal particles, such as dirt, stones, pyritic sulfur, and other contaminants. Physical cleaning can remove about 60 percent of ash, as well as about 10 to 30 percent of sulfur.[73] Fly ash must be disposed of, usually by land application, though ash can be mixed into construction material. About 40 percent of coal used for electricity in the United States receives physical cleaning, most of which occurs at the minemouth.[74] In India, none of the coal used for electricity generation is cleaned, although there are plans to implement some washing of coal for electricity.[75] In China, about 18 percent of all coal is cleaned.[76] Costs for simple physical cleaning are estimated at about $6/ton, and by one estimate is cost-effective for the transport savings alone, whenever transport distances are greater than about 600 kilometers (km).[77]

Advanced cleaning processes include chemical and biological cleaning, both of which use various processes to remove organic sulfur and other contaminants. These processes may be able to remove up to 99 percent of ash and 90 percent of sulfur,[78] but their commercial availability and costs are unclear.

As the largest industrial producer of coal, the United States has developed extensive coal production and combustion technologies. Although many of the advanced technologies, such as chemical beneficiation, are too costly for developing countries, there may be a future market. DOE has a new program to explore these export opportunities in developing countries for American coal-based technologies, including clean coal technologies.

BIOMASS RESOURCES

Biomass provides about 14 percent of the world's energy and 35 percent of the total energy supply in developing countries. Overall, Africa obtains about two-thirds of its energy from biomass, Asia about a third or more, and Latin America about one-quarter. Some developing countries are almost wholly dependent on biomass for energy; Ethiopia, Nepal, and Tanzania, for example, rely on biomass for 90 percent or more of their energy supply. In rural and poor urban areas of most developing countries, biomass is often the only accessible and affordable source of energy. As discussed elsewhere, harvesting biomass for use by small industries and in urban areas plays an important role in the traditional rural economy, particularly in employment for the poorest.[79]

Demand for biomass will rise in the future. The rural and urban poor populations in developing countries are growing rapidly and are unlikely to make a quick transition to cleaner, higher quality fuels. Therefore, subsistence-level populations will require increasing biomass supplies to meet household and small industry needs.

Biomass, as it is currently used (see chs. 3 and 4), is often a dirty and low efficiency fuel. In the future, an increasing amount of biomass could be converted

---

[69] IDEA, op. cit., footnote 63, p. 34.

[70] Availability is the amount of time a power plant is operating divided by the amount of time the plant would operate if it worked perfectly.

[71] IDEA, op. cit., footnote 63, p. 34.

[72] IDEA, op. cit., footnote 63, p. 35.

[73] U.S. Department of Energy, Assistant Secretary for Fossil Energy, *Clean Coal Technology*, DOE/FE-0217P, revised January 1991, p. 13.

[74] Ibid.

[75] Tata Energy Research Institute, *Teri Energy Data Directory and Yearbook 1988* (New Delhi, India: Urhnak and Arvind, 1988) p. 41.

[76] Ministry of Energy, People's Republic of China, *Energy in China*, 1989, op. cit., footnote 60, p. 17.

[77] IDEA, op. cit., footnote 63, p. 36.

[78] U.S. Department of Energy, Assistant Secretary for Fossil Energy, op. cit., p. 14.

[79] U.S. Congress, Office of Technology Assessment, *Energy in Developing Countries*, OTA-E-486 (Washington, DC: U.S. Government Printing Office, January 1991).

to a cleaner, higher quality fuel using the processes described in chapter 6. Biomass could be burned directly for process heat and electricity, fermented or hydrolyzed to form alcohol fuels for transport, thermochemically gasified or anaerobically digested to produce gas for direct use or to generate electricity, or converted to charcoal for high temperature process heat for industries. These uses could increase commercial demand for biomass resources.

The potential biomass resource base is enormous. The biomass energy theoretically available worldwide from residues alone is equal to about 85 exajoules (81 quads), 25 exajoules (24 quads) from crop residues, and 30 exajoules (29 quads) each from forest residues and animal dung.[80] Dedicated energy crops could add significantly to the resource base.

There are a number of advantages to increasing the use of biomass. If produced on a sustainable basis, biomass would not add to greenhouse emissions and could improve local environments through reduced emissions of SO_x. Fuel switching to biomass could also reduce energy import dependence and stimulate rural development.

Increased use of biomass for energy also has potential disadvantages. Use of these resources will require careful consideration of environmental impacts, especially the risks of negative effects on soil and water availability and quality. Increased use of biomass for energy also raises social equity issues, including the possibility of growing fuel for the rich rather than food for the poor and exacerbating inequalities between landowners and landless.

Increasing biomass availability on a sustainable basis requires considerable effort. In light of all their needs, developing countries may not view the enhancement of biomass resources as a priority. Therefore, development of these natural resources cannot occur independent of the larger context of development.

Nonetheless, there are a number of opportunities for expanding the use of biomass resources in an environmentally and economically supportable fashion. In the near term, greater use of industrial wood wastes, some agro-processing wastes, and some agricultural residues are feasible options. In the mid term, careful management and use of natural forest resources could provide additional energy supplies. In the long term, there are opportunities for growing woody or herbaceous crops for energy production.

Agricultural/Industrial Residues

In the near term, agricultural and industrial residues[81] are the most promising bioenergy resource. The biomass residues from agricultural and industrial (e.g., forest products industry) processing can be used to provide power in the form of process heat or electricity. These resources already are used to a limited extent for such purposes in most industrialized and developing countries, though they usually could be used more extensively and efficiently (see also ch. 4 for the pulp and paper industry and ch. 6).

The availability of field wastes raises the issue of "determining when a waste is really a waste."[82] In developing countries, the potential availability of field residues varies widely. Worldwide estimates for annual residue production are 3.1 billion tonnes of food crop residues and 1.7 billion tonnes of animal dung,[83] but it is not known how much of this amount is actually available for use as fuel. In wood-scarce areas such as China, Pakistan, Bangladesh, and parts of India, agricultural residues are often already heavily used as cooking fuels in rural households. Agricultural residues also can be plowed in to fertilize fields. Decomposing crop and animal wastes add both nutrients and fibrous matter to the soil. As soil degradation and erosion are chronic problems in many parts of the world, collection of agricultural residues has to be managed so that fields are not deprived of these resources.

In many cases, however, there is room for multiple uses of these residues. The portion of crop residues recycled for fertilizer and soil stability may actually be small, due to the physical difficulty of replowing. Farmers using hand implements may burn or landfill residues, particularly dry residues such as coconut shells, rather than replowing them. If this is so, then using some of the residues for fuel

---

[80] D.O. Hall, "Biomass Energy," *Energy Policy*, October 1991, pp. 711-737.

[81] Forest residues will be discussed under the "forest management" heading.

[82] William Ramsay, *Bioenergy and Economic Development: Planning for Biomass Energy Programs in the Third World* (Boulder, CO: Westview Press, 1985), p. 65.

[83] G.W. Barnard, "Use of Agricultural Residues as Fuel," *Bioenergy and the Environment* (Boulder, CO: Westview Press, c. 1990), p. 87.

will not hurt the productivity of the soil.[84] However, availability of residues for combustible fuel will still be site specific. For example, in some places field residues may have other uses such as fiber or fodder. Residues might be used for thatching, or sold to small industries, such as paper manufacturers.[85]

Dung is important to agricultural productivity both as a fertilizer and in maintaining soil structure. Using dung for fuel (or occasionally as a building material) raises issues similar to those concerning field residues. Collecting dung from grazing animals is time-consuming and much of the dung may just be left where it falls. Moreover, dung that sits in the sun before being plowed under can lose as much as 80 percent of its nitrogen content, a key soil nutrient in dung.[86] In contrast to direct burning of dung for fuel, biogas digesters allow use of dung for fuel while still retaining (or even improving) its value as a fertilizer if applied to fields and plowed in.[87]

Densification of agricultural and forestry residues to briquettes or pellets increases the energy content per unit volume, reducing transport and handling costs. These reduced costs can help to open commercial markets for residues and therefore stimulate demand for these resources. For example, the cost of delivering briquettes to Addis Ababa in Ethiopia from a farm 300 km away would be about one-third the cost of delivering baled wheat straw. The processing costs can be considerable, however, limiting the use of densified fuels to rural and urban industries and middle to high income households in countries where other fuel prices are high.

In Ethiopia, for example, agricultural residues from small farms are currently used as a fuelwood substitute throughout the country. At least 3.3 million tons of surplus coffee, cotton, wheat, and maize residues are produced annually, including surpluses from State-owned lands. Although costs for densifying and transporting residues are site specific, costs for Ethiopia are summarized in table 7-4. The "ready-to-burn" costs at the market are equivalent to unprocessed crude oil prices of $15 to $20 per barrel. For industrial use, agricultural residue briquettes can be produced and delivered to users in Addis Ababa at a lower cost than most other industrial fuels.

In Sri Lanka, on the other hand, the delivered cost of coir dust (derived from coconut husks) briquettes is more than three times that of fuelwood per unit weight and the energy cost per unit of energy content is twice that of fuelwood. Sri Lanka has, however, large coir dust resources with an energy potential of 84 million GJ (80 trillion Btus).[88] Fuelwood resources are predicted to dwindle in Sri Lanka, so coir dust briquettes may become an economically viable alternative. In general, briquette and pellet costs will vary considerably according to the densification process, the scale of processing, and the original biomass feedstock.

Expanded use of agricultural residues and dung requires careful planning, whether for household use or when briquetted for commercial use. The availability of some agro-processing residues, however, is less constrained by competing uses and environmental concerns. Residues resulting from industrial processes, such as bagasse from sugar refining or sawdust from sawmills, present a "ready" resource.[89] Many developing countries already use these resources for energy, though often inefficiently. Improvements in system efficiencies could enable industries to sell to the grid some of the power they produce from residues.

In summary, the overall picture for increased use of agricultural and industrial residues is mixed. Field residues are a promising resource, especially when briquetted, but they must be carefully collected to protect agricultural and livestock productivity. Other competing uses of the field resources also must be taken into account. Waste-to-energy schemes for using residues are, however, an immediate opportunity for developing countries. In fact, many developing countries already take advantage of this econom-

---

[84]Ibid., p. 93.

[85]Ibid., p. 94.

[86]Ibid., p. 93.

[87]Many families in the developing world can not afford sufficient livestock to supply their own biogas needs, however. In order to meet cooking needs with biogas, a typical family would need three head of cattle. Few families in the developing world can afford to own or feed so many cattle. See ch. 6 for more detail. David Pimentel, personal communication, Apr. 23, 1991.

[88]V.R. Nanayakkara, *Wood Energy Systems for Rural and Other Industries* (Bangkok: Food and Agriculture Organization of the United Nations), p. 15 and p. 40.

[89]William Ramsay, op. cit., footnote 83, p. 66.

Table 7-4—Production Cost Estimates for Commercial Scale Crop Residue Briquetting in Ethiopia
(US$ (1983)/ton of product)

	Residue		
Stage of production	(1) Cotton stalks	(2) Corn and sorgum stover	(3) Wheat and barley straw
Harvesting	7.23	19.03	10.85
Capital charges	(4.22)	(10.40)	(2.39)
Energy and lube	(1.35)	(4.11)	(1.64)
Maintenance and other	(1.50)	(4.32)	(6.40)
Labor	(0.16)	(0.20)	(0.42)
Grinding	—	1.44	1.44
Briquetting	11.80	8.54	8.54
Capital charges	(5.56)	(2.37)	(2.37)
Energy and lube	(1.76)	(5.25)	(5.25)
Maintenance and other	(4.37)	(0.80)	(0.80)
Labor	(0.11)	(0.12)	(0.12)
Storage, etc.	1.0	0.88	0.88
Financial cost ex-plant	20.05	29.89	21.71
Economic cost ex-plant	25.02	32.15	27.35
Economic costs of transport and tagging, etc.	19.41	19.41	19.41
Bagging (40 kg sacks)	(3.38)	(3.38)	(3.38)
Transport[a]	(14.03)	(14.03)	(14.03)
Handling at each end	(2.01)	(2.01)	(2.01)
Economic cost delivered to market ("ready to burn")	44.43	51.56	46.76
Net heating value: MJ/kg	17.3	15.0	17.4
Moisture content: % (wb)	(12)	(15)	(15)
Economic cost per energy unit delivered to market: US$/GJ	2.57	3.44	2.69

[a]Transport: 22 ton trucks over 300 km of deteriorated paved roads to Addis Ababa.

SOURCE: K. Newcombe, "The Commercial Potential of Agricultural Residue Fuels: Case Studies on Cereals, Coffee, Cotton and Coconut Crops," World Bank Energy Department Paper No. 26 (Washington, DC: World Bank, June 1985).

ically viable resource, though in an inefficient manner in many instances. Overall, residues present both near and long term resource expansion possibilities.

Forest Management

Forests are an important resource for developing countries. Woodlands furnish fuelwood, timber, fiber, food products, fodder, pesticides, and medicine. Forest products are the main industrial base for a number of developing countries, and many subsistence populations depend on natural woodlands for their livelihood. Forests are also an integral part of the ecosystem, promoting healthy soils, maintaining diverse plant and animal life, regulating the flow of water, and controlling flooding and other potential hazards.[90] Moreover, tropical forests in developing countries are critical to global environmental quality; these forests store a significant share of the world's carbon stock.

Despite the importance of forests to developing-country economies and environments, many natural woodlands are being depleted faster than they can regenerate. Management techniques can slow the degradation of forests, however, and still allow for commercial and small-scale harvesting.

Commercial and subsistence wood harvesters need different forest management strategies. Generally, rural subsistence farmers cause relatively little damage to forests when they are collecting wood. These farmers tend to take dead wood or cut branches and leaves from trees, often lacking even the tools for cutting whole trees. In some cases, farmers may use hedges and other vegetation growing on their farms for fuel. A study of West Java found, for example, that three-fourths of all the fuel collected came from within family courtyards and gardens, and two-thirds of this fuel was branches and twigs.[91]

---

[90] U.S. Congress, Office of Technology Assessment, *Technologies to Sustain Tropical Forest Resources*, OTA-F-214 (Springfield, VA: National Technical Information Service, March 1984), p. 10.

[91] M. Hadi Soesastro, "Policy Analysis of Rural Household Energy Needs in West Java," *Rural Energy to Meet Development Needs: Asian Village Approaches*, M. Nurul Islam, Richard Morse, and M. Hadi Soesastro (eds.) (Boulder, CO: Westview Press, 1984), p. 114. See also, U.S. Congress, Office of Technology Assessment, *Energy in Developing Countries*, OTA-E-486 (Washington, DC: U.S. Government Printing Office, January 1991), p. 119.

In cases where farmers are cutting down whole trees, a number of techniques are available to minimize damage. "Thinning" (cutting trees selectively) may actually stimulate growth in the remaining forest as there is less competition for nutrients, light, and water. "Pollarding" (cutting off branches and twigs, rather than felling whole trees) can also stimulate forest growth by reducing competition for nutrients, water, and light.

In contrast to the rural foragers, commercialized fuelwood and charcoal harvesters routinely fell whole trees, damaging or even clear-cutting entire forests. Even so, there are techniques, such as thinning or cutting in strips rather than clear-cutting a whole area, to minimize damage. After harvesting, forest lands can be reseeded or replanted. Adequate maintenance or upgrades of cutting and harvesting equipment also minimizes damage to trees and soil surrounding the logging site.[92] Better charcoal conversion processes could decrease the number of trees cut. In the long run, though, commercial logging for fuel may sometimes be demand-constrained as consumers switch to other fuels when wood becomes scarce and prices climb.[93] There is no guarantee, though, that this will occur early enough to prevent serious damage to forest resources in some areas.

Putting forest management techniques into practice can be difficult. Intervention by governments or nongovernmental organizations often is required. Governments can encourage forestry techniques through taxes on wood and charcoal, tax incentives, laws, and enforcement of laws.[94] Governments and nongovernmental organizations can teach forest management techniques through extension staff and forestry projects. These measures have proven difficult to instill in the past. Taxes or laws requiring certain forestry practices tend to be difficult and costly to enforce and forestry projects often fail when overseas funds dry up.[95] Management techniques may have limited success in conserving

Photo credit: Jennifer Cohen

In many developing countries, land may be cleared to make way for agriculture and livestock.

forests because forests are usually cleared to make room for migrants, agriculture, and livestock,[96] rather than by fuelwood gatherers.

Energy Crops

Some developing countries already successfully grow crops dedicated to energy production. Energy crops can be divided into two broad categories; herbaceous field crops and woody biomass plantations. These crops can be planted as separate lots or fields devoted to energy production or as stands "intercropped" between other crops and around residential areas. Five key variables govern the viability of woody and field energy crops: technical feasibility; availability of suitable land; economic viability; implementation; and environmental impacts.[97]

Technical Feasibility

Research and development on plant species and methods of planting have greatly enhanced the technical feasibility of energy cropping. Energy crops are highly site specific, however; species that respond well to test conditions may be vulnerable to

---

[92] U.S. Congress, Office of Technology Assessment, *Changing By Degrees: Steps to Reduce Greenhouse Gases*, OTA-O-482 (Washington, D.C.: U.S. Government Printing Office, February 1991).

[93] U.S. Congress, Office of Technology Assessment, *Energy in Developing Countries*, op. cit., footnote 79, p. 119.

[94] Keith Openshaw and Charles Feinstein, World Bank, Industry and Energy Department, "Fuelwood Stumpage: Considerations for Developing Country Energy Planning," Industry and Energy Department Working Paper, Energy Series Paper No. 16 (Washington, DC: World Bank, June 1989).

[95] Paul Kerkhoff, *Agroforestry in Africa: A Survey of Project Experience* (London: Panos Publications, Ltd., 1990).

[96] For more detail on deforestation, see U.S. Congress, Office of Technology Assessment, *Energy in Developing Countries*, op. cit., footnote 79, ch. 5.

[97] David Hall, "Biomass, Bioenergy and Agriculture in Europe," *Biologue*, vol. 6, No. 4, September/October 1989.

actual site conditions. Flexibility of a given species in different environments is one of the most important technical characteristics for energy crops in developing countries.[98]

For woody biomass plantations, other desirable characteristics include fast growth; high density (high heat value per unit of volume); robustness (ability to withstand weather, pests, and disease); nitrogen fixing capability (a trait that reduces the need for fertilizer); and good potential for coppicing (ability to regrow from stumps). Since 1978, the U.S. DOE has supported research and development of Short Rotation Intensive Culture (SRIC) woody energy crops that incorporate most of these features. SRIC plantations consist of a fast-growing single tree[99] species (a "monoculture"), planted in carefully spaced patterns, harvested in 3 to 10 year cycles, and intensively managed.[100]

DOE and other researchers have tested a wide variety of species and found them to be appropriate for SRIC plantations.[101] Although some species native to the region in question can be used, especially in the tropics, exotic species with particularly favorable characteristics are often brought in. Especially versatile species include acacia, grevillea, eucalyptus, calliandra calothyrus, populus (poplar), salix (willow), sesbania, and leucaena.[102] Leucaena, native to Central America and Hawaii, is especially popular for energy crops because it is drought tolerant, coppices readily, has the highest measured yields of any tree species, has a very high density, and is a good source of wood and foliage for fuel, fodder, construction, or pulp.[103] Even leucaena does not universally grow well, however. A project supported by the Cooperative for American Relief Everywhere to grow leucaena leucocephala, calliandra calothyrus, and sesbania sesban in Ethiopia:

. . . showed abysmal growth. It was a painful lesson in the risks of putting too much trust in the textbook. One project worker said, "I was devastated when leucaena didn't work. We inoculated it, we did everything to make it grow, and it was like we were growing a dwarf variety."[104]

Even the most versatile, fast-growing, and resilient species may not grow well in any given environment. Monocultured short-rotation forests are susceptible to a variety of hostile environments. Poor soil characteristics, harsh microclimates, pests, fires, weeds, and diseases may all devastate or stunt plantation growth. There is some evidence that monocultured plots are more susceptible than the more diverse natural stands, in which some trees may have characteristics that ward off disease or fix nitrogen in the soil, protecting the surrounding trees. In many cases, monocultured trees perform better in trials because they are nurtured under very controlled situations on small plots, conditions that are hard to replicate in the field.[105] Other technical problems resulting from monoculture SRIC plantations include possible exhaustion of the land and loss of biodiversity.[106]

Bioengineering has the potential to overcome some of these problems, such as vulnerability to pests and environmental stress. Desirable characteristics, such as nitrogen fixation and fast growth, also may be augmented through bioengineering.[107] Bioengineering technologies that have proven successful in some cases are cloning and hybridization.[108] For example, the U.S. DOE has produced hybrid black cottonwoods that have yields that exceed those of the parent stock by a factor of 1.5 to 2.[109] Genetic engineering of trees is a relatively new field, however. In comparison, agricultural biotechnology is much further advanced. Technology transfer from agricul-

---

[98] W. Ramsay, op. cit., footnote 82, p. 58.

[99] In this section, references to trees include shrubs.

[100] U.S. Congress, Office of Technology Assessment, op. cit., footnote 79, p. 214.

[101] W. Ramsay, op. cit., footnote 82, pp. 71-87.

[102] I. Stjernquist, "Modern Wood Fuels," *Bioenergy and the Environment* (Boulder, CO: Westview Press, c. 1990), pp. 61-65.

[103] Ramsay, op. cit., footnote 82, p. 83.

[104] P. Kerkhof, op. cit., footnote 95, p. 83.

[105] Edward A. Hansen, "SRIC Yields: A Look to the Future," unpublished manuscript, 1990, p. 3.

[106] U.S. Congress, Office of Technology Assessment, op. cit., footnote 92, p. 223.

[107] Ramsay, op. cit., footnote 82, p. 88.

[108] Edwin H. White, Lawrence P. Abrahamson et. al., "Bioenergy Plantations in Northeastern North America," paper presented at Energy from Biomass and Wastes XV in Washington, DC, Mar. 25, 1991, p. 10.

[109] Philip A. Abelson, "Improved Yields of Biomass," *Science*, vol. 252, No. 5012, June 14, 1991, p. 1469.

ture will speed SRIC genetic engineering, but only to a point; trees and shrubs have unique characteristics, including a long breeding cycle.[110] Nonetheless, the potential to increase yields through biotechnology is enormous—according to one researcher, even more significant than the successes already achieved in agricultural genetic engineering.[111]

Field crops dedicated to energy production in the developing world have focussed in part on starches or sugars that can be fermented to alcohol fuels. For example, Brazil grows a large amount of sugarcane exclusively for production of ethanol. In temperate climates, other crops that can be grown for energy include cereals, pasture plants, cassava, sugar beet, sweet sorghum, potatoes, oilseed crops, Jerusalem artichoke, and Japanese knotwood. Crops with high oil contents can be grown for conversion to vegetable oils, and some experimentation has occurred with hydrocarbon-producing species, such as euphorbia lathyrus, in arid or semi-arid regions.[112] "Energy grasses," often indigenous grasses that will grow on marginal land, are also feasible. These grasses generally have high nitrogen requirements, however, and would need to be "intercropped" between nitrogen-fixing trees or crops.[113]

A technical barrier for developing countries is that much of the research and development on energy crops does not reflect developing-country needs and conditions. For example, there is little research and development on tree species indigenous to developing countries.[114] In practice, determining the viability of indigenous species for energy crops may be a trial and error process. More demonstration and in-the-field testing of SRIC and engineered species would increase their effective deployment in developing countries. Even though more research and development could certainly enhance energy cropping, however, current knowledge suggests that prospects for high-yield SRIC plantations and field crops in developing countries are encouraging.

Availability of Suitable Land

To be a reliable source of energy, biomass plantations would require significant amounts of land, especially as biomass conversion is energy intensive. For example, with short-rotation intensive culture trees, about 8 percent of the energy provided would have to be used to plant, harvest, and dry the trees. About one-third of the energy then would be lost if the wood were converted to a liquid or gaseous fuel. Conversion to other forms, such as charcoal, involve even higher losses (the majority of charcoal kilns in Minas Gerais, Brazil have conversion efficiencies of about 50 percent, for example).[115] To break even economically, plantations would have to be large and may require fertilizer and other inputs to achieve a sufficiently high level of productivity.

In many countries, plantations of this size and productivity would have to compete with agriculture for the same land resources and inputs. According to the United Nations Food and Agriculture Organization (FAO), 46 percent of the developing countries surveyed already lack sufficient land resources to support their populations at low levels of agricultural inputs.[116] With projected population growth and the contined low agricultural inputs, 55 percent of the countries would lack sufficient land resources to feed their populations by 2000.[117] According to the FAO's analysis, Central and South America are the regions most likely to be able to support energy crops. Many of the most populous African countries, however, lack sufficient land resources to sustain their populations, let alone energy crops given current low levels of agricultural inputs. For the countries with insufficient land resources for energy cropping, small scale agroforestry projects, discussed below, may still be viable.

For countries with adequate land resources, a growing market for biomass resources could have national and local benefits. Developing countries could save foreign exchange dollars now spent on oil

---

[110]Edward A. Hansen, op. cit., footnote 105, p. 12.

[111]Ibid.

[112]I. Stjernquist, "Modern Wood Fuels," *Bioenergy and the Environment* (Boulder, CO: Westview Press, 1990), p. 70.

[113]Ramsay, op. cit., footnote 82, pp. 59-60.

[114]Cynthia C. Cook and Mikael Grut, *Agroforestry in Sub-Saharan Africa: A Farmer's Perspective* (Washington, DC: The World Bank, 1989), p. 37.

[115]J. Warren Ranney, Lynn L. Wright, et al., "Hardwood Energy Crops: The Technology of Intensive Culture," *Journal of Forestry*, vol. 85, pp. 17-28.

[116]United Nations Food and Agriculture Organization, *Potential Population Supporting Capacities of Lands in the Developing World*, Technical Report of Project INT/75/P13 (Rome: United Nations Food and Agriculture Organization, 1982), Table 3.4, pp. 134-136.

[117]Ibid.

imports and reinvest the money within the country. Local economies could benefit by growing biomass as a cash crop, stimulating rural development. There are, however, risks. New technologies could lead to an increase in deforestation if they rely solely on existing biomass resources.

Economic Viability

The expenses involved in establishing energy crops include land opportunity costs, the cost of inputs, and the costs of maintaining, harvesting, and transporting the crops. For energy crops to be viable, the price at which the fuel can be sold must exceed the sum of these costs. If this is not the case, energy crops may still be viable if other benefits justify the costs, or if there is an external source of financial support, such as an international aid organization.

In rural areas, with woodlots meant for local fuel uses, trees may be planted on farms without many inputs and with almost no mechanization, particularly where there is subsistence farming. Transportation costs may also be irrelevant as many of the products (poles, fodder, fuelwood) will be used locally, rather than transported and sold in markets. In these cases, economic viability may be determined by number of hours spent rather than monetary gain. The World Bank estimates that an attractive time savings to farmers would involve a rate of return greater than 30 percent.[118]

Large scale woody plantations have mixed economic results. Many plantations, particularly community woodlots, fail economically for a number of reasons. The returns on planting trees for energy are inherently long term; the time between planting and harvesting is 3 to 10 years. Farmers, community members, and other investors are generally reluctant to make the investment due to this long lead time, particularly with current low prices for other forms of energy, including gathered wood.[119] On the other hand, tree plantations with multiple end-uses—trees that provide fruit or timber—may well be economically viable, as is the case in Gujarat, India.[120]

Another plantation strategy could be large industries growing trees specifically to support industrial processes. There are examples of such industrially dedicated plantations. A study of Sri Lankan industries using wood for fuel concluded that 49 percent of the total fuelwood consumed came from manmade rubber plantations. Another 2 percent came from other forest plantations.[121]

Markets are the key to economic viability, and yet, according to one analyst, "it is probably a safe guess . . . that more renewable energy projects have come to grief through omitting a thorough investigation of potential markets than through any other cause."[122] Generalizing about markets for energy crops is extremely difficult, however. There are scant data available and there are significant differences between commercial and noncommercial markets, rural and urban markets, and markets in different regions and nations.

In rural areas, "non timber products tend to be marketed locally and in a decentralized manner, [so] their value is generally hard to recognize and assess."[123] Wood is gathered from the forests for "free," so developing a market for marginal small scale industrial and residential users would be difficult. In some rural areas with scarce wood resources, however, there is a small market for wood, and in such areas, wood is an important source of income. Even with scarcity, local subsistence populations are likely to substitute other forms of energy—such as crop residues—rather than pay cash for wood fuel. Some communities have been encouraged to pursue social forestry, or community woodlots, but returns often are insufficient to attract investment—either time or money.[124]

Markets also can be created, as is the case with field energy crops for ethanol fuels in Brazil (see ch. 6). Through a mixture of price controls, subsidies, and tax breaks, the Brazilian government has been

---

[118]C. Cook and M. Grut, op. cit., footnote 114, p. 13.

[119]Ramsay, op. cit., footnote 82, p. 167.

[120]Ibid.

[121]V.R. Nanayakkara, *Wood Energy Systems for Rural and Other Industries: Sri Lanka* (Bangkok: United Nations Food and Agriculture Organization), pp. 20-21.

[122]Ramsay, op. cit., footnote 82, p. 161.

[123]U.S. Congress, Office of Technology Assessment, op. cit., footnote 92, p. 221.

[124]Samuel F. Baldwin, *Bioenergizing Rural Development*, unpublished manuscript, February 1988, p. 1.

able to create and maintain a market for fuel ethanol.¹²⁵ Few developing countries can afford to run programs on this scale, though. Other smaller scale market interventions are possible, however. For example, a tobacco company in Kenya in effect creates a market for woodfuels by placing the farmers under a contractual obligation to plant trees. Farmers must demonstrate that they grow a certain number of trees before the tobacco company will purchase tobacco from those farmers.¹²⁶

Urban areas of developing countries often depend on a commercialized fuelwood and charcoal market. Many urban residents still rely on wood and charcoal for fuel. As fuelwood generally is not located close enough to cities for collection, supplies are purchased. Often, these supplies originate in distant areas. A study of fuelwood use in two Indian cities revealed that in one Indian city, Bangalore, one-half of the fuelwood consumed was transported from between 120 and 300 km away and 17 percent from as far as 700 km from the city borders. In another Indian city, Hyderabad, about 55 percent of all fuelwood supplies were brought in from forests over 100 km away.¹²⁷ Transportation infrastructure is therefore an important precursor for an urban fuelwood market. In Thailand, for example, transportation systems are fairly well developed and a large segment of the urban population derives its energy from tropical forests located far from the cities.¹²⁸

The introduction of the technologies discussed in chapter 6 could improve the economic viability of both field and woody energy crops, by making biomass resources competitive with fossil fuels in the provision of such desirable energy carriers as gas and electricity. As these biomass conversion technologies become available, they may well create a market for biomass resources.

Implementation

There are two different types of energy crop planting schemes; dedicated energy crops and plants or trees intercropped between agricultural crops. In the former case, crops or trees are planted exclusively for energy production. In the latter case, also referred to above as "agroforestry," the provision of fuel is a secondary reason for planting the trees or crops. Implementation issues differ for these two scenarios.

In the past, tree-planting schemes (both for large scale plantations and for smaller scale agroforestry) in developing countries sometimes have been introduced without adequate attention to local needs and local environments. As a result, these schemes have needed extensive infrastructures to initiate and maintain and, with some exceptions, have resulted in widespread failures.¹²⁹

Experience suggests that dedicated energy crops and agroforestry have a better chance of success if the local farmers are intrinsic to the process. Tree schemes that cost the farmers time, money, and land without any clear benefits require extensive, costly, and long-term intercessions from governmental or nongovernmental organization extension staff. Even with incentives and proper training, farmers may not be attracted to energy cropping.

Given the land availability and economic viability problems discussed above, farmers will weigh the advantages of biomass against the other possible uses for the limited land available. Energy projects designed by outside experts often fail because they do not respond to local priorities.¹³⁰ Sometimes, afforestation strategies are based on wrong assumptions. For example, it has been assumed that local people would use wood resources until the resource was completely, irreversibly depleted. In many

---

¹²⁵Marcia Gowen, "Biofuel v. Fossil Fuel Economics in Developing Countries," *Energy Policy*, vol. 17, No. 5, October 1989, p. 460.

¹²⁶P. Kerkhof, op. cit., footnote 95, p. 63. Although this program has resulted in successful tree growth, the farmers continue to cut the natural growth for wood, preferring to use the cultivated trees for poles.

¹²⁷Manzoor Alam, Joy Dunkerley, Amulya Reddy, "Fuelwood Use in the Cities of the Developing World: Two Case Studies from India," Natural Resources Forum, United Nations, 1985, p. 207.

¹²⁸Norman Myers, *The Primary Source: Tropical Forests and our Future* (New York: W.W. Norton and Co., 1984), p. 119.

¹²⁹P. Kerkhof, op. cit., footnote 95; C. Cook, and M. Grut, op. cit., footnote 114.

¹³⁰Steven Meyers and Gerald Leach, U.S. Department of Energy, Oak Ridge National Laboratory, Office of Scientific and Technical Information, *Biomass Fuels in the Developing Countries: An Overview* (Oak Ridge, TN: U.S. Department of Energy, 1989).

developing countries, however, local populations respond of their own accord to woodfuel shortages.[131] In these cases, building on the foundation already provided by local practices will speed and ease project implementation. Increasingly, developing-country governments and foreign donors are realizing that local knowledge and aspirations must be included if these projects are to be successful and that the problem of biomass energy must be considered in the broader context of development.

Environmental Impacts

With sound management, environmental impacts of energy cropping can be minimal or even favorable.[132] Environmental problems associated with energy crops are very similar to those associated with agricultural crops, with the degradation of soils being a key risk. Although specific problems and benefits vary, the general environmental benefits and costs are valid not only for the developing countries, but also for the industrial nations.

If grown sustainably, biomass energy does not make a net contribution to the buildup in atmospheric CO_2 from anthropogenic sources. When trees are harvested, a certain amount of stored carbon is released. If vegetation with comparable carbon storage is replanted, carbon emissions resulting from the decomposition, combustion, or harvest of the original trees may be offset.[133]

Large-scale woody plantations or energy field crops can further offset carbon emissions if they replace fossil fuels.[134] Biomass also contains little sulfur, so combustion produces less sulfur oxides than are emitted during fossil fuel (especially coal) combustion.[135] Sulfur oxides are a component of local air pollution and acid rain, and a serious problem in many cities in developing countries[136] (see figure 7-1).

Negative local environmental impacts of biomass depend on the crop grown, the site characteristics, and the circumstances. Impacts can range from "minor changes, easily accommodated within the existing ecosystem," to "major changes drastically altering the site or its surroundings."[137] Effects of energy cropping can include soil erosion, increased water runoff, loss of soil nutrients and organic matter, negative effects on hydrologic systems, and reduction in biological diversity as fast-growing monocultures are substituted for indigenous vegetation.[138]

Although these effects are site and crop specific, changes in the soil are very likely to occur. Soil is composed of mineral matter and organic matter, largely from decaying trees or animals. Removing the original trees will therefore change the composition of the soil by changing the organic component. In most forest ecosystems, the decaying vegetation contributes organic matter and helps stabilize soils. Soil is extremely slow to reform; forests are estimated to take 1,000 years or more to reform 2.5 centimeters of soil.[139] Clearing a site for an energy crop and then leaving the soil exposed before or during planting can result in significant soil erosion. Removing forest debris and interrelated ecosystems also can lead to erosion. The soil and water runoff can hurt local hydrological systems by increasing the levels of sedimentation in the water. Runoff polluted by herbicides, pesticides, and fertilizers can exacerbate the damage to aquatic ecosystems and foul drinking water.

Soil conservation techniques, such as terrace and contour planting, can moderate or mitigate these

---

[131] U.S. Congress, Office of Technology Assessment, op. cit., footnote 79, p. 119.

[132] James Pasztor and Lars A. Kristoferson, "Bioenergy and the Environment—the Challenge," *Bioenergy and the Environment* (Boulder, CO: Westview Press, 1990), p. 28.

[133] For more on the carbon cycle and the variables that effect carbon releases by trees, see U.S. Congress, Office of Technology Assessment, *Changing By Degrees: Steps to Reduce Greenhouse Gases*, op. cit., footnote 92, ch. 7.

[134] D.O. Hall, H.E. Mynick, and R.H. Williams, "Cooling the Greenhouse with Biomass Energy," *Nature*, vol. 353, No. 6339, Sept. 5, 1991, pp. 11-12; D.O. Hall, H.E. Mynick, and R.H. Williams, "Alternative Roles for Biomass in Coping with Greenhouse Warming," *Science and Global Security*, vol. 2, 1991, pp. 1-39.

[135] Samuel F. Baldwin, op. cit., footnote 126, p. 2.

[136] For more information on pollution from biomass combustion, see chs. 2 and 3 and U.S. Congress, Office of Technology Assessment, op. cit., footnote 79.

[137] William Ramsay, op. cit., footnote 82, p. 91.

[138] David Pimentel, personal communication, Apr. 23, 1991.

[139] David Pimentel, Alan F. Warneke, et al, "Food Versus Biomass Fuel: Socioeconomic and Environmental Impacts in the United States, Brazil, India, and Kenya," *Advances in Food Research* (New York, NY: Academic Press, Inc, 1988), p. 216.

Figure 7-1—Sulfur Dioxide Levels in Selected Cities, 1980-84

Shown is the range of annual values at individual sites and the composite 5-year average for the city.

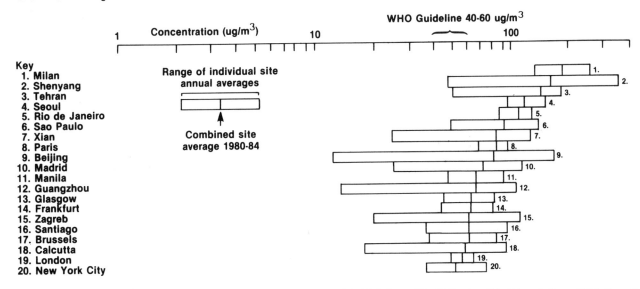

SOURCES: World Health Organization and United Nations Environment Fund, *Global Pollution and Health* (London: Yale University Press, 1987), figure 2.

effects. However, "because of their high labor or capital costs, many of these techniques are only of interest to large-scale biomass energy production schemes."[140] For small scale biomass planting, agroforestry can be an alternative. Soil has been estimated to erode 12 to 150 times faster than it is formed world wide, with the bulk of losses due to agricultural crops.[141] Trees, bushes, and grasses planted among crops and around houses can stabilize soils, slow erosion rates, and provide fuel. Intercropped vegetation also can enhance the organic makeup of the soil since some tree and bush species contribute or "fix" important soil nutrients, such as nitrogen.

Energy cropping and agroforestry also can alter or destroy ecosystems. To make plantations economically viable, fast growing crops generally have to be planted, particularly if biomass is to be a fossil fuel substitute. Monoculture short-rotation intensive culture plantations usually best meet this need. These monoculture plantations often replace diverse tree and plant species, however, which were a habitat for diverse animal and insect populations. Not only will the various flora be lost, most of the fauna may not be able to survive in the new habitat.[142] Planting strategies exist that maintain biological diversity, such as multiple species cropping or retaining patches of old growth within the plantation. These strategies are likely to mean lower crop yields, though.[143] Although agroforestry also may rely on exotics, there is better potential to maintain the indigenous species. Since the trees usually are planted for multiple purposes, fast growth is not essential to economic viability.

Energy cropping must be carefully planned and implemented in order to minimize negative environmental impacts. With such careful planning, crops and agroforestry have the potential to benefit local and global environmental quality by stabilizing soils, adding soil nutrients such as nitrogen, and acting as a net carbon sink.

---

[140] D.R. Newman and D.O. Hall, "Land-Use Impacts," *Bioenergy and the Environment* (Boulder, CO: Westview Press, 1990), p. 242.

[141] D. Pimentel, op. cit., footnote 141, p. 216.

[142] J.H. Cook and J. Beyea, "Preserving Biological Diversity in the Face of Large-Scale Demands for Biofuels," paper presented at the Energy from Biomass and Wastes Conference XV, Washington, DC, Mar. 25, 1991, p. 3.

[143] Ibid., p. 5.

Conclusions

Biomass resources are extremely important to developing countries. Urban and rural subsistence populations rely on biomass for fuel and are likely to continue relying on biomass in the future. Small and large scale industries already use biomass and could use it more in the future. The present resource base may not be able to support this increasing demand from a growing population.

There are a number of strategies for enhancing the biomass resource base. In the near term, use of processing wastes is the most promising opportunity. United States industries have extensive experience in this area and could provide assistance and exports to developing-country industries. In the future, briquetting field residues, managing existing forest resources better, and growing crops dedicated to energy hold out the possibility of resource expansion. The United States already encourages developing countries to pursue these strategies through the Agency for International Development, the Department of Energy, the Peace Corps, and other bilateral aid programs. All of these strategies must be planned, however, with the larger developmental context in mind. Larger issues of deforestation, land ownership, rural development, and food production need to be considered in allocating project funds.

Chapter 8
Issues and Options

Photo credit: U.S. Agency for International Development

Contents

	Page
INTRODUCTION AND SUMMARY	261
EXISTING U.S. PROGRAMS AND POLICIES FOR ENERGY TECHNOLOGY TRANSFER AND DIFFUSION	263
The U.S. Policy Framework	263
AID Programs	264
Multilateral Bank Activities	265
IMPROVING TECHNOLOGY TRANSFER AND DIFFUSION	269
Technology Transfer Support	269
Improving the Institutional Framework	274
Providing Economic Incentives	276
Financing Issues	276
REDEFINING PRIORITIES	279
Enhancing U.S. Trade and Investment Opportunities	279
Reducing Poverty and Improving Social Welfare	283
Protecting the Environment	284
Setting a Good Example	285

Boxes

Box	Page
8-A. AID Office of Energy and Infrastructure	266
8-B. Multilateral Development Banks	267
8-C. The Global Environmental Facility (GEF)	269
8-D. The Department of Energy	272
8-E. Programs for U.S. Trade and Investment	281
8-F. Examples of Import and Other Regimes Affecting the Transfer of Technology in Selected Developing Country Markets	282
8-G. Tied Aid	284

Chapter 8
Issues and Options

INTRODUCTION AND SUMMARY

Developing countries need energy services to raise productivity and improve their standard of living. But the traditional way of meeting these energy needs, through increasing energy supplies with little attention to the efficiency of energy use, raises serious financial, institutional, and environmental problems. The magnitude of these problems underlines the need for improving the efficiency with which energy is currently used and produced in developing countries.

This report has shown the potential for improving energy efficiency through the adoption of proven cost-effective technologies. For a wide range of electricity using services, energy savings of nearly 50 percent and life cycle cost savings to consumers of nearly 30 percent are possible with current or near-commercial energy technologies. Energy efficiencies could be further improved by technologies expected to be commercial in the near to midterm.

While the energy saving advantages of these technologies are usually recognized, the common perception is that their widespread adoption will not occur because of their high initial cost, an important consideration for poor, heavily indebted countries. This study shows, however, that when the capital requirements of both supply and end use technologies are combined—on a systemwide basis—higher efficiency technologies reduce overall capital costs. The higher capital costs of the energy efficient end use equipment are more than offset by the lower investment required for electricity generation.

Even with major end use efficiency gains, the rising demand for energy services means that increases in supply will also be required. Improved supply technologies offer opportunities to augment the efficiency of energy production and distribution, while moderating environmental impacts. Recent cost reductions in renewable technologies (e.g., wind turbines, photovoltaics, and biomass conversion) will encourage their increased use, and also help improve rural energy supplies.

The rapid adoption of these technologies is, however, being held back by a variety of technical, institutional, and economic and financial barriers that occur throughout the entire technology transfer and diffusion process.

- **Technical.** Many improved technologies, though well established in industrial countries, may not be well adapted to conditions in developing countries. People in developing countries may not be aware of new technologies or have access to the training necessary to make effective use of new technologies
- **Institutional.** Both public and private developers are organized to fund large scale conventional energy supply expansion rather than demand side energy efficiency projects. Rules and practices in the critically important electric power sector often do not give efficiency and renewable energy equal weight with conventional large scale supply options in providing energy services.
- **Economic and Financial.** Energy prices are frequently subsidized in developing countries and therefore provide neither the economic incentives for energy efficient equipment, nor adequate revenues for system expansion. Particularly in poor countries, consumers may not have access to the capital needed for the higher initial costs of energy efficient equipment (even though these technologies reduce costs to the user over the product's lifetime, and lower overall capital costs for the nation).

The United States has strong reasons for taking a leadership role in policy changes to overcome these barriers to rapid diffusion of improved energy technologies.

International political stability depends on steady broad-based economic growth in the developing countries, which in turn requires economic and reliable energy services. The developing countries are of growing importance in global energy markets and global environmental issues. Because of their rapid population increase, the developing countries are projected to account for over one half of the increase in both global energy consumption and associated CO_2 emissions over the coming decades, despite their low levels of per capita consumption. Sharply rising demand for oil from the developing countries contributes to upward pressure on international oil prices. Developing-country debt, often energy related, affects the stability of U.S. and international banking systems. At the same time, developing countries offer the United States major trade opportunities in their large and expanding market for energy technologies.

This chapter explores ways in which the United States could contribute to more rapid technology diffusion in developing countries, the policy mechanisms that could be used, and their relative costs and benefits. A large number of U.S. programs already influence or have the potential to influence technology adoption in developing countries, and many are undergoing change—often at the instigation of Congress. This chapter describes those programs that appear, on the basis of our analysis of developing-country needs and U.S. policy interests, to provide the greatest opportunities for effective action. These areas establish a framework for more detailed program planning and analysis.

This is a timely moment for reviewing U.S. policies towards the adoption of improved energy technology in the developing countries. These countries are themselves demonstrating interest in seeking alternative ways of meeting the demand for energy services, despite the difficulties inherent in changing entrenched systems. Increased attention is being given to politically sensitive questions of energy price reform, improved management, and operations efficiency in State-owned energy supply industries. Several developing countries are taking steps to encourage private investment in electricity, and in oil and gas exploration and development. Many developing countries now have capable energy research and policy institutions. Progress is also being made on the environmental front. At present, efforts are largely directed at local rather than global conditions, though developing countries have also participated in international environmental protection treaties. There is also evidence of change in donor institutions. The bilateral donor agencies and the multilateral development banks (MDBs) (of which the most influential is the World Bank) are beginning, often under pressure from Congress and nongovernmental organizations (NGO), to incorporate environmental planning into their projects, develop energy conservation projects, and encourage a larger role for the private sector. While many problems remain, this momentum for change offers a good opportunity for U.S. initiatives.

An analysis of possible U.S. actions to augment the process of improved technology adoption and diffusion provides a long list of policy options. The following merit priority consideration:

- Additional attention to energy efficiency, and the environmental impacts of energy developments in current bilateral and multilateral programs.
- Encouragement of energy price reform in developing countries, including assistance in adjusting to the impacts of higher prices. Rational energy pricing both encourages the adoption of energy efficient equipment and helps finance needed supply expansion. The United States has a number of policy levers (through bilateral and multilateral aid agencies, and in debt negotiations) to achieve this objective.
- Promotion of Integrated Resource Planning (IRP) and associated regulatory reform to provide incentives for investments in energy efficiency as part of electricity projects. The United States has experience with this approach in the electricity sector on which to base technical assistance to developing countries.
- Provision of technical assistance in a variety of other areas of potential large impact. Examples include environmental protection (an area of growing concern in developing countries), appliance efficiency, utility management, and transportation planning.
- Encouragement of private sector participation (including the private sector of both the United States and the developing countries) in energy development. The United States has policy levers through membership in MDBs and programs for export and investment support. The developing countries are indicating interest in further private sector development.
- Help in building institutions in developing countries, especially for technology development, adaptation, and testing.
- Expansion of U.S. trade and investment programs, recognizing that U.S. energy related exports and investment are an important channel for energy technology transfer to developing countries.
- Setting a good example to the rest of the world in energy efficiency and environmental protection. This provides credibility to U.S. policy advice to others.

Congress has already taken action in several of these areas by promoting IRP, efficient energy pricing, and consideration of environmental impacts of projects. The implementation of these Congressional directives may need to be monitored.

Since efficient energy technologies often reduce systemwide capital investment, redirecting capital funds from supply expansion to energy efficiency projects could free resources for additional investment in energy services. Even so, the rapid rise in demand for energy services may require more investment than that projected to be available. High levels of support will continue to be needed from the MDBs and other bilateral and multilateral financial institutions. At the same time supporting actions (e.g., debt negotiations, macroeconomic reform, and privatization) will be required to encourage increased private sector participation.

Several of the options for accelerating the adoption of energy efficient technology imply an increase in U.S. bilateral assistance. While increases in bilateral aid run counter to efforts to control budget expenditures, the share of bilateral aid (particularly in the U.S. Agency for International Development (AID)) directed to energy is at present low in relation to: 1) total bilateral assistance; 2) the share of energy in the aid efforts of other donors; 3) the potential importance of developing countries as markets for U.S. exports of improved energy technologies; and 4) the contribution of developing country energy use to global warming. The current geographical distribution of AID energy expenditures is concentrated in the Near East. This distribution may not adequately reflect the totality of U.S. policy interests.

EXISTING U.S. PROGRAMS AND POLICIES FOR ENERGY TECHNOLOGY TRANSFER AND DIFFUSION

Technology transfer can be influenced through a variety of channels, both public and private. (See table 8-1.) These channels are:

- capital projects financed by development agencies,
- intergovernmental agreements on research and development,
- foreign investment and direct export sales, and
- training and technical assistance carried out by both the private sector and the government.

The U.S. Policy Framework

A substantial number of U.S. agencies already have programs and activities for energy technology transfer. These include: AID; the Trade and Development Program (TDP); the Departments of Energy, Commerce, and the Treasury; the Overseas Private Investment Corporation (OPIC); the Export-Import Bank (Eximbank); the Small Business Administration; and the U.S. Trade Representative (USTR). Through membership in international organizations, notably the multilateral development banks (MDBs) and United Nations (UN) programs, the United States exercises additional influence. Further, a number of industry groups and NGOs are also active in this field.

These agencies and organizations cover a wide range of technology transfer and diffusion activities: research, development, and demonstration; project loans and grants; education, training, technical assistance; information services; policy advice; and support to exports and private investment. Although the U.S. policy infrastructure for promoting energy technologies is largely in place, it has not in the past focused on efficiency and is only now beginning to accept efficiency as an important theme. In general, the U.S. energy technology transfer "program" lacks a consistent mandate. Efforts in energy are normally a small, low priority component of agency budgets.

Efforts have been made to coordinate some aspects of these numerous programs through formal and informal channels. For example, Congress in the Renewable Energy Industry Development Act of 1983 initiated a multiagency committee—the Committee on Renewable Energy Commerce and Trade (CORECT) to promote exports of U.S. renewable energy technologies (see box 8-D below). A U.S. General Accounting Office (GAO) evaluation of CORECT'S activities, currently underway, will provide guidance on whether CORECT can serve as a model for other areas such as energy efficiency and the environment.

Despite the large number of programs, the current level of U.S. bilateral aid for energy (see table 8-1) is modest. The main bilateral aid agency is the U.S. Agency for International Development (AID). Annual AID energy expenditures, largely grants, amount to $200 million, compared with MDBs' energy loans of $5 billion annually. The relatively small scale of

Table 8-1—Overview of Organizational Functions Relating to Energy Technology Transfer

Organization	Capital project support	Technology support and dissemination	Training and human resource development	Policy persuasion and advice	Private section support, export promotion	Level of resources (million $)[a]
U.S. agencies:						
Agency for International Development	√	√	√	√	√	$214 (1990 est.)
Department of Energy	—	√	√	—	√	$1-$10
Trade & Development Program	√	—	—	—	√	$8
Department of Commerce	√	—	—	—	√	NA
Environmental Protection Agency	—	√	√	√	—	NA
Treasury	√	—	—	√	√	NA
Export-Import Bank	√	—	—	√	√	NA
Overseas Private Investment Corporation	√	√	—	—	√	NA
Small Business Administration	—	—	—	—	√	NA
Multilaterals:						
World Bank	√	—	√	√	√	$3,704 (1987)
International Finance Corporation	√	—	√	√	√	NA
Multilateral Investment Guarantee Agency	√	—	—	—	√	NA
InterAmerican Development Bank	√	—	√	√	√	$405 (1988)
Asian Development Bank	√	—	√	√	√	$567 (1988)
African Development Bank	√	—	√	√	√	$185
UN Development Program	—	√	√	√	√	$25-30 (1988)
International Energy Agency	—	√	√	—	—	—
UN Industrial Development Organization	—	√	√	—	—	$5-10 (1988)

NA = Not available or not applicable.
[a]Lawrence Berkeley Laboratory, *Energy Technology for Developing Countries: Issues for the U.S. National Energy Strategy*, U.S. Department of Energy, December 1989, p. 17.

U.S. bilateral assistance for energy suggests that the sums available will continue to be used to greatest effect by:

- using grant monies to promote technical assistance and institution building for technology transfer and diffusion;
- introducing energy efficiency and related environmental considerations into broader international policy discussions where the U.S. voice carries considerable weight;
- bringing influence to bear on the activities of the multilateral development banks, whose expenditures represent a major force in developing country energy decisionmaking; and
- developing cooperative approaches with other bilateral donors and lending agencies, and the private sectors in both the United States and developing countries.

AID Programs

AID is the major conduit for U.S. bilateral energy support. However, energy accounts for a relatively small share, about 3 percent, of AID's total annual economic assistance. Obligations have been declining in real terms in recent years—from $214 million in fiscal year 1990 to $177 million in fiscal years 1991 and 1992.[1] Beginning in fiscal year 1991, assistance for East European countries is included in this total. Less than one-half of 1 percent of Agency staff are full-time, direct-hire energy experts.

AID assistance is concentrated in projects in Egypt and Pakistan,[2] which are the only two

---

[1] U.S. General Accounting Office, *Foreign Assistance: AID Energy Assistance and Global Warming*, GAO/U.S.AID-91-221 (Washington, DC: U.S. General Accounting Office, July 1991), p. 2.

[2] Funding to Pakistan has been suspended because of that country's nuclear weapons program.

countries to have major efficiency projects. In 1980, annual AID energy funding to Africa was just over $1 million. Latin America also receives little attention from AID for energy (under 6 percent of AID's total energy expenditures) despite strong U.S. interest in trade and investment markets in this area.

The remainder of AID energy expenditures is devoted to a centralized Office of Energy and Infrastructure that provides energy planning and policy advice on an agency-wide basis, supporting country missions with advice and training. In fiscal year 1990, the Office had a $15.4 million budget, which was increased to $20 million in fiscal year 1991 or about 8 percent of total AID energy assistance.

The Office of Energy and Infrastructure conducts a variety of policy, training, technical assistance, and institution building programs (see box 8-A). The Office of Energy and Infrastructure also coordinates a number of programs with other government agencies (e.g., the Department of Energy National Laboratories), multilateral aid donors, and private organizations (e.g., Bechtel and the American Wind Energy Association).

Congress has expressed a high level of interest in AID energy and environmental activities. For example, Congress has directed AID to encourage energy pricing reform, end use energy efficiency, Integrated Resource Planning (termed "least cost planning"), and renewable energy; and to increase the number and expertise of personnel devoted to these areas. AID has also been directed to include global warming considerations in its energy assistance activities. In particular, Congress requested AID to identify key developing countries in which changes in energy and forestry policies might significantly reduce greenhouse gas emissions.

Multilateral Bank Activities

The multilateral development banks (MDBs) operate major energy programs mainly devoted to large-scale conventional energy supply projects. Through their lending policies, the MDBs influence energy technology transfer to developing countries and in general play a major role in determining the types of energy projects that will be developed.

Among the multilateral donors (see box 8-B), the World Bank has the largest single energy program, providing three quarters of total multilateral energy lending. The World Bank exercises considerable influence in the energy sector of developing countries through its own loans and through the leveraging effect of its lending activities. For example, in 1989, World Bank energy sector loans of $3.8 billion together with Inter-American Development Bank (IDB) loans of $407 million formed part of the financing of 33 projects whose total cost was $25 billion. In addition, the World Bank has a large lending program in areas (e.g., transportation, industry, and urban development) that closely affect the way energy is used. Through these activities the World Bank has a good opportunity to act in an integrated fashion over many relevant sectors. It is also influential in setting policy guidelines.

The World Bank and other development banks have traditionally concentrated on large scale conventional energy supply projects. These include centralized electricity generation projects, and coal, oil, and gas production and processing. Few energy projects have been devoted to efficiency or the development of renewable energy resources, other than large scale hydroelectric. In their 1989 lending program, for example, MDB support for solar, geothermal, and wood-based energy projects accounted for less than 1 percent of total energy lending.[3]

Recently, often under pressure from Congress and NGOs, the World Bank has devoted more attention to energy efficiency considerations (mainly in the electricity sector) and has set up a special task force to examine efficiency efforts within the World Bank. The World Bank/UNDP Energy Sector Management Assistance Program (ESMAP) has been reorganized. As a program outside the main geographical departments that are largely responsible for project development, ESMAP could be well placed to develop innovative energy efficiency and renewable energy programs of broad application throughout the bank lending program. On the other hand, its isolation from the project development process could also diminish its effectiveness. By recently joining ESMAP, the United States has increased its voice in this potentially innovative program and could bring its influence to bear on improving the

---

[3] U.S. Export Council for Renewable Energy, *Energy Lending at the World Bank and Inter-American Development Bank* (Arlington, VA: January 1990), p. 20.

Box 8-A—AID Office of Energy and Infrastructure

The Office of Energy and Infrastructure has five basic goals: increased consideration of environmental criteria; increased technical efficiency and financial performance of energy systems; greater private enterprise involvement in energy development and management; expanded use of suitable indigenous energy resources; and enhanced availability of energy for sustained rural development. To implement these goals, the Office currently supports the following projects.

Energy Policy Development and Conservation Project.

This project includes the following elements: advancing the Multi-Agency Group on Power Sector Innovation (MAGPI); promoting least-cost investment planning; improving efficiency and performance of electric power systems in developing countries; encouraging price reform policies; developing institutions to promote technology innovation and commercialization programs such as Program for the Acceleration of Commercial Energy Research (PACER) in India; conducting technology assessment and prefeasibility studies, including options for rural power delivery; and developing a program in environmental management.

Energy Conservation Services Project

This project includes: promotional and planning activities related to energy efficiency as a response to global warming; energy efficiency in electric power systems; and efficiency in the industry, buildings and transportation sectors. Through this program, the Office of Energy is participating in the design and implementation of the Global Energy Efficiency Initiative and a PACER building energy efficiency project.

Renewable Energy Applications and Training (REAT) and Biomass Energy Systems and Technology (BEST) Projects

These programs are both aimed at stimulating the use of renewable energy in development projects. REAT addresses the commercialization of renewable energy projects through the following elements; preinvestment studies; feasibility studies; training and informational activities in cooperation with the U.S. Export Council for Renewable Energy; participation in the FINESSE (Financing of Energy Services for Small-Scale Energy Users) program; and coordination with multilateral banks and bilateral donors through the MAGPI mechanism. BEST focuses on promoting biomass energy development, programs of applied biomass research; a Venture Investment Program; and information dissemination, including workshops.

Private Sector Energy Development (PSED) Project

The PSED Project is intended to catalyze policy changes that will enhance private sector involvement in the energy sector through the following means; conferences and workshops in assisted countries and in the United States; technical assistance; dissemination of Private Power Reporter, a publication based on an in-house database of project opportunities; financial support for feasibility studies; and coordination with other bilateral donors, multilateral development banks, and the private sector.

Conventional Energy Technical Assistance (CETA) Project

The CETA project is directed at transferring U.S. advances in energy technology to developing countries in the areas of resource assessment and development, and technology innovation. In the latter case, CETA has targeted coal combustion technologies, particularly atmospheric fluidized-bed combustion and integrated coal gasification combined cycle. CETA will be phased out in 1991, and the Energy Technology Innovation Project (ETIP) will take its place. ETIP will add a Clean Energy Technology Feasibility Study Fund; a focus on energy efficiency improvements in supply and distribution in the power sector (complementing the Energy Conservation Service Project); transfer of rehabilitation technologies; and workshops aimed at institutional reform in the power sector.

Energy Training Program

The Energy Training Program offers short-term (2 to 7 months) training to governmental, parastatal, and private employers in developing countries in the following categories: energy policy and analysis; indigenous fossil fuel development; power industry development; energy conservation and efficiency; alternative energy systems; environmental policy and regulation; pollution-control systems; and data collection and analysis. Through the Energy Training Program, the Office of Energy also trains USAID personnel (in Washington and in overseas missions) in environmental topics, including the relevance of least-cost planning, efficiency, and renewable resources. In addition, the Office of Energy and Infrastructure works closely with AID missions to implement country specific programs and projects.

SOURCE: U.S. Agency for International Development, Bureau for Science and Technology, Office of Energy, "Program Plan," for fiscal years 1990-1992.

Box 8-B—Multilateral Development Banks

Multilateral agencies are international organizations with a consortium of member nations or contributors. The United States participates in several multilateral organizations, principally as a member or majority contributor to four multilateral development banks (MDBs). Most of these agencies are associated with the United Nations. U.S. membership in each organization resulted from congressional legislation, and Congress annually passes new legislation to renew subscriptions of funds to the banks or to suggest policy direction to the U.S. representative. The Department of the Treasury has oversight of U.S. directors of the MDBs, and the Department of State has oversight of the U.S. participation in the UN, through the U.S. Ambassador to the United Nations.

The World Bank

The World Bank Group is an independent organization under the United Nations and is the largest multilateral development bank. It supports the majority of developing country loans, accounting for around 75 percent of all MDB loans made to developing countries. There are four interrelated organizations that are part of the World Bank Group: the International Bank for Reconstruction and Development, the International Development Association, the International Finance Corporation, and the Multilateral Investment Guarantee Agency.

The International Bank for Reconstruction and Development (IBRD) finances its loans through its own borrowings on the world capital markets, through retained earnings (approximately $1 billion in 1990) and through repayments on loans. IBRD loans carry near-commercial interest rates, repayable over a 15 to 20 year period. IBRD loan decisions are based on economic considerations and "prospects for repayment."[1] In 1990, IBRD loans amounted to a total of about $15.2 billion, comprised of 121 projects.

The International Development Association (IDA) targets countries that cannot afford to pay the near-market interest rates of the IBRD loans. Currently, countries with $580 or less annual per capita GNP (1987 dollars) are eligible for IDA loans. Internationally, IDA is the largest source of multilateral development bank lending on "concessional" terms. In 1990, IDA credits totaled over $5.5 billion, distributed through 101 projects. Credits were granted most heavily in Africa, with $2.7 billion distributed among 67 projects.

The top four lending sectors—agriculture, energy, structural adjustment, and transportation—account for over 50 percent of IBRD/IDA (referred to subsequently as "World Bank") loans. In fiscal year 1990, 16 percent of the World Bank's loans went to the energy sector, making energy second only to agriculture. Over two-thirds of World Bank energy sector loans are to the power sector (notably hydropower), and most of the rest to oil and gas exploration and development. Structural adjustment is the fastest growing sector, accounting for 15 percent of Bank lending in recent years. Typical conditions of structural adjustment loans relate to macroeconomic reforms, such as pricing policies, including energy prices.

The International Finance Corporation (IFC), an independent affiliate of the World Bank, promotes private sector investment by providing long term loans and risk capital without government guarantees to private sector companies. Of the 125 IFC projects in fiscal year 1990, 6 were energy related, and amount to an IFC investment totaling $140 million in loans and $40.8 million in equity and syndications. Three projects were in oil exploration, two in electricity transmission and/or distribution, and one in hydro. The majority of IFC's investments are in local financial institutions, such as banks and credit unions.

The Multilateral Investment Guarantee Agency (MIGA) aims to stimulate foreign direct investment in developing countries by providing guarantees to foreign investors against losses from noncommercial risks. MIGA was initiated in 1988 and has guaranteed a total of $132.3 million for four projects, with the projects totaling $1.04 billion in direct foreign investment. One project is energy related, an investment by GE of the United States in a lighting product manufacturer in Hungary.

The Regional Development Banks

The regional development banks, the Inter-American Development Bank, the Asian Development Bank, and the African Development Bank, all conform to the World Bank model. Each makes loans to developing country members and each has a concessional arm that makes grants and loans to the poorest member countries.

(continued on next page)

> **Box 8-B—Multilateral Development Banks—Continued**
>
> The Inter-American Development Bank (IDB) annually makes loans amounting to about one-tenth of the World Bank's total, with lending in fiscal year 1989 around $2 billion. The United States has the single largest subscription. About 25 percent of the IDB's loans, about $407 million in fiscal year 1988, go to the energy sector. IDB's energy lending closely parallels that of the World Bank; large hydropower projects usually receive the most funding, followed by fossil fuel generation and exploration, electricity transmission and distribution, and oil and gas development.
>
> Japan and the U.S. contribute equal amounts to the Asian Development Bank. In fiscal year 1988, ADB made $2.8 billion in loans, with $392.14 million going to energy projects. The projects included oil and gas exploration; transmission and distribution; and one hydropower project.
>
> The African Development Bank with its concessional arm, the African Development Fund (AfDB/AfDF), dispensed about $2.2 billion in loans and grants in fiscal year 1989. AfDB/AfDF lends for energy projects as part of the public utilities sector. In fiscal year 1989, the public utilities sector comprised 18.2 percent of the Bank's loans and grants. The near-commercial loans for energy totaled $185.4 million in fiscal year 1989, with concessional loans totaling $69.7 million, and one grant of $900,000 for a joint water supply/power project in Sudan.
>
> [1] World Bank, *Recent World Bank Activities in Energy* (Washington, DC: October 1992), p. 2.
> SOURCE: Office of Technology Assessment, 1992.

interaction between ESMAP and the Bank operational department.

World Bank energy project lending usually contains requirements to increase energy prices to cover long term marginal costs, as well as other policy and institutional reform conditions. A more active role in energy efficiency, however, may be useful. Past experience in removing market distortions in power sector pricing and increasing competition and operating efficiency indicate that it may take many years to improve the functioning of the energy market.

The new Global Environmental Facility (see box 8-C) provides an opportunity for additional energy efficiency projects. The criteria governing lending under this fund, however, appear to exclude energy efficiency projects that would be cost effective if energy prices reflected long term production costs. Such an interpretation risks excluding many energy efficiency projects from consideration. It is as yet too early to assess the operational impacts of these rules.

Despite these initiatives, energy efficiency still receives minor attention. This appears to be the result of three factors: First, energy efficiency projects are more diverse and complex than conventional energy supply projects and harder to put into a project format for lending. Second, results of energy efficiency initiatives are hard to forecast and incorporate in supply plans. Third, past emphasis in favor of traditional supply-side projects is difficult to change. Given the importance of efficiency, it is of particular concern that there is no clear organizational center of expertise within the World Bank to support implementation of energy efficiency projects.

Even with energy efficiency improvements, energy supplies in developing countries will need to increase, and MDBs will continue to be influential in the choice of technology. Increased utilization of natural gas offers both a cost effective and environmentally attractive opportunity for developing countries. However, a variety of economic and institutional factors (see ch. 7) have constrained its development. Oil exploration in many developing countries has also lagged. There is a role here for the MDBs to stimulate hydrocarbon development by helping provide information on geological prospects, insuring foreign investors, and providing assistance in developing satisfactory long term agreements between oil companies and developing countries.

In recognition of the MDBs role in energy lending, Congress has taken an active interest in their activities. Congress has instructed the U.S. Executive Directors to the MDBs to take into account end use energy efficiency and renewable energy in making decisions about new energy projects. Congress has also addressed the issue of bundling (or combining) small energy projects into large projects on the financial scale usually handled

> **Box 8-C—The Global Environmental Facility (GEF)**
>
> The GEF is a pilot program administered by the World Bank intended to provide grants or concessional loans to developing countries to help implement programs of global environmental protection. Four areas are included in the GEF:
>
> 1. Protection of the Ozone Layer,
> 2. Limiting Emissions of Greenhouse Gases,
> 3. Protection of Biodiversity, and
> 4. Protection of International Waters.
>
> The second area, Limiting Emissions of Greenhouse Gases, is particularly related to energy supply and use. Energy efficiency and renewable energy are to be included under certain circumstances.
>
> The World Bank, United Nations Environment Program, and United Nations Development Program jointly implement the GEF. The World Bank takes primary responsibility for project definition, evaluation, and lending. The initial 3 years of the GEF is considered a pilot program period, with an expected $1.5 billion budget. The United States has committed $150 million for GEF "parallel" financing through AID for the initial 3-year period.
>
> While the GEF is intended to address areas of global concern, substantial overlap with issues of substantial national environmental self-interest are also eligible. Eligibility criteria incorporate three types of projects: a) Projects economically viable on the basis of domestic benefits and costs to the country itself, but which would not proceed without GEF involvement; b) investment that is not justified in economic terms if full costs are borne by the country itself; and c) investment that is justified in the country context, but the country would have to incur additional costs to derive the full global benefits. It was intended that the GEF operations complement but not substitute for actions that could be supported under existing programs.
>
> SOURCE: Office of Technology Assessment, 1992.

by the large development banks. It has done this by directing the Treasury Department to work with borrowing countries to develop loans for bundled projects on end use energy efficiency and renewable energy. In further action to facilitate development of small scale projects, the 1989 International Development and Finance Act (PL 101-240) requires the U.S. Executive Directors of MDBs to promote increased assistance and support for nongovernmental organizations.

This survey of current policies and programs with respect to the accelerated diffusion of improved energy technologies to developing countries suggests two broad types of policy options (options 1 to 3 of table 8-2). The first is to give the U.S. program greater cohesion by establishing and strengthening coordinating bodies in selected areas such as energy efficiency and environmental protection activities. The second is to monitor the implementation of existing congressional directives to AID and the executive directors of the MDBs with regard to energy efficiency, energy pricing, renewables, the bundling of small scale projects, and the role of NGOs in project lending. It may be that there is less need for additional directives than effective implementation of existing ones.

IMPROVING TECHNOLOGY TRANSFER AND DIFFUSION

Three main categories of barriers to the accelerated adoption of improved energy technologies in the developing countries have been identified. These are technical barriers relating to the technologies themselves, institutional barriers, and economic and financial barriers. The following sections examine current U.S. activities and policies that help to remove these barriers and discuss options for improving their effectiveness.

Technology Transfer Support

Technology consists not just of the appropriate hardware, but also the knowledge and training necessary to use it effectively. In addition to technology research, development, and demonstration, the dissemination of information about technology and training in its use is also important to technology transfer.

Research, Development, and Demonstration

Since scale of production, skills, relative prices, and raw materials often differ among the developing countries, technology developed in the industrial

Table 8-2—Options

Structure of existing programs for energy technology transfer and diffusion
1. To increase the cohesion and effectiveness of U.S. efforts, the Committee on Renewable Energy Commerce and Trade (CORECT) could be considered as a model for energy conservation and environmental programs.
2. The effects of previous congressional directives regarding energy efficiency, energy pricing, integrated resource planning, renewables, the bundling of small scale projects, and the role of non-governmental organizations in project lending could be monitored. It may be that there is less need for additional directives than effective implementation of existing ones.
3. Recognizing the need for continued conventional energy supply development, the MDBs could further encourage natural gas development in the developing countries.

R&D and demonstration
4. The Department of Energy and associated National Laboratories could increase efforts to develop or adapt energy efficient technologies for developing country use in close cooperation with the U.S. technical community.
5. In order to integrate industrial country research, development and demonstration more closely with the needs, conditions, and expertise of the developing countries, special institutes (analogous to the Consultative Group for International Agricultural Research institutions) for that purpose could be established in the developing world.

Information dissemination
6. Ways could be examined of improving developing country access to, coordinating, and expanding, technology databases, especially in defined areas of priority need (e.g., in electrical power generation and transmission and distribution system efficiency; renewables; oil and gas exploration and development; clean coal power generation; energy environmental control technology; and transport system efficiency).

Training and technical assistance
7. AID could be encouraged to hold more regionally based or in-country training, thus stretching available budgets, extending the language coverage of the courses, and strengthening the growing number of energy institutions in the developing countries.
8. The U.S. executive directors of MDBs could be requested to ensure that energy sector loans, as appropriate, incorporate provisions for training.
9. The donor agencies could be encouraged to pay particular attention to training needs in the electricity sector of developing countries.

Institutional reform in developing countries
10. DOE could provide additional support for Integrated Resource Planning and related regulatory reform to the bilateral and multilateral donor agencies, including drawing on the substantial experience of the various State and regional authorities (e.g., the California Energy Commission and Pacific Northwest Coordinating Council) and utilities, which have been at the forefront of this effort.
11. The growing interest in privatization in the electric utility sector of developing countries could be encouraged, again drawing on U.S. experience.

Providing incentives for the adoption of improved energy technologies
12. The multilateral donors, AID, and U.S. negotiators in debt negotiations could be encouraged to include energy pricing issues in their economic reform efforts. This is an opportune time as many of the developing countries are expressing interest in moving towards market-based systems.
13. AID could be requested to make greater efforts to provide technical assistance on pricing reform and the development of energy efficiency standards, drawing on U.S. experience.

Financing issues
14. Consideration could be given to increasing the share of energy in total bilateral aid, which is low in relation to other major donors.
15. AID and the MDBs and other bilateral agencies could be requested to investigate ways of overcoming the initial higher costs of energy efficient equipment to end users, and to consider mechanisms for bundling a large number of small projects into larger projects consistent with MDB scale of operation.

U.S. trade and investment
16. Existing trade and investment support programs (including those of the United States and Foreign, Commercial Service, Trade and Development Program, and U.S. Trade Representative for energy could be expanded.
17. Additional efforts could be made in existing trade support programs to respond to the special needs of small to medium scale companies.
18. The extent to which the present trade and investment support agencies are able to support new forms of finance such as project and nonrecourse finance could be investigated.

Poverty and rural energy needs
19. Bilateral and multilateral donor agencies could undertake additional activities to improve access of both the rural and urban poor to improved forms of energy.
20. Donor organizations could examine ways to seek more effective means to deliver technical assistance to the urban and rural poor, including through NGOs and nonprofits with demonstrated ability to deliver technical assistance and financing at the local level.

Protecting the environment
21. AID and EPA could be requested to provide additional technical assistance to developing countries for environmental planning. Efforts could be made to ensure an effective degree of coordination under EPA's leadership of U.S. environmental activities in the developing countries.
22. In connection with longer term issues of global sustainability, the appropriate level of funding for family planning, how these funds should be distributed, and under what restrictions, if any, they should be distributed could be further examined.

Setting a good example
23. The United States could set a good example to the rest of the world by improving energy efficiency at home.
24. The State Department, as requested by Congress, could ensure that U.S. facilities abroad, particularly those in developing countries, are highly energy efficient and take advantage of renewable energy sources where feasible. U.S. embassies, office buildings, and residences could serve as show pieces for advanced U.S. energy efficient technologies, such as highly efficient air conditioners, heat pumps, household appliances, insulation standards, solar systems, and cogeneration.

countries may need considerable adaptations to developing country conditions. Examples of potential areas for RD&D include high reliability rural off-grid power using indigenous fuels and renewable energy, and more efficient end use devices such as lighting systems and refrigerators.

The Department of Energy (DOE), with the associated National Laboratories, is the center of energy research and development in the U.S. Government (see box 8-D). DOE is currently active in fossil energy technology, notably clean coal technology. DOE has agreements with a number of countries for cooperative R&D carried out through exchanges of information and research personnel. In addition, DOE's National Laboratories work directly with public and private organizations in developing countries. For example, Oak Ridge National Laboratory has undertaken studies on institution building, technical assessments, energy efficiency improvements, and fossil energy options in a wide range of developing countries. Lawrence Berkeley Laboratory focuses on global climate change and energy efficiency in buildings, working especially with China and Association of Southeast Asian Nations (ASEAN) countries. Argonne National Laboratory is active in energy system planning. Los Alamos has specialized in geothermal energy R&D, directed at Central America and the Caribbean nations. Sandia National Laboratory, working through its Design Assistance Center, has worked in transferring photovoltaic technologies to Latin America. The National Renewable Energy Laboratory (formerly the Solar Energy Research Institute) conducts research in a wide range of renewable energy technologies.

Further cooperation between U.S. Government agencies and between these agencies and the MDBs is possible (see option 4 of table 8-2). DOE, for example, could develop simplified procedural mechanisms and an experience/technology inventory, whereby other agencies and MDBs might more readily access DOE laboratories' resources.

In order to integrate industrial country energy RD&D more closely with developing country needs and to give the developing countries greater influence in defining their R&D agendas, consideration could be given to establishing or supporting special institutes for that purpose (see option 5 of table 8-2). An example of an analogous effort for agriculture is the Consultative Group for International Agricultural Research (CGIAR). This is an the agricultural research network which links a series of research institutes in different parts of the world. An energy "CGIAR" could set up centers of research and development, technology demonstration and evaluation, and pilot projects in developing countries.

Such centers could be staffed by scientists from both the industrial and developing countries, and draw on existing institutions with relevant experience. Feedback between the developing countries and technology research institutions in such a system could be much closer than is often the case at present, taking advantage of developing countries' greater local knowledge and growing expertise in energy technologies. These centers could help create in the developing countries an institutional ability to conduct successful R&D and achieve its commercialization. The centers could be funded through a variety of sources—bilateral and multilateral donors, foundations interested in energy and rural development, and some industries who might find work on the adaption of their products to developing country conditions a stimulus to future sales. Debt swaps could also be used as a source of funds.[4]

Information Dissemination

Despite efforts to improve access to information on energy technology, there are still reports from developing countries of inability to obtain, assess, and utilize technical information.

Responsibility for information dissemination is now divided among several U.S. agencies, including the Departments of Commerce and Energy. The Trade and Development Program has an "Industrial Equipment Familiarization Program" that assists foreign buyers to locate U.S. technology. Several private sector organizations are also active. For example, the U.S. Export Council for Renewable Energy (US/ECRE) provides information on small scale, renewable energy systems, and the International Institute for Energy Conservation (IIEC) provides information on energy efficiency technologies. Several U.S. utilities have provided information and assistance to "sister" utilities in developing

---

[4] George E. Brown, Jr. and Daniel R. Sarewitz, "Fiscal Alchemy: Transforming Debt Into Research," *Issues in Science and Technology*, vol. VIII, No. 1, Fall 1991.

Box 8-D—The Department of Energy

The Department of Energy has a number of international activities. These include over 30 direct agreements with private or public organizations in developing countries (mostly the Newly Industrialized Countries)[1] as part of its international energy research and development cooperation activities. These bilateral agreements usually consist of collaborative R&D and exchanges of technical information, accounted for almost entirely as "in-kind contributions of staff effort." Although there is joint research on geothermal energy (in Mexico, for example) and renewables, most of the present bilateral agreements involve fossil fuels. In addition, through the Assistant Secretary for Fossil Energy, DOE supports a $1.1 million program aimed at stimulating the export of coal-based technologies to developing countries, emphasizing clean coal technologies. Of this total, $750,000, plus funds from AID and TDP, support prefeasibility studies by U.S. companies. The Assistant Secretary for Fossil Energy also chairs a Coal Export Initiative Program for the coordination of governmental agencies involved in coal export issues. Agreement to share information on clean coal was reached with Chile and Costa Rica and all 75 AID countries were assessed as possible candidates for cooperative efforts in the field of power generation and/or industrial utilization for clean coal.

A recent initiative—the Export Assistance Program—was announced to add to the Department's expertise in promoting exports of U.S. energy goods, services, equipment and technology, working with ongoing activities of the Departments of Commerce and State, and the U.S. Trade Representative. The Program will be managed by a new office within the Office of International Affairs and Energy Emergencies, headed by a Deputy Assistant Secretary who will serve as a U.S. energy industry international advocate and contact for information and action. Promotional trips to Venezuela and Chile have already been made, and an annual Latin American Energy Forum is in the planning stage.

DOE's National Laboratories work directly with public and private organizations in developing countries as well as in cooperation with U.S. governmental and multilateral groups. The National Labs spend about $10 million annually to provide energy assistance to developing countries, and from 1986-1988, worked on 36 diverse projects. In addition to some DOE funds, the laboratories receive substantial support from AID, the World Bank, the International Atomic Energy Agency, the Environmental Protection Agency and other outside sources.

The Department of Energy also leads the Committee on Renewable Energy Commerce and Trade (CORECT), a multi-agency committee initiated by an Act of Congress[2] that includes AID, the Departments of Commerce, State, and Treasury, the Export-Import Bank, Overseas Private Investment Fund, the Trade and Development Program, the U.S. Trade Representative and the Small Business Administration. CORECT promotes trade of U.S. renewable energy technologies through a variety of activities: (see below). CORECT's current funding level is about $1.5 million. CORECT helps fund the U.S. Export Council for Renewable Energy, formed by 9 national renewable energy trade associations to promote exports, which in turn provides industry direction for the Committee.

CORECT's activities include:

- Financing Energy Services for Small-Scale Energy Users (FINESSE)-Financing Task Force. The World Bank and CORECT are working closely together on this effort to improve credit accessibility for small-scale energy users who may be unable to attract the attention of multilateral donors and are unable to support the high interest rates of commercial banks.
- Pacific Rim Initiative. This initiative has involved identification of markets and was followed by identification of specific project opportunities, initially focusing on Indonesia and the Philippines.
- One-Stop Application. CORECT is coordinating the U.S. government agencies necessary to develop a simplified form and procedures for obtaining support from Eximbank (team leader), the Trade and Development Program, Small Business Administration, Overseas Private Investment Corporation and AID.
- Integrated Electric Utility Program. The goal of this program is to encourage the consideration and use of decentralized energy systems for delivering electric power in developing countries.
- Design Assistance Center of Sandia National Laboratory. The Design Assistance Center at Sandia works with industry and users in the photovoltaics area to provide feasibility studies, system specifications, procurement document preparation, design evaluation, and education and training assistance.

[1] Lawrence Berkeley Laboratory, *Energy Technology for Developing Countries: Issues for the U.S. National Energy Strategy* (Berkeley, CA: December, 1989), p. 20. LBL lists Brazil, India, South Korea, PRC, Venezuela and Egypt as Newly Industrialized countries.

[2] The Renewable Energy Industry Development Act of 1983.

SOURCE: Office of Technology Assessment, 1992.

countries. The United Nations serves an informational function in development countries. For example, the UN Industrial Development Organization provides information on industrial energy conservation.

Ways of improving access to and coordinating these databases and of expanding these services could be explored. Of particular importance are areas such as electrical power generation, transmission and distribution system efficiency, small scale renewables, oil and gas exploration and development, clean coal power generation, energy environmental control technology, and transport system efficiency (see option 6 of table 8-2).

Training and Technical Assistance

Training and technical assistance are integral parts of technology transfer. Without good training, efforts to transfer technology hardware are frequently ineffective. The transfer of renewable energy and energy efficiency technologies requires skills to evaluate their appropriateness, cost-benefit, applications, and local manufacturing prospects. Implementation of rural and small scale energy technology transfer, in particular, requires sustained and concentrated attention to improve local human capability: foreign technical assistance is generally both less effective and too expensive to be a practical long term vehicle for support.

AID sponsors an Energy Training Program that includes academic advanced degrees and in-service and industry fellowships. A recent evaluation of the project[5] found widespread satisfaction with the program among former attendees, while also suggesting several new directions. These include training in environmental assessment and management, general management training, local/indigenous resource utilization, coal resource development and contracting, and energy project financing.

Energy training programs might be expanded to meet the needs of non-English speaking participants. Neither the French-speaking West African countries nor the Latin American countries appear to be adequately supported in AID training. One option is to hold more regionally based or in-country training. This would stretch available budgets and stimulate the growing number of energy institutions in the developing countries. Private voluntary organizations (PVOs), many of which have long histories of operations in developing countries, can play a role here. In recent years, AID has enrolled PVO support on an increasing scale, taking advantage of their experience and generally lower costs. Training programs might benefit from the wide experience of U.S. companies in training associated with their foreign operations in developing countries.

Increased emphasis on training can yield important benefits to the United States. It creates awareness of U.S. technology and services. At times when capital project aid is severely constrained, training may be one of the more cost-effective, long term trade strategies.

In the MDBs, training is largely associated with specific projects or, as in the case of the World Bank's Economic Development Institute, as part of specially designed economic development courses. Considering the importance of investment in human capital, much greater attention to this function appears warranted. It has been observed in many projects, for example, that training funds are often either not used fully by governments or not used effectively, due to a lack of government priority on training or lack of a training plan. Most existing loans could incorporate a more comprehensive treatment of training needs.

The deteriorating performance of the electric utility sector is attributed in part to poor management and maintenance. This suggests the need for greater attention to training across the entire spectrum of activities, from power operational efficiency, to billing and collection of revenues, to diesel maintenance and lineman training.

While promotion of operational efficiency improvement may appear more appropriate for agencies with large capital programs such as the World Bank, areas that require intense technical assistance, training, and institutional development may in fact be much better suited to an aid agency with grant funds like AID. Expanded cooperation in these areas between AID the World Bank, and other MDBs, could be beneficial (see options 7, 8, and 9 of table 8-2).

---

[5] Development Sciences, Inc., "An Evaluation of the Conventional Energy Training Program and the Energy Training Program for the U.S. AID Office of Energy," Washington, DC, Jan. 12, 1990, pp. 23-24, 31-34.

Improving the Institutional Framework

In many developing countries there are numerous opportunities for improving the economic and institutional framework for energy sector development. The present system, particularly in the electricity sector, often discourages energy efficiency and small scale decentralized energy facilities. The decision process is usually biased towards large scale supply options even when efficiency improvements and small scale renewables are more beneficial. Furthermore, private sector investment has had little role in electricity sector development. These issues are of growing concern because of the serious deterioration in the electricity sector in many developing countries and the prospect of a major shortfall in the financial resources available for supply expansion.

The Regulatory Framework

Decisions on how to provide energy services are made independently by suppliers and users. This leads to a "disconnect" in the system, which discourages energy efficiency. On the one hand, purchasers faced by high interest rates and low incomes opt for low first cost, typically low energy efficiency equipment. Moreover, frequently subsidized electricity prices provide users with little incentive to invest in efficiency. Suppliers of electricity, on the other hand, face a number of incentives, financial or otherwise, that encourage supply expansion rather than efficiency improvements.

One means of overcoming the system obstacles to energy efficiency is the use of Integrated Resource Planning (IRP)—an expanded form of least cost planning—in project design.[6] Traditionally, least cost planning techniques focused on a limited range of large scale conventional supply technologies, such as coal-or oil-fired generation. Forecasts of electricity demand were seen as a given, not subject to influence. This process has now been expanded in many utility service areas in the United States to include consideration of demand. Thus, energy efficiency investments are balanced on a cost-benefit basis with new supply additions, including nonconventional supply options such as privately generated renewable energy. This system can be used in both publicly and privately owned utilities, and its value is being increasingly recognized by analysts and policymakers in developing countries.

It is important to realize that Integrated Resource Planning systems will require substantial reform in the regulatory structures if they are to be effective. Furthermore, support activities such as data gathering, metering, demonstration, standards, financing, and incentives will need to be developed.

United States agencies are already providing advice on the long term planning of the energy sector, often at the instigation of Congress. For example, legislation has directed AID to focus on IRP (termed "least cost energy planning"), end use energy efficiency, and renewable energy. These elements are contained in two of the AID Office of Energy Programs: the Energy Policy Development and Conservation Project and the Energy Conservation Services Project. The implementation of these congressional directives could be examined.

The United States could make further contributions to establishing appropriate regulatory systems in the developing countries. U.S. State and regional utility regulatory apparatus have played a leading role in the development and application of IRP (as in California and the Pacific Northwest). Programs of demand-side management have been operating for several years. In these programs, utilities give financial incentives (rebates and special rates) to consumers to make greater investments in high efficiency equipment.

The Department of Energy has an Integrated Resource Planning Program, whose annual appropriations were recently increased from $1 to $3 million. This program works closely with the national laboratories to provide U.S. utilities with data and analysis. The results of these activities could be further incorporated into bilateral energy programs, and could contribute to MDB programs (see option 10 of table 8-2).

Realistically, effective implementation of IRP will require additional resources. These include grant or technical assistance funds to perform the necessary surveys, studies, and pilot demonstrations and provide guidance of regulatory reform. Such support is of particular importance as developing-country utilities do not have prior experience with

---

[6] Integrated Resource Planning should be distinguished from least-cost supply-side planning that includes supply-side options only. The term Integrated Resource Planning, covering both the supply and end use of energy, will be used here.

this approach to planning, and are likely to have very limited experience with efficiency programs in general. They may also be highly risk averse, with very limited financial and staff resources available for experimentation, and no established regulatory structure to oversee and stimulate full implementation of these innovations. Risk sharing by the MDBs or bilateral donors could also be an effective means to stimulate demand-side actions.

The role of the World Bank is critical in this area. Recognizing this key role, Congress has already instructed the executive directors of the MDBs to promote end use energy efficiency and renewable energy in decisions on new projects, and to incorporate IRP in their project planning. The implementation and effectiveness of these directive may need to be examined.

Role of the Private Sector

Much of the energy sector, particularly the electricity sector, is government owned, controlled, and operated in the developing countries. There is increasing interest in developing countries, however, in opening up part or all of the energy sector to private investors (see option 11 of table 8-2). This interest is based on a number of considerations:

- the need to attract funds both from domestic and foreign investors for system expansion,
- the need to introduce an element of competition into this sector and achieve better utilization of existing capacity, and
- the desire to gain access to new technologies.

A considerable effort has already been made to increase the role of the private sector in electricity.[7] These efforts vary from outright sale of State owned facilities to the private sector (as in Chile) to the more typical opening of the grid to small private generators as was done in the United States under the Public Utility Regulatory Policies Act (PURPA)[8].[9]

U.S. and multilateral agencies are assisting in this process. The AID Office of Energy, for example, supports private sector energy initiatives by providing information on PURPA type legislation, helping to develop private energy policies and projects, cost-sharing feasibility studies, and collecting and disseminating information.[10] There may however, be a gap in complementary in-country or AID mission programs, where the actual delivery of technical assistance, training, and other resources is most important.

The United States is a leader in the field of private power production, with valuable experience under PURPA of providing access to the grid to small private producers. Like IRP, PURPA-type legislation changes utility rules to help correct the bias in favor of large conventional generating facilities. The benefit of this experience could be made readily available to developing countries under AID or other programs.

The World Bank also has programs to promote privatization. A privatization group, applicable to all industries but of particular importance to energy, assists in developing the legal and statutory framework for regulation of private investments. The program helps in the formulation of divestiture policies for the public sector and provides technical assistance as needed to support privatization. Bank lending and cofinancing arrangements are also designed to support privatization as appropriate. Assistance is given in structuring agreements that serve as the security package for investors, lenders, and governments. Critical to the success of privatization and private capital mobilization are programs that guarantee payments for investors if their power purchase contracts are not honored. Agencies such as MIGA can play a role here.

A major element of the broad private sector power strategy is development of the local capital market,

---

[7] James B. Sullivan, "Private Power in Developing Countries: Early Experience and a Framework for Development," *Annual Review of Energy 1990*, vol. 15, pp. 335-363.

[8] The Public Utilities Regulatory Policies Act of 1978 (PURPA) requires utilities to purchase electricity from qualifying facilities at avoided cost.

[9] The appropriate system and the mix of public and private elements in it will, however, be country specific. A recent World Bank review found that public ownership could be as effective as private ownership, and that good performance was generally found to be related to management capacity and the degree of autonomy. Common features of successful power companies were found to be their relative autonomy of management, procurement, and staffing functions; sound professional and nonpolitical organizational structure; appropriate tariff policy that takes economic costs into account; and operation on sound commercial principles. W. Teplitz-Sembitzky, "Regulation, Deregulation or Reregulation: What Is Needed in the LDCs," Energy Development Division, World Bank, Energy Series Paper #30, World Bank, July 1990.

[10] The major centrally funded programs within AID for implementing these objectives are the Bureau for Science and Technology, Office of Energy, Private Sector Energy Development (PSED) project, and some aspects of AID centrally funded projects such the Renewable Energy Applications and Training Project, Energy Conservation Services Project, and Biomass Energy Applications and Training.

both to supply needed local currency inputs and also to vest continued control in local hands. The International Finance Corporation, part of the World Bank family, is one of the few donor programs devoted to the encouragement of developing-country private sector activities. The International Finance Corporation is a natural vehicle to stimulate the development of private sector energy supply and end use energy service initiatives.

Providing Economic Incentives

In addition to institutional barriers, economic systems frequently do not provide sufficient incentives to encourage the rapid diffusion of energy efficient technology or small scale renewables. High interest rates faced by consumers discourage purchase of initially high cost yet efficient equipment. Heavy reliance on indirect taxes often means high tariffs on imported equipment. The operation of energy markets is hampered by price controls, price distortions, and cross subsidies. Subsidized prices may make energy more affordable to low income groups, but do not encourage efficient use of energy or provide adequate revenues for system maintenance or expansion.[11]

Issues of energy pricing are frequently addressed in programs of general macroeconomic and energy sector reform by international and bilateral agencies. In recent years, the World Bank has increased the scale of its structural adjustment programs, which now account for 15 percent of total World Bank lending. While designed to improve the health of a country's economy generally, such programs often include specific requirements to reform energy and other pricing systems and to improve the working of capital markets. World Bank energy sector loans typically include energy pricing reform.

Congress has also instructed AID to pay attention to energy pricing in its energy programs. The renegotiation of developing country public and commercial debt can be used as a vehicle for leveraging economic and energy sector reform (see option 12 in table 8-2). In so far as these actions reduce inflation and real interest rates and make energy prices reflect their long term supply costs, they improve the incentives to adopt energy efficient equipment and renewables.

Efficiency standards have played an important role in U.S. energy policy, but have received less attention in developing countries. U.S. experience could be useful to developing countries as they formulate energy efficiency (and environmental) standards. Efficiency standards could also be useful to developing countries to offset the impact of energy price reforms—a politically sensitive issue. If higher efficiency equipment can be introduced at the same time that energy prices are raised to reflect their true cost, the total cost of the service to the consumer need not rise (see option 13 in table 8-2).

Financing Issues

The lack of finance is often perceived to be a major barrier to the introduction of energy efficient and renewable energy equipment as well as to supply expansion. On the supply side, funds for energy sector development come from both public (about two-thirds) and private (one-third) sources. On average, about one-half of total investments for energy supply is in the form of foreign exchange (see table 8-2).

Bilateral aid from all sources accounts for only a small part of total developing-country energy investments. Although AID budgets have remained steady in real terms in recent years, the U.S. share of global foreign aid has declined. Any increase in overall aid budgets would run counter to efforts to control Federal spending, but the continuation of current trends will reduce U.S. influence in the development assistance forum and in related trade issues. Increased aid could reverse this trend and possibly leverage increased contribution from other bilateral donors. Alternatively, the share of energy in the existing AID budget could be increased, reflecting rising concern about global environmental issues, the reduction of global poverty, and the increasing demands on current budgets from new programs directed to Eastern Europe (see option 14 of table 8-2).

The MDBs already operate large energy programs in which they are presently, partly in response to congressional initiative, raising the energy efficiency component and incorporating environmental considerations. In addition, the Global Environmental Facility could, if interpreted more broadly to include efficiency projects, increase the resources

---

[11] Indeed, insofar as inadequate revenues lead to shortages or unreliable supplies of energy and consequently reduce job opportunities, subsidies can ultimately damage those elements of the population they were designed to protect.

Table 8-3—Estimate of Investment in Commercial Energy Supplies in Developing Countries, Early 1980's (billions of 1982$)

	Electricity	Oil and gas	Coal	Total
Foreign exchange: external borrowing	11	4.2	0.5	19.7[a]
Foreign exchange: other	0	16.3	0.5	16.8[b]
Local	19	10.5	3.0	32.5
Total	30	31.0	4.0	65.0

[a]Of which:
export related	5.1
multilateral	3.9
bilateral	1.5
financial institutions	5.2
	15.7

[b]Residual, assumed private investment and expenditure of countries own foreign reserves.

SOURCE: Based on data in World Bank, *Energy Transition in Developing Countries*, Washington, DC, 1983, pp. 68, 69.

available for energy efficiency and small renewable energy projects. The increasing demands on World Bank and AID for funds for the countries of Eastern Europe and the Soviet Union, however, could come at the cost of assistance to developing countries.

Though public sector funding has provided about one-third of the foreign exchange costs of overall energy supply investments and a large part of electricity investments, there is general agreement in many developing countries that the private sector must play a larger role in the future. The interest in more private sector participation stems from both the shortage of public funds to finance development and the belief that competition from privately owned facilities will improve energy sector performance.

The United States has a number of long standing programs to encourage U.S. private sector investment and trade. In addition, the U.S. supports a variety of activities to bolster investment reforms and build local capital markets in developing countries. Activities in this area include: investment reform developed by the Inter-American Development Bank under the Enterprise for the Americas Initiative, IFC programs, and World Bank and International Monetary Fund economic adjustment projects. While designed for broader purposes, progress in this area will also benefit the energy sector.

High levels of debt in developing countries (currently about $1.3 trillion)[12] limit the resources available for investment in energy and other development sectors, and discourage private sector investment. Official debt traditionally accounts for a high share (75 percent or more) of total debt of the poorest countries. Most OECD governments, including the United States, have provided, and are continuing to provide, debt relief to poor African countries.

Among the middle income highly indebted countries, however, the share of official debt rose from 22 percent in 1982 to 37 percent in 1988 as private borrowing fell sharply in the 1980s.[13] Private debt still, however, accounts for the largest part of developing-country debt. In recent years, this debt has been reduced somewhat either under the Brady Plan or market based conversion procedures.[14]

Debt forgiveness, renegotiation, or restructuring provide an opportunity for conditionality—debt written down and interest forgiven to the extent that debtor governments undertake to fulfill certain agreed conditions such as increasing domestic energy prices. Debt for nature conversions are often a part of these transactions. The debt for nature model could equally be applied to debt for energy efficiency. Similarly, debt swaps could be used to fund support of energy technology research, development, and demonstration.[15] Additional require-

---

[12]Karin Lissakers, "Debt and Energy," contractor report prepared for the Office of Technology Assessment, January 1991, Table 1.

[13]K. Lissakers, Ibid.

[14]The Institute of International Finances estimates that debt obligations were reduced by $18 billion in 1988 though debt equity swaps, other local currency conversions and private sector restructuring. Institute of International Finance, Inc., "Improving the Official Debt Strategy: Arrears Are Not the Way," Washington, DC, 1990, figure 2.

[15]Brown, op. cit., footnote 4.

ments could include establishing domestic revolving funds for energy conservation investment. Debt restructuring, by improving the investment climate, encourages the repatriation of capital (as has recently taken place in Mexico) and the resumption of new foreign investment (as in both Mexico and Brazil). This is increasing resources available for investment in energy.

The financing problem concerns not only the total amount of resources available, but also the tailoring of these resources to energy sector needs and opportunities. For example, World Bank lending usually consists of a small number of large scale conventional supply projects rather than a large number of small projects, as typical of energy conservation and small scale renewables projects.[16] Changes in funding procedures will be needed to better match the requirements of new types of projects.

The total life cycle costs of energy (particularly electricity) efficient equipment are generally lower than those of inefficient equipment. The first costs of efficient equipment to end users, however, are usually higher, even though total systemwide capital costs—including both utility and end user capital investment—are lower. This is an important deterrent in countries where capital is scarce and expensive. Further, adoption of electricity-using energy efficient equipment shifts investment from utilities that usually have easy access to relatively low cost capital to the individual consumer who usually faces higher interest rates. This situation calls for innovative financing mechanisms. These could include the development of intermediary institutions or service companies to finance the initial higher capital cost to the end user through the subsequent electricity savings; or action by the utilities (through loans, rebates, etc.) to assume part of the higher initial costs to the consumer (see option 15 of table 8-2).

A second characteristic of energy efficiency is the need to influence a large number and variety of consumers, from large energy intensive industries to individual householders. This wide range underlines the importance of broad incentive measures such as energy price reform, equipment or appliance standards, and innovative financial mechanisms. There will also be new supply side actors—promoters of private power schemes and small scale decentralized renewables. Some of these actors are likely to be smaller in scale then traditional suppliers, and will lack the traditional suppliers knowledge of the special conditions of developing country markets, financial resources, and established relations with large banking systems. Special banking facilities grouping a series of small projects together in a way that would make them suitable for traditional banking procedures would be useful (see option 15 of table 8-2). There may also be a need to redistribute funds within the usual project cycle, with a larger share of the resources in the prefeasibility and feasibility stage than is usual in well-established systems.

Ultimately, it is the developing countries themselves that provide the bulk of resources for energy sector investment. At present, these resources come from the public sector as most energy supply facilities (the electricity sector, oil and gas in many countries) are public sector enterprises. Many developing countries are experiencing severe financial stringency, however, and are unlikely to be able to meet the increasing financial demands of the energy supply sector from traditional sources. This is particularly acute in the electricity sector, which in many developing countries is running at heavy financial losses. Given the scale of investment requested, the developing countries will need to both improve the revenue situation of public supply enterprises and attract private capital into the energy sector. Many of the new initiatives (e.g., energy efficiency improvements, new power generation, and development of new natural gas or renewable energy resources) might best be undertaken by the private sector.

---

[16]Multilateral organizations find it easier to fund large rather than small projects because: 1) A loan for a given amount consisting of a large number of small projects is inherently more information and management intensive, and therefore collectively have higher staffing requirements than a similar loan covering only one project. 2) Small projects require budgets and staff from multiple program areas, which is often difficult to achieve in existing organizational structures. 3) Many new smaller-scale technologies may not be known to staff or, even if known, may be felt to be too risky given the existing state of knowledge and experience to recommend to recipient countries. 4) Small scale technology generally does not involve large scale projects and therefore may not be politically attractive to borrowers. 5) Project appraisal methodologies are incompletely developed and have a bias against renewable technologies and conservation. The lack of integrated least cost energy planning at the national level results in bias towards supply investments. 6) Conservation involves direct intervention into specific industry production processes, which runs counter to World Bank funding policies.

REDEFINING PRIORITIES

Priorities in foreign policies are constantly changing, especially in the present climate of global political and economic uncertainty. In recent years, a number of issues of U.S. policy concern have emerged that are closely connected with energy. These include: U.S. trade and international competitiveness; environmental problems; and growing poverty in many developing countries despite rapid rates of economic growth.

Enhancing U.S. Trade and Investment Opportunities

Much U.S. technology transfer to developing countries occurs through private sector exports of machinery and equipment and the transfer of equipment and know-how associated with private foreign investment. U.S. based companies account for a significant part of the oil and gas exploration in developing countries. They are also taking a keen interest in new opportunities for foreign investors in gas and electricity development in the developing countries.

Expanding markets for energy technology in developing countries offer an important opportunity for U.S. industry and exporters.[17] In recent years, however, the U.S. share of global markets for major items of energy equipment has been declining (see table 8-4). Several factors have contributed to this decline. U.S. investors and exporters frequently encounter difficulties operating in developing countries. Some of the smaller U.S. companies lack experience in international markets. A lack of open markets in energy technology and market distortions, including anticompetitive practices by OECD competitors and greater direct financial support by other governments, compounds the difficulty. These practices hurt not only the United States, but may also be disadvantageous to the developing country. In some cases, the developing country may be obliged to accept less than optimal technologies or may accept a wide range of incompatible equipment, complicating training and spare part problems.

Trade and Investment Support Programs

A number of programs have been developed to support the U.S. private sector in technology transfer. These programs cover: trade and investment support through market information, finance, and insurance facilities; trade policies to establishing fair and open competition in export markets and protection for U.S. exporters; and removal of restrictions on certain commodities and markets (see box 8-E).

One option for Congress would be to expand such programs (see option 16 of table 8-2). Support could be increased for the overseas activities of the United States and Foreign Commercial Service (US&FCS), which is responsible for much of the overseas export promotion undertaken by the Federal Government.[18] Compared with many of its trading partners, the United States devotes only modest resources to export promotion abroad. For example, Japan has about 5,000 overseas commercial officers, the UK and France 400 or more, compared with the United States' 200, despite its much larger economy.

The Trade and Development Program could also be increased in size to better match the efforts of U.S. competitors. There are several areas where TDP might expand their promotion of energy projects. For example, it appears that the demand for TDP financial support for definitional missions and prefeasibility studies is substantially in excess of TDP resources, even without a substantial outreach and promotion effort. Given the high multiplier in trade and/or investment benefits from TDP expenditures and the substantial industry support for these activities, funding could be increased.

These trade support programs have in general responded to the increased interest in conservation technologies and renewables, and environmental impacts of energy technologies, but could do more (see option 17 of table 8-2). Many programs were originally set up to promote exports made in connection with large conventional supply projects.

---

[17]One report, "Opportunities in the Worldwide Overseas Power Generation Market," prepared for the U.S. Agency for International Development, Office of Energy by RCG/Hagler Bailly Inc., estimates that over the next 20 years, the United States could sell about $94 billion of power equipment to the developing world (page 4). A companion report, "U.S. Exports of Oil and Gas Exploration and Production Equipment and Services," RCG/Hagler Bailly, Inc., for U.S. Agency for International Development, draft, Exhibit 2, suggests a market for U.S. exports of oil and gas exploration and production equipment and services of an additional $110 billion, of which a substantial share would occur in developing countries. Expanding markets in energy conservation equipment could also result in large U.S. exports.

[18]This option is also discussed in U.S. Congress, Office of Technology Assessment, *International Competition in Services*, OTA-ITE-328 (Washington, DC: U.S. Government Printing Office, July 1987), ch. 10.

Table 8-4—Share of United States in Selected Global Exports
(as a percent of the total world market)

	1976	1980	1985	1987
Steam boilers:				
U.S.	17.8	19.2	14.9	11.3
Japan	11.0	21.8	36.0	29.0
Germany	25.1	18.0	9.6	14.5
Heating, cooling equipment:				
U.S.	23.0	20.7	17.9	14.1
Japan	10.8	14.9	18.2	15.9
Germany	17.6	15.5	15.6	17.3
Pumps for liquids:				
U.S.	21.3	18.8	19.3	13.5
Japan	6.0	7.9	10.2	8.8
Germany	25.5	23.2	23.8	29.1
Pumps:				
Other U.S.	24.1	20.4	16.4	12.4
Japan	5.8	9.1	14.8	13.2
Germany	22.9	19.8	22.8	22.2
Switchgear:				
U.S.	14.2	13.4	15.9	13.1
Japan	10.0	12.3	15.8	17.1
Germany	24.2	22.6	20.6	22.7
Electric distributing equipment:				
U.S.	11.6	10.0	16.2	16.3
Japan	13.8	17.5	15.1	11.5
Germany	15.7	15.0	12.3	13.9
Electrical machinery:				
U.S.	17.2	16.0	16.6	13.0
Japan	12.7	15.5	21.5	21.4
Germany	23.7	18.4	15.8	18.2

SOURCE: United Nations, Department of International Economic and Social Affairs, Statistical Office, *Statistical Yearbook* (New York, NY: United Nations, 1988), various issues.

These were primarily done by large corporations accustomed to dealing in foreign markets and aware of the official programs available. Exports of renewables and conservation equipment and services, however, may in many cases be made by small companies, unfamiliar with foreign markets and government programs. Program responses to this new clientele include providing increased prefeasibility and feasibility funding needed by small companies, and simpler application procedures. CORECT for example, is promoting a standardized application form for all of its member agencies.

Another development, which may require changes in existing programs, is the growth of project or limited recourse financing arrangements (see option 18 of table 8-2), projects in which the U.S. investor does not receive formal sovereign guarantees. This is an attractive form of financing for developing countries as it does not add to their official debt burden. As purchases of energy equipment and services looking increasingly to sellers to organize financing, it is important to establish a framework within the U.S. system that facilitates this financing. Both TDP and OPIC appear to be moving in this direction. OPIC, for example, has taken the initiative in establishing several equity investment funds. These could serve as model mechanisms to stimulate commercial capital flows to the developing countries. Broadening Eximbank's authority to support project finance could also assist in stimulating new U.S. energy sector technology transfer and investment.

Ensuring Competitive Markets for U.S. Exports

U.S. exporters and investors may encounter uncompetitive practices in developing-country markets. The U.S. Trade Representative conducts bilateral and multilateral trade negotiations to reduce or eliminate trade and investment barriers. These include: lack of guarantees for intellectual property rights, restrictive import licenses, and high tariffs (see box 8-F). USTR resources could be expanded to better respond to these and many other trade related issues now facing the nation. Additional responsibil-

Box 8-E—Programs for U.S. Trade and Investment

The *Agency for International Development (AID)* supports prefeasibility funding studies and sponsors reverse trade missions and an energy and environmental training program for host country nationals. AID maintains a private power database, which includes a compilation of information relevant to private sector power activities for selected countries.[1] It consists of information in five categories: government policy, private project opportunities, project commitments, country specific points of contact, and a general country private power related bibliography. The primary intended beneficiaries of this information are U.S. project developers.

The *Department of Energy* supports exports through its Export Assistance program and CORECT and by other means (see Box 8-D).

In the *Department of Commerce*, the Office of International Major Projects (OIMP) functions as a facilitator for U.S. architectural, engineering, and construction industries in the promotion of exports for major projects overseas, mainly in developing countries, in addition to the export assistance activities of the United States and Foreign Commercial Service. The *Trade and Development Program* (TDP) in the U.S. International Development Cooperation Agency (whose focus is primarily on large public sector projects) provides support for definitional missions, feasibility studies, technical symposia, orientation visits by high-level government officials, and plans to initiate training activities. In addition, TDP operates an Investor Assistance Program, which offers feasibility study support on a cost-sharing basis, and the State Initiative Program, which is authorized by Congress up to $5 million to develop a cooperative program with State agencies. The program has concentrated on information dissemination, including, for example, support to the California Energy Commission in its export promotion activities. The *U.S. Trade Representative* formulates overall trade policy and conducts bilateral and multilateral trade negotiation, including the removal of barriers to U.S. exports.

U.S. private investment and exports are also supported by a number of autonomous agencies. The *Export-Import Bank (Eximbank)* is an independent U.S. Government agency, chartered under the Export-Import Bank Act of 1945, that helps finance and facilitate the sale of U.S. goods and services to foreign buyers, particularly in developing countries through direct loans, guarantees, and insurance. The 1990 Foreign Operations Appropriations Act (Public Law 101-167) instructed Eximbank to direct not less than 5 percent of its financial assistance in the energy sector to renewable energy projects. This goal has probably been exceeded. In fiscal year 1989, the Eximbank provided final commitments to support $2.1 million in renewable energy projects (i.e., hydroelectric, photovoltaics) and had pending commitments for an additional $11.8 million. Assuming pending commitments are finalized, Eximbank's fiscal year 1990 support for renewable energy projects would represent 7.4 percent of its total energy sector support. The *Private Export Funding Corp.* (PEFCO) is closely related to Eximbank although it is a private corporation owned by 62 investors, mostly commercial banks. It makes medium and long term loans to borrowers in foreign countries for U.S. goods and services, using unconditional Eximbank guarantees.

The *Overseas Private Investment Corporation (OPIC)* is an independent corporation created by Congress. It directly finances projects sponsored by U.S. private investors in over 100 developing countries and provides insurance against political risks for U.S. private investments in those countries. It can provide direct loans of up to $6 million to small and medium-sized firms and investment guarantees for up to $50 million. OPIC is developing a privately owned and managed Environmental Investment Fund for business enterprises in developing countries (and Eastern Europe) that involve renewable energy, ecotourism, sustainable agriculture, forest management, and pollution prevention. OPIC hopes to capitalize the fund with $60 million of equity raised from U.S. businesses and institutional investors and $40 million in OPIC-guaranteed long-term debt.

In addition to federally sponsored programs, several States have initiated programs centered on export promotion. The geographical focus varies from State to State. New York tends to pursue trade in Europe and Japan, California and the Pacific Northwest are turning more to the Pacific Basin, and Florida focuses on Latin America. California has been especially active in export promotion through various programs, including one that guarantees export loans given by commercial banks and supports trade missions, and others through the California Energy Commission. Since 1986, the California Energy Commission (CEC) has supported an Energy Technology Export Program, designed to facilitate exports by California firms.

---

[1] Philippines, Pakistan, Costa Rica, Jamaica, and India, with plans to add Thailand and Indonesia.

SOURCE: Office of Technology Assessment, 1992.

> **Box 8-F—Examples of Import and Other Regimes Affecting the Transfer of Technology in Selected Developing Country Markets**
>
> *Brazil*: U.S. merchandise exports to Brazil in 1989 were $4.8 billion, with imports of $8.4 billion. Brazil's import policies have been characterized as forming substantial pervasive barriers to U.S. exports. Restrictive import licensing policies have been particularly effective barriers. Tariffs have recently been reduced, from about the range of 40 to about 25 percent ad valorem, taking into account exemptions. Government procurement in many cases is still restricted to national firms; the United States has in response prohibited U.S. Government awards to Brazilian suppliers. Brazil has various export subsidy programs, and lacks intellectual property protection for some classes of goods. Provision of services by non-Brazilian firms is also restricted in some areas, for example in construction engineering and architectural, data processing, and telecommunications areas. Foreign investment is prohibited in several areas including petroleum production and refining, public utilities, and various other "strategic" industries.
>
> *India*: In 1989, U.S. merchandise exports to India were $2.5 billion, versus imports of $3.3 billion. India is characterized as having a complex and comprehensive web of market access barriers, which are a serious barrier to U.S. firms. Indian tariffs are exceptionally high, with a weighted average level of 118 percent. Import licensing requirements are considered particularly effective trade barriers. They are limited to end users, not distributors, and require in many cases a certification that the product is not available from domestic sources. Government procurement discriminates against foreign suppliers. India appears to have expanded its export subsidy program. The USTR has judged that intellectual property protection in India is inadequate. Policies also severely limit potential U.S. investment. Criteria in many cases are unpublished, and decisions are made on a case-by-case basis.
>
> *Pakistan*: The United States had a trade surplus with Pakistan in 1989, with merchandise exports of $1.1 billion, versus imports of $523 million. Both import bans and high tariffs inhibit U.S. exports. Export subsidies of various types are common. Intellectual property protection is judged to be inadequate, with the USTR continuing to seek improvements. Investment barriers face foreign investors, both in terms of a location policy promoting dispersal, an indigenization policy requiring a gradual deletion of imported components, and cumbersome approval policies. Technology licensing restrictions also exist through limits on license fees and approval procedures.
>
> SOURCE: Office of the U.S. Trade Representative "Foreign Trade Barriers" 1990 Trade Estimate Report, US Government Printing Office, 1990.

ities arising out of the proposed U.S.-Mexican Free Trade Area, and the Enterprise for the Americas reinforce this argument.

The issue of "level playing fields" in international energy technology export markets and international competitiveness has become of major concern to Congress and the American people in recent years. There are widespread perceptions and reports of our frequent inability to offer terms in export markets as attractive as those of our competitors (Japan, Germany, France, Italy, the UK) because of the use of "mixed credits" and the practice of tied aid. Mixed credits occur when a part of the total costs of a commercial export transaction is covered by concessional rates, thus resulting in a lower average cost to the recipient country. The advantage of mixed credits to the donor country is that it frequently secures an export order; the disadvantage is that this order is partly subsidized by the taxpayer. The recipient country benefits from goods that are cheaper than they would be in the absence of mixed credits. The disadvantage to the recipient country is that it may accept the cheapest offer even if for technical, maintenance, or other reasons another choice would have been preferable. Tied aid occurs when donors provide aid to developing countries on condition that the necessary goods and services are bought from the donor country rather than through open competitive bidding on the world market.

While there appears to be general agreement on the competitive advantage that mixed credit and tied aid has given other nations vis-a-vis U.S. exporters, the extent of this impact is controversial (see box 8-G). Various limiting terms and conditions governing tied-aid credits have been negotiated by the United States to reduce the incidence of trade-distorting credits by competitors. The "Arrangement on Guidelines for Officially Supported Export Credits" agreement of March 1987, amended in July 1988, governs OECD member practices. The "arrangement" requires, for example, that tied-aid credit packages should include a substantial grant element (a minimum 35-percent grant for middle-income LDCs and a 50-percent grant for low-income

LDCs). The OECD rules also require advanced notification of tied-aid credit deals. The purpose of these arrangements is to discourage mixed credit financing by increasing their cost to the donor country.

The United States has recently initiated a more active policy. In 1986 Congress authorized a 2-year, $300 million tied-aid War Chest as part of the Export-Import Bank Act Amendments (Public Law 99-472). Due to the failure to establish clear violations, a limitation of the fund to defensive purposes only, and a generally negative view of use by the Administration, the War Chest was seldom used up to 1989. In September 1990, in support of negotiations on tied-aid credits, the Administration indicated that it had decided to use all available budgetary resources, including the War Chest. In fiscal year 1991, $500 million[19] has been allocated to a tied-aid pool for enhancing U.S. industrial competitiveness. The focus is to be on Indonesia, Pakistan, the Philippines, and Thailand. Project sectors targeted include electric power as well as telecommunications, transportation, and construction equipment. Two recent energy projects for which the War Chest has been used are the Uruguay Power Project (versus a French 35 percent grant tied-aid credit offer, with the U.S. bidder ultimately winning this contract); and the Philippines Barge-Mounted Power Plant, where the offer is opposing Japanese and British concessional offers.

Even in the absence of tying, U.S. exports of energy technologies might still encounter difficulties. U.S. competitors allocate a large share of their aid funds to the energy sector and to capital infrastructure. In contrast, the United States directs a larger fraction of aid for structural development and policy reform. AID capital project support has largely disappeared except for the programs in Egypt and Pakistan. In these two cases, the large-scale use of Economic Support Funds occurs without tying to U.S. procurement.

Effective tying may also require the development of complementary institutional structures. In its current form, tying has been found to be ineffective in some cases in stimulating new markets for U.S. goods, and has sometimes complicated and hindered project operations.[20] A recent GAO[21] report found evidence of U.S. source and origin requirements requiring ''countless waivers in Egypt for lack of interest on the part of U.S. suppliers or unavailability of certain commodities from U.S. sources. Furthermore, the Cargo Preference Act to ship in U.S. flag vessels cost up to 5 times that of non-U.S. flag vessels and at times caused lengthy delays while shipments were consolidated for shipment in U.S. vessels.'' Increasing the project element in economic security funds also requires substantially more up-front planning and analysis.

Reducing Poverty and Improving Social Welfare

The reduction of poverty is a central aim of U.S. foreign assistance. Despite rapid rates of economic growth experienced in the developing countries from the 1950s to the end of the 1970s, the numbers of people living in poverty in the developing world continued to increase. In recognition of the need for special programs targeted at the poor, Congress directed AID to pay increased attention to the development problems of the poorest of the poor. Fulfilling this policy objective has implications for energy, which is an essential component in meeting basic needs. While reliable supplies of high grade commercial fuels are essential for raising productivity and living standards, they are currently used primarily by the modern urban sector of the economy. However, many of the developing world's population live in rural areas with low standards of living based largely on low resource farming. Rural populations have little access to commercial fuels and technologies and only limited connection with the modern economy.

Additional attention to energy for rural development could be considered (see option 19 of table 8-2). Rural industry is a major user of biomass energy in rural areas and an important source of rural employment and cash income; its energy needs also

---

[19] $100 million from EXIM's tied-aid credit fund, $300 million in EXIM-guaranteed commercial loans, and $100 from AID Economic Support Funds.

[20] ''Delivery time of U.S. equipment has been long, and inoperative U.S. vehicles, pumps, and other equipment litter the Sahel for want of spare parts, maintenance skills, or operating funds... In addition, these ''buy American'' requirements have led to use of inappropriate capital-intensive technologies ... [and] greatly diminished its value to them [less developed countries].'' U.S. Congress, Office of Technology Assessment, OTA-F-308, *Continuing the Commitment; Agricultural Development in the Sahel* (Springfield, VA: National Technical Information Service, 1986), p. 105.

[21] U.S. General Accounting Office ''Foreign Assistance; AID Can Improve Its Management of Overseas Contracting'' GAO/NSIAK-91-31, October 1990, pp. 23-24.

> ### Box 8-G—Tied Aid
>
> According to one study[1] the tied-aid affected market for telecommunications, power generation, computers, and transportation is considerable, estimated to be about $10 to $12 billion per year, with losses to U.S. exporters on the order of $2.4 to $4.8 billion. Furthermore, according to this report, long run costs could be much greater. An earlier Eximbank report,[2] on the other hand, estimates a substantially lower figures, $400 to $800 million per year.
>
> Other factors, however, suggest that losses may be greater than those estimated by Eximbank:
>
> - Tied-aid credits (which include a 35 percent grant element) have substantial distributive impacts since they go heavily to middle-income developing countries. Given limited total aid dollars, this allocation subtracts from the concessional resources available for poorer countries.
> - The closure of the European and Japanese markets for heavy electrical equipment to foreign competition as matters of national policy, appears to produce financial surpluses to the domestic supply industries of these countries, which are then available to underwrite export sales to the United States and developing-country markets.
> - Use of tied engineering and consultancy services, a practice widely used by the Japanese, results in specifications favoring donor's suppliers. Partly as a result, some 85 percent of partially untied development assistance (about 37 percent of Japanese ODA loans) results in procurement from Japan, and for fully untied loans (62 percent of Japanese ODA loans) about 60 percent of procurement is in Japan. Furthermore, Japanese ODA is largely devoted to project assistance and relies heavily on loans (about two-thirds at near market rates).
> - Frequent failure of OECD competitors properly to follow notification procedures for tied-aid credits leads to a competitive advantage.
>
> ---
> [1] Ernest H. Preeg, "The Tied Aid Credit Issue," Center for Strategic and International Studies, Washington, DC, 1989.
>
> [2] Export-Import Bank, "Report to the U.S. Congress on Tied Aid Credit Practices," Washington, DC, April 1989.

merit attention. The improved economics of decentralized renewables, such as photovoltaics or mini-hydro, suggest renewed emphasis on the institutional obstacles that deter diffusion. The domestic energy needs of the urban poor are currently receiving little attention, even though energy used for cooking contributes heavily to poor urban air quality in many areas.

AID funding of energy projects targeted at the poor is modest. In the decade of the 1980s, for example, total AID energy funding to Africa was just over $1 million annually. Application of the technologies suggested above will require a variety of complementary inputs, which AID could facilitate, including policy reform, credit programs, and technical assistance.

Rural energy receives relatively little attention from the MDBs as well. The World Bank, which earlier in the 1980s had substantial lending programs for rural electrification, has reduced the number of projects in recent years. The recent reorganization of ESMAP may lead to lessened interest in energy projects for the rural and urban poor. As previously constituted, ESMAP contained a special household energy unit that addressed the technical, economic, and social issues related to energy use and supply in lower income urban and rural households and in rural industries. The reorganized ESMAP will deal only with commercial fuels and confine itself to fewer "priority" countries. This reorganization is expected to promote better energy sector planning in the selected countries and be more instrumental in increasing investments, but could result in less of a role than before in rural and household energy supplies.

One of the problems in rural energy projects is securing the high level of technical assistance needed. Donor agencies could examine ways of delivering such technical assistance through non-governmental oranizations with demonstrated ability to deliver technical assistance and financing at the local level (see option 20 of table 8-2).

Protecting the Environment

Congress has become increasingly concerned with the problem of achieving sustainable economic growth in developing countries and the growing role of developing countries in global environmental problems. This concern is expressed in directives to AID and MDBs to take into account environmental impacts in their developing country programs. Progress in these efforts may need to be monitored (see option 21 of table 8-2). There may also be a need to coordinate the activities of the many U.S. agencies and NGOs that have environmental activities in developing countries. The CORECT program or others could serve as a model. The Environmental Protection Agency may be a logical host for such a program (see option 21 of table 8-2).

Environmental protection in many countries is still in its infancy due to competing economic and social priorities, limited resources, and lack of knowledge and experience in environmental management and protection. The policy infrastructure for environmental protection is as yet poorly developed. There is a need for technical assistance and advice to support environmental impact monitoring and mitigation strategies and to establish the government policies and institutions for environmental protection (see option 21 of table 8-2). The United States could help provide technical assistance in these areas.

One of the obstacles to securing environmental protection in all countries is that the cost of environmental damage is typically not included in prices, nor is use of natural resources included in the current system of national accounts. Unsustainable exploitation of forests for timber exports, for example, appears in the current system of national accounts as increases in the gross domestic product, while ignoring any environmental damage associated with this exploitation and the reduction of the natural resource stock. Congress has already instructed the U.S. Ambassador to the United Nations to adopt economic accounting procedures that include natural resource values and services in the UN accounting system. The UN has agreed to incorporate these concepts to the extent possible in its next revision of its standardized national accounts system, which serves as the national accounting model for all market or mixed economies.

Concern over global warming focuses particular attention on large carbon dioxide emitters including both industrial and developing countries—China, India, Brazil, Indonesia, and Mexico. The United States, however, does not have a strong bilateral relationship with all of these countries. Congress recently directed AID to help key low and middle-income countries likely to become major emitters of greenhouse gases whether or not they are currently AID countries. The United States is also cooperating with the Global Environmental Facility through a special AID program.

Population plays a major role in long term environmental sustainability. Under the Foreign Assistance Act as amended in 1965, the U.S. position was that family planning could be considered an important contributor to economic development and improved health and nutrition.[22] Most nations now firmly support family planning assistance. At the World Population Conference held in Mexico City in 1984, however, the United States reversed this position and introduced new restrictions on family planning assistance in AID. Two important international population assistance programs lost U.S. funds—the International Planned Parenthood Federation, and the United Nations Population Fund (UNFPA). According to the UNFPA, more assistance, particularly for family planning services, is needed to ensure the stabilization of the world's population. The additional direct cost of providing contraceptive services would be under $1 billion per year, but several billion dollars per year would also be needed for a range of supporting activities such as education women's programs, and research and development (see option 22 of table 8-2).

Setting a Good Example

U.S. commercial energy consumption per capita is almost 15 times higher than the average per capita consumption for all developing countries, and over 60 times higher than in the poorest developing countries. Strong efforts to improve efficiencies, and the use of renewable energy at home are needed to provide credibility to U.S. policies in developing countries (see options 23 and 24 of table 8-2).

Evidence of energy efficiency gains in the United States is provided by the leveling of energy consumption after 1973—a marked change from the previous steady annual increases. Since 1986, however, energy consumption has been rising steadily, in virtual "lockstep" with the gross national product

---

[22] P.J. Donaldson and C.B. Keely, "Population and Family Planning: An International Perspective," Family Planning Perspectives 20(6):307-311,320 (November/December 1988; U.S. Congress, Library of Congress, Congressional Research Service, "International Population and Family Planning Programs: Issues for Congress," IB85187 (Washington, DC: Mar. 30, 1990); U.S. Congress, Office of Technology Assessment, *Changing by Degrees: Steps to Reduce Greenhouse Gases*, OTA-O-482 (Washington, D.C.: U.S. Government Printing Office, February 1991).

and posting substantial increases in per capita consumption. Further efforts to achieve perceptible and measurable improvements in energy efficiencies in the United States would reinforce policies designed to improve energy efficiencies in developing countries. Such actions to improve efficiency at home would have collateral economic, energy security, environmental, and competitiveness benefits for the United States. Also, an expanding market for energy efficient technologies lowers manufacturing costs, thus adding to U.S. competitiveness. A particularly useful initiative could be to ensure, as requested by Congress, that U.S. facilities in developing countries are highly energy efficient and take advantage of renewable energy sources where feasible. U.S. embassies, office buildings, and residences could serve as show pieces for advanced U.S. energy efficient technologies.

Appendixes

Appendix A
Capital and Life Cycle Costs for Electricity Services

This appendix calculates the capital and life cycle costs of providing electricity services to users. This is done in three stages. First, a general framework for calculating costs is presented. Second, the capital cost of providing electricity supplies is determined. Third, the total systemwide—including both electricity supply and end-use equipment—capital costs and life-cycle operating costs are determined for standard and energy efficient equipment.

General Framework for Calculating Costs

All costs are in constant 1990 U.S.$. Where necessary, other currencies are converted to U.S.$ in the year cited and then the U.S.$ are deflated to 1990 values using GNP deflators. The factors used to convert foreign currencies to U.S.$ are generally market exchange rates as these are capital goods.

Capital costs are calculated using a simple capital recovery factor (CRF) method.[1] This method divides the capital cost into an equal payment series—an annualized capital cost—over the lifetime of the equipment. For example, the CRF for a 30-year lifetime and a discount rate of 7 percent is 0.080586. A $1,000 widget then has an annualized capital cost of $80.586, including interest payments and in constant dollars, each year of its 30-year expected lifetime.

In the base case, a uniform real discount rate of 7 percent is used for both the supply and end-use sectors. The discount rate used is intended to be the equivalent of a societal discount rate. In comparison, the real cost of capital in the United States averaged about 3 percent between 1950 and 1980, rising briefly to nearly 10 percent in 1983, before dropping back to about 6 percent and below from 1987 onward.[2] Capital costs in developing countries vary widely. Sensitivity analyses, discussed below, examine the effect of varying the discount rates and other parameters for both the supply and end-use sectors.[3]

The total systemwide capital cost of each energy service and each choice of equipment to deliver that service is determined by adding: 1) the annualized capital costs of the end-use equipment, and 2) the corresponding annualized capital costs for the electricity supply system needed to power the end-use equipment. This latter value, (2), is determined by averaging the number of kWh used over the year by the equipment to get an equivalent average kW demand and then multiplying by the corresponding annualized cost of electricity supply per kW of delivered power (as determined in detail below).

Note, however, that the capital cost of delivering a kW of power is substantially greater than the cost of a kW of supply capacity. This is due to the additional capital costs of, e.g., coal mining equipment, transmission and distribution equipment, etc. that are needed on the supply side; because supply equipment can deliver only a fraction of its rated capacity due to maintenance needs, breakdowns, imperfect matching of the demand to the available capacity, the need to maintain reserve capacity, etc.; and because of losses in the system before the power is delivered to the consumer.

More sophisticated analyses of capital and operating costs are possible.[4] These include, for example, taking into account the higher cost of delivering electricity during the peak of the system load. Such refinements are avoided here in order to make the presentation as simple and transparent as possible, while still presenting reasonable estimates of the relative costs of different means of delivering desired energy services.

Finally, the following estimates of capital and lifecycle operating costs have a number of highly conservative factors built in. The cost of electricity supply is estimated on the low side. In particular, factors that lower the estimated cost of electricity supply include low assumed values for the capital costs of coal mining, utility generation plants, transmission and distribution equipment, and other capital investments; high assumed values for the capacity utilization levels of generation equipment and Transmission and Distribution (T&D) equipment; and low assumed losses in T&D systems; among others. These are detailed in the following section.

In contrast, the cost of end-use efficiency is intentionally estimated to be higher than it is likely to be in practice, specifically:

---

[1]The CRF = $\{i(1+i)^n\}/\{(1+i)^n-1\}$ where i is the discount rate and n is the lifetime or period of capital recovery of the system.

[2]Margaret Mendenhall Blair, "A Surprising Culprit Behind the Rush to Leverage," *The Brookings Review*, Winter 1989/90, pp. 21.

[3]Additional factors—such as levelizing increasing real costs of inputs like labor, energy, etc.; shadow pricing/opportunity costs; economies of scale and learning; byproduct credits; environmental costs; etc.—are not included in order to keep the model calculations as simple as possible so as to clearly show the overall driving financial forces in the system.

[4]Electric Power Research Institute, "Technical Assessment Guide: Electricity Supply," and "End-Use Technical Assessment Guide," various volumes, Palo Alto, CA various years; Jonathan Koomey, Arthur H. Rosenfeld, and Ashok Gadgil, "Conservation Screening Curves to Compare Efficiency Investments to Power Plants," *Energy Policy*, October 1990, pp. 774-782.

- **Direct Substitutions Only.** Only direct substitutions of more efficient for less efficient end-use equipment are considered. This excludes many highly capital-conserving and life-cycle cost-effective investments, such as insulating buildings, using improved windows, daylighting, improved industrial processes, and many others.

- **No Synergisms.** Synergisms between efficient equipment are excluded. For example, high efficiency lights, refrigerators, and other equipment reduce heat loads in buildings that must otherwise be removed by ventilation and air conditioning systems. For each kW of internal heat load that is avoided, roughly one-third kW of electricity required by an air conditioner to remove this heat is saved.

- **No Downsizing Credits.** With more efficient lights and refrigerators, etc. lowering internal heat gains, the size of the needed air conditioner to cool the space can be reduced. Similarly, with more efficient pumps/fans, ASDs, etc. the size of the motor needed could be reduced. Such downsizing is not considered here.

- **No Credit for Manufacturing Volume.** Margins for efficient equipment are sometimes larger than those for standard equipment. Manufacturing costs for efficient equipment may also be higher than for standard equipment due to the smaller production volumes. The impact of such learning can be substantial. As shown in chs. 2 and 3, the real cost of refrigerators in the United States declined by a factor of 5 between 1950 and 1990 due to improved materials and manufacturing methods. No credits were given for expected cost reductions at higher volume production or for reduced manufacturer margins.[5]

- **Capacity Increments.** Capacity increments are assumed to be added as needed rather than in large lumps as is the case in reality. This reduces the effective cost of supply.

Together, these low-side supply costs and high-side end-use costs are intended to bias the case against efficient equipment in order to be as conservative as possible. Even under these circumstances, energy efficient equipment shows systemwide capital savings, life-cycle operating savings, and energy savings as illustrated in various figures in chs. 1, 2, and 3.

Capital Costs of Electricity Supplies

This section calculates the cost of delivering a kW of electric power to the end-user. Capital costs for electric power systems are usually cited in terms of $/kW of electricity generation **capacity**. Such figures are substantial understatements of the full capital costs of delivering electricity **supplies**.

Typical capital costs for generation capacity range from $500/kW for a conventional gas turbine used for providing peak power to over $2000/kW for current nuclear power plants and even higher for current photovoltaic systems. The World Bank estimates that developing countries, under current power expansion plans, will invest $775 billion (1990 U.S.$) during the 1990s for 384 GW of additional capacity, including generation, T&D and other capital expenditures or $2,018/kW total. This cost is divided into 60-percent generation, 31-percent T&D, and 9-percent general.[6]

This overall figure of $2,018/kW total assumes that existing plant and equipment can be used more intensively than at present,[7] a highly desirable opportunity. Without this credit, the World Bank estimates the capital cost of new capacity at $2,618, including generation, T&D, and other capital expenditures, or $1,568/kW for generation equipment alone (assuming the same percentage breakdown as above). Corresponding estimates of the capital cost of new generation equipment from the World Energy Conference are $2,310-$2,770 (1990 U.S.$), shown in table A-1,[8] and costs for the United States are shown in table A-2. It should be noted that the capital costs of fossil steam and gas turbine plants listed in table A-2 are "recommended best practice" estimates and do not include any contingency for unexpected costs (or savings) incurred in actual field construction. The costs of capacity for various generation technologies are examined in more detail in chapter 6 and appendix B.

Estimates of generating capacity alone do not, however, reflect the full capital cost of delivering electricity supplies to users. First, the capacity to produce electricity is not the same as actually producing it. Typical baseload coal-fired plants, for example, might be available for operation 70 percent of the time. The remaining 30 percent of the time they are shut down for routine maintenance or due to breakdowns. Additional generating capacity is needed to make up this shortfall.

In operation, electric power systems normally maintain a "spinning reserve" of perhaps 5 to 7 percent of the

---

[5]Note, however, that a constant retail markup of 100 percent over estimated manufacturing cost was assumed for refrigerators.

[6]Edwin A. Moore and George Smith, "Capital Expenditures for Electric Power in the Developing Countries in the 1990s," Washington, DC World Bank, February 1990, Industry and Energy Department Working Paper, Energy Series Paper No. 21 (Washington, DC: World Bank, February 1990).

[7]Specifically, it assumes that reserve capacities can be reduced from the 1989 level of 42.5 percent to a 1999 level of 36.3 percent. See Moore and Smith, op. cit., annex tables 2.1 and 2.2.

[8]World Energy Conference, "Investment Requirements of the World Energy Industries, 1980-2000," London, 1987.

Table A-1—Capital Costs of Delivered Electricity Supply

	World Bank parameters		World Energy Conference		
	Low[a]	High[b]	Low	High	OTA base case
Coal mining and transport[c]	$55/kW	$126/kW	$55/kW	$126/kW	$55/kW
Generation	$1,211/kW	$1,568	$2,310/kW	$2,770	$1,536[d]
Capacity factor	50%	50%	50%	50%	60%
T&D loss	15%	15%	15%	15%	10%
Capacity needed	2.36 kW	2.36 kW	2.36 kW	2.36 kW	1.85 kW
Firm kW	$2,858	$3,700	$5,451	$6,537	$2,844
T&D	$625/kW	$812	$1,897/kW	$2,770/kW	$625
T&D loss	15%	15%	15%	15%	10%
Capacity factor	75%	75%	75%	75%	75%
Capacity needed	1.57 kW	1.57 kW	1.57 kW	1.57 kW	1.48 kW
T&D Capacity	$981	$1,275	$2,978	$4,349	$926
Other capital[e]	$363	$471	NA	NA	$337
Total cost ($/delivered kW)	$4,257	$5,572	$8,484	$11,012	$4,162

NOTE: Optimistic values are used throughout this table for the generation capacity factor (50%) and for the T&D capacity factor (75%), among others. The assumed values for these and other key parameter—generation capacity factor (60%), T&D capacity factor (75%), and T&D losses (10%)—are substantially more optimistic in the OTA base case than those observed in practice.

[a] The $ values from the World Bank include a credit for improving the utilization of existing capacity and so are not strictly comparable to the capital cost of new electricity supply capacity.
[b] These $ values simply do not include a credit for better utilization of existing capacity as in the "low" case and use high values for coal mining. In other respects, these $ values are the World Bank baseline case.
[c] Based on World Energy Conference estimates cited in text.
[d] Electric Power Research Institute. See table A-2.
[e] This assumes that the other capital requirements—buildings, other equipment, administrative support, etc.—increase directly with the generation capacity, as indicated by the World Bank.

SOURCE: Edwin A. Moore and George Smith, "Capital Expenditures for Electric Power in the Developing Countries in the 1990s," Washington, DC World Bank, February 1990, Industry and Energy Department Working Paper, Energy Series Paper No. 21; World Energy Conference, "Investment Requirements of the World Energy Industries, 1980-2000," London, 1987.

Table A-2—Typical Capital Costs and Capacity Factors for Existing U.S. Electricity Generating Plants, 1990 U.S.$

Prime mover	Capital cost, $/kW capacity	Capacity factor, percent	T&D loss[a] percent
Fossil steam	$1,536[b]	50	6
Gas turbine	$ 500[c]	7	6
Nuclear	$2,580[d]	62[e]	6
Hydroelectric	—[f]	33	6

[a] T&D loss is the U.S. average of about 6 percent.
[b] Average cost of 300 MW coal-fired steam plants in the United States under EPRI recommended practice, table 7-4, EPRI.
[c] Average cost of conventional simple and combined cycle gas turbines operating on distillate or natural gas, exhibits 15-19, EPRI.
[d] Note that this is the average cost for the 63 nuclear power plants put into operation in the United States since 1975. Dollars are mixed current dollars over the construction period and then discounted from the date of operation to 1990 U.S.$. Consequently, the costs are somewhat underestimated. The comparable average cost for all 108 U.S. nuclear power plants in operation is $1,834. The estimated cost for an advanced reactor design has been estimated by EPRI at approximately $1,667.
[e] Note that a 10-year unweighed average, 1980-1990, capacity factor for nuclear plants is 58.8%. The figure shown is for 1989. Energy Information Administration, "Monthly Energy Review," May 1991, U.S. Department of Energy, DOE/EIA-0035(91/05)
[f] Too variable to be readily quantified here.

SOURCE: Steam and gas turbine capital costs are from the Electric Power Research Institute, "Technical Assessment Guide: Electricity Supply, 1989," EPRI P-6587-L, vol. 1, Rev.6, September 1989, Palo Alto, CA.; Nuclear power costs are from Energy Information Administration, "Nuclear Power Plant Construction Activity, 1988," DOE/EIA-0473(88); capacity factors are from: Energy Information Administration, "Annual Energy Review, 1989," U.S. Department of Energy, 1989, tables 89, 93.

system load in order to handle normal short-term fluctuations in demand.

Excess generation capacity is also built into the overall system in order to handle peak loads—for example, on exceptionally hot summer days when everyone with an air conditioner turns it on. Typical reserve margins on well-run systems might be 20 to 30 percent greater than the maximum peak load including the spinning reserve. That reserve margins aren't larger than this—corresponding to the plant availability cited above, or 43 percent (1/0.7)—is because some of the routine maintenance of the plants can be scheduled during nonpeak times and because some of the peak is met by using additional equipment specifically designed just for peak loads, such as low capital cost—but high fuel cost—gas turbines.

Finally, because generation equipment comes in large units, there is typically a stairstep increase in system capacity above overall demand.

As a consequence of these considerations, the typical power system produces electricity at only a fraction of the capacities of its individual plants. For example, the average (weighted by country) generation capacity factor—measured as the ratio of actual gross generated kWhs divided by the potential kWhs generated by the power plant if it ran at full capacity all the time—across 98 developing countries is 36 percent. (This very low generation capacity factor reflects serious institutional and operational problems in many of these power systems. These issues are discussed in ch. 6.) In comparison, the generation capacity factor of the United States[9] was 46 percent in 1989. Only 10 of the developing countries reviewed by the World Bank had generation capacity factors of greater than 50 percent. For example, Brazil had 50 percent, China 55 percent, Colombia 57 percent, Egypt 59 percent, and Kenya 52 percent.[10]

This average generation capacity factor includes both peaking and baseload power plants. Actual operating experience for different types of prime movers is shown in table A-2 for the United States. Generation capacity factors ranged from 7 percent for gas turbines, to 50 percent for fossil steam, to 62 percent for nuclear plants. Corresponding generation capacity factors for prime movers in India are shown in table A-3. Low generation capacity factors increase the capital cost of delivering electricity to users.

Table A-3—Capacity Factors for Electricity Generating Plants in India

Prime mover	Capacity factor
Coal-steam	42.2%
Nuclear	44.8%
Oil and gas	28.5%
Hydro	37.5%

SOURCE: Ashok V. Desai, "Energy, Technology and Environment in India," contractor report for the Office of Technology Assessment, December 1990.

Second, to determine the capital cost of supplying electricity to users, the large losses of electricity between the plant and the customer must be considered. Transmission and distribution losses in developing countries include both technical losses and nontechnical losses due to billing errors, unmetered use (theft), and other factors. Technical losses of 15 percent are typical. This relatively high level of loss is due to poor system power factors, long low voltage lines to dispersed consumers, inefficient equipment, and other factors. System improvements can reduce this high level of loss, but it will probably remain higher than the 6 percent typical for the United States due to the large dispersed rural demand likely in the future. Nontechnical losses are not considered here.[11] To deliver 1 kW of power to a consumer, then, requires 1.18 kW of power to be generated at the utility when 15-percent transmission and distribution losses are included. Assuming an optimistic 50-percent generation capacity factor and with 15-percent T&D losses, to deliver 1 kW of power on average to consumers, a generation capacity of 2.36 kW would be needed.

Third, estimates of the capital cost of generation capacity alone ignore the cost of transmission and distribution equipment to deliver it to customers. The World Bank estimates the cost of T&D capacity in current developing country expansion plans at $625 to $812/kW. The World Energy Conference estimates the cost of new T&D capacity at $1,900-$2,770/kW. Costs in developing countries are likely to be particularly high because of the extension of electric power grids into rural areas, an expensive undertaking. This capacity, too, must be augmented by the T&D losses of 15 percent. In addition, utilization capacity factors for T&D systems are often quite low. Lines to residential areas are used primarily in the evenings and on hot summer afternoons and the T&D capacity is substantially underused the rest of the time. Lines to commercial areas are used primarily during weekdays, but carry little load during evenings and

---

[9]Note that the generation capacity factor for the United States is net generation—not including the electricity consumed in operating the power plant—rather than gross generation.

[10]Jose R. Escay, World Bank, Industry and Energy Department, "Summary Data Sheets of 1987 Power and Commercial Energy Statistics for 100 Developing Countries," Industry and Energy Department working paper, energy series paper No. 23 (Washington, DC: World Bank, March 1990).

[11]Mohan Munasinghe, Joseph Gilling, Melody Mason, "A Review of World Bank Lending for Electric Power" (Industry and Energy Department working paper energy series paper No. 2 (Washington, DC: World Bank, March 1988), p. 33.

weekends, etc. This results in an increased capital investment per kW of electricity supply delivered to the user.

Fourth, there are other capital costs associated with electricity supplies, such as investment in buildings, administrative support offices, etc. The World Bank estimates these capital costs at 9 percent of the total required investment, or $182 to $236/kW of capacity.

Fifth, the capital costs of producing the fuels to run thermal power plants must also be included if the capital costs of supply expansion is to be accurately depicted. The World Energy Conference estimates the capital investment in coal mining at $0.38 to $1.00/GJ (1990 U.S.$)[12] plus $0.20 to $0.33 per GJ[13] for transportation equipment. For thermal power plants with conversion efficiencies of 33 percent, 1 GJ produces 92.6 kWh or 0.01057 kW-yr. This is equivalent to $55 to 126/kW of annual electric power output.

The corresponding capital costs for oil production are $0.80 to $1.10/GJ[14] for exploration and production, and $0.18/GJ[15] for downstream investment in storage tanks, refineries, and transportation equipment. This is equivalent to $93 to $121/kW of annual electric power output.

For gas, the corresponding capital costs are $0.04 to $0.16/GJ[16] for exploration and production, and $0.58 to $1.12/GJ[17] for natural gas transport and distribution infrastructure. For a thermal power plant with a conversion efficiency of 33 percent, this gives a capital cost of $59 to $121/kW of annual electric power output. Obviously, for coal, oil, and gas there can be wide variations from these estimates based on local conditions.

These capital costs and capacity factors can now be used to estimate the approximate total capital cost of supplying electricity, as shown in table A-1. Estimated costs of a firm kW of power range from $4200 to over $11,000. These values are comparable to those found in more detailed analyses of electricity supply options in India[18] and in Brazil.[19]

To be as conservative as possible, OTA uses more optimistic values for its base case than those found in or estimated for developing countries or, indeed, the United States (table A-1). The extent of this conservatism should be noted.

First, OTA assumes a low capital cost for coal production. This value is in part based on the World Energy Conference assumption that intensive energy efficiency improvements will reduce the elasticity of energy use with economic growth in developing countries to just 0.7—that is, that energy use will increase at just 70 percent of the rate at which developing country economies grow. In turn, this results in greater availability of low cost coal supplies, according to the World Energy Conference.

Second, OTA optimistically assumes that generation capacity factors can be raised from the current average of 36 percent (by country) to 60 percent—a level higher than those found in all but two of the 98 developing countries reviewed by the World Bank.[20] This is also better, for example, than the 56-percent capacity factor projected for Brazil by Eletrobras—the federal utility holding company—for the year 2010.[21] The cost of new capacity was chosen to be $1,536 corresponding to the estimates by the U.S. Electric Power Research Institute for new coal-fired steam plants constructed under "good" practice conditions.[22]

A more comprehensive analysis would examine the capacity factors and costs for each component of the electricity supply system, including base load and peak load generation capacity.

---

[12] Table 5.4 World Energy Conference, 1987, op. cit., footnote 8. Excludes South Africa.

[13] Table 5.7 divided by Table 5.2 World Energy Conference, 1987, ibid.

[14] Table 3.2 of World Energy Conference, 1987, ibid.

[15] Table 3.5 of World Energy Conference, 1987, ibid.

[16] Table 4.3 World Energy Conference, 1987, ibid.

[17] Table 4.7, World Energy Conference, 1987, ibid.

[18] Amulya Kumar N. Reddy et al., "Comparative Costs of Electricity Conservation: Centralized and Decentralized Electricity Generation," *Economic and Political Weekly* (India), June 2, 1990, pp. 1201-1216.

[19] Jose Goldemberg and Robert H. Williams, "The Economics of Energy Conservation in Developing Countries: A Case Study for the Electrical Sector in Brazil," in David Hafemeister, Henry Kelly, and Barbara Levi, "Energy Sources: Conservation and Renewables," American Institute of Physics, New York, NY, 1985.

[20] Note that Cape Verde achieved a generation capacity factor of 76 percent and Madagascar achieved a level of 64 percent. These high capacity factors, if correctly reported, are probably unique to these very small systems and may be due to the lack of reserve margins, not meeting peak loads, having little or no backup capacity, or other unusual features. See Jose R. Escay, op. cit., footnote 10.

[21] Howard S. Geller, "Electricity Conservation in Brazil: Status Report and Analysis," Contractor Report for the Office of Technology Assessment, November, 1990.

[22] For conceptual clarity and again to be as conservative as possible in estimating costs, this analysis assumes a high generation capacity factor for a baseload coal plant operating under near ideal conditions of almost constant load. This should be contrasted with a typical electric power system which includes a variety of different baseload and peaking plants with differing capital, operating and maintenance, and fuel costs; availabilities and effective capacity factors; and efficiencies.

Third, OTA assumes that nontechnical losses can be ignored and that even with extensive rural electrification T&D losses can be lowered by one-third, from the current 15 percent to 10. In the much longer term, increasing urbanization and various technical improvements may make further reductions possible.

Fourth, OTA assumes that T&D capacity is used at 75 percent of its limit and ignores the frequently low level of utilization in many parts of a typical T&D system, as noted above.

Together, these considerations indicate that the OTA base case is likely to be substantially conservative in its estimate of the capital costs of electricity supply systems.

Calculating Operating Costs for Electric Power Systems and End Use

To determine life-cycle operating costs, it is necessary to know the full cost of electricity, including capital, operations and maintenance, fuel costs, and other factors. Electricity costs for new supplies are estimated at $0.09/kWh for the residential and commercial sectors (1990 U.S.$) and $0.07/kWh for the industrial sector. The lower cost for industry reflects the greater concentration of use—allowing the purchase of bulk supplies, reduced T&D costs, lower administrative overheads, and other benefits.

These values correspond to the 1987 OECD weighted average electricity price of $0.087/kWh (1990 U.S.$). Electricity prices charged in developing countries, however, have a weighted (by electricity sales) average price of $0.048/kWh (1990 U.S.$) but have a marginal cost of production of $0.094/kWh assuming a high 60 percent average system capacity factor.[23] Expected costs of $0.09 to $0.13/kWh (1990 U.S.$) are listed by Jhirad for some 18 commercially available technologies running at a high capacity factor of 75 percent corresponding to baseload applications. Gadgil and Januzzi give marginal costs of production of $0.09/kWh and $0.12/kWh for India and Brazil respectively.[24] OTA has thus chosen the current average cost of electricity or lower in order to be conservative. A more detailed examination of electricity costs are given in ch. 6.

Systemwide Capital and Operating Costs

To complete the analysis, systemwide capital and life-cycle operating costs can now be calculated for standard and energy efficient equipment. A level of services corresponding to a U.S. or Western European standard of living is assumed, as described below and summarized in table A-4. Parameters for each energy service are shown in tables A-5 through A-14 together with notes providing context. The corresponding summary values of systemwide capital, life-cycle operating costs, and electricity use are shown in tables A-15 and A-16.

Residential households are assumed to have five persons; capital costs of end-use equipment are allocated equally among them to get per capita capital and life-cycle costs.

The following discussion of costs is not inclusive. There are many related costs that are not included as they are assumed to be the same for both the standard and the energy-efficient cases. Examples include the wiring, switches, and related capital components within the home, business, or industry—note, particularly, that for industry these related components such as switchgear, pipes and ducts, and related process equipment are a very substantial part of the total systemwide costs; labor and other costs associated with actually putting equipment into service; and many other costs. There are also many other cases where energy efficient equipment is not considered or where it is left out due to it not being cost effective. Several examples are listed below.

Cooking[25]

Cooking levels are scaled by efficiency factors from those currently observed in developing countries—6 GJ per person per year when using wood with a stove thermal efficiency of 17 percent. This corresponds to an electricity consumption in the all electric household of about 2250 kWh/household-year. This is slightly lower than the 2500 kWh/yr observed in, for example, Guatemala (see ch. 2), but is substantially higher than the 700 kWh/household-year observed in the United States. The dramatically lower residential household energy consumption for cooking in the United States is due to a number of factors, including: 1) smaller households—e.g. two people—than assumed here; 2) extensive dining out or purchase of

---

[23] Calculated from Annex 9 of A. Mashayekhi, ''Review of Electricity Tariffs in Developing Countries During the 1980s,'' Industry and Energy Department working paper, energy series paper No. 32, World Bank, Washington, DC November 1990.

[24] A. Mashayekhi, Ibid. David J. Jhirad, ''Innovative Approaches to Power Sector Problems: A Mandate for Decision Makers,'' (New Delhi, India: PACER Conference, Apr. 24-26, 1990, U.S. Agency for International Development, Washington, DC) A more general discussion of electric power pricing issues can be found in Mohan Munasinghe, ''Electric Power Economics'' (London, England: Butterworth & Co, 1990). Ashok Gadgil and Gilberto De Martino Januzzi, ''Conservation Potential of Compact Fluorescent Lamps in India and Brazil,'' Lawrence Berkeley Laboratory, report No. LBL-27210, July 1989.

[25] Additional detail on cooking can be found in, Samuel F. Baldwin, ''Cooking Technologies,'' Office of Technology Assessment, staff working paper, 1991.

Table A-4—Assumed Levels of Electricity Services and Other Parameters

Energy service	Level of service provided
Residential/commercial	Five people per household.
Cooking	Comparable to cooking energy requirements in developing countries today, scaled by efficiency. This results in per capita consumption levels somewhat above those observed in the United States (see text for details). Cooking energy is allocated 75 percent to electric resistance/gas stoves in the standard/high-efficiency cases, respectively, and 25 percent to microwave ovens in both cases.
Water heating	50 liters of water heated 30 °C per day, corresponding to the U.S. level of hot water usage.
Lighting	Lighting levels at the midrange of industrial countries. Residential lighting is the equivalent of six hours of lighting by 60 W incandescent bulbs (four hours with one bulb and two hours with a second) per capita per day. Commercial and industrial lighting is equivalent to 10 hours, 260 days per year by 4 standard 40 W fluorescents per capita.
Refrigeration	One 510-liter top-mount freezer, automatic defrost refrigerator, corresponding to the most popular type in use in the United States.
Air conditioning	1,200 kWh of electricity for cooling (SEER=8) annually per capita. This is slightly lower than the 1,400 kWh/year used per capita in the United States.
Electronic information services	One color TV used about 5 hours per day.
Industrial motor drive	Industrial motor drive power consumption of 300 W/capita in the base case, comparable to the 308 W used in the United States.

NOTE: This does not cover all energy services, nor all costs associated with a particular energy service.
SOURCE: Office of Technology Assessment, 1992.

Table A-5—Cooking, OTA Base Case Parameters

Stove/fuel	Stove parameters		Efficiency	Fuel parameters	
	Lifetime	Capital cost		Capital cost	Total cost
Standard case:					
Electric resistance	15	$ 75	63%	$335/kW[a] ($10.6/GJ)	$0.09/kWh ($25/GJ)
Microwave	15	$250	58%	same	same
High efficiency case:					
LPG	20	$ 50	58%	$1/GJ/yr	$7/GJ
Microwave	15	$250	58%	above	above

SOURCE: Note that the same discount rate of 7 percent real is used to calculate a capital recovery factor for the given stove lifetime. The CRF is then used to annualize all capital costs. Energy consumption is scaled by the relative efficiencies of the stoves from a baseline value of 6 GJ/capita (30 GJ/household) with a 17 percent efficient wood stove. The electric resistance stove then uses the kW equivalent of 1.6 GJ.
[a] This is the annualized cost per kW over the 30-year lifetime of the utility power plant, using a total cost per delivered kW of $4,162 as in table A-1.
[b] Note that this is the estimated thermal efficiency of a microwave; in practice, a microwave can realize overall cooking efficiencies substantially higher than conventional cooking devices, particularly for baking.
SOURCE: Office of Technology Assessment, 1992.

"take-out" meals rather than cooking at home; 3) extensive use of highly processed foods such as "minute" rice or TV dinners, etc. rather than cooking unprocessed grains for long periods at home; 4) greater use of high-value foods such as meats that typically do not require as much cooking energy to prepare as unprocessed grains; and others.

In the standard efficiency case, an electric resistance stove provides 75 percent of the required cooking energy and a microwave oven provides the remaining 25 percent. In the high efficiency case, an LPG stove substitutes for the electric resistance stove—reducing upstream capital costs and cutting primary energy consumption by two-thirds, and a microwave oven again provides the remaining 25 percent. Natural gas could readily substitute for LPG, but upstream capital costs for installing and maintaining a pipeline distribution system would vary, depending on the total demand. In the industrial countries, the large winter space heating requirements help justify the high capital costs of a natural gas distribution system.

The high efficiency case summary values for electricity consumption do not include the LPG used for cooking (capital and operating costs for the LPG system are included in the totals, however). If the total systemwide

Table A-6—Costs, Efficiencies, and Lifetimes for Alternative Cooking Technologies

Technology	Efficiency		Stove capital cost $	Lifetime years	Fuel cost $/GJ
	Stove percent	System percent			
Traditional stoves					
Dung	11-15	10-14	0.00	—	0.00
Agricultural residues	13-17	12-16	0.00	—	0.00
Wood	15-19	14-18	0.00	—	0.00
Wood (commercial)	15-19	14-18	0.00	—	1.50
Charcoal	19-23	8-12	3.00	2	4.00
Improved biomass stoves					
Wood	27-32	26-31	6.00	2	1.50
Charcoal	29-34	13-17	8.00	3	4.00
Wood II	40-44	38-42	10.00	3	1.50
Liquid stoves					
Kerosene wick	40-45	36-41	20.00	10	5.00
Kerosene pressure	45-50	41-45	40.00	10	5.00
Alcohol wick	40-45	33-37	20.00	10	10.00
Alcohol pressure	45-50	37-42	40.00	10	10.00
Gas stoves					
Central gasifier	55-60	39-42	20.00	10	1.50
Site gasifier	40-45	39-44	50.00	4	1.50
Biogas	55-60	54-59	20.00	10	0.00[a]
LPG	55-60	48-53	50.00	20	7.00
Natural gas	55-60	53-58	20.00	20	1.50
Electricity					
Resistance	60-65	17-21	75.00	15	25.00
Microwave	55-60	16-20	250.00	15	25.00
Solar					
Solar box oven	25-30	25-30	25.00	5	0.00

[a]There are substantial capital costs for the fuel system, as well as large amounts of labor involved in collecting the biomass and dung to be put into the digester. For a detailed discussion of this data, including fuel cycle capital costs, see Baldwin, below.

SOURCE: Samuel F. Baldwin, "Cooking Technologies," Office of Technology Assessment, U.S. Congress, staff working paper, 1991.

energy consumption values are converted to their primary energy equivalents, using a fuel to end-user conversion efficiency of 33 percent, then the high efficiency case primary energy consumption—including LPG—increases to 34.3 GJ. The corresponding ratio of primary energy use between the efficient and standard cases (efficient case divided by standard case) is 59.4 percent, compared to their ratio for electricity consumption of 57.4 percent.

Water Heating

The OTA base case assumes that each household will use 250 liters of 50 °C water daily for a per capita consumption of 50 liters per day. The standard efficiency equipment is an electric resistance storage water heater; the efficient case is a solar water heater with a storage tank and electric resistance heater backup. The solar water heater is assumed to provide 85 percent of the household water heating requirements on an annual average.

Lighting

The OTA base case assumes that residential, commercial, and industrial lighting services will total about 30 million lumen hours per person per year. This is in the middle of the range of lighting levels currently provided in the industrial countries (see ch. 3). No additional use of daylighting or other such techniques is considered. This lighting is assumed to be provided by two 60 W incandescents—one burning 4 hours and one 2 hours each day—within the home for each person and by, on average, a bank of four standard 40 W fluorescent tubes with conventional core-coil ballasts in the commercial and industrial sectors for each person that are used 10 hours per day, 5 days per week, 52 weeks per year. Obviously, there will also be other lights that are used for short periods of time—such as in a hall closet, etc.—that are not included in the analysis here.

The energy efficient case assumes the use of 15 W compact fluorescent lamps to directly replace the 60 W incandescents; and the use of two 32 W high efficiency fluorescent lamps with electronic ballast and a mirrored glass reflector to directly replace the bank of four 40 W standard fluorescent lamps. Data for the efficient fluorescent lamp case is shown in table A-9. To the extent that the assumed utilization rates are relatively low—for example, using one incandescent/compact fluorescent for 2 hours per day and the commercial/industrial fluorescent just 10 hours per day (particularly in industry it might be

Table A-7—Water Heating, OTA Base Case Parameters

	Water heater parameters			
Water heater	Lifetime years	Capital cost $	Efficiency percent	Solar fraction percent
Standard case:				
Electric resistance	13	360	100[a]	—
Intermediate case:				
Heat pump water heater	13	800	200[b]	—
High efficiency case:				
Solar water heater	13	1,125	100[c]	85

[a] It is assumed that 100 percent of the electric energy is converted into heat in the water. Standby losses are included in all cases.
[b] The heat pump water heater is assumed to heat water using half the electricity used by the electric resistance heater.
[c] The solar water heater obtains 85 percent of water heating needs from sunlight; the remaining 15 percent is provided by a backup electric resistance heating coil with 100 percent efficiency. The solar water heater is a thermosyphon type with a flat plate collector and a storage tank.

SOURCES: Sunpower, Ltd., Barbados, installed cost October 15, 1990; Howard S. Geller, "Residential Equipment Efficiency: A State-of-the-Art Review," American Council for an Energy-Efficient Economy, Washington, DC, Contractor Report for the Office of Technology Assessment, May 1988; and Howard S. Geller, "Efficient Electricity Use: A Development Strategy for Brazil," American Council for An Energy Efficient Economy, Washington DC, and Contractor Report for the Office of Technology Assessment, 1991.

Table A-8—Lighting, OTA Base Case Parameters

	Lamp Parameters			
	Lifetime hours	Capital cost $	Efficiency lumens/W	Useful output lumens
Residential:				
Standard case:				
Incandescent	1,000	0.5	12	—
Intermediate case:				
Halogen	3,500	1.50	16	—
High efficiency case:				
Compact fluorescent	10,000	15.00	48	—
Commercial:				
Standard case:				
fluorescent	10,000	43.00	56	2,260
High efficiency case:				
Advanced fluorescent	10,000	68.00	83	2,100

SOURCES: Residential lamp costs are from Gilberto De Martino Januzzi et al., "Energy-Efficient Lighting in Brazil and India: Potential and Issues of Technology Diffusion," Apr. 28, 1991, draft, and from manufacturer data. Commercial and Industrial lamp data is from table A-9, below.

used for longer periods than that)—this increases the effective capital and life-cycle operating costs of the efficient equipment relative to the standard lighting equipment.

Refrigeration

The OTA base case assumes a U.S. style (18 cubic feet or, equivalently, 510 liter adjusted volume) top-mounted freezer with automatic defrost that consumes 955 kWh/yr. This is a much larger refrigerator and has more features (particularly automatic defrost) than those generally in use in developing countries today, but is likely to become more popular in the future as the economies of developing countries grow. It is also much more efficient (taking into account its larger size and added features) than refrigerators commonly sold in developing countries today, but is comparable in size and efficiency to new refrigerators sold in the United States. The average U.S. refrigerator, however, has much lower efficiency than this one. This biases the case against more efficient equipment relative to actual existing conditions.

The energy-efficient refrigerator chosen for comparison is technology "I" listed in ch. 3, which uses evacuated panel insulation and higher efficiency compressors, evaporators, and fans than the base case. Much larger and cost-effective improvements in refrigerator performance are possible as discussed in ch. 3.

The capital cost of these refrigerators is assumed to have a retail markup of 100 percent over the factory cost. This is somewhat lower than the 124 to 133 percent

Table A-9—Cost and Performance of Commercial Lighting Improvements, Brazil

	Standard[a]	Efficient[b]
Performance		
Power input	192 W	60 W
Rated light output	10,800 lm	5,000 lm
Useful light output	2,260 lm	2,100 lm
Capital costs		
Lamps	$ 9.70	$ 5.80
Ballasts	$ 33.30	$28.65
Reflectors	NA	$33.95
Subtotal	$ 43.00	$68.40
Annualized capital costs	$ 12.66	$20.18
Annual energy use		
Direct electricity use	507 kWh	158 kWh
Air conditioning energy[c]	142 kWh	44 kWh
Total electricity use	649 kWh	202 kWh
Utility costs[d]		
Capital investment	$296.00	$92.00
Annualized capital cost	$ 23.85	$ 7.41
Annual electricity costs	$ 58.41	$18.18
System wide costs		
Total annual capital cost	$ 36.50	$27.60
Total annual operating cost	$ 71.10	$38.40

NA = not applicable.
[a] Based on four 40 W tubes with conventional core-coil ballast in a standard fixture with completely exposed lamps.
[b] Based on two 32 W high efficiency tubes with electronic ballast and with a mirrored glass reflector. Useful output is so high because of: (1) the narrow 32 W tubes trap less light in the fixture; (2) the electronic ballast operates at high frequencies and raises nominal output of the tube; (3) the mirrored reflector increases useful light output, etc.
[c] This is the amount of air conditioning power needed to remove the heat generated by the lights. This synergism is not included in the OTA analysis.
[d] Utility costs are set at estimated marginal prices rather than prevailing average prices in Brazil which may be undervalued. Thus, capital investment is based on $4,000 per delivered kW and electricity prices are set at $0.09/kWh.
SOURCE: Adapted from Howard S. Geller, "Efficient Electricity Use: A Development Strategy for Brazil," American Council for an Energy Efficient Economy, Washington, DC 1991. Contractor report to the Office of Technology Assessment, November 1990.

markup assumed by the U.S. Department of Energy,[26] but is believed representative of the lower overheads that can be expected in a developing country.

Commercial refrigeration systems are not considered in the OTA scenarios, but information on efficiency improvements in these systems is presented in ch. 3.

Table A-10—Refrigeration, OTA Base Case Parameters

	Refrigerator parameters		
Technology[a]	Lifetime years	Retail capital cost $	Annual energy consumption kWh
Standard case:			
A	20	495.00	955
Intermediate case:			
B	20	495.20	936
C	20	498.40	878
D	20	506.00	787
E	20	514.20	763
F	20	534.00	732
G	20	550.50	706
H	20	561.20	690
Efficient case:			
I	20	635.70	577
More advanced[b]			
J	20	746.20	508
K	20	781.56	490

NOTE: A constant retail markup of 100 percent over factory prices was assumed. This is somewhat lower than the retail markups assumed for the United States by the Department of Energy, but corresponds to lower overheads in developing countries.
[a] Specific descriptions of these technologies are listed in chapter 3, table 3-13.
[b] These were not considered in the OTA high efficiency scenario because, even though they appear to be cost effective on a lifecycle basis, they have substantially higher capital costs due to their projected use of two compressor systems.
SOURCE: *Technical Support Document: Energy Conservation Standards for Consumer Products: Refrigerators and Furnaces* (Washington, DC: U.S. Department of Energy, November 1989) publication DOE/CE-0277. See also Table 2-13.

Air Conditioning

The OTA base case averages the use of two air conditioners for both residential and commercial cooling over the five persons per household. One is smaller, at 2 tons equivalent capacity, and uses a relatively low 4000 kWh/year; the other is larger, at 3 tons, and uses a higher 8000 kWh/yr.[27] Combined, these might correspond to a household and a small office demand, respectively. Larger offices, however, would have substantially lower per occupant air conditioning costs and higher efficiencies than those assumed here due to economies of scale in the equipment and much higher capacity factors than those assumed here. The assumed base case efficiencies were, in both cases, an SEER of 8, which is comparable

---

[26] U.S. Department of Energy, "Technical Support Document: Energy Conservation Standards for Consumer Products: Refrigerators and Furnaces; Including: Environmental Impacts and Regulatory Impact Analysis," report No. DE90-003491, November 1989.

[27] A ton of air conditioning is the cooling power provided by melting 1 ton—2000 pounds—of ice over a 24-hour period. This is equivalent to 200 Btu per minute or 12000 Btu/hour of heat removal, equivalent to 0.211 (MJ)/minute or 12.66 MJ/hour. The energy efficiency ratio of an air conditioning unit is its cooling capacity in Btu per hour divided by its required power input in watts. Thus, an air conditioner with an EER of 8.0 requires 426.5 watts of energy input to remove 3412 Btu/hour of heat from a building, or equivalently, to remove 1 kW of heat input. An air conditioner with a 2-ton capacity operating continuously consumes (2 tons)*(12000 Btu/hr)/8=3 kW of power, or 26,280 kWh/year. At an annual energy consumption of 4000 kWh/yr, it is then operating at an annual average capacity factor of 15 percent. A 3-ton air conditioner consuming 8,000 kWh per year has a capacity factor of 20 percent.

Table A-11—Air Conditioning, OTA Base Case Parameters

	Air Conditioner Parameters			Average power consumption watts
	Lifetime years	Retail capital cost $	SEER	
Low load: 2 ton				
Standard case	15	1,400	8	456
Intermediate case	15	1,700	10	365
	15	2,000	12	304
High efficiency case	15	2,300	14	261
High load: 3 ton				
Standard case	15	2,100	8	913
Intermediate case	15	2,400	10	730
	15	2,700	12	608
High efficiency case...........	15	3,000	14	521

NOTE: Costs are based on a flat rate of $700 per ton of cooling power and $150 per SEER above an SEER of 8, corresponding roughly to U.S. retail prices installed. Power consumption is in watts—averaged over the year—and is based on the average energy use of 4000 kWh in the low-load case, scaled by SEER using an SEER of 8 for the baseline, and 8000 kWh annual energy use in the high load case, also scaled by SEER.

SOURCE: Office of Technology Assessment, 1992.

Table A-12—Color Television, OTA Base Case Parameters

	Color Television Parameters		
	Lifetime years	Retail capital cost $	Annual energy consumption kWh/yr
Standard case..................	10	$316.00	205
Intermediate case 1	10	$320.30	184
Intermediate case 2..............	10	$322.90	176
High efficiency case..............	10	$323.50	171

NOTE: The assumed lifetimes of 10 years are somewhat longer than the observed lifetime in the U.S. of 7 years. Efficiency improvements include reducing standby power by replacing voltage dropping resistor with a transformer; replacing the surge protection resistor and adding output taps on the power supply; and in the high efficiency case, replacing the picture tube with a slightly higher efficiency picture tube. Much larger efficiency improvements may be readily and cost-effectively achieved.

SOURCE: *Technical Support Document: Energy Conservation Standards for Consumer Products: Refrigerators, Furnaces, and Television Sets* (Washington, DC: U.S. Department of Energy, November 1988) publication DE89-002738.

to the average for new room air conditioners sold in the United States in 1988, and slightly lower than the SEER of 9 for new central air conditioners (see ch. 3).

These levels of air conditioning are then divided by five (per household) to get the corresponding per capita costs and energy use; these values are then divided in half to reflect an overall average air conditioning penetration of 50 percent of households and offices. In the base case, per capita electricity consumption for air conditioning is then about 1,200 kWh/year. This is comparable to the 1,400 kWh/year used per capita in the United States for cooling residential and commercial buildings (this includes buildings with no air conditioning and cooler climates as well) and is in the same range as that observed in some developing countries for those with air conditioners (see ch. 3).

In hot, humid climates, however, air conditioning loads can be substantially higher than those assumed here. In Florida, for example, air conditioning loads on uninsulated concrete block houses were 8200 kWh/yr—twice the assumed levels here. In large buildings, there is typically a cooling demand year around in order to remove internal heat gains—from lights, people, etc.—which increases the capacity utilization of the air conditioner and improves the cost effectiveness of high efficiency units compared to the values assumed here.

Overall, air conditioning loads could easily dwarf most other electricity demands in hot, humid climates. They are intentionally kept comparable to other demands here, because there is little available data to project future air conditioning demand in developing countries, and because the case presented was intended to be as conserva-

Table A-13—Industrial Motor Drive, OTA Base Case Parameters

	Industrial motor drive parameters				Annual energy consumption kWh/yr-motor[a]	Weighting by motor	Power consumption by size class watts
	Motor lifetime years	Motor capital cost $/hp	Pump capital cost $/hp	ASD capital cost $/hp			
Standard case:							
1 hp	15	218.00	75	NA	3,621	0.036283	15
10 hp	19	56.30	75	NA	30,390	0.007782	27
30 hp	22	41.03	75	NA	86,341	0.003956	39
100 hp	28	43.53	75	NA	283,116	0.002506	81
200 hp	29	37.5	75	NA	560,752	0.002155	138
Efficient case:							
1 hp	15	294.00	90	543.00	70%*	0.036283	10.5
10 hp	19	71.40	90	359.70	70%	0.007782	18.9
30 hp	22	49.70	90	203.60	70%	0.003956	27.3
100 hp	28	46.96	90	135.75	70%	0.002506	56.7
200 hp	29	41.24	90	111.07	70%	0.002155	96.6

NA = Not applicable.

SOURCES: Motor lifetimes and weighting factors are derived from table A-14. Motor costs are from Marbek Resource Consultants, Ltd., "Energy Efficient Motors in Canada: Technologies, Market Factors and Penetration Rates," Energy Conservation Branch, Energy, Mines and Resources, Canada, November 1987; Pump (fan) costs are very rough estimates from various manufacturers data sheets—note that these costs can vary dramatically depending on the particular type of pump and application; ASD costs are from Steven Nadel, et al., "Energy Efficient Motor Systems: A Handbook on Technology, Programs, and Policy Opportunities," (Washington, DC: American Council for an Energy Efficient Economy, 1991). System engineering and installation costs are assumed to be the same for both standard and energy efficient cases, corresponding to the situation where there is considerable practical experience with high efficiency systems and the development of effective design rules and procedures.

Table A-14—Characteristics of U.S. Motor Population, 1977

	Electric motor size, horsepower					
	<1	1-5	5.1-20	21-50	51-125	>126
Total number, millions	660.0	55.0	10.5	3.3	1.7	1.0
Average life, years	12.9	17.1	19.4	21.8	28.5	29.3
Weighted average size, hp	0.3	2.1	11.9	32.5	86.7	212.0
Average efficiency, %	65.0	77.0	82.5	87.5	91.0	94.0
Average load, %full load	70.0	50.0	60.0	70.0	85.0	90.0
Average capacity factor, %	3.5	7.0	17.4	27.7	37.5	43.2
Total annual use, 10^9 kWhr	30.5	33.9	103.4	155.2	337.7	573.3

SOURCE: Samuel F. Baldwin, "Energy Efficient Electric Motor Drive Systems," Thomas B. Johansson, Birgit Bodlund, and Robert H. Williams, eds., *Electricity: Efficient End-Use and New Generation Technologies and Their Planning Implications* (Lund University Press, Lund Sweden, 1989).

tive as possible—higher air conditioning loads weight the case even more heavily in favor of more efficient air conditioning equipment. Finally, it must be noted that many techniques, such as building insulation, improved window technologies, and many others, are generally much more cost effective than even the high efficiency air conditioner case presented here (see ch. 3). Again, these alternatives were not examined, both to keep the standard and efficient cases strictly comparable and to be as conservative as possible.

Electronic Information Services: Color Television

The OTA Base Case assumes the use of 19-inch to 20-inch color TVs for about 5 hours per household per day. The standard TV uses about 109 W of power; the efficient TV uses about 91 W of power. As discussed in ch. 3, much greater efficiency improvements are possible using Complementary Metal Oxide Semiconductor (CMOS) electronic devices and power management techniques and, in the future, converting to flat panel displays.

Industrial Motor Drives

The OTA base case assumes the use of standard efficiency motors, pumps, fans, and other equipment. The size class of motors is weighted by U.S. data, as shown in table A-14. Motor costs and efficiencies are discussed in ch. 4.

The energy-efficient equipment case assumes that average savings of 30 percent are realized (a 30 percent reduction in energy consumption), compared to the base case through the use of energy-efficient motors, high efficiency pumps/fans, adjustable speed drives, and the elimination of throttling valves.[28] The corresponding capital costs of motors and Adjustable Speed Drives (ASDs) is listed by size; the capital costs of high efficiency pumps/fans were assumed to be 20 percent greater than standard equipment. The same weighting by size class is used as for the Base Case. No credit is given for potential reductions in the size (and cost) of efficient equipment. No consideration is given to the potential for optimizing the sizes of pipes or ducts, etc., or for improved design rules or other changes. Again, these various assumptions combine to make the case for energy efficient equipment conservative.

Summary

The results of these cases can now be summed as shown in tables A-15 and A-16. These data form the basis of the summary graphs shown in chs. 1 through 4 for the systemwide capital and life-cycle costs of electricity services.

Sensitivity Analysis

Each of the various parameters can now be varied to determine the sensitivity of the analysis to the input values. The results of such a sensitivity analysis are shown in table A-17. This sensitivity analysis shows that the above estimates of systemwide capital costs, life-cycle operating costs, and electricity consumption are fairly insensitive to the input values.

In order to erase the overall capital savings advantage of more efficient equipment: the discount rate for end-users would have to be raised to 2.3 times that for utilities; the marginal cost of efficient equipment would have to be increased by 70 percent over observed values;

---

[28] A more conservative assumption would be that two-thirds of the motor drive systems could be retrofit and achieve such 30 percent energy savings while one-third could not be usefully retrofit in terms of cost or energy efficiency (the motor systems might already operate at full constant load with high efficiency). Such an assumption obviously does not reduce the economic or energy savings for the two-thirds of the motors that could be retrofit, but does change the total economy-wide energy and lifecycle and capital cost savings realizable (the numerator is decreased by the change in motor drive systems retrofit while the denominator remains the same). The overall impact is to reduce society-wide energy savings from 47 percent in the case of all motors retrofit to 43 percent if two-thirds of the motors are retrofit; lifecycle cost from 28 percent to 25 percent; and capital savings from 13 percent to 11 percent.

Table A-15—Summary Results for Standard Equipment

Standard system	Capital costs			Operating cost $/capita	Power Watts	Fuel GJ
	End-user $/capita	Utility $/capita	System $/capita			
Cooking	7.14	17.74	24.87	48.82	52.9	0.0
Water heating	6.80	25.72	32.52	67.25	76.7	NA
Lighting	14.31	24.15	38.45	71.06	72.0	NA
Refrigeration	9.34	7.31	16.66	26.53	21.8	NA
Air conditioning	38.43	45.95	84.38	146.43	137.0	NA
Information services	9.00	1.57	10.57	12.69	4.7	NA
Industrial motor drive	9.80	100.63	110.43	193.76	300.0	NA
Total	94.81	223.07	317.87	566.54	665.0	0.0

NA = not applicable.
NOTE: Many related capital costs, particularly for industrial motor drive, are not included as they are assume to be the same in both the standard and the energy efficient cases.
SOURCE: Office of Technology Assessment, 1992.

Table A-16—Summary Results for Energy Efficient Equipment

Energy efficient system	Capital costs			Operating cost $/capita	Power Watts	Fuel GJ/year
	End-user $/capita	Utility $/capita	System $/capita			
Cooking	6.43	6.05	12.48	26.83	14.1	1.3
Water heating	21.23	3.78	25.02	30.12	11.3	NA
Lighting	25.34	7.23	32.58	42.34	21.6	NA
Refrigeration	12.00	4.42	16.42	22.39	13.2	NA
Air conditioning	58.19	26.26	84.45	119.91	78.3	NA
Information services	9.21	1.31	10.52	12.29	3.9	NA
Industrial motor drive	25.44	70.44	95.88	154.21	210.0	NA
Total	157.87	119.48	277.34	408.09	352.24	1.3

NA = not applicable.
NOTE: Many related capital costs, particularly for industrial motor drive, are not included as they are assumed to be the same in the standard and energy-efficient cases.
SOURCE: Office of Technology Assessment, 1992.

the demand for a given piece of efficient equipment would have to be cut nearly in half; or the marginal efficiency gain of more efficient equipment would have to be reduced by one-third.[29]

Similarly, to erase the overall life-cycle cost savings advantage of more efficient equipment: the discount rate for end-users would have to be raised to over five times that for utilities; the cost of electricity would have to be reduced to less than one-third of its marginal cost of production; the marginal cost of efficient equipment would have to be increased by 2.5 times; or the marginal efficiency gain of more efficient equipment would have to be reduced by over two-thirds.[30]

Of particular interest in this sensitivity analysis is that real consumer discount rates must be raised to 2.3 times—or 16 percent real—that for utilities (at 7 percent) in order to erase the systemwide capital cost advantage of energy efficient equipment; and the discount rate must be raised to over five times—or 38 percent real—that for utilities to erase the life-cycle cost advantage of efficient equipment. Some will respond that observed capital costs to end-users can be that high in developing countries. This is true. Such high rates are not primarily due to the difficulty of administering large numbers of loans or other such factors, however, but rather are due to institutional mechanisms that route, often intentionally, capital from end-users to large capital users such as utilities. These mechanisms include taxes on end-users, but tax breaks for utilities; low-interest loans or special financial bonds for utilities; or other such proactive mechanisms. Unintentional impacts of these mechanisms include capital shortages in end-user markets due to the large demand for capital by utilities and other public or favored sectors. As noted in chs. 1 through 4, however, even if consumers had access to capital at rates comparable to those available to

---

[29] The one-third figure does not include cooking, for which the shift from electric resistance burners to LPG burners provides particularly large capital savings. If cooking is included, the marginal efficiency gain of energy-efficient equipment would have to be reduced by two-thirds overall in order to erase the capital saving advantage of energy efficient equipment.

[30] Again, this does not included cooking, as above.

Table A-17—Sensitivity Analysis

Parameter	Baseline	Required value to erase system capital cost advantage of efficient equipment	Required value to erase lifecycle cost advantage of efficient equipment
Discount rate for end-use sectors[a]	7%	16%	38%
Electricity cost:			
Residential/commercial	$0.09/kWh	NA	$0.028/kWh
Industrial	$0.07/kWh	NA	$0.021/kWh
Marginal cost of efficient end-use equipment	Tables A-5 to A-14	70% increase over baseline	250% increase over baseline
Load factor	Tables A-5 to A-14	55% of baseline load	25% of baseline load
Marginal efficiency Advantage of efficient end-use equipment	Tables A-5 to A-14	66% of baseline efficiency advantage[b]	29% of baseline efficiency advantage[b]

NA = not applicable.
[a] While keeping utility discount rate at 7 percent.
[b] This does not include cooking as the shift from electric resistance to gas burners results in efficiency gains and corresponding capital and lifecycle savings irrespective of the relative efficiency factor.
SOURCE: Office of Technology Assessment, 1992.

utilities, there are a number of market failures which would impede the purchase of efficient end-use equipment by end-users.

If real discount rates increase for both the utility and the end-user, the capital savings realized by installing efficient equipment increase. For example, increasing the real discount rate for both utility and end-user from 7 to 15 percent increases the relative capital savings of efficient equipment from 12.75 percent to 17 percent.

From a societywide perspective, failure to invest in energy efficient equipment thus results in substantially higher systemwide capital costs, life-cycle operating costs, and energy consumption—with all its related environmental impacts—than necessary. A more optimal allocation of capital to end-use efficiency would require changes in institutional mechanisms. Certainly, administrative overheads associated with oversight of large numbers of small loans will lead to higher discount rates than those for the utility sector, but they are unlikely to be twice as high, let alone the five times as high needed to erase the overall cost savings of efficient equipment found in this analysis. Such institutional mechanisms might range from channeling capital through utilities for purchase of efficient end-use equipment by end-users, to mandated efficiency standards. Combinations of these are being used in the United States and other industrial countries.

It is also useful to put the marginal cost of energy efficient equipment in the context of the overall decline in the real cost of consumer goods. As shown in chs. 1 and 2, the real cost of refrigerators in the United States declined by a factor of five between 1950 and 1990 due to improvements in materials and in manufacturing technologies. Averaged over this 40-year period, this is a 12.5 percent annual decline in real cost. In comparison, the energy-efficient equipment costs end users about 67 percent more than standard equipment, or the equivalent of about 5 years worth of manufacturing improvements.

Finally, it must be noted that these changes to erase the systemwide capital cost and life-cycle cost advantages of energy efficient equipment are on top of the highly conservative choices of parameters used in the OTA base case standard and energy efficient equipment scenarios. Based on these considerations, it appears that the overall conclusion—that energy efficient equipment can reduce systemwide capital costs and lifecycle operating costs—is robust.

Appendix B
Electricity Supply Technologies

On-Grid Electricity Supply Technologies

Table B-1 summarizes key characteristics for on-grid electric generating technologies. Representative values are shown; specific cases may vary considerably from these estimates.

Factors listed in this table include the following:

- **Application:** Base, intermediate, or peaking. Base load plants provide the slowly varying baseline power demanded by the grid and account for the bulk of the power supplied. Because of the large amount of power they supply, low cost fuels such as coal or—in some cases—nuclear are preferred for these plants. Because of the difficulties inherent in using these fuels, base load plants are generally large and capital intensive. Intermediate and peaking plants are chosen to be successively less capital intensive as they are used for shorter periods, such as for the afternoon peak demand due to air conditioning loads or the early evening peak due to residential lighting demand. To minimize capital costs, more expensive fuels, such as oil or gas, are generally used, and the plants are installed in smaller units.
- **Capital Costs.** Costs shown are nominal values; these costs will vary widely depending on local conditions, specifics of the technology, and other factors. Many of these cost estimates are from Electric Power Research Institute, *Technical Assessment Guide (TAG)*, Report No. EPRI P-6587-L, Palo Alto, CA, September 1989; other cost estimates have been developed by OTA from various sources. All costs are deflated to December 1990 U.S. dollars. Note that the capital costs listed here do not include mining, transmission and distribution, administration, or other overhead capital costs, and are not adjusted for capacity factors as is necessary to account for full systemwide costs as done in appendix A.
- **Operating Requirements.** Estimated heat rates (Btu/kWh) are from the Electric Power Research Institute (TAG) where available; other heat rate or thermal efficiency (in percent) estimates have been developed by OTA from various sources. Heat rates or efficiencies will vary depending on specifics of the technology, fuel quality, and other factors.
- **Fuel.** Plants listed as using "distillate" can often use natural gas as well. Fuel costs are from the Electric Power Research Institute (TAG), and are for U.S. delivery in 1990. Availability of petroleum-based fuels varies with world market conditions; if a domestic source is used then availability may be much improved.
- **Total Costs.** Levelized capital costs assume the discount rate, lifetime, and capacity factors shown; results are sensitive to all these assumptions. O&M costs are estimated by OTA based on information from manufacturers, consultants, utilities, and others. Both fixed ($/kW-yr) and incremental ($/kWh) O&M costs are included; fixed O&M costs are levelized using the capacity factor. Fuel costs are based on given fuel prices and heat rates. Note that these costs are lower than the total systemwide costs estimated in appendix A due to assumed high capacity factors here (70 percent) for individual plants rather than typical system capacity factors of 60 percent or less as assumed in appendix A, and because other cost components such as coal mining, transmission and distribution, and operating overhead are included in appendix A but are not included here.
- **Time Requirements.** Installation lead times are typical values for the time from decision to build to actual operation. These values are strongly affected by site-specific permitting and other regulatory considerations.
- **Environmental Impacts.** Air indicates the relative quantity of NO_x, SO_x, and particulate emission, per kWh; it does not include CO_2 emissions, which are highest for coal-based technologies, followed by oil, natural gas, and others. Water requirements indicate the relative quantity of water needed to operate the plant. Solid and liquid waste products indicate the volume of waste products which must be handled.
- **Infrastructure Requirements.** In general, large (over 100 MW) plants will require river or rail access.

The technologies evaluated in table B-1 include the following with the listed parameters:

- **Conventional Combustion Turbine.** Based on an 80 MW unit.
- **Conventional Steam Plants.** Values for coal plant are based on a 300 MW subcritical plant using West Virginia bituminous pulverized coal without flue gas desulfurization (FGD).
- **Hydroelectric Plants.** Costs may vary widely, depending on local conditions.
- **Advanced Combustion Turbine.** Based on a 140 MW unit.
- **Combined Cycle.** Based on a 120 MW unit using distillate fuel.

Appendix B—Electricity Supply Technologies • 305

Table B-1—Nominal Parameters for Selected On-Grid Generating Technologies—Part 1

	Conventional combustion turbine	Existing reference technologies				
		Conventional steam, fueled by:			Diesel engine	Hydropower
		Gas	Oil	Coal		
Application (base, intermediate, peaking)	I,P	B,I	B,I	B,I	I,P	B,I,P
Capital cost ($/kW)	400	1,100	1,200	1,500	1,000	1,500
Operating requirements:						
Efficiency (Btu/kWh or percent)	14,500	10,000	10,000	10,000	8,800	NA
Fuel:	natural gas	natural gas	residual	coal	diesel	water
Cost, $/million Btu (1990)	$2.72	$2.72	$2.72	$2.00	$4.10	$0.00
Total costs:						
Discount rate (percent)	7	7	7	7	7	7
Lifetime (years)	30	30	30	30	30	30
Capacity factor	15	70	70	70	40	40
Levelized capital costs (cents/kWh)	2.50	1.50	1.60	2.00	2.30	3.50
O&M costs (cents/kWh)	0.90	0.50	0.60	1.00	0.80	0.30
Fuel costs (cents/kWh)	3.90	2.70	2.70	2.00	3.50	0.00
Total (cents/kWh)	7.30	4.70	4.90	5.00	6.60	3.80
Sizes of available units (MW)						
Minimum	4	20	20	50	4	1
Typical	80	300	300	300	8	200
Time requirements						
Installation lead time (years)	2-3	6-8	6-8	7-9	2	10
Expected year of commercial availability	available	available	available	available	available	available
Environmental impacts						
Air	low	low	medium	high	medium	none
Water requirements (gallons/MW-day)	low	medium	medium	medium	low	high
Solid and liquid waste products	low	low	low	high	low	none
Land requirements (per MW)	low	medium	medium	high	medium	high
Infrastructure requirements: rail or river access	useful	required	required	required	none	useful

NA = not available, not applicable, or too variable to quantify.
SOURCE: Office of Technology Assessment, 1992.

Table B-1—Nominal Parameters for Selected On-Grid Generating Technologies—Part 2

	Advanced combustion turbine	Improvements in existing technologies				
		Combined cycle	Fluidized bed	Life extension	Municipal solid waste	Steam-injected gas turbine
Application (base, intermediate, peaking)	I,P	B,I	B,I	NA	NA	B,I,P
Capital cost ($/kW)	400	600	1,700	50-1,000	5,000	600
Operating Requirements						
Efficiency (Btu/kWh or percent)	13,800	8,100	10,000	NA	17,000	9,400
Fuel:	natural gas	distillate	coal	NA	MSW	natural gas
Cost, $/million Btu (1990)	$2.72	$4.10	$2.00	NA	(varies)	$2.72
Total costs:						
Discount rate (percent)	7	7	7	7	7	7
Lifetime (years)	30	30	30	30	30	30
Capacity factor (percent)	15	70	70	NA	70	40
Levelized capital costs (cents/kWh)	2.50	0.80	2.20	NA	11.50	1.40
O&M costs (cents/kWh)	0.90	0.50	1.30	NA	NA	0.80
Fuel costs (cents/kWh)	3.70	3.20	2.00	NA	NA	2.50
Total (cents/kWh)	7.10	4.50	5.50	NA	NA	4.70
Sizes of available units (MW)						
Minimum	4	50	100	NA	40	NA
Typical	80	150	300	NA	50	100
Time requirements						
Installation lead time (years)	2-3	5-6	7-8	1-4	7-8	2-3
Expected year of commercial availability	1992	available	1994	available	available	available
Environmental impacts						
Air	low	medium	medium	NA	NA	low
Water requirements (gallons/MW-day)	low	medium	medium	NA	high	medium
Solid and liquid waste products	low	low	high	NA	NA	low
Land requirements (per MW)	low	low	medium	none	NA	low
Infrastructure requirements: rail or river access	useful	required	required	NA	useful	none

NA = Not available, not applicable, or too variable to quantify.
SOURCE: Office of Technology Assessment, 1992.

Table B-1—Nominal Parameters for Selected On-Grid Generating Technologies—Part 3

	Advanced batteries	Advanced small nuclear	New or innovative technologies			Integrated gasification combined cycle
			Binary geothermal	Compressed air energy storage	Fuel cell	
Application (base, intermediate, peaking)	P	B	B,I	P	B,I,P	B
Capital cost ($/kW)	600	1,900	2,000	650	3,000	1,650
Operating requirements						
Efficiency (Btu/kWh or percent)	86%	10,500	29,000	52%	8,500	9,200
Fuel:	electricity	nuclear	warm brine	electricity	distillate	coal
Costs, $/million Btu (1990)	NA	$0.80	variable	NA	$4.10	$2.00
Total costs						
Discount rate (percent)	7	7	7	7	7	7
Lifetime (years)	30	30	30	30	30	30
Capacity factor (percent)	15	70	70	15	40	70
Levelized capital costs (cents/kWh)	3.80	2.50	2.60	4.00	6.90	2.20
O&M costs (cents/kWh)	NA	1.40	1.60	NA	0.80	1.20
Fuel costs (cents/kWh)	NA	0.80	NA	NA	3.40	1.80
Total (cents/kWh)	NA	4.70	4.20	NA	11.10	5.20
Sizes of available units (MW)						
Minimum	NA	150	5	50	10	50
Average	20	350	50	110	25	75
Time requirements						
Installation lead time (years)	2	10-12	5-6	6-8	NA	8-9
Expected year of commercial availability	1997	2002	available	1993	NA	1994
Environmental impacts						
Air	low	low	high	low	low	medium
Water requirements (gallons/MW-day)	low	medium	high	low	low	medium
Solid and liquid waste products	medium	medium/high	varies	low	low	high
Land requirements (per MW)	low	medium	low-medium	medium	low	high
Infrastructure requirements: rail or river access	none	required	required	none	none	required

NA = not available, not applicable, or too variable to quantify.
SOURCE: Office of Technology Assessment, 1992.

Table B-1—Nominal Parameters for Selected On-Grid Generating Technologies—Part 4

	Pumped hydro	New or innovative technologies		
		Solar photovoltaic	Wind turbines	Solar-thermal electric/natural gas hybrid
Application (base, intermediate, peaking)	P	NA	NA	NA
Capital cost ($/kW)	1,100	6,000	1,100	3,000
Operating requirements				
Efficiency (Btu/kWh or percent)	75%	10%	NA	3,300
Fuel:	electricity	sunlight	wind	sunlight/Gas
Costs, $/million Btu (1990)	NA	0	0	$2.72
Total costs				
Discount rate (percent)	7	7	7	7
Lifetime (years)	30	30	30	30
Capacity factor (percent)	15	20	15	40
Levelized capital costs (cents/kWh)	6.80	36.80	6.80	6.90
O&M costs (cents/kWh)	NA	0.70	1.50	1.50
Fuel costs (cents/kWh)	NA	0.00	0.00	0.90
Total (cents/kWh)	NA	37.50	8.30	9.30
Sizes of available units (MW)				
Minimum	NA	NA	0.25	NA
Average	1,000	NA	75	80
Time requirements				
Installation lead time (years)	11	2	2	2
Expected year of commercial availability	available	available	available	available
Environmental impacts				
Air	low	none	none	low
Water requirements (Gallons/MW-day)	high	none	none	high
Solid and liquid waste products	low	none	none	low
Land requirements (per MW)	high	high	high	high
Infrastructure requirements: rail or river access	useful	none	none	none

NA = not available, not applicable, or too variable to quantify.
SOURCE: Office of Technology Assessment, 1992.

Table B-2—Nominal Parameters for Selected Off-Grid Generating Technologies

	Diesel	Micro-hydro	Photovoltaics	Wind
Capital cost ($/kW)	700	2,000	10,000[a]	5,000[b]
Lifetime (years)	10	20	30	20
Discount rate (percent)	7	7	7	7
Levelized capital cost ($/kW)	100	190	800	470
Capacity factor[c] (percent)	20	20	20	20
Capital cost ($/kWh)	0.06	0.11	0.46	0.27
O&M ($/kWh)	0.02	0.02	0.005	0.01
Fuel ($/kWh)[d]	0.23	0.00	0.00	0.00
System losses[e] (percent)	5	5	10	10
Total cost ($/kWh)	0.33	0.14	0.51	0.31

NOTE: Numbers may not add due to extensive rounding.
[a] Photovoltaic costs include $6,000 per peak kW for the panel, $2,000 for the balance of system costs, and $2,000 for batteries and their replacements (every 5 years over the 30 year life of the system).
[b] Wind costs include $3,500 per peak kW for the turbine and balance of system, and $1,500 for batteries and their replacements (every 5 years over the 20 year life of the system).
[c] The capacity factor is listed as percent but might equally well be given in terms of annual kWh output per kW capacity. A capacity factor of 20 percent then corresponds to 1,750 kWh/kW.
[d] For diesel priced at $0.50 per liter.
[e] System losses include battery losses for PV and wind systems, and other system losses for diesel and hydro equipment.
SOURCE: Office of Technology Assessment, 1992. Sources for costs are discussed in chapter 6 and are based primarily on retail quotes from manufacturers and distributors, where available. Developing country costs may be higher due to increased transportation costs, and due to duties and taxes. Costs shown here are for low-volume production, however, and higher volumes may allow for lower per unit costs.

- **Fluidized Bed.** Based on a 200 MW unit with a circulating atmospheric bed using Illinois bituminous coal.
- **Life Extension.** Costs and other attributes vary widely.
- **Municipal Solid Waste.** Based on a 40 MW mass burn technology.
- **Steam Injected Gas Turbine.** Based on a 100 MW unit.
- **Advanced Batteries.** Based on a 5-hour, 20 MW unit.
- **Advanced Nuclear.** Based on a 600 MW light water reactor with passive safety features.
- **Binary Geothermal.** Based on a 54 MW unit.
- **Compressed Air Energy Storage.** Based on a rock formation cavern, using an electric compressor and a 110 MW combustion turbine. Approximately 10 hours of storage provided. Each kWh of output requires 0.76 kWh electric input plus 4,000 Btu of fuel.
- **Fuel Cell.** Based on a phosphoric acid fuel cell with 4 units at 25 MW each.
- **Integrated Gasification Combined-Cycle.** Based on a 400 MW plant using bituminous coal.
- **Pumped Hydro.** Based on a conventional above-ground 3 by 350 MW unit.
- **Solar Photovoltaics.** Based on a flat plate technology. Note that the costs of solar photovoltaic systems are largely scale independent.
- **Wind Turbine.** Based on a farm of three hundred 250 kW units.
- **Solar Thermal.** Based on a parabolic trough/natural gas hybrid. Note that the relatively high capacity factor is due to use of natural gas cofiring to supplement insolation.

Off-Grid Electricity Generating Technologies

A variety of technologies are available that can provide electricity at (remote) sites that are not connected to the electric power grid. Two sets of calculations are of interest. First, how do the costs of these various technologies compare. Four technologies—diesel engine generator sets, micro hydroelectric plants, flat panel photovoltaics, and wind turbines—were selected to illustrate this representative calculation. Second, how do the costs of these technologies (including battery storage, as needed) compare with the cost of extending the grid and providing on-grid generation.

In order to compare the cost of power from these various technologies, estimates were made for the life-cycle costs of a 10 kW peak capacity electricity generating system with storage. Nominal parameters and results are shown in table B-2 and a sensitivity analysis is shown in table B-3. These results are also shown in figures in chapter 6.

As can be seen in table B-2, where available, micro-hydroelectric power can be relatively lower in cost and wind power comparable in cost to diesel systems. Photovoltaic systems tend to be higher in cost for the baseline parameters chosen. The costs of power are highly sensitive, however, to fuel costs in the case of diesel

Table B-3—Sensitivity Analysis for the Cost of Selected Off-Grid Generating Technologies

	Diesel	Micro-hydro	Photovoltaics	Wind
Baseline: total cost ($/kWh)	$0.33/kWh	$0.14/kWh	$0.51/kWh	$0.31/kWh
Variable: fuel cost				
Baseline: $0.50/liter	0.33	0.14	0.51	0.31
$0.75/liter	0.45	0.14	0.51	0.31
$1.00/liter	0.57	0.14	0.51	0.31
Variable: discount rate				
3 percent	0.32	0.10	0.33	0.22
Baseline: 7 percent	0.33	0.14	0.51	0.31
10 percent	0.33	0.16	0.67	0.38
15 percent	0.35	0.21	0.96	0.51

NOTE: Parameters are based on table B-2.
SOURCE: Office of Technology Assessment, 1992.

Table B-4—Nominal Costs and Break-Even Distances for Grid Extension

	Diesel	Micro-hydro	Photovoltaics	Wind	Grid + extension
Baseline: total cost ($/kWh)[a]	$0.33/kWh	$0.14/kWh	$0.51/kWh	$0.31/kWh	$0.07/kWh + $0.032/kWh-km
Break-even distance (kilometers)					
Variable: grid extension					
Baseline: $ 4,500/km	8.1 km	2.1 km	13.9 km	7.5 km	NA
$ 7,000/km	5.2	1.3	8.9	4.8	
$ 9,000/km	4.0	1.0	6.9	3.8	
$10,000/km	3.6	0.9	6.3	3.4	
$13,000/km	2.8	0.7	4.8	2.6	
Variable: discount rate					
3 percent	10.4	1.8	10.8	6.6	NA
Baseline: 7 percent	8.1	2.1	13.9	7.5	
10 percent	6.8	2.2	15.7	8.0	
15 percent	5.3	2.4	17.8	8.6	

NA = not available or not applicable.
[a]Baseline values for remote generation technologies are listed in table B-2; baseline parameters for on-grid generation are $1500/kW capacity, 30 year lifetime, 7 percent discount rate, $0.01/kWh for O&M, $0.02/kWh for fuel (fuel priced at $2.00 per million Btu and a heat rate of 10,000 Btu/kWh), and an actual capacity factor of 60 percent in operations, with power routed to the site at the same rate as generated by the remote technologies; baseline values for grid extension were $4500/km—corresponding to a single-wire earth return (a very low cost system), a lifetime of 20 years, a discount rate of 7 percent, and an annual O&M cost of 3 percent of the initial capital cost.
SOURCE: Office of Technology Assessment, 1992.

systems, and to discount rates in the case of photovoltaic, wind, and hydro systems (see table B-3).

Actual costs may vary considerably from the parameters chosen in tables B-2 and B-3. Other factors are also important in choosing a system for remote generation. Many regions will not have access to good micro-hydro or wind resources. Diesel systems require timely delivery of fuel, spare parts, and competent maintenance for reliable operation; in many areas these factors are not available. Finally, photovoltaic systems, though expensive, may have substantial advantages over diesel systems by not using fuel, having few or no moving parts, and requiring little maintenance.

These systems can also be compared to the cost of extending the electric power grid and providing on-grid generation capacity. There are two components of cost that must then be considered: the cost of conventional on-grid generation as detailed in table B-1, and the cost of the grid extension itself.

As detailed in Chapter 6, the cost of grid extension ranges from $4,600 to nearly $13,000 per kilometer for single phase systems, depending on the terrain, the precise type of system, and a host of other factors. O&M costs range from 2 to 4 percent of the capital cost.

By equating the cost per kWh of the remote generation to the cost of the grid technology plus grid extension, an approximate "break-even" distance can be calculated. For distances less than the break-even distance, it will then be lower cost to extend the grid; for distances greater than the break-even distance, it will be lower cost to install a remote generation technology. Note, however, that there are many caveats to this highly simplified analysis. The capacity of grid extension may be much greater than that assumed here for the remote generation technologies, or may be upgraded more easily. On the other hand, remote generation technologies may be more reliable than extending the grid with lines easily downed during storms. A much more detailed analysis is required for actual implementation of a real system.[1]

Results of such a calculation are shown in table B-4. As seen there, the most sensitive factor is the cost of the grid extension itself. Breakeven distances are not particularly sensitive to the discount rate as both grid extension and remote generation require large upfront capital investments.

---

[1] For a slightly more detailed analysis, see Chandra Shekhar Sinha and Tara Chandra Kandpal, "Decentralized v Grid Electricity for Rural India," *Energy Policy* June 1991, pp. 441-448.

Appendix C
Conversion Factors, Abbreviations, and Glossary

Conversion Factors

Area

1 square kilometer (km^2)=
 0.386 square mile
 247 acres
 100 hectares
1 square mile=
 2.59 square kilometers (km^2)
 6.4×10^2 acres
 2.59×10^2 hectares

1 acre=
 0.405 hectare (ha)
 1.56×10^{-3} square miles
 4.05 square kilometers (km^2)
1 hectare=
 0.01 square kilometer (km^2)
 3.86×10^{-3} square miles
 2.47 acres

Weight

1 kilogram (kg)=
 2.20 pounds (lb)
1 pound (lb)=
 0.454 kilogram (kg)

1 metric ton (mt) (or "long ton")=
 1,000 kilograms or 2,200 lbs
1 short ton=
 2,000 pounds or 907 kg

Energy

1 quad (quadrillion Btu)=
 1.05×10^{18} Joules (J)
 1.05 exajoules (EJ)
 3.60×10^5 metric tons, coal
 1.72×10^6 barrels, oil
 2.36×10^5 metric tons, oil
 2.83×10^{10} cubic meters, gas
 1.07×10^{12} cubic feet, gas
 2.93×10^2 terawatthours

1 kilowatthour=
 3.41×10^3 British thermal units (Btu)
 3.6×10^6 Joules (J)
1 Joule=
 9.48×10^{-4} British thermal unit (Btu)
 2.78×10^{-7} kilowatthours (kWh)
1 British thermal unit (Btu)=
 2.93×10^{-4} kilowatthours (kWh)
 1.05×10^3 Joules (J)

Volume

1 liter (l)=
 2.64×10^{-1} gallons (liquid, U.S.)
 6.29×10^{-3} barrels (petroleum, U.S.)
 1×10^{-3} cubic meters (m^3)
 3.53×10^{-2} cubic feet (ft^3)
1 gallon (liquid, U.S.)=
 3.78 liters (l)
 2.38×10^{-2} barrels (petroleum, U.S.)
 3.78×10^{-3} cubic meter (m^3)
 1.33×10^{-1} cubic feet (ft^3)
1 barrel (bbl) (petroleum, U.S.)=
 1.59×10^2 liters (l)
 42 gallons (liquid, U.S.)
 1.59×10^{-1} cubic meters (m^3)
 5.61 cubic feet (ft^3)

1 cubic meter (m^3)=
 1×10^3 liters (l)
 2.64×10^2 gallons (liquid, U.S.)
 6.29 barrels (petroleum, U.S.)
 35.3 cubic feet (ft^3)
1 cubic foot (ft^3)=
 2.83×10^1 liters (l)
 7.48 gallons (liquid, U.S.)
 1.78×10^{-1} barrels (petroleum, U.S.)
 2.83×10^{-2} cubic meters (m^3)
1 cord wood=
 128 cubic feet (ft^3) stacked wood
 3.62 cubic meters (m^3) stacked wood
 1 dry (i.e., no moisture) ton of wood

Temperature

From Centigrade to Fahrenheit:
$((9/5) \times (°C)) + 32 = °F$

From Fahrenheit to Centigrade:
$(5/9) \times (°F - 32) = °C$

Temperature changes:
—To convert a Centigrade change to a Fahrenheit change:
 $9/5 \times$ (change in °C) = change in °F
—To convert a Fahrenheit change to a Centigrade change:
 $5/9 \times$ (change in °F) = change in °C
—Example: a 3.0 °C rise in temperature = a 5.4 °F rise in temperature

Abbreviations

AC	—Alternating current
ACEEE	—American Council for an Energy Efficient Economy
ADB	—Asian Development Bank
ADF	—African Development Foundation
AFBC	—Atmospheric fluidized bed combustion
AfDB	—African Development Bank
AfDF	—African Development Fund
AID	—Agency for International Development
ASD	—Adjustable speed drive
ASEAN	—Association of South East Asian Nations
ASHRAE	—American Society of Heating, Refrigeration and Air-Conditioning Engineers
ASTRA	—Centre for the Application of Science and Technology to Rural Areas
BEST	—Biomass Energy Systems and Technology (AID)
BIG/GT	—Biomass gasifier/gas turbines
BOF	—Basic oxygen furnace
BOS	—Balance-of-system
Btu	—British thermal unit
CAEX	—Computer aided exploration and development
CAFE	—Corporate average fuel efficiency
CEST	—Condensing-extraction steam Turbine
CETA	—Conventional Energy Technical Assistance (AID)
CFCs	—Chlorofluorocarbons
CGIAR	—Consultative Group on International Agricultural Research
CH_4	—Methane
CKD	—Completely knocked down kits
CNG	—Compressed Natural Gas
CO_2	—Carbon dioxide
CORECT	—Committee on Renewable Energy Commerce and Trade
DC	—Direct current
DOE	—Department of Energy
DRI	—Directly reduced iron
DSM	—Demand side management
EAF	—Electric arc furnace
EAI	—Enterprise for the Americas Initiative
EDI	—Economic Development Institute (World Bank)
EPA	—Environmental Protection Agency
EPDCP	—Energy Policy Development and Conservation Project (AID)
EPRI	—Electric Power Research Institute
ESCO	—Energy service companies
ESMAP	—Energy Sector Management Assistance Program
ETIP	—Energy Technology Innovation Project (AID)
Eximbank	—Export-Import Bank
FAO	—Food and Agriculture Organization
FBC	—Fludized bed combustion

International System of Units (SI): Prefixes

Prefix	SI symbol	Multiplication factor
exa	E	10^{18} (1,000,000,000,000,000,000)
peta	P	10^{15} (1,000,000,000,000,000)
tera	T	10^{12} (1,000,000,000,000)
giga	G	10^{9} (1,000,000,000)
mega	M	10^{6} (1,000,000)
kilo	k	10^{3} (1,000)
hecto	h	10^{2} (100)
deca	da	10

EXAMPLES: 1 Teragram or Tg (10^{12} or 1,000,000,000,000 or 1 trillion grams); 1 megawatt-electric or MWe (10^6 or 1,000,000 or 1 million watts-electric).

EXCEPTION: 10^{15} (1,000,000,000,000,000) British thermal units (Btu) is not generally referred to as a PBtu. Instead it is known as a quad, or one quadrillion Btu's.

FCIA	—Foreign Credit Insurance Association
FGD	—Flue gas desulfurization
FINESSE	—Financing of Energy Services for Small Scale Energy Users
GAO	—General Accounting Office
GATT	—General Agreement on Tariffs and Trade
GDP	—Gross domestic product
GEEI	—Global Energy Efficiency Initiative
GEF	—Global Environmental Facility
GNP	—Gross national product
HVAC	—Heating, ventilation, air-conditioning equipment
IAF	—Inter-American Foundation
IBRD	—International Bank for Reconstruction and Development
IDA	—International Development Association
IDB	—Inter-American Development Bank
IEA	—International Energy Agency
IFC	—International Finance Corporation
IGCC	—Integrated gasification combined cycle
IIEC	—International Institute for Energy Conservation
IPCC	—Intergovernmental Panel on Climate Change
ISTIG	—Intercooled steam injected gas turbine
JIT	—Just-in-time (inventory control)
LCP	—Least cost planning
LDC	—Lesser developed country
LPG	—Liquefied petroleum gas
LWR	—Light water reactor
MAGPI	—Multi-Agency Group for Power Sector Innovation
MDB	—Multilateral development bank
MFN	—Most favored nation
MIGA	—Multilateral Investment Guarantee Agency
MWD	—Measurement while drilling
NGL	—Natural gas liquids
NGO	—Non-governmental organization
NIC	—Newly industrializing country
O&M	—Operations and maintenance
ODA	—Overseas Development Assistance

OECD	—Organization for Economic Cooperation and Development
OIMP	—Office of International Major Projects (Dept of Commerce)
OPEC	—Organization of Petroleum Exporting Countries
OPIC	—Overseas Private Investment Corporation
PACER	—Program for the Acceleration of Commercial Energy Research (AID)
PC	—Pulverized coal
PEFCO	—Private Export Funding Corporation
PPP	—Purchasing power parity
PROCEL	—National Electricity Conservation Program (Brazil)
PSED	—Private Sector Energy Development (AID)
PURPA	—Public Utilities Regulatory Policy Act
PV	—Photovoltaic
PVO	—Private volunteer organizations
R&D	—Research and development
RD&D	—Research, development and demonstration
REAT	—Renewable Energy Applications and Training (AID)
REDAC	—Renewable Energy Design Assistance Center (SANDIA)
SRIC	—Short rotation intensive culture
STIG	—Steam injected gas turbine
T&D	—Transmission and distribution
TDP	—Trade and Development Program
UHP	—Ultra high power furnace
UNDP	—United Nations Development Program
UNEP	—United Nations Environment Program
UNFPA	—United Nations Family Planning Association
UNIDO	—United Nations Industrial Development Organization
USDA	—United States Department of Agriculture
US/ECRE	—United States Export Council for Renewable Energy
US&FCS	—United States and Foreign Commercial Service
USTR	—United States Trade Representative
VHF/FM	—Very high frequency/frequency modulated
WB	—World Bank

Glossary

Appliance: Any household energy-using device.

Biodiversity: Biological diversity, i.e., the variety of species in a given area.

Biomass: Technically, the total dry organic matter or stored energy content of living organisms in a given area. As used by OTA, biomass refers to forms of living matter (e.g., grasses, trees, shrubs, agricultural and forest residues) or their derivatives (e.g., ethanol, timber, charcoal, dung) that can be used as a fuel.

Btu (British thermal unit): The amount of heat needed to raise the temperature of 1 pound of water by 1 °F at a specified temperature.

Capacity factor: The actual output of the generating technology in kWh, divided by the theoretical maximum output of the technology operating at peak design resource levels.

Capital cost: The investment in plant and equipment. This includes construction costs, but does not include operations, maintenance, or fuel/electricity costs.

Chlorofluorocarbons: Compounds containing chlorine, fluorine, and carbon; they generally are used as propellants, refrigerants, blowing agents (for producing foam), and solvents. They are identified with numbered suffixes (e.g., CFC-11, CFC-12). They are known to react with and deplete stratospheric ozone and also are ''greenhouse'' gases in that they effectively absorb certain types of radiation in the atmosphere.

Cogeneration: The simultaneous production of both electric power and heat for use in industrial or commercial/residential or other applications.

Commercial energy: Usually refers to coal, oil, gas, and electricity on the basis that they are widely traded in organized markets. These fuels are distinguished from other fuels such as firewood, charcoal, and animal and crop wastes, which are mostly described as ''biomass'' in this report.

Deforestation: Converting forest land to other vegetation or uses (i.e., cropland, pasture, dams).

Demand side management: The planning, implementation, and monitoring of utility activities designed to encourage customers to modify their pattern of electricity usage.

Discount rate: The rate at which money grows in value (relative to inflation) if it is invested.

Efficiency: For electricity generating technologies, efficiency is the actual output in kWh divided by the energy consumed or used to produce that output. For end use technologies, efficiency is often defined as the ratio of output to input, but for some end uses (such as transportation), more complex definitions are used.

Emissions: Flows of gases, liquid droplets, or solid particles into the atmosphere. Gross emissions from a specific source are the total quantity released. Net emissions are gross emissions minus flows back to the original source. Plants, for example, take carbon from the atmosphere and store it as biomass during photosynthesis, and they release it during respiration, when they decompose, or when they are burned.

End use: Any of the services or processes (e.g., lighting, refrigeration, mechanical drive) made possible through the provision of energy (also see energy services).

Energy carrier: A fuel—liquid, gaseous, or solid—or electricity to provide energy.

Energy conversion: The process of converting energy from one form to another. Often involves transforming primary or raw energy to a high quality carrier, such as gas or electricity.

Energy intensity: The amount of energy required per unit of a particular product or activity. Often used interchangeably with "energy per dollar of GNP."

Energy services: The service or end use ultimately provided by energy. For example, in a home with an electric heat pump, the service provided by electricity is not to drive the heat pump's electric motor but rather to provide comfortable conditions inside the house (also see end use).

Feedback: When one variable in a system (e.g., increasing temperature) triggers changes in a second variable (e.g., cloud cover) which in turn ultimately affect the original variable (i.e., augmenting or diminishing the warming). A positive feedback intensifies the effect. A negative feedback reduces the effect.

Fossil fuel: Coal, petroleum, or natural gas or any fuel derived from them.

Generating capacity: The capacity of a powerplant to generate electricity, typically expressed in watts-electric (e.g., kWe or MWe).

Greenhouse effect: The effect produced as certain atmospheric gases allow incoming solar radiation to pass through to the Earth's surface, but prevent the (infrared) radiation, which is reradiated from the Earth, from escaping into outer space. The effect responsible for warming the planet.

Greenhouse gas: Any gas that absorbs infrared radiation in the atmosphere.

Integrated Resource Planning (IRP): In energy planning, a cost-based ranking of all of the supply and end use technologies that could provide an energy service, beginning implementation with the lowest cost opportunities. Integrated Resource Planning changes the regulatory framework in order to encourage utilities and others to implement the least-cost demand and supply options. Among other changes, regulators allow utilities to earn income based on the net benefits from investments in energy efficiency improvements.

Least cost planning: Often used interchangeably with Integrated Resource Planning, though a more limited frame of reference. The practice of basing investment decisions on the least costly option for providing energy services. It is distinguished from the more traditional approach taken by utilities, which focuses on the least costly ways to provide specific types of energy, with little or no consideration of less costly alternatives.

Life cycle or lifecycle operating cost: The cost of a good or service over its entire life cycle.

Methane: A compound consisting of one carbon atom and four hydrogen atoms; it occurs naturally, often in association with coal and petroleum and as a byproduct of the metabolic activities of some microorganisms; it can also be synthesized artificially.

Monoculture: The exclusive cultivation of single species (e.g., corn or soybeans), a common practice in modern agriculture and energy forestry.

Natural gas: A naturally occurring mixture of hydrocarbons (principally methane) and small quantities of other gases found in porous geological formations, often in association with petroleum.

OECD: Organization for Economic Cooperation and Development, an organization that includes most of the world's industrialized market economies. Members include Australia, Austria, Belgium, Canada, Denmark, Finland, France, Germany, Greece, Iceland, Ireland, Italy, Japan, Luxembourg, Netherlands, New Zealand, Norway, Portugal, Spain, Sweden, Switzerland, Turkey, United Kingdom, and United States.

Primary energy: The term "primary energy" includes fossil fuels (e.g., coal, crude oil, gas) and biomass in their crude or raw state before processing into a form suitable for use by consumers.

Reliability: Measured by the actual output of the generating technology in kWh, divided by the output that would occur if the technology worked perfectly all the time. As used in this report, reliability does not reflect resource limits. The term **availability** is also used.

Reserves: The portion of a resource base that is proven to exist and can be economically recovered (i.e., the value of the product exceeds the production and transportation costs).

Residues: Agricultural or agroindustrial byproducts (e.g., sawdust, coconut shell, bagasse from sugar cane) that can be used as fuel.

Resources: The total existing stock of a given resource—including discovered and not yet discovered portions—regardless of the economic feasibility of recovering the resource. Also refers to subset of resources that have been proven to a degree of certainty, which are likely to be proved recoverable in the future based on a defined set of technical and economic specifications.

Retrofit: To update an existing structure or technology by modifying it, as opposed to creating something entirely new from scratch. For example, an old house can be retrofitted with advanced windows or insulation to slow the flow of heat energy into or from the house.

Sectors: Categories of end users or suppliers. The sectors included in this report are residential, commercial, industrial, agricultural, transportation, conversion, and resources.

Sustainable: A term used to characterize human activities that can be undertaken in such a manner as to not adversely affect the environmental conditions (e.g., soil, water quality, climate) necessary to support those same activities in the future.

Systemwide: In the context of this report, an analytic point of view that accounts for each interdependent aspect of the process of producing, providing and using energy.

Traditional energy: Typically, fuels that are gathered and burned by individuals with little or no processing. Some processed forms, such as charcoal, are included in this definition.

Transport mode: The different means of transporting people and freight within a system. This includes, road, rail, and maritime transport.

Watt (W): A common unit used in measuring power (i.e., as the flow of energy over time), equivalent to 3.41 Btu per hour. Where an ''e'' follows the unit (as in kWe or MWe), the watt is in the form of electrical energy. Where a ''t'' follows the unit (as in kWt or MWt), the watt is in the form of thermal energy.

Index

Index

Acid rain, 4-5, 10, 11, 17, 37, 246
Africa. *See also specific countries*
   agriculture, 136, 140, 141
   AID assistance, 265
   biomass, 247
   economic growth, 22, 103
   electricity systems, 189
   energy consumption, 22
   energy crops, 253
   industry, 103
   microhydroelectric power, 206
   oil imports, 26, 234
   oil reserves, 234
   paper mill capacity, 132
   recycling aluminum, 130-131
   residential electricity use, 48
   steel production, 120
   transport energy use, 23
   urban population, 22
   wind power potential, 202
Agency for International Development
   congressional directives, 276, 283
   energy-related projects, 13, 30, 258, 264-265, 276, 281, 284, 285
   family planning, 285
   geographical concentration of aid, 14
   Office of Energy and Infrastructure, 265, 266
   training, 273-274, 281
Agriculture
   energy use, 33-37, 103, 136, 139
   environmental damage, 37, 38-39
   improving efficiency, 103, 136-137
   irrigation, 137-140
   residues as fuel, 248-250
   traction, 140-141
AID. *See* Agency for International Development
Air conditioning, 79-83, 114
Algeria, 22, 241-242
Alternative fuels, 171-176
Amoco, 238
Analytical framework of OTA study, 19-20
Angola, 141, 242
Appliances
   barriers to purchase of efficient appliances, 8, 83-84, 87
   demand, 29, 48
   energy use, 23, 47, 65
Argentina, 22, 151, 194, 241
Argonne National Laboratory, 190, 271
Asia. *See also specific countries*
   biomass, 247
   coal consumption, 25
   economic growth, 22
   energy consumption, 22
   irrigation, 137
   oil reserves, 234
   paper mill capacity, 132
   residential electricity use, 48
   transport energy use, 23, 155
   wind power potential, 202

ASTRA. *See* Center for the Application of Science and Technology to Rural Areas
Automobiles. *See also* Transport
   electric, 176
   environmental impact, 145, 148, 167, 172
   improving efficiency, 145-150, 161-165
   role of, 167
   scrappage, 148, 161
Bangladesh, 36, 248
Benin, 189
Bicycles, 155-156
Biogas, 215, 219-222
Biomass Energy Systems and Technology Project, 266
Biomass fuels
   barriers to new technologies, 12, 183, 214, 233
   consumption, 23, 27, 91, 214, 247
   conversion to gas and electricity, 179, 183, 215-222
   conversion to liquid fuels, 179, 183, 222-227
   energy crops, 233, 251-257
   environmental impact, 233
   forage v. purchase, 34-35
   forest management, 250-251
   improving efficiency, 10-11, 247-248
   residues as energy, 233, 248-250, 258
   sales and rural economies, 36-37, 39
   shift from, 26, 28, 47, 59, 179-180
   stoves, 58, 64-65, 131
   supply, 25, 34-35, 36-37, 247, 248, 258
   use in industry, 91
Bolivia, 91, 161
Botswana, 36
Brazil
   agriculture, 103, 136
   air conditioning, 79, 80
   capital cost constraints, 136
   cement production, 124, 125
   charcoal production, 253
   coal reserves, 244, 245
   debt restructuring, 278
   electric water heating, 60
   electricity system, 48, 187
   electronic ballasts, 71
   energy consumption, 27
   ethanol production, 174, 175, 224, 225, 254-255
   foreign trade, 282
   hydroelectric power, 190
   industry, 91, 136
   lighting, 69, 70
   motor efficiency, 107
   nuclear power, 194
   oil production, 239, 241
   per capita income, 22
   PROCEL, 85, 135
   producer gas, 215
   refrigerators, 72, 73-74
   steel production, 117, 118, 119, 120-121, 122, 123
   transport, 151, 158, 163, 166, 167, 169
   urbanization, 22

–319–

Brookhaven/Stony Brook Energy Management Training Program, 190
Burkina Faso, 35, 189
Burma, 241
Buses, 145, 151, 152, 166, 167-168

CAEX. *See* Computer-Aided Exploration and Development
California, 199, 200, 202
Cameroon, 161, 235, 241
Canada, 114, 194
Capital Flight, 30
Capital investment
　for efficient technologies, 9, 52-53, 76, 133-134, 136, 261, 263, 278
　in energy supply systems, 180, 186, 201, 208, 231, 232, 278-279
　projected, 4, 7, 17, 29-30, 48-49, 231
Carbon dioxide
　coal and, 191, 193, 244
　deforestation and, 40, 42
　fossil fuels and, 4, 31, 37
　gas turbines and, 197
　geothermal power and, 199
　natural gas combustion and, 241
　vehicle, 172
Carbon monoxide, 145, 171, 172, 174
Cement, 5, 28, 32, 94, 95, 101
　production, 123-127
Center for the Application of Science and Technology to Rural Areas, 217, 221
Central African Republic, 245
Central America. *See also specific countries*
　energy crops potential, 253
　oil imports, 26
CFCs. *See* Chlorofluorocarbons
CGIAR. *See* Consultative Group for International Agricultural Research
Chad, 242
Chemicals, 95, 101, 129-130
Chevron, 242
Chile, 214, 244
China
　agricultural residues, 248
　agriculture, 103, 129, 138, 140-141
　biogas, 219-222
　capital cost constraints, 136
　cement production, 125, 126
　coal consumption and production, 10, 19, 25, 179, 231-232, 245-246, 247
　cogeneration, 197
　electricity system, 187
　energy consumption, 19, 23, 26-27, 32, 77
　fluidized-bed combustion, 191-192
　HVDC projects, 186
　hydroelectric power, 190, 193-194, 206
　improving efficiency, 85, 135
　industry, 22, 23, 91
　irrigation, 138
　nuclear power, 182, 195, 196
　quality of life, 21-22
　solar water heating, 60
　space heating, 77, 78
　steel production, 24, 117, 118, 121, 122, 123, 134
　tractors, 140-141
　transport, 155, 157-158, 159, 160, 161-163, 170
　wind turbines, 202
Chlorofluorocarbons, 39
Coal
　clean, 190-191, 246-247
　consumption and production, 10, 25, 179, 190, 231-232, 244-246
　for electricity generation, 181, 191-193
　environmental impact, 10, 11, 37, 190-191, 231, 245
　and rail freight, 159
　reserves, 244-245
Cogeneration, 197-198
Colombia, 60, 244
Committee on Renewable Energy Commerce and Trade, 13, 263, 280
Competitiveness, 4, 17, 282-283, 286
Compressed natural gas, 151, 174-175
Computer-Aided Exploration and Development, 237
Congo, 241
Consultative Group for International Agricultural Research, 271
Consumer goods
　demand, 3, 29, 48, 100-101
　energy for manufacture, 100
　first cost, 8, 50, 67, 81, 148
　quality control, 131
Context of OTA study, 18-19
Conventional Energy Technical Assistance Project, 266
Conversion technologies. *See also specific technologies or energy sources*
　comparing, 180-181, 200-201, 209-211
　improving efficiencies, 9-10, 183-190
Cooking, 23-24, 47, 50-60, 214
Cooperative for American Relief Everywhere, 252
CORECT. *See* Committee on Renewable Energy Commerce and Trade
Costa Rica, 136
Cote d'Ivoire, 186, 189, 235
Crops for energy, 251-257

Debt, 3, 11, 26, 30, 262, 277-278
Deforestation, 4, 31, 37, 38, 39, 40-42
Demand for energy
　difficulties in meeting, 4, 29-31, 181-182
　factors increasing, 3, 27-29, 47-48, 179, 181
Developing countries
　compared with industrialized countries, 21-22
　defining, 21
　developmental differences, 18, 22
Diesel fuel
　consumption, 155, 172
　environmental impact, 145, 172
　pricing, 148, 158, 164
Dominican Republic, 207, 210

Eastern Europe, 264, 276, 277
Economic development, 28, 31-33
Ecuador, 60, 241, 245
Egypt, 244, 265, 283
Electric motor drive systems. *See* Motor drive systems
Electric vehicles, 176

Electricity. *See also specific production methods and power sources*
  capital investment, 6-7, 48-49, 180, 186, 201, 208
  cogeneration, 197-198
  demand, 47-48, 181
  efficiency improvements, 5-8, 182, 183-190
  environmental issues, 189-190
  generating system rehabilitation, 183-184
  management improvements, 188-189, 273
  options for new on-grid generation, 190-201
  policy options, 182
  private sector role, 275
  rural electrification, 182, 201-212
  system interconnections, 187-188, 189
  system planning procedures, 188
  transmission and distribution systems, 185-187
Electronic ballasts, 70, 71
Electronic equipment, 47, 82-83
Employment, 11
Energy Conservation Services Project, 266
Energy Policy Development and Conservation Project, 266
Energy Sector Management Assistance Program, 265-269, 284
Energy supply. *See also specific energy sources*
  analytical framework, 19, 20
  biases, 9
  reliability, 11, 93, 180
  sources, 10-11, 25-26, 179-180, 231-233
Energy use
  analytical framework, 19-20
  developing countries, 3-4, 17-19, 26-27
  patterns, 22-25
Environmental issues
  benefits of energy supply technologies, 11
  energy services and, 17, 37-42
  environmental planning, 12-13
  global, 4-5
  policy issues and options, 261, 263, 284-285
Environmental Protection Agency, 70
Equatorial Guinea, 235
ESMAP. *See* Energy Sector Management Assistance Program
Ethanol, 174, 175, 223-227, 255
Ethiopia, 164, 235, 245, 247, 249, 252
Export-Import Bank, 13, 243, 263, 281
Export promotion, 32, 33, 279-280
Exxon, 242

Family planning, 285
FAO. *See* Food and Agriculture Organization
FBC. *See* Fluidized-bed combustion
FGD. *See* Flue gas desulfurization
Financial issues, 279-280. *See also* Capital investment; Debt; Financial markets
Financial markets, 5, 17, 262
Finland, 111, 124
Flue gas desulfurization, 191
Fluidized-bed combustion, 9, 181, 191-192
Food and Agriculture Organization, 253
Forced outage rate, 184
Ford Motor Co., 131
Foreign Credit Insurance Association, 243
Forest management, 250-251
Formaldehyde, 174

France, 282
Freight transport
  improving efficiencies, 156
  modal shifts, 159-161
  rail, 159
  trucks, 156-159
Fuel cells, 200
Fuels, alternative, 171-176

Gabon, 235, 241
Gas turbines, 9-10, 11, 12, 181-182
Gasoline, 145, 148, 155, 164, 172
GEF. *See* Global Environment Facility
General Motors, 164
Geothermal energy, 10, 198-199
Ghana, 188, 189
Global Environment Facility, 268, 269, 277, 285
Global warming, 5, 11, 17, 37, 39-42
Greenhouse gases, 4, 17, 191
  global climate changes, 39-42
  transport systems and, 145, 172, 174, 175
Grid extension, 211-212
Guatemala, 60

Haiti, 245
Heating, 77-78
High voltage direct current, 186-187
H_2S emissions, 199
Humanitarian issues, 4, 17, 283-284
HVDC. *See* High-voltage direct current
Hydrocarbons, 171
Hydroelectric power, 10, 37, 182, 193-194, 199, 201
Hydrogen, 176

IAEA. *See* International Atomic Energy Agency
IFC. *See* International Finance Corporation
IGCC. *See* Integrated Gasification Combined Cycle systems
IIEC. *See* International Institute for Energy Conservation
Import duties, 148
Import substitution, 32
Imports, 25, 148, 180, 231, 234
India
  agricultural residues, 248
  agriculture, 103, 137-138, 141
  appliances, 72
  biogas, 222
  biomass, 254, 255
  capital cost constraints, 136
  cement production, 125, 133
  coal consumption and production, 10, 19, 25, 179, 231-232, 245-246, 247
  cogeneration, 197-198
  electricity system, 187, 217
  energy consumption, 19, 23, 26-27
  fluidized-bed combustion, 191-192
  foreign trade, 282
  gasifier-engine systems, 217, 219, 220
  HVDC projects, 186-187
  hydroelectric power, 190, 193
  industry, 11, 22, 23, 180
  irrigation, 137-138
  lighting, 63, 65-66
  nuclear power, 194, 196

oil production, 234, 238
pricing, 134, 140
steel production, 24, 117, 118, 120, 121, 122, 123, 131
thermal efficiency, 183
tractors, 141
transport, 145, 155, 157, 159, 160, 161-163, 165, 169, 170-171
water heating, 60
women's role in rural areas, 36
Indonesia
   biomass, 91
   building efficiency standards, 135
   capital cost constraints, 136
   coal reserves, 244
   energy consumption, 27
   lighting, 69
   oil development, 241
   oil imports, 234
   refrigerators, 73
   steel production, 120
Industrial residues, 248-250
Industrialization, 32-33
Industry
   barriers to efficiency, 132-134
   energy use, 23, 24, 91
   growth, 91
   material usage, 130-132
   modern, 94-103, 117-130
   small scale, 91-94, 132-133
Infrastructure, 8, 95, 117, 133, 148, 157, 158
Institutional issues, 11-12, 30-31, 261, 274-276
Integrated Gasification Combined Cycle systems, 192-193
Integrated Resource Planning, 7-8, 9, 13, 87-88, 188
   Department of Energy, 274-275
   promotion of, 262-263
Intergovernmental Panel on Climate Change, 40, 41
International Atomic Energy Agency, 190
International Development Bank, 265, 277
International Finance Corporation, 214
International Institute for Energy Conservation, 273
International Monetary Fund, 277
International Planned Parenthood Federation, 285
IPCC. *See* Intergovernmental Panel on Climate Change
Iraq, 196
IRP. *See* Integrated Resource Planning
Irrigation, 137-140
Israel, 60
Italy, 282
Ivory Coast, 186, 189, 235

Japan, 60, 118, 123, 124, 132, 282
Just-in-time (JIT) inventory control, 132

Kenya
   agriculture, 103, 136
   biomass, 37
   electricity system, 189
   energy crops, 255
   industry, 135
   solar water heating, 60-61
   thermal plant rehabilitation, 186
   women's role in rural economy, 36

Latin America. *See also* specific countries
   agriculture, 141
   AID assistance, 265
   biomass, 247
   energy consumption, 22
   energy crops potential, 253
   hydroelectric power, 190
   industry, 103
   microhydroelectric power, 206
   oil exploration, 242
   oil reserves, 234, 235
   paper mill capacity, 132
   residential use of energy, 48
   transport energy use, 23
   wind power potential, 202
Lawrence Berkeley Laboratory, 71, 271
Liberia, 91
Lighting, 23, 47, 63-71, 72
Liquefied natural gas, 175-176
Liquid fuels from biomass, 222-227
Los Alamos, 271

Madagascar, 235
Malawi, 36-37, 242
Malaysia, 69, 234, 241
Mali, 245
Material usage, 130-132
MDBs. *See* Multilateral development banks
Mechanical drive, 23, 24. *See also* Motor drive systems
Methane, 4, 10, 41
Methanol, 151, 172-174, 214, 222-223
Mexico
   air conditioning, 79
   automobile production, 163
   coal reserves, 244
   debt restructuring, 278
   energy use, 26, 27
   nuclear power, 194
   oil reserves, 234, 235, 239
   steel production, 120, 121, 122
   transport, 167
Microhydroelectric power, 206-208
Middle East, 234
Mongolia, 203, 244
Morocco, 202
Motor drive systems
   adjustable speed drives, 107, 112-115
   design, 104-107
   energy use, 103-104
   improving efficiency, 101, 117
   motors, 107-108
   pipes and ducts, 115-117
   pumps and fans, 108-112, 138-140
   systems, 117
   test standards, 108
Mozambique, 235, 242
Multilateral development banks, 262, 263, 264, 285
   activities of, 265-269
   issues and options for the U.S., 12, 13, 14
   technology transfer and, 269-279
Multilateral Investment Guarantee Agency, 243

Namibia, 235
National Renewable Energy Laboratory, 174, 271

National Rural Electric Cooperative Association, 190
Natural gas
　compressed, 151, 174-175
　environmental impact, 10, 11, 38, 241
　liquefied, 175-176
　production, 10, 190, 240-241
　reserves, 10, 235-236, 243
　use in developing countries, 23, 25
Nepal, 247
Newly Industrializing Countries, 22. *See also specific countries*
Nigeria, 189, 236, 241, 242
Nitrogen oxides, 4, 9, 233, 244
　air pollution, 37, 39, 41
　coal, 190, 191, 193
　diesel vehicles, 145
　gas turbines, 197
North Korea, 244, 245
Norway, 242
NRECA. *See* National Rural Electric Cooperative Association
Nuclear energy, 10, 11, 37-38, 182, 194-196, 201

Oak Ridge National Laboratory, 271
OECD. *See* Organization for Economic Cooperation and Development
Ohio, 191
Oil
　consumption, 5, 17, 25, 37, 145, 155, 179, 232, 233-234
　electricity generation, 190, 201
　environmental impact of exploration, 236, 237
　exploration and development, 10, 12, 234, 236-241
　global markets, 3, 5, 17
　imports, 17, 25-26, 32, 148, 231, 234
　institutional issues, 241-243
　offshore production, 239-240
　policy options, 243-244
　refining, 182, 212-214
　reserves, 10, 234-236
OPIC. *See* Overseas Private Investment Corporation
Organization for Economic Cooperation and Development, 28, 282-283
Overseas Private Investment Corporation, 13, 243, 263, 281

Pakistan
　agricultural residues, 248
　AID assistance, 265, 283
　capital cost constraints, 136
　electricity system interconnections, 189
　foreign trade, 282
　gas exploration, 244
　generating system rehabilitation, 184
　microhydroelectric power, 208
　nuclear power, 194
　oil imports, 234
　oil reserves, 235
　reliability of supply, 11, 180
　thermal efficiency, 183
Paper, 95, 101, 127-129, 132
Papua New Guinea, 62
Particulates, 172, 190, 244, 246-247
Peace Corps, 258
Peru, 235
Philippines, 69, 125, 135, 194, 217, 219, 234, 283
Photovoltaics, 10, 204-206, 207, 210, 211

Policy issues and options, 12-14, 261-286
　biomass, 183, 258
　electricity systems, 182
　environment, 12-13, 284-285
　humanitarian issues, 4, 17, 283-284
　importance to U.S., 4-5, 17
　improving efficiency, 9, 269-279
　industry, 134-136
　oil and gas, 243-244
　programs for policy implementation, 263-269
　residential and commercial energy use, 84-88
　trade, 279-283
　transport, 152-155
　U.S. example, 285-286
　U.S. policy framework, 263-264
Population control, 285
Population growth, 3, 27-28, 36, 38-39
Poverty. *See* Humanitarian issues
Pricing policies, 8, 30
　cost-plus pricing, 134
　energy price reforms, 262-263, 276
　subsidies, 29, 67, 84, 86, 140, 181, 261
Private sector role, 273, 275-276
PROALCOOL (Brazilian Fuel Alcohol Program), 225
PROCEL (Brazilian National Electricity Conservation Program), 85, 135
Producer gas, 183, 215-222
Public Utility Regulatory Policies Act, 197, 198, 275
PURPA. *See* Public Utility Regulatory Policies Act

Quality control, 131-132
Quality of life, 21-22, 33

Rail systems, 150, 151, 159-160, 168
Recycling, 130-131. *See also* Waste-to-energy technologies
Refrigeration, 29, 47, 71-77, 114, 206
Reliability, 11, 93, 180
Renewable Energy Applications and Training Project, 266
Royal Dutch/Shell, 242
Rural areas
　economy of, 33-37
　electrification, 182, 201-212
　energy use, 11, 23, 24-25, 34-35
　policy options, 283-284
　traditional industries, 93
　transport needs, 170

Sandia National Laboratory, 271
Scrubbers, 191
Senegal, 235
Short Rotation Intensive Culture crops, 252-253, 257
Singapore, 69, 135, 151, 167
Small Business Administration, 13, 263
Solar Energy Research Institute. *See* National Renewable Energy Laboratory
Solar power
　cooking, 58
　thermal-electric technology, 200
　water heating, 60-62
Somalia, 138, 235, 242
South Africa, 27, 120, 244, 245
South Korea
　appliances, 74, 78, 83

automobiles, 163
building efficiency standards, 135
coal consumption, 231, 245
energy efficiency activities, 85
energy use, 27
industry, 91
nuclear power, 182, 194
quality of life, 21, 22
steel production, 118, 120, 121, 122, 123
Southern California Edison, 114
Soviet Union, 234, 244
Space conditioning, 47, 77-82
Sri Lanka, 249, 254
SRIC. See Short Rotation Intensive Culture crops
Steel, 24, 28, 94-102, 130
industry development, 32-33
industry importance, 117-123
Stoves, 24, 56-60, 61-62, 64-65, 131
Subsidies. See Pricing policies
Sudan, 138, 235, 242
Sulfur dioxide, 9, 37, 39, 172, 233
biomass, 248
coal, 190, 191, 193, 244, 246
gas turbines, 197
Swaziland, 244
Sweden, 111, 191

Taiwan
appliance efficiency standards, 135
automobiles, 163
energy efficiency activities, 85
nuclear power, 182, 194
steel production, 33, 120, 121, 122
Tanzania, 126, 134, 189, 235, 247
Taxes and tax credits, 9, 24, 136, 164, 276
Technology transfer, 3, 9, 13, 261, 263-279, 282
Television, 29, 82, 83, 84
Thailand
air conditioning, 79, 81
automobiles, 161, 163
biomass, 91
building efficiency standards, 135
capital cost constraints, 136
energy crops, 255
energy use, 79
lighting, 69
microhydroelectric power, 208
oil imports, 234
tractors, 140-141
Thermal efficiency, 183
Tied-aid, 282-283, 284
Togo, 189
Toyota, 165
Trade and Development Program, 13, 263, 279, 281
Trade issues, 3, 4, 17, 279-283
Traffic management, 151, 152, 167, 169
Training programs, 188-189, 190, 266, 273-274
Transport. See also specific modes of transport
alternative fuels, 171-176

demand, 169, 176
in developing countries, 8, 23, 145, 155-156
environmental impact, 145, 172
freight, 156-161
improving efficiency, 8, 145-152, 156
land-use planning, 151, 167, 169, 176
nonmotorized, 151, 169-171
passenger, 161-169
policy options, 152-155
Trinidad and Tobago, 91, 241
Trucks, 145, 156-159
Tunisia, 125, 244
Turkey, 60, 125
Two- and three-wheelers, 165-166

United Kingdom, 194, 242, 282
United Nations, 13, 244, 253, 263, 273, 285
United Nations Population Fund, 285
University of Pennsylvania, 190
Urbanization, 22, 28, 167, 169, 176
Uruguay, 136, 283
U.S. and Foreign Commercial Service, 279
U.S. Department of Agriculture, 233
U.S. Department of Commerce, 281
U.S. Department of Energy, 71, 232, 233, 252, 258
information dissemination, 271, 273
Integrated Resource Planning Program, 274
U.S. Export Council for Renewable Energy, 271
U.S. Trade Representative, 13, 263, 280-282
US/ECRE. See U.S. Export Council for Renewable Energy
United States and Foreign Commercial Service (US&FCS), 279
Utility planning, 188-189, 190

Vaccine refrigerators, 206
Venezuela, 12, 22, 27, 123, 234, 239
Ventilation, 78, 79, 116-117
Volkswagen, 165
Volta River Authority, 188
Volvo, 164

Waste-to-energy technologies, 233, 248-250, 258
Water heating, 47, 60-63
West Germany, 124, 282
West Java, 250
Wind power, 10, 202-204, 210
Women
demand for appliances as women enter workforce, 29, 48
role in rural labor, 35-36
World Bank, 4, 17, 186, 243, 274, 277. See also Multilateral development banks
creation of Natural Gas Unit, 244
financing of preinvestment studies, 232-233
methanol export project, 214
World Energy Conference, 17, 27, 28, 231, 235
World Population Conference, 285

Zaire, 193
Zambia, 188
Zimbabwe, 188, 244

Superintendent of Documents **Publications** Order Form

Order Processing Code:
***5246**

Charge your order.
It's Easy!

To fax your orders (202) 512–2250

☐ **YES**, please send me the following:

_____ copies of ***Fueling Development: Energy Technologies for Developing Countries*** *(336 pages)*
S/N 052-003-01279-1 at $15.00 each.

The total cost of my order is $ _____ . International customers please add 25%. Prices include regular domestic postage and handling and are subject to change.

_____ (Please type or print)
(Company or Personal Name)

_____
(Additional address/attention line)

_____
(Street address)

_____
(City, State, ZIP Code)

_____
(Daytime phone including area code)

_____
(Purchase Order No.)

May we make your name/address available to other mailers?   YES ☐   NO ☐

Please Choose Method of Payment:

☐ Check Payable to the Superintendent of Documents

☐ GPO Deposit Account ☐☐☐☐☐☐☐–☐

☐ VISA or MasterCard Account

☐☐☐☐ ☐☐☐☐ ☐☐☐☐ ☐☐☐☐

☐☐☐ (Credit card expiration date)   *Thank you for your order!*

_____
(Authorizing Signature)   4/92

Mail To: New Orders, Superintendent of Documents
P.O. Box 371954, Pittsburgh, PA 15250–7954

P3